Prem K. Goel
N. Sreenivas Iyengar (Eds.)

# Bayesian Analysis in Statistics and Econometrics

Springer-Verlag
New York Berlin Heidelberg London Paris
Tokyo Hong Kong Barcelona Budapest

Prem K. Goel
Department of Statistics
The Ohio State University
Columbus, OH 43210-1247
USA

N. Sreenivas Iyengar
Economic Analysis Unit
Indian Statistical Institute, Bangalore
8 Mile Mysole Road
R.V. College - P.O.
Bangalore 560059
India

---

Mathematics Subject Classification: 62-06

---

Library of Congress Cataloging-in-Publication Data
Bayesian analysis in statistics and econometrics /
[edited by] Prem K. Goel, N. Sreenivas Iyengar.
p. cm. — (Lecture notes in statistics ; v. 75)
Based on the papers presented at the Indo-U.S. Workshop on Bayesian Analysis in Statistics and Econometrics, Dec. 19-23, 1988 in Bangalore, India.
Includes bibliographical references and indexes.
ISBN 0-387-97863-1
1. Econometrics—Congresses. 2. Bayesian statistical decision theory—Congresses. I. Goel, Prem K., 1943- . II. Iyengar, N. S. III. Indo-U.S. Workshop on Bayesian Analysis in Statistics and Econometrics (1988 : Bangalore, India) IV. Series: Lecture notes in statistics (Springer-Verlag) ; v. 75.
HB139.B3934 1992
330'.01'5195—dc20 92-11138

Printed on acid-free paper.



Camera ready copy provided by the editors.
Printed and bound by Edwards Brothers, Inc., Ann Arbor, MI.
Printed in the United States of America.

9 8 7 6 5 4 3 2 1

ISBN 0-387-97863-1 Springer-Verlag New York Berlin Heidelberg
ISBN 3-540-97863-1 Springer-Verlag Berlin Heidelberg New York

Lecture Notes in Statistics 75

Edited by J. Berger, S. Fienberg, J. Gani, K. Krickeberg, I. Olkin, and B. Singer

# Preface

This volume is based on the invited and the contributed presentations given at the Indo-U.S. Workshop on Bayesian Analysis in Statistics and Econometrics (BASE), Dec. 19-23, 1988, held at the Hotel Taj Residency, Bangalore, India. The workshop was jointly sponsored by The Ohio State University, The Indian Statistical Institute, The Indian Econometrics Society, U.S. National Science Foundation and the NSF-NBER Seminar on Bayesian Inference in Econometrics. Profs. Morrie DeGroot, Prem Goel, and Arnold Zellner were the program organizers.

Unfortunately, Morrie became seriously ill just before the workshop was to start and could not participate in the workshop. Almost a year later, Morrie passed away after fighting valiantly with the illness. Not to find Morrie among ourselves was a shock for most of us. He was a continuous source of inspiration and ideas. Even while Morrie was fighting for his life, we had a lot of discussions about the contents of this volume and the Bangalore Workshop. He even talked about organizing a Second Indo-U.S. workshop some time in the near future. We are dedicating this volume to the memory of Prof. Morris H. DeGroot. We have taken a conscious decision not to include any biography of Morrie in this volume. An excellent biography of Morrie has appeared in Statistical Science [(1991), vol. 6, 1-14], and we could not have done a better job than that.

The BASE workshop was attended by 80 scientists from four countries; 29 from the United States, two from Canada, one from Switzerland and 48 from India. Most of the Indian participants were young faculty members from universities across India. It provided them an opportunity to learn about the state of the art in Bayesian Statistics and Econometrics. We believe that the workshop was a grand success in its mission. It also opened doors for collaboration between some young Indian scientists and the visiting U.S. scientists.

The workshop presentations covered the entire spectrum of current research efforts in Bayesian Statistics and Econometrics. Some presentations produced lively comments from invited discussants. The papers based on the invited talks and contributed poster sessions have undergone an extensive refereeing and editing process. Whenever applicable, the invited paper is followed by the Comments from the invited discussants followed by a rejoinder from the author(s). The invited papers have been organized into six parts, namely: A. Computational approaches; B. Contingency tables and nonparametric Bayes; C. Hierarchical and empirical Bayes inference; D. Priors and Utility; E. Reflections on Bayesian paradigm; F. Reliability and dose-response modeling; G. Time series modeling, forecasting and robustness. Finally, part H consists of some of the contributed papers.

The financial support for the Indo-U.S. Workshop was provided by the *National Science Foundation*, under grant INT- 8913294, to The Ohio State University, the Seminar on Bayesian Inference in Econometrics and a grant from the University Grants Commission, Government of India, to the Indian Statistical Institute. Without their generous support, it would have been impossible to organize the workshop. The editing effort on this volume is partially supported by the National Science Foundation under Grants DMS-8906787 and DMS-9008067.

We are grateful to the other members of the Coordination Committee, Prof. C. R. Rao, the honorary Chair, and Prof. J. K. Ghosh, Director, Indian Statistical Institute, Calcutta. The editors of this volume were the co-organizers of the workshop arrangements. We highly appreciate the efforts of the invited speakers and discussants, who created an excellent learning environment at the workshop in addition to a high standard of scientific content of the presentations. We also convey special thanks to the referees of the papers submitted for publication in the volume. Finally, this volume would not have been possible without the long hours put in by Peg Steigerwald in the text processing of the manuscripts. Her effort is most genuinely appreciated.

N. Sreenivas Iyengar
Bangalore, India

Prem K. Goel
Columbus, Ohio
February 28, 1992

## Morris H. DeGroot

June 8, 1931 to November 2, 1989

The following are excerpts from my remarks at the memorial service for Prof. Morris H. DeGroot, held at the Carnegie Mellon University on November 18, 1989. They came from the bottom of my heart. I have decided to include them in this volume as a dedication to Morrie.

Professor Morris H. DeGroot was the first person I met on arriving at Carnegie Mellon University in 1968. He was a *Guru* in the real sense in that he was an excellent teacher both in and outside the classroom and was always eager to share ideas about everything. To celebrate my graduation in 1971, Prof. DeGroot invited me to a luncheon. We discussed everything that had happened during my three years stay at CMU. That day I realized that Morrie hasn't taught me everything he knows. If I wished to learn everything that he knew, I must cultivate our relationship. That was the start of our friendship. He was a friend *in need* and a friend *in deed.*

During these years, Morrie and I sat together numerous times for discussions about all aspects of statistics, science, and life in general. Three people can take the credit for most of my statistical education: Morrie DeGroot; Herman Rubin, who I had the good fortune to learn from, only because I decided to join Purdue University's faculty on Morrie's advice; and D. D. Joshi, my statistics professor at the Institute of Social Sciences, Agra University.

John F. Kennedy was my hero as well as one of the two heroes of Morrie. I believe that JFK must be describing Morrie when he described an American as follows:

> The American by nature, is optimistic. He is cxperimental, an inventor and a builder who builds best when called upon to build greatly.

The following observations validate this point:

1. Morrie was called upon to build a Department of Statistics at the Carnegie Mellon University in a great tradition. He created it to be among one of the best in a short duration.

2. He was called upon to experiment and conceive a great journal which will help in boosting the morals within the Institute of Mathematical Statistics. He created the best one- *Statistical Science.*

Another hero of Morrie was Robert F. Kennedy, who once said that *"Progress is a nice word. But change is its motivator and change has its enemies."* According to Morrie, this is the situation with the Statistical Science community. He once explained to me that a full changeover from the classical statistics to Bayesian statistics won't be an easy thing for a long time, because in the words of RFK *"One fifth of the people are against everything all the time."*

However, Morrie was not a risk averse person. During my last meeting with him at his home, I asked for his advice on how to deal with difficult things in life. After some pause, he told me *"Take risk if you want to be successful in life."* I will always remember these words of wisdom.

For a long time after Morrie's death, I felt lost. Finally, I was led by the following verse in Bhagavad Gita (The Celestial Song), the most sacred Hindu scripture, on the essence of life and death:

> As a man casting off worn-out garments puts on new ones, so the embodied, casting off worn out bodies enters into others that are new. [2:22, Bhagavad Gita]

This verse awakened me to the fact that Morrie has not gone anywhere, that he has merely changed his body which was worn out, that memories of all the good times with Morrie are still with me, and that his ideas and advice will keep on leading me on the path ahead.

Prem K. Goel

## Table of Contents

1

# LaPlace Approximation for Curved Surfaces


## Parthasarathy Bagchi & Joseph B. Kadane [1]



**Abstract**

The computation of posterior moments and marginal distributions in non-conjugate cases is an important research area for Bayesians. This paper reviews general progress in the area, with special emphasis on Laplacean approximation, especially in the fully exponential form. Then the paper considers the extension of these methods to curved surfaces (circles, spheres, cylinders, etc.). First, we address why a separate theory for curved surfaces is convenient, since Euclidean Laplace approximations apply. After a review of recent work by the authors that gives a Laplacean approximation for curved surfaces, a number of examples are considered in depth.



*Key words and Phrases:* Laplacean; Curved surfaces; Bayesian aproximation; non-conjugate priors; exponential family.


## 1. INTRODUCTION

For a Bayesian method to become widely applicable, it is obvious that convenient methods for integrating non-conjugate cases must be readily available. Recently, interesting theoretical results have been developed showing that an unrestricted prior can be approximated arbitrarily closely by convex combinations of conjugate priors (Dallal and Hall 1983, Diaconis and Ylvisaker 1985). However, these results apply only when conjugate priors exist, essentially only in the exponential family. Furthermore, as a computational method, it is far from clear that where these results apply, it is efficient to express the prior of interest as such a convex combination, update each of the summands, and then compute the posterior accordingly.

Usually a simple grid method suffices for easy cases, in which the integrals required are at most three-dimensional. Thus attention focuses on the multidimensional cases. At present there are three different research approaches to such integrals. The first is Monte Carlo integration with importance sampling (Geweke, 1986, Van Dijk, Hop, and Lowter 1987). These methods require an artful choice of importance function that both well approximates the density being integrated and is easy to compute.


[1] Carnegie Mellon University

A second strand of research relies on classical polynomials to do sophisticated numerical analysis (see Smith, et al. 1985 for a review). These methods benefit from a judicious choice of parameterization. Finally, the third strand is recent work on Laplacean methods. This paper is in the latter tradition.

Laplace (1776, see also Stigler, 1986) was apparently the first to put an integral in exponential form, and then expand the exponent in a Taylor series. For this reason, his name is attached to the methods. Interest in these methods was revived by Lindley (1961, 1980) and Mosteller and Wallace (1964). Recent work of Leonard (1982) and Tierney and Kadane (1986) have also contributed. These methods also benefit from a judicious choice of parameterization.

## 2. LAPLACE APPROXIMATIONS IN THE FULLY EXPONENTIAL FORM

Suppose $g(\theta)$ is positive and bounded away from zero. Then the posterior expectation of $g$ can be expressed as

$$E(g(\theta) \mid x) = \frac{\int g(\theta)\ell(\theta \mid x)\pi(\theta)d\theta}{\int \ell(\theta \mid x)\pi(\theta)d\theta} \tag{1}$$

$$= \frac{\int e^{\log g(\theta)+\log \ell(\theta \mid x)+\log \pi(\theta)}d\theta}{\int e^{\log \ell(\theta)+\log \pi(\theta)}d\theta} \tag{2}$$

$$= \frac{\int e^{L^*(\theta)}d\theta}{\int e^{L(\theta)}d\theta} \tag{3}$$

where $\pi(\theta)$ is the prior, $\ell(\theta \mid x)$ is the likelihood, $L(\theta) = \log \ell(\theta \mid x) + \log \pi(\theta)$, and $L^*(\theta) = L(\theta) + \log g(\theta)$.

Now $L(\theta)$ and $L^*(\theta)$ may be expanded in a Taylor series around their respective maxima, $\hat{\theta}$ and $\hat{\theta}^*$, as follows

$$L(\theta) = L(\hat{\theta}) + (\theta - \hat{\theta})L'(\hat{\theta}) + \frac{(\theta - \hat{\theta})^2}{2}L''(\hat{\theta}) + \text{HOT} \tag{4}$$

(HOT stands for Higher Order Terms), and similarly for $L^*(\theta)$. Since $\hat{\theta}$ and $\hat{\theta}^*$ maximize $L(\cdot)$ and $L^*(\cdot)$, respectively, we have

$$E\{g(\theta \mid x)\} \doteq \frac{e^{L^*(\hat{\theta}^*)} \int e^{\frac{(\theta - \hat{\theta}^*)^2}{2} L^{*\prime\prime}(\hat{\theta}^*)} d\theta}{e^{L(\hat{\theta})} \int e^{\frac{(\theta - \hat{\theta})^2}{2} L''(\hat{\theta})} d\theta} \tag{5}$$

When the space of integration is the real line, then these integrals may be recognized as normal, with respective means $\hat{\theta}^*$ and $\hat{\theta}$, and respective variances $\sigma^{*2} = -[L^{*\prime\prime}(\hat{\theta}^*)]^{-1}$ and $\sigma^2 = -[L''(\hat{\theta})]^{-1}$. Hence we have the approximation

$$E\{g(\theta \mid x)\} \doteq \frac{\sigma^* e^{L^*(\hat{\theta}^*)}}{\sigma e^{L(\hat{\theta})}}. \tag{6}$$

Tierney and Kadane (1986) show that this approximation is $O(1/n^2)$, which is a bit surprising since both numerator and denominator approximations are $O(1/n)$. The reason for this is that the first order error terms in the numerator and denominator are the same, and hence cancel. The consequence of this is that if the denominator were known exactly, a better approximation to the ratio results from approximating both numerator and denominator than would result from using the exact denominator.

The extension of this approximation to the multivariate case is immediate: Equation (2) applies in vector form, again the constant order terms do not depend on $\theta$, again $L^{*\prime}(\hat{\theta}^*) = L'(\hat{\theta}) = 0$, and again the second order terms give normal integrals with means $\hat{\theta}$ and $\hat{\theta}^*$ respectively. Now, however, they are multivariate normal integrals, as $\sigma^*$ and $\sigma$ are redefined as minus the square-root of the determinant of the Hessian matrix of $L^*$ and $L$ respectively. Consequently this approximation satisfies the basic requirements for a successful computational method for Bayesian inference; it is multivariate, simple and has a claim to high-order accuracy.

Computer programs based on these ideas have been developed (Tierney, Kass and Kadane 1987, Kass, Tierney, and Kadane, 1988). The claims of asymptotic accuracy are proved with an attendant set of regularity conditions in Kass, Tierney and Kadane (1990).

The principle requirement on the fully exponential method as exposited so far in this paper, is its limitation to function $g$ bounded away from zero. However, Tierney, Kass and Kadane (1989c) show that application of (4) to the moment generating function and taking derivatives suffice to obtain approximations accurate to $O(1/n^2)$ for general functions $g$.

The question of computing marginal distributions, discussed by Tierney and Kadane (1986) is returned to in Tierney, Kass and Kadane (1989a), who demonstrate an efficient method to find an approximation to the marginal posterior distribution of an arbitrary smooth function of the parameters without having to find a 1-1 transformation, take the Jacobean and integrate out the unwanted dimensions. Finally, Tierney, Kass and Kadane (1989b) demonstrate how to use Laplace approximations to compute efficiently the effect on a posterior expectation of dropping a single observation, or of changing the prior.

## 3. LAPLACE'S METHOD ON CURVED SPACES

The first issue to address about Laplace's method as applied to curved surfaces is why a special treatment is desirable. After all, application of

(4) will still be order $O(1/n^2)$, since asymptotically a curved surface looks flat. There are two responses to this. To apply (4), which will be redubbed "rectangular asymptotics" to, say, the circle, requires treating the angle, say $\theta$, as being in the space $[0, 2\pi)$, without recognizing that 0 and $2\pi$ are the same angle. Thus if the maxima $\hat{\theta}$ or $\hat{\theta}^*$ were too close to 0 or $2\pi$, it would be desirable to reparameterize by rotation before applying the rectangular asymptotics. While this would be easy enough on the circle, in higher dimensional cases such rotation could easily become burdensome. Furthermore, an analysis of this sort seems the wrong mathematics: these problems are being forced into an awkward and unnatural framework. Second, our ambitions are to apply the asymptotics to problems with fairly small sample sizes (we will later discuss examples with sample sizes of nine and fifteen). In this context, the argument given above for asymptotics may be relying too heavily on large samples. For both these reasons, arithmetic and pragmatic, a treatment that pays special attention to the curved nature of the surface is desirable.

For these reasons, Bagchi and Kadane (1991) recently addressed the question of Laplace approximations on circular spaces and generalizations. The reminder of this section gives the ideas in this work, without repeating the details.

Reconsider equation (3), now taking the space of integration in both integrals to be the circle instead of the real line. The resulting integral, instead of being normal, is a Fisher-von Mises integral (see Mardia 1972), whose value for the numerator is given by $I_0(\gamma^*)\exp\gamma^*$, where $\gamma^* = L^{*\prime\prime}(\hat{\theta})$ and $I_0(\cdot)$ is the modified Bessel function of purely imaginary argument and zero order.

For spheres and hyperspheres, a similar analysis applies, only the integrals are Bingham integrals (Bingham 1964), with value ${}_1F_1(\frac{1}{2}, \frac{3}{2}, A)$, where ${}_1F_1$ is the confluent hypergeometric function of matrix argument and $A$ is diagonal.

For cylinders and hypercylinders (i.e., spaces of the form $\Re^q \times S_p$, when $S_p = \{\mu \mid \mu'\mu = 1$ and $\mu$ is of dimension $p\}$, the analysis depends on the order of integration. When the real part can be integrated out first, the usual partitioning of the quadratic form in the exponent into a conditional and a marginal applies to give a normal integral, and then a Bingham integral. However, there are problems involving marginal distributions over the $\Re^q$ space in which it is convenient to integrate first over $S_p$. This presents a bit of a delicate problem in that the partitioning into the marginal and conditional sums of squares would require the inverse of a singular matrix. However, a Lagrange multiplier technique, using the constraint $\mu^T\mu = 1$, yields a positive definite matrix.

Finally, the same methods apply to more general spaces of the form $\Omega = \Re^q \times \otimes_{i=1}^I S_{p_i}$. For example, when $q = 0, I = 2$, and $p_1 = p_2 = 2$, $\Omega$ in the surface of a torus. Similarly, when $q = 1, I = 2$ and $p_1 = p_2 = 2$, $\Omega$ is a solid torus.

Again, the details for all these are given in Bagchi and Kadane (1991).

## 4. EXAMPLES

For the following examples, we assume that the data follows a von Mises distribution with mean direction $\mu$ and concentration parameter $\kappa$. That is, we have a set of $n$ observation from a distribution with density given by

$$f(\theta \mid \mu, \kappa) = \frac{1}{2\pi I_0(\kappa)} \exp(\kappa \cos(\theta - \mu)).$$

The likelihood is given by

$$l(\mu, \kappa) \propto I_0^{-n}(\kappa) \exp\{\kappa R \cos(\mu - \bar{\theta})\},$$

where $R$ and $\bar{\theta}$ are the resultant and the mean direction of the sample respectively. That is, they are solutions to equations

$$\begin{aligned} \textstyle\sum_{i=1}^{n} \cos\theta_i &= R\cos\bar{\theta} \\ \textstyle\sum_{i=1}^{n} \sin\theta_i &= R\sin\bar{\theta}. \end{aligned}$$

Assuming a flat prior (purely for convenience and illustrative purposes), $p(\mu, \kappa) \propto \kappa$, the posterior distribution of the parameters is such that

$$p(\mu, \kappa \mid \text{ data }) \propto \kappa I_0^{-n}(\kappa) \exp\{\kappa R \cos(\mu - \bar{\theta})\}.$$

This joint distribution can now be used to compute the various marginal posterior distribution. We do so and plot them for the following data sets obtained from Mardia (1972).

Pigeon Data 1 consists of a set of 10 observations on vanishing angles of pigeons. For this data set we plot the exact marginal posterior distribution of directional parameter and the approximation obtained by the Laplace method and the first order approximation. Whereas the Laplace Approximation is almost the same as the exact, the marginal posterior using the first order approximation has fatter tails. This is to be expected since this approximation is valid for large values of the concentration parameter $\kappa$.

The approximate marginal posterior distribution of the concentration parameter $\kappa$, however, is the same as the exact computed numerically (see Bagchi and Kadane (1991)). The performance of the first order approximation in this case is even worse than the previous case as is very clear from the plot shown.

The same conclusions are reached for the other three data sets analyzed, namely the Pigeon Data 2 and 3, which are from similar experiments sample sizes $n = 15$. The Rock Data is a set of 10 "strikes" from a geological formation known as the "Upper Whitcliffians". (The "strikes" are the dip directions of the rock specimens.)

## 5. CONCLUSION

The methods given in this paper extend Laplacean methods to many curved surfaces. They replace posterior expectations over these spaces with two maximazations and two evaluations of special functions. They give good accuracy for all the examples considered. Bayesian analysis on such spaces should be feasible, provided the hypergeometric functions of matrix argument can be computed.

## ACKNOWLEDGMENT

We thank Professors Rob Kass, Luke Tierney, James Press, and J. Sethnuraman for their valuable comments. This research was supported by the National Science Foundation under Grant No. NSF/DMS- 8705646.

## REFERENCES

[1] ARNOLD K.J. (1941) "On Spherical Probability distributions", *Ph.D. thesis*, Massachusetts Institute of Technology.

[2] BAGCHI P. (1987) "Bayesian Analysis of Directional Data", *Ph.D. Thesis*, University of Toronto.

[3] BAGCHI P. and KADANE J.B. (1991) "Laplace Approximations to Posterior Moments and Marginal Distributions on Circles, Spheres and Cylinders". To appear in the *Canadian Journal of Statistics.*

[4] BAGCHI P. and GUTTMAN I. (1988) "Theoretical Considerations of the von Mises-Fisher Distribution", *Journal of Applied Statistics*, **15**, 149-169.

[5] BAI Z.D., RAO C.R, and ZHAO L.C. (1988) "Kernel Estimators of Density Function of Directional Data" *Journal of Multivariate Analysis*, **27**, 24-39.

[6] BINGHAM C. (1964) "Distributions on the Sphere and on the Projective Plane, *Ph.D. Thesis*, Yale University.

[7] DALLAL S. and HALL W.J. (1983) "Approximating Priors by Mixtures of Natural Conjugate Priors", *Jour. of Royal Stat. Society*, Ser. B, **45**, 278-286.

[8] DIACONIS P. and YLVISAKER D. (1985) "Quantifying Prior Opinion", in *Bayesian Statistics 2*, Bernardo J.M., Degroot M.H., Lindley D.V., Smith A.F.M., eds., North-Holland Publishing Co., 133-156 (with discussion).

[9] GEWEKE J. (1986) "Exact Inference in the Inequality constrained Normal Linear Regression model", *Jour. of Appl. Econ.*, **1**, 127-141.

[10] KASS R., TIERNEY L., and KADANE J.B. (1990) "The Validity of Posterior Expansions Based on Laplace's Method" in *Essays in Honor of George Barnard*, eds. Geisser S., Hodges J., Press J., and Zellner A. North Holland, 473-488.

[11] KASS R., TIERNEY L., and KADANE J.B. (1988) "Asymptotics in Bayesian Computation", *Bayesian Statistics 3*, eds. Bernardo J.M., Degroot M.H., Lindley D.V., and Smith A.F.M. Oxford University Press, 261-278.

[12] LAPLACE P.S. (1776) "Recherches, sur l'intégration des équations différentielles aux différenece finies, and sur leur usage dans ls theéorie des hasards, *Mémoires de mathématique et de physique presentés á l'Académie royale des sciences, par divers savans, and lus dans ses assembleées*, 1773 (published 1776) 37-162. (Reprinted in Laplace's *Oeuvres complétes 8*, pp. 69-197). Translated by S. Stigler in *Statistical Sciences*, (1986) **1** 364-378.

[13] LEONARD T. (1982) Comments on "A Simple Predictive Density Function", *Jour. of the Amer. Stat. Assoc.*, **77**, 657-658.

[14] LINDLEY D.V. (1961) "The use of the Prior Distribution in Statistical Inference and Decision", *Proc. of the Fourth Berkeley Symposium*, **1**, 453-468.

[15] LINDLEY D.V. (1980) "Approximate Bayesian Methods", in *Bayesian Statistics*, eds. J.M. Bernardo, M.H. Degroot, D.V. Lindley, and A.F.M. Smith, Valencia, Spain: University Press.

[16] MARDIA K.V. (1972) *Statistics of Directional Data*, Academic Press, New York.

[17] MARDIA K.V. and EL-ATOUM S.A.M. (1976) "Bayesian Inference for the von Mises-Fisher Distribution", *Biometrika*, **63**, 203-206.

[18] MOSTELLER F. and WALLACE D.L. (1964) *Inference and Disputed Authorship: The Federalist*, Reading, MA: Addison-Wesley.

[19] RAIFFA H. and SCHLAIFER R. (1961) *Applied Statistical Decision Theory*, Cambridge: Harvard University Press.

[20] SMITH A.F.M., SKENE A.M., SHAW J.E.H., NAYLOR J.C., and DRANSFIELD M. (1984) "The Implementation of the Bayesian Paradigm", *Commun. Statist. Theor. Meth.*, **14**, 1079-1102.

[21] STIGLER S.M. (1986) "Laplace's Memoir on Inverse Probability", *Statistical Science*, **1**, 359-362.

[22] TIERNEY L. and KADANE J.B. (1986) "Accurate Approximations for Posterior Moments and Marginal Densities", *Jour. of the Amer. Stat. Assoc.*, **81**, 82-86.

[23] TIERNEY L., KASS R., and KADANE J.B. (1987) "Interactive Bayesian Analysis using Accurate Asymptotic Approximations", *Computer Science and Statistics: Proceedings of the 19th Symposium on the Interface*, Heiberger R.M., ed., American Statistical Association, Alexandria, Va., 15-21.

[24] TIERNEY L., KASS R., and KADANE J.B. (1989a) "Approximate Marginal Densities of Nonlinear Functions", *Biometrika*, **76**, 425-433.

[25] TIERNEY L., KASS R., and KADANE J.B. (1989b) "Approximate Methods for Assessing Influence and Sensitivity in Bayesian Analysis", *Biometrika*, **76**, 663-674.

[26] TIERNEY L., KASS R., and KADANE J.B. (1989c) "Fully Exponential Laplace Approximation to Expectations and Variances of Non-positive Functions", *Jour. of Amer. Stat. Assoc.*, **84**, 710-716.

[27] VAN DIJK H.K., HOP J.P., and LOUTER A.S. (1987) "An Algorithm for the Computation no Posterior Moments and Densities using simple importance sampling", *The Statistician*, **36**, 83-90.

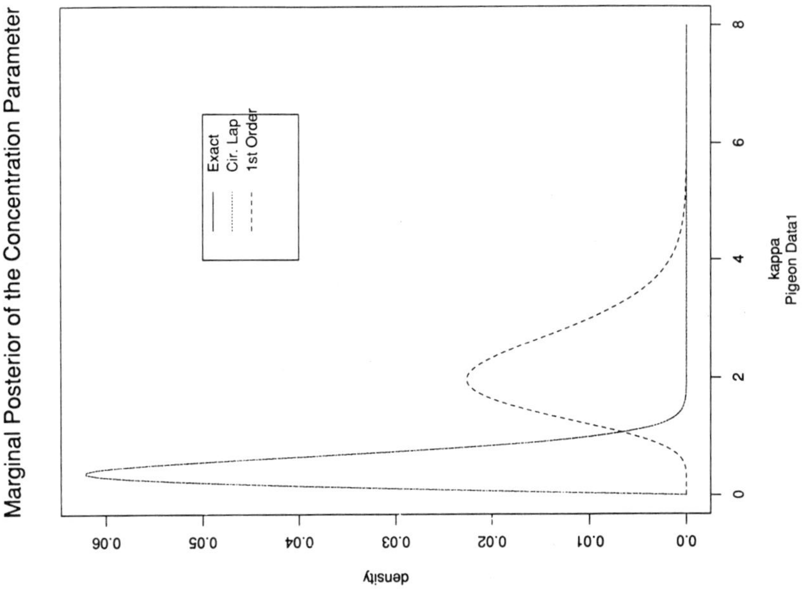
Marginal Posterior of the Concentration Parameter
Exact
Cir. Lap
1st Order
kappa
Pigeon Data1
density
0.0
0.01
0.02
0.03
0.04
0.05
0.06
0
2
4
6
8

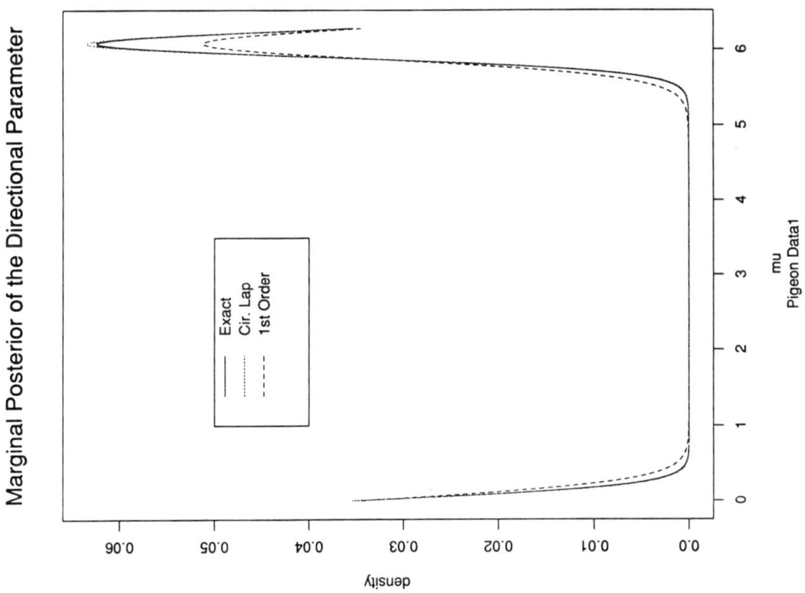
Marginal Posterior of the Directional Parameter
Exact
Cir. Lap
1st Order
mu
Pigeon Data1
density
0.0
0.01
0.02
0.03
0.04
0.05
0.06
0
1
2
3
4
5
6

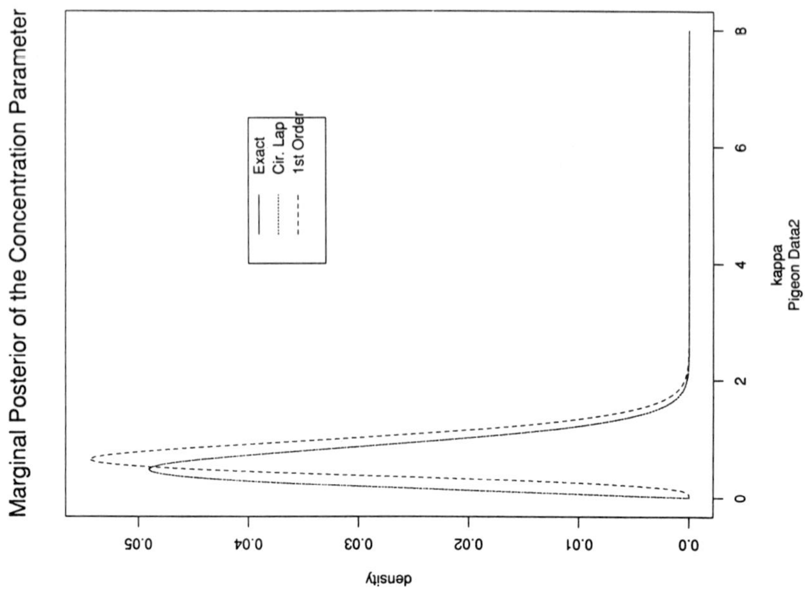
Marginal Posterior of the Concentration Parameter
Exact
Cir. Lap
1st Order
density
0.0
0.01
0.02
0.03
0.04
0.05
0
2
4
6
8
kappa
Pigeon Data2

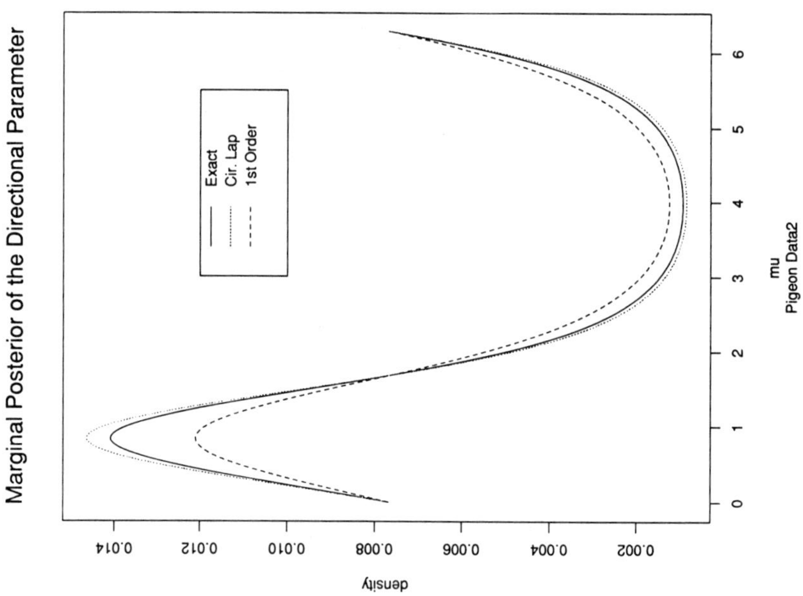
Marginal Posterior of the Directional Parameter
Exact
Cir. Lap
1st Order
density
0.002
0.004
0.006
0.008
0.010
0.012
0.014
0
1
2
3
4
5
6
mu
Pigeon Data2

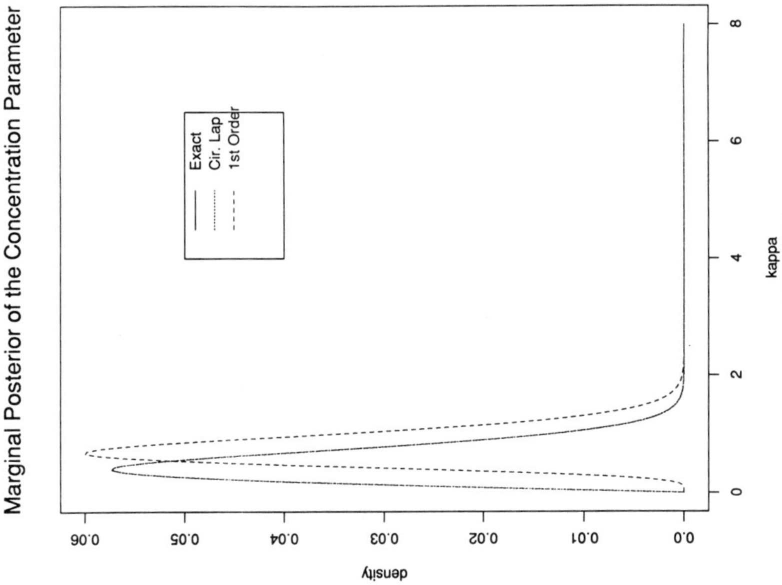
Marginal Posterior of the Concentration Parameter
Exact
Cir. Lap
1st Order
density
0.0
0.01
0.02
0.03
0.04
0.05
0.06
0
2
4
6
8
kappa
Pigeon Data3

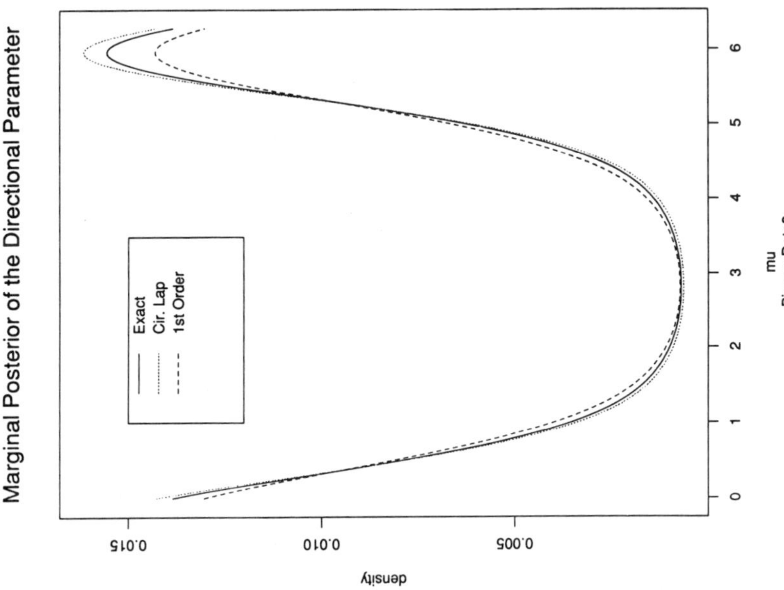
Marginal Posterior of the Directional Parameter
Exact
Cir. Lap
1st Order
density
0.005
0.010
0.015
0
1
2
3
4
5
6
mu
Pigeon Data3

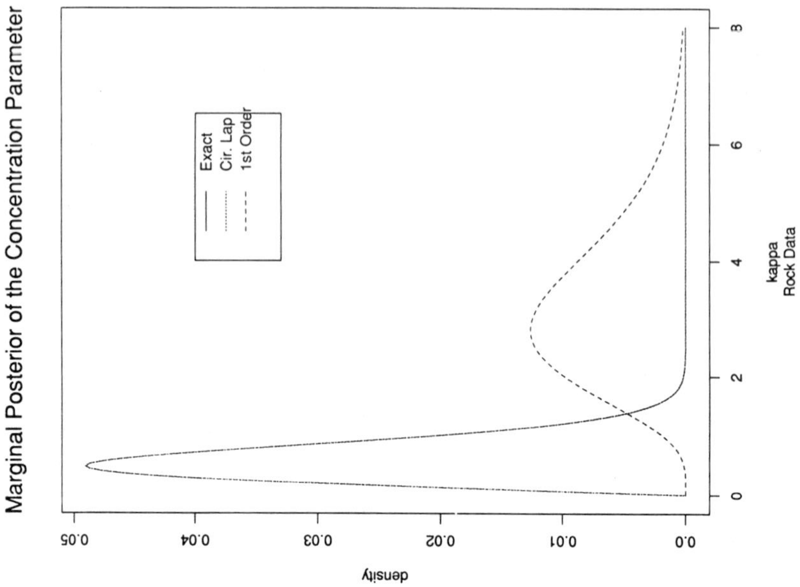
Marginal Posterior of the Concentration Parameter
Exact
Cir. Lap
1st Order
density
0.0
0.01
0.02
0.03
0.04
0.05
0
2
4
6
8
kappa
Rock Data

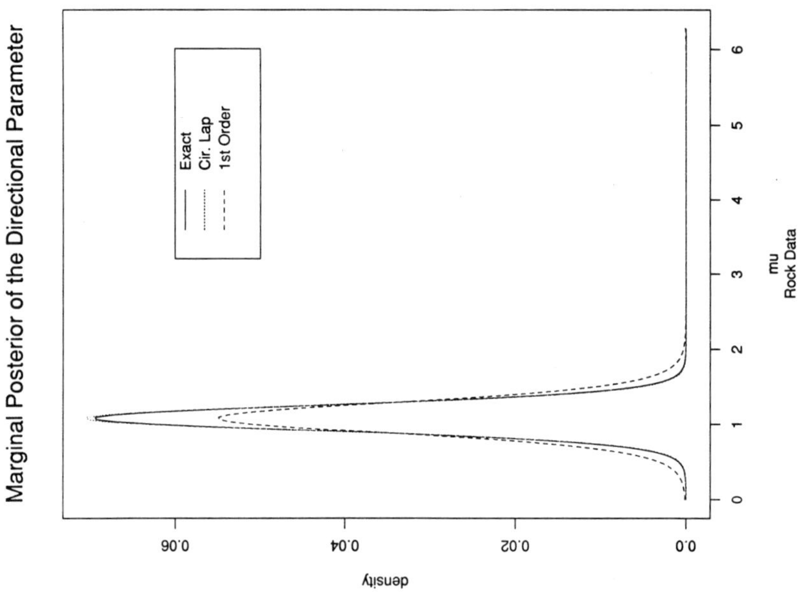
Marginal Posterior of the Directional Parameter
Exact
Cir. Lap
1st Order
density
0.0
0.02
0.04
0.06
0
1
2
3
4
5
6
mu
Rock Data

# 2

# Designing a Bayesian Object-Oriented Computing Environment

## William DuMouchel[1]

**Abstract**

This paper begins to work out the implications of providing software to support a fully Bayesian approach to a very general class of estimation problems. It focuses on continuing estimation problems, in which past projects contribute information about current parameters, and for which the hierarchical Bayesian approach is especially powerful. There are three key features of the proposed approach:

First, simple $t$-distributions approximates each separate likelihood function, so that the software design can be independent of the particular application and require only the first and second moments of classical parameter estimates. Second, similar $t$- distributions approximate the prior and posterior distributions of the parameters and hyperparameters. These have the two-fold advantage of being robust against misspecification of the model or prior distribution at each stage of the estimation, and of being closed under the Bayesian updating process. Third, an object-oriented software design allows users to construct, visualize, manipulate, and otherwise work with the distributions and likelihoods of the parameters of interest. Especially important is the ability to extend the distributions to include new parameters which may arise after a project has begun.

*Key words and Phrases:* Hierarchical Bayes; t-priors; Hyperparameters; Object-oriented software; Likelihood; informative priors; robustness.

## 1. INTRODUCTION

To fix ideas, consider the case of a toxicologist in an industrial or pharmaceutical laboratory who is repeatedly engaged in dose-response studies. Over time, many different chemical substances are under test, and many biological and physical systems are involved in the different testing protocols. Projects come and go, and although the scientist has expectations

[1]BBN Software Products Corporation, Cambridge, MA

about the effect of different classes of chemicals on different test systems, a principal goal may be to find the unexpected. Most protocols involve the estimation of a dose- response curve, and the parameters of these curves are constantly being compared from study to study. Sometimes information from other laboratories is to be used, as in DuMouchel and Harris (1983). For any particular study, the scientist has trustworthy classically-oriented software for estimating the dose-response parameters and the standard errors and covariance matrix of the estimates, and for exploring the data and the goodness of fit of the models. How can Bayesian software help this researcher?

To take another example, consider a quality control technician testing incoming samples of different types of products purchased from different vendors over a period of time. Routine statistical procedures, perhaps a straight-line regression or a paired comparison, are applied to the data from each sample. Can we design simple, easy-to-use, and robust general-purpose Bayesian software that increases the productivity of such technicians?

**Types of Bayesian Analyses**

When designing a Bayesian computing environment, it is important to distinguish between the many kinds of Bayesian analyses. This paper discusses Bayesian analyses that are intended to handle a very common, "bread and butter" statistical situation. The purpose of choosing this particular type of analysis is to encourage the widest possible use of Bayesian methodology by providing software support where it can be most useful.

The type of Bayesian analyses to be discussed here might be termed *standard, informative, hierarchical, continuing, and parametric.* The implications of each of these terms will be discussed next.

**Standard Applications vs Special Purpose Models**

The general linear model is the workhorse of statistics. The simple regression and analysis of variance setups comprise the majority of data analyses. Software for the classical approach to these problems is easy to use and has a standardized format so that nonstatisticians are comfortable using it to answer questions about their data. To be competitive, Bayesian software will have to build on this base and allow users to take advantage of what they know already. The methods discussed here will rely on standard models and the usual first and second order approximations which are often associated with them.

Of course, one of the chief advantages of the Bayesian paradigm is that it often supplies a direct solution for nonstandard problems for which there is no natural classical approach. Although these special-purpose applications will always play a large role in the Bayesian scheme of things, this paper focuses on the more common problems.

**Informative vs Noninformative Priors**

This paper also assumes that the prior distributions to be used will usually be informative, and incorporate whatever knowledge the user may possess about the problem. Specific attention is paid to the problems of eliciting prior knowledge, and of being robust against possible misspecification of the prior distribution. The assumption here is that if Bayesian methods merely mimic the classical analysis by choosing a prior which is designed to reproduce the classical results, then users have little incentive to switch to them. The true power of Bayesian methods comes from the fact that it can solve problems which are very hard to attack within the classical framework. The key contribution of the Bayesian approach to many difficult statistical problems is its ability to integrate into the analysis the user's background knowledge of the process being studied. The user must have tools designed to make informative prior distributions easy and natural to work with.

**Hierarchical Priors vs Conjugate or Special-Purpose Priors**

The use of hierarchical prior distributions, as discussed in Lindley and Smith (1972), Rubin (1980), DuMouchel and Harris (1983), and DuMouchel (1988, 1989), is emphasized. These Bayesian models have many advantages, and are useful in a great variety of data analysis problems.

**Continuing Estimation Problems vs Isolated Data Analyses**

The discussion here assumes that the data analyst is employed for a medium-to-long period of time trying to solve problems which have some degree of commonality. In order for informative prior distributions to be appropriate, there must be some background information available, usually from previous analyses of similar problems. A hierarchical model can usually incorporate such knowledge into an analysis.

**Parametric Formulations vs Predictive Formulations**

The Bayesian methods advocated here involve parametric models, freely defining parameters which are not observable directly. Of course the software should also allow the prediction of future observables, and the investigation of their posterior distributions. But the use of parameters is well established in scientific methodology, and most hierarchical models encourage the specification of parameters at various levels of abstraction.

## 2. DESIGN GOALS

Robust Bayes software should attempt to provide a computing environment for Bayesian applied statistics which is flexible enough, powerful enough, friendly enough, useful enough, and safe enough to be used regularly by data analysts. Several strategies are involved in striving for these goals:

**Simplicity**

The software should not hesitate to use approximations whenever they are useful, and most interactions with the user involve only estimation of means and 95% ranges of uncertainty, in the same spirit as the general linear model is usually used in classical statistics. Although the mathematics behind the hierarchical Bayesian approach can seem complicated, these complications can be hidden from the casual user.

**Object Orientation**

The prior and posterior distributions and likelihood functions involved in the Bayesian paradigm are objects in the computer system which the user interacts with directly. An environment for creating, visualizing, storing, editing, and otherwise manipulating these objects is provided. These objects interact with each other in accordance with the Bayesian paradigm.

**Modularity**

The statistician should not have to give up all the other tools of data analysis in order to benefit from the Bayesian paradigm. On the other hand, it would be a mammoth and wasteful project for Bayesians to reimplement most statistical computing algorithms just to give them a Bayesian flavor. This paper describes a more realistic strategy – a simple but general purpose Bayesian computer module which fits smoothly into the data analysts' usual computing environment. It supports the two specifically Bayesian operations in an analysis: the construction of a prior distribution and the updating of the prior distribution to form the posterior distribution. The computations to form the likelihood function, which are the same in a Bayesian or classical analysis, are performed by whatever software the user prefers to use.

**Robustness**

Since the Bayesian approach involves choosing the functional form for both the likelihood and the prior distribution, there is twice the chance for misspecification as in the classical approach to data analysis. Scientists who use informative prior distributions can accidentally overstate their certainty about a parameter. If a statistic diverges greatly from its prior expectation, the posterior uncertainty about the corresponding parameter should be greater than before, not less. Unfortunately, this does not happen in the standard Bayesian setup with conjugate priors. We cannot expect the scientific or industrial community to use Bayesian methodology routinely unless its users are routinely protected from the potentially severe effects of such misspecification.

## 3. THE $t$-DISTRIBUTION COMPUTER OBJECT

The probabilistic building-block of the proposed Bayesian objects is the multivariate Student's-$t$- distribution. Just as the most common uses of classical inferential statistics are based on statistics assumed to have an approximate $t$-distribution, a similar approximation is proposed here for both likelihood functions and prior and posterior distributions.

## Likelihood Object for a Vector of Parameters

The proposed Bayesian computer environment enables the creation and manipulation of a special class of data structures, or objects, called *likelihood objects.* Each likelihood object has a name and a set of instance variables which describe the results of a statistical estimation of a vector of parameters. These results may have been obtained from some standard estimation method such as least squares or maximum likelihood.

The instance variables would include the following information:

Names for each parameter
Estimates for each parameter
Estimated covariance matrix for the sampling distribution of the estimates
Degrees of freedom to be associated with the estimated covariance matrix
Text describing the background of the estimation

For a particular likelihood object, suppose that the vector of estimates has the value $\mathbf{y}$ and that the estimated covariance matrix has the value $\mathbf{C}$. Then the interpretation of the likelihood object is that the vector $\mathbf{y}$ is a value of a random variable that, conditional on the vector $\boldsymbol{\theta}$ and the scalar $\tau$, has the $N(\boldsymbol{\theta}, \tau^2\mathbf{C})$ distribution, where $\tau^{-2}$ has a chi-squared over degrees of freedom distribution, and is independent of $\boldsymbol{\theta}$.

The degrees of freedom instance variable might be the degrees of freedom of the residual sum of squares in a traditional least squares estimation problem, or it may be subjectively assigned as a way of describing the reliability which the user ascribes to $\mathbf{C}$.

## Distribution Object for a Vector of Parameters

The other major type of object in the proposed Bayesian computing environment is a *distribution object.* A distribution object is similar in structure to a likelihood object, but it has a somewhat different set of instance variables. Each distribution object has a name as well as the following instance variables:

Names for each parameter
Means for each parameter
Hyperparameter information
  Names for each hyperparameter
  Means for each hyperparameter
  Matrix describing the relationship between the hyperparameters and the parameters
Precision matrix for the vector of parameters and hyperparameters
Degrees of freedom to be associated with the precision matrix
Text describing the background of the distribution
List of names of likelihood objects, if any, for which Bayes rule has been applied to form this distribution object, and time stamps of such update operations

For a particular distribution object, suppose that the vector of means of parameters (including the hyperparameters) has the value $m$ and that their precision matrix has the value $\mathbf{V}^{-1}$. Then the interpretation of the distribution object is that the parameter vector $\boldsymbol{\theta}$, conditional on the scalar $\sigma$, has the $N(\mathbf{m}, \sigma^2\mathbf{V})$ distribution, where $\sigma^{-2}$ has a chi-squared over degrees of freedom distribution. The relationship between parameters and hyperparameters in a distribution object is discussed in the next subsection.

**Constructing a Multivariate $t$-Distribution Object**

Suppose that the parameters are denoted by the vector $\boldsymbol{\theta}$, of length $K$, and the hyperparameters are denoted by the vector $\boldsymbol{\mu}$, of length $p$. Then, in the absence of any knowledge about $\boldsymbol{\theta}$ specifically, the hierarchical prior distribution assumes that, conditional on $\boldsymbol{\mu}$ and $\sigma^2$,

$$\boldsymbol{\theta} = \mathbf{X}_{\mu} + \boldsymbol{\delta},$$

where $\mathbf{X}$ is an $n$-by-$p$ known matrix and $\boldsymbol{\delta}$ is a vector of error terms with means $\mathbf{0}$ and variances $\sigma^2\mathbf{V}_{\delta}$, where $\mathbf{V}_{\delta}$ is a user- specified diagonal matrix. The user must also specify a prior mean vector, $\mathbf{m}$, and prior variances, $\mathbf{V}_{\mu}$ for the hyperparameter vector $\boldsymbol{\mu}$. The proposed design restricts $\mathbf{V}_{\mu}$ to be a diagonal matrix, so that conditional on $\sigma$, all the elements of $\boldsymbol{\mu}$ and $\boldsymbol{\delta}$ are independent and normally distributed *apriori*.

DuMouchel (1987) discusses one possible method for using interactive graphics to elicit the prior distributions for $\mu$ and $\delta$, and the matrix $\mathbf{X}$ in the simple situation in which $\mathbf{X}$ consists of the design matrix for a one-way layout. The proposed software will extend that methodology.

Once the values of $\mathbf{X}$, $\mathbf{m}$, $\mathbf{V}_{\mu}$, and $\mathbf{V}_{\delta}$ are given, the moments of $\boldsymbol{\theta}$ are

$$\begin{aligned}
&E(\boldsymbol{\theta}) = \mathbf{Xm} \\
&V(\boldsymbol{\theta}|\sigma^2) = \sigma^2(\mathbf{XV}_{\mu}\mathbf{X}^t + V_{\delta}), \\
&\mathrm{Cov}(\boldsymbol{\theta}, \boldsymbol{\mu}|\sigma^2) = \sigma^2\mathbf{XV}_{\mu}.
\end{aligned}$$

In addition, the proposed software will handle the common situation in which some of the elements of $\mathbf{V}_{\mu}$ are arbitrarily large, so that the corresponding elements of $\boldsymbol{\mu}$ have a diffuse distribution.

Finally, $\mathbf{V}$ is the $(K+p)$-by-$(K+p)$ covariance matrix of $(\boldsymbol{\theta}, \boldsymbol{\mu})$, conditional on $\sigma = 1$, and $\mathbf{V}^{-1}$ is the precision matrix of these parameters.

**Examining a Multivariate $t$-Distribution Object**

Each of the $t$-distribution objects in the user's computing environment can respond to requests for various descriptive displays which show the corresponding likelihood function or distribution. For example, measures of location and scale and graphs of densities and distributions. In addition, distribution objects can display numerical and graphical summaries of the distributions of linear combinations of their parameters and hyperparameters.

## 4. BAYES RULE FOR MULTIVARIATE $t$- DISTRIBUTIONS

The mathematical framework for updating a $t$-distribution object for a parameter vector, using a $t$-distribution likelihood object, is as follows.

Assume that the likelihood identifies a parameter vector $\boldsymbol{\theta}$, and that conditional on $\boldsymbol{\theta}$ and $\tau$, a statistic $\mathbf{y}$ has the approximate likelihood

$$\mathbf{y} \sim N(\boldsymbol{\theta}, \tau^2\mathbf{C}),$$

where $\tau^{-2}$ has a chi-squared/df distribution with $df_\tau$ degrees of freedom. This information is contained in a likelihood object. Suppose that a distribution object contains the information that the parameter vector $(\boldsymbol{\theta}, \boldsymbol{\mu})$ has, conditional on $\sigma^2$, the distribution

$$(\boldsymbol{\theta}, \boldsymbol{\mu}) \sim N(\mathbf{m}, \sigma^2\mathbf{V}),$$

where $\mathbf{m} = (\mathbf{m}_\theta, \mathbf{m}_\mu)$, and where $\mathbf{V}$ is a similarly partitioned covariance matrix, and where $\sigma^{-2}$ has a chi-squared/df distribution with $df_\sigma$ degrees of freedom. Assume that the distributions of $\tau$ and $\sigma$ are independent at every update operation. The parameter vector $\boldsymbol{\mu}$ contains all the parameters in the distribution object which are not referenced in the likelihood object, including the hyperparameters. (Assume that that the distribution object contains all the parameters referenced in the likelihood. If not, the user must extend the distribution to include the missing parameters, before using the likelihood.)

The derivation of the posterior distribution of $(\boldsymbol{\theta}, \boldsymbol{\mu})$ is only sketched here. Although the notation in this paper implies that $\mathbf{V}$ is nonsingular and finite, the theory and the software can be extended to allow some components of $(\boldsymbol{\theta}, \boldsymbol{\mu})$ to have a diffuse distribution. See DuMouchel (1988, 1989) for details.

### Posterior Distribution of Variance Ratio

To derive the posterior distribution, note that if one conditions on the two variance-component hyperparameters, $\sigma^2$ and $\tau^2$, the conjugate normal theory applies and the posterior distributions are easily derived. In fact, one merely needs to condition on the ratio, $\phi = \sigma^2/\tau^2$, and the posterior distribution of $(\boldsymbol{\theta}, \boldsymbol{\mu})$, conditional on $\phi$, is an easily derived multivariate $t$-distribution with $df_\tau + df_\sigma + K$ degrees of freedom, where $K$ is the dimensionality of $\boldsymbol{\theta}$.

Therefore the first step in the calculations is to compute the posterior distribution of $\phi$. Denote the posterior distribution $p(\phi|\mathbf{y})$, which has the form

$$p(\phi|\mathbf{y}) \propto \phi^{-[df_\sigma+1]/2}|W(\phi)|^{.5}\{df_\tau + df_\sigma/\phi + S(\mathbf{y},\phi)\}^{-[df_\tau+df_\sigma+K]/2},$$

where

$$\mathbf{W}(\phi) = (\phi\mathbf{V}_{\theta\theta} + \mathbf{C})^{-1},$$
$$S(\mathbf{y},\phi) = (\mathbf{y} - \mathbf{m})^t\mathbf{W}(\phi)(\mathbf{y} - \mathbf{m}).$$

**Computation of Posterior Means and Covariances**

Once $p(\phi|\mathbf{y})$ has been computed, the computation of moments of the posterior distribution is a matter of integrating conditional moments with respect to it. For example,

$$E(\boldsymbol{\theta}|\mathbf{y}) = \int E(\boldsymbol{\theta}|\phi, \mathbf{y})p(\phi|\mathbf{y})d\phi,$$

and similar formulae are used to compute $E(\boldsymbol{\mu}|\mathbf{y}), \mathrm{Cov}(\boldsymbol{\theta}, \boldsymbol{\mu}|\mathbf{y}), E(\sigma^2|\mathbf{y})$, $E(\sigma^{-2}, |\mathbf{y})$, and $E(\tau^2|\mathbf{y})$. See DuMouchel (1988, 1989) for details.

**Approximation of Posterior by a Multivariate *t*- Distribution**

The posterior distribution of $\boldsymbol{\theta}, \boldsymbol{\mu}$ is thus a mixture of multivariate $t$-distributions, where the mixing distribution is $p(\phi|\mathbf{y})$. In order to continue using the multivariate $t$-distribution objects for continuing problems, a method of moments approximation is used. The means and covariances of $(\boldsymbol{\theta}, \boldsymbol{\mu}|\mathbf{y})$ are computed as mentioned above. It only remains to choose the degrees of freedom for the posterior distribution of $\sigma^2$. Remember that the prior distribution of $\sigma^{-2}$ is a chi-squared/$df$ distribution, so that

$$E(\sigma^{-2}) = 1$$
$$E(\sigma^2) = df_\sigma/(df_{\sigma^{-2}}), \qquad (\text{ if } df_\sigma > 2).$$

This suggests that the posterior degrees of freedom, say $df^\star_\sigma$ can be chosen so that

$$E(\sigma^{-2}|\mathbf{y})E(\sigma^2|\mathbf{y}) = df^\star_\sigma/(df^\star_\sigma - 2).$$

Thus, after every Bayesian update of the $t$-distribution object, the mean of $\sigma^{-2}$ is rescaled to unity, and the $df_\sigma$ value is reset to measure the relative uncertainty in $\sigma$, as measured by the product $E(\sigma^{-2}|\mathbf{y})E(\sigma^2/y)$, which, by Shwartz' inequality, is at least one.

The posterior expectations $E(\tau^2|\mathbf{y})$ and $E(\tau^{-2}|\mathbf{y})$ can also be computedand used to approximate the posterior distribution of $\tau^{-2}/E(\tau^{-2}|\mathbf{y})$ by a chi-squared over $df$ distribution. The model does not assume any relationship between the $\tau$ parameters for different likelihood functions, and the software design has no provision for updating the distribution of a previous $\tau$ parameter to make use of information from later likelihood functions. We assume that the variance hyperparameters $\sigma$ and $\tau$ are of little intrinsic interest to the user, except in how they affect certainty about the other parameters and hyperparameters.

**Updating the Distribution of Hyperparameters**

During the construction of a hierarchical prior distribution, before it has been updated with a likelihood function, the roles of parameters, $\boldsymbol{\theta}$, and hyperparameters, $\boldsymbol{\mu}$, are quite distinct. As discussed in Section 3, the hyperparameters are used to define the conditional means of the parameters, with the matrix $\mathbf{X}$ defining this relationship. Frequently the hyperparameters are assigned a diffuse distribution, while the distribution of

$\boldsymbol{\theta}$ conditional on $\boldsymbol{\mu}$ may be quite informative. But after a Bayesian update using an informative likelihood function, the posterior distributions of both types of parameters are usually similarly dispersed. Also, the relation $E(\boldsymbol{\theta}|\mu) = \mathbf{X}\boldsymbol{\mu}$ no longer holds. Therefore, in the continuing Bayesian scenario envisioned here, the distinction between parameters and hyperparameters becomes blurred. During each successive Bayesian update, the formulas merely distinguish between the subset of parameters for which the current likelihood carries information, and the remaining parameters, which usually include the original hyperparameters. (Because these remaining parameters are usually correlated with the parameters referenced in the likelihood, their distributions are affected by the Bayesian update procedure.)

## 5. EXTENDING HIERARCHICAL DISTRIBUTIONS

### Today's Posterior Distribution is Tomorrow's Prior Distribution

The purpose of approximating the posterior distribution by a multivariate $t$-distribution is to make the update process repeatable. The next likelihood can be included in the same way as the last. At each stage the approximation preserves the first and second moments of the distribution of the parameters, and the degrees of freedom of the distribution provides a data-dependent robustness tuning constant. If $df_\sigma$ is large, the updating equations lead to results similar to that of a normally distributed prior distribution. But when $df_\sigma$ is small, the posterior distribution is more robust to misspecification of $\mathbf{m}$ and $\mathbf{V}$. If the data $\mathbf{y}$ differ greatly from the prior mean $\mathbf{m}_\theta$, the elements of $\mathbf{V}$ can increase rather than decrease with the update calculations, and the value of $df_\sigma$ can decrease, to reflect greater uncertainty about the process. See DuMouchel (1989) for an example.

### Extendability of Distributions to Estimate New Parameters

In order to have the distribution object remain useful over a period of time, it is necessary to allow for the definition and use of new parameters as new, but related, data analysis problems arise. The hierarchical aspect of the model now comes into play. In order to add a new element of $\boldsymbol{\theta}$ to the distribution, it is only necessary to add a new row to $\mathbf{X}$, the matrix introduced in Section 3 which relates the parameters to the hyperparameters, assuming that the new parameter can be defined in terms of the same hyperparameters. The new mean and covariances are determined by those of $\boldsymbol{\mu}$:

$$E(\boldsymbol{\theta}_{new}) = \mathbf{x}_{new}\boldsymbol{\mu},$$
$$\mathbf{V}(\boldsymbol{\theta}_{new}|\sigma^2) = \sigma^2(\mathbf{x}_{new}\mathbf{V}_{\mu\mu}\mathbf{x}^t_{new} + \nu_\delta),$$

where $\mathbf{x}_{new}$ new is the new row of $\mathbf{X}$, $\nu_\delta$ is the user-specified variance of the new error term and $\mathbf{V}_{\mu\mu}$ is the current covariance matrix of the vector $\boldsymbol{\mu}$.

Sometimes it may be necessary to add new hyperparameters to the distribution in order to analyze new data. For example, in the quality control example mentioned in Section 1, a vendor may come along who has not been considered previously. In such cases the user may have to elicit means, variances and covariances for new hyperparameters. It is also necessary to add a column to $\mathbf{X}$ for each new hyperparameter, whose values are zero for the already-defined parameters.

## 6. THE FUTURE OF BAYESIAN STATISTICS

The article by Efron (1986) and its accompanying discussion discuss the question "Why isn't everyone a Bayesian?" This paper argues that the primary factor is the lack of simple and useful software to implement the Bayesian paradigm. The review of currently available Bayesian software by Goel (1988) makes it painfully clear that the only software powerful enough to help users solve realistic problems either requires a Ph. D.- level of knowledge of the ins and outs of Bayesian statistics, or else encourages the use of noninformative prior distributions to such a degree that the answers hardly differ from the classical approach. We have to have simple-to-use software, and it has to encourage the use of informative prior distributions.

Unless such software becomes generally available soon, it is hard to imagine Lindley's (1975) prediction of a Bayesian 21st century coming to pass. Consider the situation today. What fraction of all currently performed data analyses use Bayesian methods? We might imagine a calculation in which the number of copies of all statistical packages in the world is multiplied by the average number of times per year each copy is accessed to perform an inferential analysis. The proportion of this total which represents the Bayesian analyses is surely infinitesimally small. I would guess that this proportion is of a similar order of magnitude as the proportion of all plays of video cassette recorders which use the Beta format. Are there still some optimistic engineers at Sony Corporation who dream of a "Beta 21st century?"

## REFERENCES

DuMouchel, William and Harris, Jeffrey (1983) Bayes methods for combining the results of cancer studies in humans and other species (with discussion). *Journal of the American Statistical Association*, **78**: 293-315.

DuMouchel, William (1988) A Bayesian model and a graphical elicitation procedure for multiple comparisons. *Proceedings of the Third Valencia Meeting on Bayesian Statistics*, (J. Bernardo et al, eds.) University Press, Valencia, Spain.

DuMouchel, W. (1989) Bayesian Metaanalysis. in *Statistical Methods for Pharmacology*, Donald Berry, ed., New York: Marcel Dekker.

Efron, B. (1986) Why isn't everyone a Bayesian? (with Discussion). *The American Statistician* 40:1-11.

Goel, P. K. (1988) Software for Bayesian analysis: Current status and additional needs-II. *Proceedings of Interface 88: Computer Science and Statistics.* Reston, Virginia.

Lindley, D. V. (1975) The future of statistics – a Bayesian 21st century. *Supp. Adv. Appl. Prob.* **7**:106-115.

Lindley, D.V., and Smith, A.F.M. (1972) Bayes estimates for the linear model (with discussion). *Journal of the Royal Statistical Society*, **Ser. B, 34**: 1-41.

Rubin, D.B. (1980) Using empirical Bayes techniques in the law school validity studies (with discussion). *Journal of the American Statistical Association*, **75**: 801-827.

## COMMENTS

Mark J. Schervish
*Carnegie-Mellon University*

Bill DuMouchel is to be applauded for attempting to design a Bayesian computing environment. Bayesian analysis will never be more than a seldom applied theoretical concept until adequate facilities are available for its routine use. I agree with the author that a chief advantage of the Bayesian paradigm is that "it often supplies a direct solution for nonstandard problems for which there is no natural classical approach." In fact, I would go so far as to say that the Bayesian paradigm is designed to supply answers to all questions concerning uncertainty. What is lacking are the tools to implement the theory in practice.

In addition to the "ability to integrate into the analysis the user's background knowledge of the process being studied", the Bayesian approach also provides answers which are more straightforward and easier to understand than the classical approach. I would even disagree with DuMouchel's claim that "since the Bayesian approach involves choosing the functional form for both the likelihood and the prior distribution, there is twice the chance for misspecification as in the classical approach to data analysis." Rather, I would say that, for the very reason that the Bayesian approach involves choosing both the likelihood and the prior, there is a greater chance of choosing a good model than in the classical approach.

Despite my agreement with DuMouchel's objective and my support for his project, I would like to mention a few points of disagreement about where the emphasis should be placed. This is not to say that I think any of Bill's efforts are misguided, but rather that I would suggest a few alterations.

With regard to the distinction between conjugate an hierarchical priors, I would point out that, in some situations (notably normal linear models) there exist conjugate hierarchical priors. Bill claims "if a statistic diverges greatly from its prior expectation, the posterior uncertainty about the corresponding parameter should be greater than before, not less. Unfortunately, this does not happen in the standard Bayesisan setup with conjugate priors." With proper hierarchical conjugate priors, the desired behavior actually does occur, but not in precisely the same manner as it does in non-conjugate priors.

Although there clearly is a need for Bayesian software to deal with problems over a medium-to-long period of time having some degree of commonality, I believe that the most urgent need is to provide Bayesian tools for those types of problems which people work on most often, namely regression and analysis of variance. Bayesian hierarchical models can help improve the analysis of data in such situations, while at the same time providing a positive impression of the usefulness of Bayesian methods to a large segment of the community of data analysts.

What I believe to be the greatest advantage of the Bayesian paradigm is its ability to treat predictive inference in a sensible fashion. For this reason, I am slightly disappointed that DuMouchel has deliberately chosen to down-play the role of prediction. It is true that "hierarchical models encourage the specification of parameters at various levels of abstraction", but statisticians have been notoriously inept at providing sensible interpretations for most parameters. For example, in an analysis of variance, the mind can often stretch to imagine arbitrary numbers of observations in each cell of a design, but it becomes increasingly more difficult to imagine arbitrary numbers of cells and arbitrary numbers of designs as we progress up the hierarchy. Thinking of the parameters as tools for expressing various levels of partial exchangeability among observables, and then providing predictive inferences for the observables seems to me, at least, to be the most promising approach for making Bayesian methodology palatable to the population at large.

There were two technical issues that I would like to raise for possible consideration in the design of the computer environment. These concern the discussion of the multivariate $t$−distribution results. First, I am slightly concerned about the approximation of the posterior of $(\theta, \mu)$ by a multivariate $t$−distribution. I would suggest an investigation of how much shape information would be lost in this approximation. This could presumably be done empirically in several examples. Secondly, there is the claim that "the software makes no provision for updating the distribution of a previous $\tau$ parameter to make use of information from later likelihood functions." Might the following approach be possible and still preserve the possibility of returning to earlier analyses with new information? Instead of making the approximation part of each likelihood object, leave the objects in the form of conditional distributions given $\sigma$. Then have a pointer to the $\sigma$ object so that the approximation could be performed when needed. This way, after the distribution of $\sigma$ is updated with later data sets, any earlier object involving a $\tau$ could then be reapproximated using the new $\sigma$ information. This would change the marginal variance of the previous $\theta$ distribution as well as the degrees of freedom and any predictive distributions with which one is concerned.

In summary, I would encourage Bill to proceed with development of his project. In addition, as already mentioned, I would suggest a few modest changes in focus. I have every confidence that DuMouchel will achieve a significant level of success, within the parameters he has set up, and I predict that the computing environment will be a most powerful tool for data analysis.

## REJOINDER

Many thanks to Professor Schervish for his careful reading of my paper and for his suggestions and encouragement. Here are my reactions to six of his comments:

(1) "for the very reason that the Bayesian approach involves choosing both the likelihood and the prior, there is a greater chance of choosing a good model than in the classical approach." With a more powerful tool, you have both a greater chance of doing better and a greater chance of misuse.

(2) "there exist conjugate hierarchical priors." Yes. I am only saying that the simplest nonhierarchical priors are often too simple to be both useful and safe in practical applications.

(3) "the most urgent need is to provide Bayesian tools for ... regression and analysis of variance." I agree that that need is urgent, since the classical approaches to hypothesis testing and simultaneous inference are so unsatisfactory from a Bayesian point of view. I would suggest taking the sufficient statistics from a standard regression or Anova problem and entering them into my proposed Bayesian object-oriented computing environment.

(4) "I am slightly disappointed that DuMouchel has deliberately chosen to down-play the role of prediction." I have no disagreement with the paragraph containing this sentence. The paper did not propose an elicitation process for the prior distributions of my proposed system. Whenever possible, elicitations should involve observables, and of course predictive distributions should be available as outputs. However, if you have to specify $\mathbf{X}$ in the relationship $\boldsymbol{\theta} = \mathbf{X}\boldsymbol{\beta} + \boldsymbol{\delta}$ in order to set up a hierarchical model, it is no use forbidding yourself to think about parameters.

(5) "I would suggest an investigation of how much shape information would be lost in this approximation" of the posterior by a multivariate $t$-distribution. Good idea.

(6) "leave the objects in the form of conditional distributions given $\sigma$. Then have a pointer to the $\sigma$ object..." Such refinements might well be feasible and more accurate than the one proposed. My inclination is to stick with the simpler method at first.

# 3

# Bayesian Estimation in Multidimensional Contingency Tables


## Leonardo D. Epstein [1]
## Stephen E. Fienberg [2]



### Abstract

This paper examines a new class of prior distributions for Bayesian estimation in contingency tables. It exploits both the conjugate Dirichlet prior when the data follow a multinomial distribution and the traditional loglinear parametrization by applying it to the means of the Dirichlet prior. The resulting prior distributions can reflect the user's uncertainty about both the applicability of a loglinear structure and the values of the cell probabilities. Connections with the maximum likelihood approach permit the user to provide initial guesses at the cell probabilities either through various margins of the cell prior means or through the "effects" of a loglinear model. We apply our method to medical data concerning skin cancer.



*Key words and Phrases:* Contingency tables; Bayesian estimation; Dirichlet prior distribution; Loglinear model; Shrinkage estimator; Mixture of distributions.


## 1. INTRODUCTION

This article addresses the problem of Bayesian estimation of the expected proportions in contingency tables and explores the properties of a new class of prior distributions. There have been two rather parallel lines of development in the Bayesian analysis of contingency tables, each one possessing its own advantages. The first line exploits the conjugate Dirichlet distribution to model prior knowledge when the data follow a multinomial distribution. This approach is best explained by Good (1965) and Good and Crook (1980). A natural extension of this work uses hierarchical priors based on mixtures of Dirichlet distributions, e.g. see Good (1967), Fienberg and Holland (1973) and more recently Albert and Gupta (1982, 1983) and Wang (1985). Leonard and Novick (1986) also use the Dirichlet approach. Their treatment remains confined to the two-dimensional case. Although they


[1] Johns Hopkins University

[2] Carnegie Mellon University

suggest it might be possible to extend it to the case of full independence among all the factors in multi-way tables they do not pursue this.

The second line of development is the loglinear approach developed independently in Fienberg and Holland (1973) and in Leonard (1975). Leonard applies in the logit scale results analogous to those of Lindley and Smith (1972) for the Bayesian analysis of linear models, and utilizes a multivariate normal approximation to study the behavior of his estimates. Other Bayesian attempts to use a loglinear approach include Nazaret (1987) and Laird (1978). While loglinear models have played a central role in the analysis of muldimensional contingency tables, the bulk of the literature has remained confined to maximum-likelihood techniques, and Bayesians have been unable to profit fully from loglinear parametrizations.

Here we combine the Dirichlet and the loglinear approaches using a two-stage prior distribution. At the first stage we use a Dirichlet distribution and a loglinear parametrization for the cell prior means. For the second stage of the prior one can either use distributions on the loglinear parameters or distributions on some suitable margins of the array of Dirichlet means means. We present several results that demonstrate how this class of priors can reflect the uncertainty about the presence of a structure in the cell probabilities as well as the uncertainty about their values in a compelling fashion.

To introduce the Dirichlet approach we consider an $r_1 \times r_2$ contingency table $\boldsymbol{x} = \{x_{ij} : i \in \bar{r}_1, j \in \bar{r}_2\}$ where $x_{ij}$ denotes the observed frequency in the $(i,j)$ cell, $\bar{r}_1 = \{1,\ldots,r_1\}$, $\bar{r}_2 = \{1,\ldots,r_2\}$, and $\theta_{ij}$ denotes the probability that an observation falls in that particular cell. We assume that $\{x_{ij}\}$ has a multinomial distribution with parameters $N$ and $\boldsymbol{\theta} = \{\theta_{ij}\}$, which we denote by $M(N,\boldsymbol{\theta})$. Under this sampling model, we consider the problem of estimating the vector of cell probabilities using prior information about the cross classification structure of the table.

The usual method of utilizing prior information assumes that $\boldsymbol{\theta}$ follows a Dirichlet distribution $d(K,\boldsymbol{\eta})$ with density

$$\frac{\Gamma(K)}{\Pi_{i,j}\Gamma(K\eta_{ij})}\Pi_{i,j}\theta_{ij}^{K\eta_{ij}-1}$$

where $K > 0$, $\eta_{ij} \geq 0$, $\sum_{ij}\eta_{ij} = 1$ and $\boldsymbol{\eta} = \{\eta_{ij}\}$. Under this parametrization the cells prior means are $E(\theta_{ij}|\boldsymbol{\eta}) = \eta_{ij}$. The parameter $\eta_{ij}$ can be interpreted as the statistician's prior guess at the probability $\theta_{ij}$, and the parameter $K$, usually called the *flattening constant* (see Good, 1965 and Fienberg and Holland, 1972), as the sureness about the set of prior guesses $\{\eta_{ij}\}$. If $\{x_{ij}\}$ follows the multinomial distribution $M(N,\boldsymbol{\theta})$ and $\boldsymbol{\theta}$ has a density $d(K,\boldsymbol{\eta})$, then the posterior distribution of $\boldsymbol{\theta}$ is a Dirichlet distribution $d(K+N, \frac{N}{K+N}\frac{\boldsymbol{x}}{N} + \frac{K}{K+N}\boldsymbol{\eta})$. The formulae presented thus far are also applicable to higher-dimensional tables.

When the statistician believes with uncertainty that there is a structure linking the row and column variables in the table but cannot specify the values of the parameters $\{\eta_{ij}\}$, then the standard Dirichlet prior is inadequate since it cannot reflect properly the uncertainty about the parameters $\{\theta_{ij}\}$. To remedy this problem we can use the Dirichlet distribution as the first stage of a two-stage hierarchical prior. A two-stage prior reflects the uncertainty about both the structure of the cell probabilities and the Dirichlet means.

Section 2 introduces, for two-way tables, a two-stage prior to reflect vague belief of independence along with information about the prior marginal means, and then briefly discusses properties of the resulting estimators. Section 3 extends the approach to multidimensional tables. Finally Section 4 presents a medical example concerning individuals affected by skin cancer who were classified according to three attributes.

## 2. ESTIMATION USING PRIOR KNOWLEDGE ABOUT INDEPENDENCE

Suppose we wish to estimate the cell probabilities in an $r_1 \times r_2$ table using prior knowledge about independence, i.e., we believe the cell probabilities satisfy $\theta_{ij} = \theta_{i\cdot}\theta_{\cdot j}$, for all $i, j$, where $\theta_{i\cdot} = \sum_{j=1}^{r_2} \theta_{ij}, i \in \bar{r}_1$ and $\theta_{\cdot j} = \sum_{i=1}^{r_1} \theta_{ij}, j \in \bar{r}_2$ are the marginal probabilities in the table. Since we want the $\eta_{ij}$'s to reflect the vague belief in independence they should satisfy

$$\eta_{ij} = \eta_{i\cdot}\eta_{\cdot j}\ , \tag{3.1}$$

where $\eta_{i\cdot} = \sum_{j=1}^{r_2} \eta_{ij}$ and $\eta_{\cdot j} = \sum_{i=1}^{r_1} \eta_{ij}$. Hence we can use the prior marginal means $\boldsymbol{\eta}_1 = \{\eta_{i\cdot}\}$ and $\boldsymbol{\eta}_2 = \{\eta_{\cdot j}\}$ as the parameters of the prior distribution for $\boldsymbol{\theta}$.

An equivalent parametrization suggested by Leonard (1973) uses the multivariate logits

$$\gamma_{ij} = \log(\eta_{ij}) + D(\{\gamma_{ij}\})\ , \tag{3.2}$$

where $D(\{\gamma_{ij}\}) = \log(\sum_{ij} \exp(\gamma_{ij}))$ so that $\sum_{ij} \eta_{ij} = 1$ for any values of the $\gamma_{ij}$. Thus

$$\eta_{ij} = e^{\gamma_{ij}} / \sum_{ij} e^{\gamma_{ij}}\,.$$

Now consider the factor 1 effects, $\{u_{1(i)}\}$, the factor 2 effects, $\{u_{2(j)}\}$, and the first-order interaction terms, $\{u_{12(ij)}\}$, in the saturated model $\gamma_{ij} = u_{1(i)} + u_{2(j)} + u_{12(ij)}$. If the $\{\eta_{ij}\}$ satisfy equation (3.1), then all the terms $u_{12(ij)}$ are zero, and

$$\gamma_{ij} = u_{1(i)} + u_{2(j)}\,. \tag{3.3}$$

Taking exponentials in (3.3) yields

$$\eta_{ij} = \frac{\exp(\gamma_{ij})}{\exp(D(\{\gamma_{ij}\}))} = \frac{\exp(u_{1(i)})}{\sum_k \exp(u_{1(k)})} \frac{\exp(u_{2(j)})}{\sum_l \exp(u_{2(l)})} = \eta_{i\cdot}\eta_{\cdot j}\,,$$

$$\eta_{i\cdot} = \frac{\exp(u_{1(i)})}{\sum_{k=1}^{r_1} \exp(u_{1(k)})}, \tag{3.4}$$

and

$$\eta_{\cdot j} = \frac{\exp(u_{2(j)})}{\sum_{l=1}^{r_2} \exp(u_{2(l)})}. \tag{3.5}$$

If we are uncertain about the values of the arrays $\{\eta_{i\cdot}\}$ and $\{\eta_{\cdot j}\}$, we can include this uncertainty in the prior distribution using a distribution on the on them, i.e. via a two-stage prior. In the first stage we use a Dirichlet distribution:

> **Stage I:** $\boldsymbol{\theta}|K, \boldsymbol{\eta} \sim d(K, \boldsymbol{\eta})$, where the $\{\eta_{ij}\}$ are reparametrized using equations (3.2) and (3.3).

In the second stage we induce a probability distribution for $\{\eta_{i\cdot}\}$ and $\{\eta_{\cdot j}\}$ using the transformations defined by (3.4) and (3.5) and by assigning distributions to $\{u_{1(i)}\}$ and $\{u_{2(j)}\}$. For the purpose of illustration we choose here normal distributions, but our approach does not depend upon this choice.

> **Stage II:** The $u_{1(i)}$ are independent, $i \in \bar{r}_1$. The $u_{2(j)}$ are independent, $j \in \bar{r}_2$ and independent of the $u_{1(i)}, i \in \bar{r}_1$. The distribution of $u_{1(i)}$ is $N(\mu_{1(i)}, \sigma^2_{1(i)})$ and the distribution of $u_{2(j)}$ is $N(\mu_{2(j)}, \sigma^2_{2(j)})$.

The resulting prior density for $\boldsymbol{\theta}$ is given by

$$\pi(\boldsymbol{\theta}|K) = \int_{\Omega_\eta} \pi_1(\boldsymbol{\theta}|\boldsymbol{g}(\boldsymbol{\eta}_1, \boldsymbol{\eta}_2), K)\pi_{2,1}(\boldsymbol{\eta}_1|\boldsymbol{\mu}_1, \boldsymbol{\sigma}_1)\pi_{2,2}(\boldsymbol{\eta}_2|\boldsymbol{\mu}_2, \boldsymbol{\sigma}_2)d\boldsymbol{\eta}_1 d\boldsymbol{\eta}_2$$

where $\pi_1$ is the $d(K, \{\eta_{ij}\})$ density, $\boldsymbol{g}(\boldsymbol{\eta}_1, \boldsymbol{\eta}_2) = \{\eta_{i\cdot}\eta_{\cdot j}\}$,

$$\boldsymbol{\mu}_1 = (\mu_{1(1)}, \cdots, \mu_{1(r_1)}), \quad \boldsymbol{\mu}_2 = (\mu_{2(1)}, \cdots, \mu_{2(r_2)}),$$
$$\boldsymbol{\sigma}_1 = (\sigma_{1(1)}, \cdots, \sigma_{1(r_1)}) \quad \text{and} \quad \boldsymbol{\sigma}_2 = (\sigma_{2(1)}, \cdots, \sigma_{2(r_2)}).$$

The densities $\pi_{2,1}$ and $\pi_{2,2}$ correspond to the distributions of $\boldsymbol{\eta}_1$ and $\boldsymbol{\eta}_2$. To estimate a cell probability $\theta_{ij}$ we focus on its posterior mean $E(\theta_{ij}|\boldsymbol{x})$. Related results are available for posterior variances.

**Fact 1** *Let $\boldsymbol{\theta}$ be distributed according to a two-stage prior distribution with a first stage as already stipulated and a second stage that consists of distributions on the margins $\{\eta_{i\cdot}\}$ and $\{\eta_{\cdot j}\}$. Then*

$$E(\theta_{ij}|\boldsymbol{x}) = \frac{N}{N+K}\frac{x_{ij}}{N} + \frac{K}{N+K}E(\eta_{i\cdot}\eta_{\cdot j}|\boldsymbol{x}).$$

Note that the posterior expectation $E(\theta_{ij}|\boldsymbol{x})$ is a weighted average of the unrestricted cell maximum likelihood estimate (MLE) and the posterior expectation $E(\eta_{i\cdot}\eta_{\cdot j}|\boldsymbol{x})$. Moreover, the MLE of $\theta_{ij}$ under independence is

$(x_{i\cdot}/N)(x_{\cdot j}/N)$ and this estimate depends on $x_{i\cdot}/N$ and $x_{\cdot j}/N$ in the same form that $\eta_{i\cdot}\eta_{\cdot j}$ depends on $\eta_{i\cdot}$ and $\eta_{\cdot j}$. The parameter $K$ can be regarded as the sample size of a table (either "real" or "imaginary") considered prior to the observation of $\boldsymbol{x}$, in which we believe that there is independence between the two factors. Thus we can regard $K$ as reflecting the strength of our prior belief in independence: the larger $K$, the stronger our belief. It is interesting to study the behavior of the posterior mean $E(\theta_{ij}|\boldsymbol{x})$ as $K$ approaches zero and as $K$ approaches infinity. Under the assumptions of Fact 1 we have,

$$\lim_{K\to 0} E(\theta_{ij}|\boldsymbol{x}) = x_{ij}/N,$$

and if in addition the second-stage consists of independent distributions on the margins $\{\eta_{i\cdot}\}$ and $\{\eta_{\cdot j}\}$ then,

$$\lim_{K\to\infty} E(\theta_{ij}|\boldsymbol{x}) = \{\lim_{K\to\infty} E(\eta_{i\cdot}|\boldsymbol{x})\}\{\lim_{K\to\infty} E(\eta_{\cdot j}|\boldsymbol{x})\}. \tag{3.6}$$

Albert and Gupta (1982) prove these results for a particular second- stage distribution. Note that the form of the dependency of $\lim_{K\to\infty} E(\theta_{ij}|\boldsymbol{x})$ on $\lim_{K\to\infty} E(\eta_{i\cdot}|\boldsymbol{x})$ and $\lim_{K\to\infty} E(\eta_{\cdot j}|\boldsymbol{x})$ is the same as that of the MLE of $\theta_{ij}$ on $x_{i\cdot}/N$ and $x_{\cdot j}/N$ under the model of independence.

## 3. EXTENSIONS TO MULTIDIMENSIONAL TABLES

Here we expand our approach to higher-dimensional tables and explain how to utilize different types of prior knowledge about the structure underlying the cell probabilities. Some results from the theory of loglinear models are relevant to our development. In presenting them we follow the notation of Andersen (1974). Theorem 1 is the key result in this section as it has important practical and theoretical implications. In particular, it explains the similarities between equations arising in our approach and maximum likelihood equations obtained when estimating the cell probabilities using loglinear models. The proofs of all the theorems in this section can be found in Epstein (1989).

Let $I = \bar{r}_1 \times \cdots \times \bar{r}_n = \{(i_1,\ldots,i_n), i_k \in \bar{r}_k, k \in \bar{n}\}$, where $\bar{n} = \{1,\ldots,n\}$ and $\bar{r}_k = \{1,\ldots,r_k\}, k \in \bar{n}$. Consider an $r_1 \times \cdots \times r_n$ contingency table $\boldsymbol{x} = \{x_{\boldsymbol{i}}, \boldsymbol{i} \in I\}$ and let $\theta_{\boldsymbol{i}}, \boldsymbol{i} \in I$, denote the probability that an observation falls in the cell $\boldsymbol{i}$. Assume that $\boldsymbol{x} = \{x_{\boldsymbol{i}}, \boldsymbol{i} \in I\}$ has a multinomial $M(N, \{\theta_{\boldsymbol{i}}\})$ distribution. This sampling model is used when a random sample of $N$ individuals is taken from an infinite population and the individuals are examined for the presence of $n$ traits, and for trait $k$ classified into $r_k$ distinct groups.

We begin with the first stage of the prior, namely the conjugate Dirichlet family indexed by $K > 0$ and $\boldsymbol{\eta}$ in an appropriate subset of $[0,1]^I$. Our approach in section 2 generalizes to $n$-way contingency tables through the multivariate logit transformation for $\boldsymbol{\eta}$. This transformation is

$$\boldsymbol{\eta} = \frac{\exp(\boldsymbol{\gamma})}{\sum \exp(\gamma_{\boldsymbol{i}})}, \ \boldsymbol{\gamma} \in M \tag{3.7}$$

for some subspace $M$ of $\Re^I$.

Next we describe the basic elements of the general loglinear parametrization. For $Y \subset \bar{n}$, we define the $Y-$*margin*, $\boldsymbol{\theta}^Y = \{\theta^Y_{\boldsymbol{i}} : \boldsymbol{i} \in I\}$, of $\boldsymbol{\theta}$ by

$$\boldsymbol{\theta}^Y_{i_1 \ldots i_n} = \sum_{\bar{n} \setminus Y} \theta_{i_1 \ldots i_n} = \sum_{i_j \in \bar{r}_j,\, j \in \bar{n} \setminus Y} \theta_{i_1 \ldots i_n}.$$

This definition of Y-margins will apply to other multidimensional arrays. For $W \subset \bar{n}$, we define $U^W$ by $U^W = \{\boldsymbol{u} \in \Re^I : u_{i_1 \ldots i_n}$ depends only on the indices in $W$ and the sum over an index in $W$ is $0\}$. The sets $U^W$ are linear subspaces of $\Re^I$. They are orthogonal and span $\Re^I$. As a consequence, $\boldsymbol{\phi}$ has a unique representation $\boldsymbol{\phi} = \sum_{W \in \bar{n}} \boldsymbol{u}^W$ with $\boldsymbol{u}^W \in U^W$, if $\boldsymbol{\phi} \in \Re^I$.

We specify a loglinear model by reparametrizing the multinomial cell probabilities $\boldsymbol{\theta} = \{\theta_{\boldsymbol{i}},\ \boldsymbol{i} \in I\}$ as

$$\log \boldsymbol{\theta} = \sum_{W \in \mathcal{A}} \boldsymbol{u}^W, \tag{3.8}$$

or equivalently

$$\log \boldsymbol{\theta} \in M = \sum_{W \in \mathcal{A}} U^W,$$

where $\mathcal{A}$ is a subclass of $\bar{n}$, but the constraint $\sum_{\boldsymbol{i} \in I} \theta_{\boldsymbol{i}} = 1$ requires that $\phi$ always be excluded from $\mathcal{A}$. Often the parameters $\boldsymbol{u}^W, W \subset \bar{n}$, are referred to as " the $u$-terms ", and the term $\boldsymbol{u}^W$ as the *interaction* between the factors in $W$. The choice of $M$ allows us in many cases to interpret the induced structure in $\boldsymbol{\theta}$ in terms of different statements of independence, equiprobability, and conditional properties (e.g., conditional independence) pertaining to a subset of the variables, given the level of some of the remaining variables. A model is said to be hierarchical if $\boldsymbol{u}^W = \mathbf{o}$ and $W \subset W'$ implies $\boldsymbol{u}^{W'} = \mathbf{o}$, where $\mathbf{o}$ is the array whose entries are all zero. Models that are not hierarchical are difficult to interpret. For a hierarchical model, let $\mathcal{A}$ be the class of subsets of $\bar{n}$ such that, if $W \notin \mathcal{A}$, then $\boldsymbol{u}^W = \mathbf{o}$.

When the subspace $M$ is built in this fashion we will say that $M$ is *hierarchical*, even outside the strict context of loglinear models, and we will also refer to $\mathcal{A}$ as a *hierarchical* class. Let $\mathcal{Y} = \{Y_1, \ldots, Y_T\}$ be the subset of $\mathcal{A}$ determined by

1. $W \in \mathcal{A}$ if and only if there exists a $Y_i \in \mathcal{Y}$ such that $W \subset Y_i$.

2. For no pair $Y_i, Y_j \in \mathcal{Y}, i \neq j$, we have $Y_i \subset Y_j$.

We call $\mathcal{Y}$ the *generator* of the model. Different models may either be characterized by the linear subspace $M$ to which $\log \boldsymbol{\theta}$ belongs, or by those interactions in the loglinear model that are 0, i.e. by $\mathcal{A}$ or by the generator $\mathcal{Y}$. The essence of a loglinear model is captured by the subspace $M$.

Since here we model the situation where the statistician believes, with uncertainty, that the cell probabilities might have a structure specified by a given loglinear parametrization, we reparametrize the cell prior means instead of the cell probabilities.

Consider a hierarchical model $M = \sum_{W \in \mathcal{A}} U^W$, with generator $\mathcal{Y} = \{Y_1, \ldots, Y_T\}$. This decomposition of $M$ allows us to write an element $\boldsymbol{\gamma} \in M$ as $\boldsymbol{\gamma} = \sum_{W \in \mathcal{A}} \boldsymbol{u}^W$. Thus a coordinate-dependent version of the suggested parametrization associates with $(\boldsymbol{u}^W, W \in \mathcal{A})$ an array $\boldsymbol{\eta}$ in $(0,1)^I$ defined by

$$\eta_{\boldsymbol{i}} = \frac{\exp(\gamma_{\boldsymbol{i}})}{\exp(D(\{\gamma_{\boldsymbol{i}}\}))}, \quad \boldsymbol{i} \in I, \tag{3.9}$$

with

$$\gamma_{\boldsymbol{i}} = \sum_{W \in \mathcal{A}} u_{\boldsymbol{i}}^W, \tag{3.10}$$

and where $u_{\boldsymbol{i}}^W$ is the $\boldsymbol{i}$–th coordinate of $\boldsymbol{u}^W$. In Leonard's notation, equation (3.9) becomes $\gamma_{\boldsymbol{i}} = \log(\eta_{\boldsymbol{i}}) + D(\{\gamma_{\boldsymbol{i}}\})$, where $D(\{\gamma_{\boldsymbol{i}}\}) = \log(\sum \exp(\gamma_{\boldsymbol{i}}))$. In section 2, we used a loglinear parametrization reflecting uncertain knowledge about independence that led to $E(\theta_{ij}|\boldsymbol{\eta}) = \eta_{i\cdot}\eta_{\cdot j}$. This illustrates how only some margins of the Dirichlet means combine to form an initial guess $\boldsymbol{\eta}$ for the entire array of cell probabilities. Theorem 1 tells how and which margins of $\boldsymbol{\eta}$ intervene when forming an initial guess at $\boldsymbol{\theta}$ in the general case.

**Theorem 1** *Let $\boldsymbol{\phi}$ be a vector in $M$ and let $\boldsymbol{\eta} = e^{\boldsymbol{\phi}} / \sum_{\boldsymbol{i} \in I} e^{\phi_{\boldsymbol{i}}}$. Assume further that $\mathbf{e} \in M$, where $\mathbf{e}$ is the vector whose entries are all 1. Define the functions $g_t(\boldsymbol{\eta}) = \boldsymbol{\eta}^{Y_t}$, for $t = 1, \ldots, T$. If the margins $g_t(\boldsymbol{\eta})$, $t = 1, \ldots, T$ are given, then $\boldsymbol{\eta}$ can be determined by solving the equations*

$$\boldsymbol{\eta}^{Y_t} = g_t(\boldsymbol{\eta}), \quad t = 1, \ldots, T. \tag{3.11}$$

Remarks:

1. Suppose the cell counts follow a multinomial model with cell means $\boldsymbol{m}$, which are to be estimated by maximum likelihood under a loglinear model. For a hierarchical model with generator $\mathcal{Y} = \{Y_1, \ldots, Y_T\}$, the minimal sufficient statistic is $\{\boldsymbol{x}^{Y_1}, \ldots, \boldsymbol{x}^{Y_T}\}$, also called *the minimal set of fitted margins.*

2. The maximum likelihood estimate $\widehat{\boldsymbol{m}}$ of $\boldsymbol{m}$, when it exists, satisfies the likelihood equations

$$\boldsymbol{x}^{Y_t} = \widehat{\boldsymbol{m}}^{Y_t} \quad t = 1, \ldots, T. \tag{3.12}$$

Note the similarities between equations (3.11) in Theorem 1 and the likelihood equations (3.12).

Under the conditions of Theorem 1, if the margins $\boldsymbol{\eta}^{Y_1}, \ldots, \boldsymbol{\eta}^{Y_T}$ of an array $\boldsymbol{\eta}$ are given, then equations (3.11) define a function

$$\boldsymbol{\eta} = \boldsymbol{g}(\boldsymbol{\eta}^{Y_1}, \ldots, \boldsymbol{\eta}^{Y_T}). \tag{3.13}$$

In other words, a family of distributions indexed by a parameter $\boldsymbol{\eta}$ defined by equations (3.9) and (3.10) can also be parametrized properly by $\boldsymbol{\eta}^{Y_1}, \ldots, \boldsymbol{\eta}^{Y_T}$. Concerning the actual evaluation of $\boldsymbol{g}$, one may use the Bishop-Goodman procedure (see Bishop, Fienberg and Holland, 1975, pp. 76-83), when $M$ is decomposable and iterative proportional fitting if $M$ is only hierarchical. For the definition of decomposable models and a description of these two algorithms in the notation of this paper see Andersen (1974).

The posterior expectation of $\boldsymbol{\theta}$ is

$$E\{\boldsymbol{\theta}|\boldsymbol{x}, \boldsymbol{\eta}^{Y_1}, \ldots, \boldsymbol{\eta}^{Y_T}\} = \frac{N}{N+K}\frac{\boldsymbol{x}}{N} + \frac{K}{N+K}\boldsymbol{g}(\boldsymbol{\eta}^{Y_1}, \ldots, \boldsymbol{\eta}^{Y_T}). \tag{3.14}$$

Suppose we believe, with uncertainty, in the structure (3.8) but we are unable to specify $\boldsymbol{\eta}^{Y_t}$, $t = 1, \ldots, T$. We can use a $d(K, \boldsymbol{\eta})$ distribution with $\boldsymbol{\eta}$ given by equation (3.13) as a first stage of a two-stage hierarchical prior. In the second stage we can use distributions on the margins $\boldsymbol{\eta}^{Y_t}, t = 1, \ldots, T$, or on the $u$- terms in expression (3.10), so that the resulting prior accounts for the extra uncertainty. In this case the posterior expectation of $\boldsymbol{\theta}$ becomes

$$E\{\boldsymbol{\theta}|\boldsymbol{x}\} = \frac{N}{N+K}\frac{\boldsymbol{x}}{N} + \frac{K}{N+K}E\{\boldsymbol{g}(\boldsymbol{\eta}^{Y_1}, \ldots, \boldsymbol{\eta}^{Y_T})|\boldsymbol{x}\}.$$

Our result on the limit as $K \to 0$ from section 2 now generalizes directly,

**Theorem 2** *Let $\pi$ be the prior distribution for $\boldsymbol{\eta}$, and suppose $\pi$ is such that the expectation $E_\pi(\boldsymbol{\eta})$ is finite. Then*

$$\lim_{K \to 0} E(\boldsymbol{\theta}|\boldsymbol{x}) = \frac{1}{N}\boldsymbol{x}.$$

To better understand the next result let us first contrast, in the particular framework of section 2, the behavior posterior means as $K \to \infty$ when we use one-stage and two-stage priors. Given the arrays $\{\eta_{i\cdot}\}$ and $\{\eta_{\cdot j}\}$ , the limiting posterior Dirichlet means are "independence estimates",

$$\lim_{K \to \infty} E(\theta_{ij}|\{\eta_{i\cdot}\}, \{\eta_{\cdot j}\}, \boldsymbol{x}) = \eta_{i\cdot}\eta_{\cdot j}, \; i \in \bar{r}_1, j \in \bar{r}_2.$$

However, if the arrays $\{\eta_{i\cdot}\}$ and $\{\eta_{\cdot j}\}$ are random then the limiting posterior means $\lim_{K\to\infty} E(\theta_{ij}|\boldsymbol{x})$ may not be independence estimates, unless the arrays $\{\eta_{i\cdot}\}$ and $\{\eta_{\cdot j}\}$ are a priori independent. To use the parameter $K$ to reflect the strength of our belief in independence and to obtain estimates that reflect our uncertainty about independence by compromising between the observed proportions and independence estimates, it is desirable that $\lim_{K\to\infty} E(\theta_{ij}|\boldsymbol{x})$ be an independence estimate. To formalize this idea note that $\lim_{K\to\infty} E(\theta_{ij}|\boldsymbol{x})$ is an independence estimate if it has the same structure as the conditional prior Dirichlet means, $E(\theta_{ij}|\{\eta_{i\cdot}\},\{\eta_{\cdot j}\}) = \eta_{i\cdot}\eta_{\cdot j}; i \in \bar{r}_1, j \in \bar{r}_2$, or,

$$E(\theta_{ij}|\{\eta_{i\cdot}\},\{\eta_{\cdot j}\}) = \boldsymbol{g}(\{\eta_{i\cdot}\},\{\eta_{\cdot j}\}),$$

where $\boldsymbol{g}(\{\eta_{i\cdot}\},\{\eta_{\cdot j}\})$ is the $r_1 \times r_2$ array $\{\eta_{i\cdot}\eta_{\cdot j}\}$. The limiting posterior means have this structure, that is,

$$\lim_{K\to\infty} E(\{\theta_{ij}\}|\boldsymbol{x}) = \boldsymbol{g}(\lim_{K\to\infty} E(\{\eta_{i\cdot}\}|\boldsymbol{x}), \lim_{K\to\infty} E(\{\eta_{\cdot j}\}|\boldsymbol{x})),$$

if $\{\eta_{i\cdot}\}$ and $\{\eta_{\cdot j}\}$ are a priori independent. This result is precisely equation (3.6) of section 2 and Theorem 3 extends it for decomposable parametrizations. A loglinear parametrization is decomposable if there exists an ordering $\mathcal{Y} = \{Y_1, \ldots, Y_T\}$ of its generator such that for each $t$, $1 \le t \le T-1$,

$$U_t \cap Y_{t+1} = Y_{k_t} \cap Y_{t+1}, \tag{3.15}$$

for some $k_t$, $1 \le k_t \le t$, where $U_t = \bigcup_{k=1}^{t} Y_k \cap Y_{t+1}$, $t = 1, \ldots, T-1$.

**Theorem 3** *Consider a loglinear parametrization for $\boldsymbol{\eta}$, with generator $\mathcal{Y} = \{Y_1, \ldots, Y_T\}$, and suppose the ordering of the elements of $\mathcal{Y}$ is such that (3.15) holds. We have*

$$\lim_{K\to\infty} E(\boldsymbol{\theta}|\boldsymbol{x}) = \boldsymbol{g}(\lim_{K\to\infty} E\{\boldsymbol{\eta}^{Y_1}|\boldsymbol{x}\}, \ldots, \lim_{K\to\infty} E\{\boldsymbol{\eta}^{Y_T}|\boldsymbol{x}\}), \tag{3.16}$$

*if the arrays*

$$\boldsymbol{\eta}^{Y_1}, \quad \frac{\boldsymbol{\eta}^{Y_t}}{\boldsymbol{\eta}^{U_{t-1}}}, \quad t = 2, \ldots, T$$

*are a priori independent.*

Thus, provided the hypotheses of Theorem 3 hold, the limiting posterior mean of $\boldsymbol{\theta}$ possesses the same structure as that for $\boldsymbol{\eta}$ in the prior. This structure is given by equations (3.13).

## 4. THREE-DIMENSIONAL TABLES: AN ILLUSTRATIVE EXAMPLE

Consider an $r_1 \times r_2 \times r_3$ table and suppose it is of interest to estimate the cell probabilities $\theta_{ijk}$ using vague prior beliefs of conditional independence of factors 1 and 2, say, given the other, say 3, that is, using the uncertain prior belief that the cell probabilities satisfy $\theta_{ijk} = \frac{\theta_{i\cdot k}\theta_{\cdot jk}}{\theta_{\cdot\cdot k}}$. In order to let the cell prior means reflect this structure we, write

$$\eta_{ijk} = \frac{\eta_{i\cdot k}\eta_{\cdot jk}}{\eta_{\cdot\cdot k}}. \tag{3.17}$$

To induce the structure in (3.17) using the multivariate logits, we take

$$\gamma_{ijk} = \log(\eta_{ijk}) + D(\{\gamma_{ijk}\}), \tag{3.18}$$

where $D(\{\gamma_{ijk}\}) = \log(\sum_{ijk} \exp(\gamma_{ijk}))$. Then we set the terms $\{u_{12(ij)}\}$ and $\{u_{123(ijk)}\}$ equal to zero in the saturated parametrization. Thus,

$$\gamma_{ijk} = u_{1(i)} + u_{2(j)} + u_{3(k)} + u_{13(ik)} + u_{23(jk)}, \tag{3.19}$$

which is a decomposable parametrization. One ordering of its generating class is $Y_1 = \{1,3\}$ and $Y_2 = \{2,3\}$. Therefore $U_1 = \{3\}$.

Any distribution such that the arrays $\{\eta_{i\cdot k}\}$ and $\{\eta_{\cdot jk}/\eta_{\cdot\cdot k}\}$ are independent satisfies the hypotheses of Theorem 2. We use a two-stage prior when we are uncertain about the values of $\eta_{i\cdot k}$ and $\{\eta_{\cdot jk}/\eta_{\cdot\cdot k}\}$, in which case the prior we adopted in the previous section becomes:

> **Stage I:** $\boldsymbol{\theta}|K, \boldsymbol{\eta} \sim d(K, \boldsymbol{\eta})$ where there $\eta_{ijk}$ are parametrized using equations (3.18) and (3.19).

In the second stage, we induce a probability distribution for $\eta_{i\cdot k}$ and $\eta_{\cdot jk}$ through equations (3.18) and (3.19) by assigning distributions to each of the $u-$terms in (3.19). Again we choose normal distributions:

> **Stage II:** The $u_{1(i)}, u_{2(j)}, u_{3(k)}, u_{13(ik)}$ and $u_{23(jk)}$ are mutually independent, for $i \in \bar{r}_1$, $j \in \bar{r}_2$, $k \in \bar{r}_3$, with distributions

$$u_{1(i)} \sim N(\mu_{1(i)}, \sigma^2_{1(i)}),\ u_{2(j)} \sim N(\mu_{2(j)}, \sigma^2_{2(j)}),\ u_{3(k)} \sim N(\mu_{3(k)}, \sigma^2_{3(k)}),$$

$$u_{13(ik)} \sim N(\mu_{13(ik)}, \sigma^2_{13(ik)}),\ u_{23(jk)} \sim N(\mu_{23(jk)}, \sigma^2_{23(jk)}).$$

The posterior mean of the cell probability $\theta_{ijk}$ takes the form

$$E(\theta_{ijk}|\boldsymbol{x}) = \frac{N}{N+K}\frac{x_{ijk}}{N} + \frac{K}{N+K}E(\frac{\eta_{i\cdot k}\eta_{\cdot jk}}{\eta_{\cdot\cdot k}}|\boldsymbol{x}), \tag{3.20}$$

while equation (3.16) takes the form

$$\lim_{K\to\infty} E(\theta_{ijk}|\boldsymbol{x}) = \frac{\{\lim_{K\to\infty} E(\eta_{i\cdot k}|\boldsymbol{x})\}\{\lim_{K\to\infty} E(\eta_{\cdot jk}|\boldsymbol{x})\}}{\lim_{K\to\infty} E(\eta_{\cdot\cdot k}|\boldsymbol{x})}. \tag{3.21}$$

The data in Table 3.1 concern a $4 \times 2 \times 2$ cross-classification of 1154 individuals whose death was caused by skin cancer. The individuals were cross-classified according to three attributes: site of the body where the cancer originally developed, sex and type of cancer. The data were collected by the US National Center for Health Statistics and previously analyzed in Nazaret (1987).

TABLE 3.1. Skin cancer data. Observed counts.

| | Melanoma | | | Non- melanoma | | |
|---|---|---|---|---|---|---|
| Site | Male | Female | Total | Male | Female | Total |
| Lips | 0 | 1 | 1 | 47 | 34 | 81 |
| Eyelids | 1 | 4 | 5 | 75 | 79 | 154 |
| Ears | 3 | 3 | 6 | 106 | 39 | 145 |
| Other | 24 | 19 | 43 | 360 | 359 | 719 |
| Total | 28 | 27 | 55 | 588 | 511 | 1099 |

In this analysis, we first specify the $\mu$ and $\sigma$ parameters reflecting our knowledge about the $u-$terms in the parametrization of the Dirichlet means. Our preliminary choice for the values of these parameters is

$$\begin{aligned}
\mu_{1(i)} &= .125, & \sigma_{1(i)} &= 1.0,\ i = 1, \ldots, 4, \\
\mu_{2(j)} &= .25, & \sigma_{2(j)} &= 1.0,\ j = 1, 2, \\
\mu_{3(k)} &= .5, & \sigma_{3(k)} &= 2.0,\ k = 1, 2, \\
\mu_{13(i,k)} &= .125, & \sigma_{13(i,k)} &= 1.0,\ i = 1, \ldots, 4,\ k = 1, 2, \\
\mu_{23(j,k)} &= .25, & \sigma_{23(j,k)} &= 1.0,\ i = 1, 2,\ k = 1, 2.
\end{aligned}$$

This specification may be seen to reflect a rather imprecise knowledge about $\{\eta_{i \cdot k}\}$ and $\{\eta_{\cdot jk}\}$. Further remarks about the relation between these margins and the $u-$terms in equation (3.19) along with suggestions to aid specification of the prior may be found in Epstein (1989). Finally, one specifies a value for the parameter $K$.

Table 2 reports the computed values for the posterior means of each of the cell probabilities for several values of $K$. Note that as the value of $K$ increases from zero to infinity the posterior means shrink the maximum likelihood estimates $x_{ijk}/N$ toward the conditional independence estimates given by equation (3.21). The estimates corresponding to finite values of $K$ reflect the imprecise prior belief in conditional independence by compromising between estimates obtained under a saturated model and estimates obtained under a conditional independence model.

In practice the posterior expectations (3.20) are difficult to compute. We used a Monte Carlo method to estimate

$$\begin{aligned}
&E\left(\frac{\eta_{i \cdot k}\eta_{\cdot jk}}{\eta_{\cdot\cdot k}} \Big| \{x_{ijk}\}\right) \\
&= \frac{\int \frac{\eta_{i \cdot k}\eta_{\cdot jk}}{\eta_{\cdot\cdot k}} f(\{x_{ijk}\}, \{\eta_{i \cdot k}\}, \{\eta_{\cdot jk}\}) d(\{\eta_{i \cdot k}\}) d(\{\eta_{\cdot jk}\})}{\int f(\{x_{ijk}\}, \{\eta_{i \cdot k}\}, \{\eta_{\cdot jk}\}) d(\{\eta_{i \cdot k}\}) d(\{\eta_{\cdot jk}\})}, \qquad (3.22)
\end{aligned}$$

where $f(\{x_{ijk}\}, \{\eta_{i \cdot k}\}, \{\eta_{\cdot jk}\})$

$$= \Pi \left\{ \frac{\Gamma(x_{ijk} + K\frac{\eta_{i \cdot k}\eta_{\cdot jk}}{\eta_{\cdot\cdot k}})}{\Gamma(K\frac{\eta_{i \cdot k}\eta_{\cdot jk}}{\eta_{\cdot\cdot k}})} \right\} \pi_{2,1}(\{\eta_{i \cdot k}\}) \pi_{2,2}(\{\eta_{\cdot jk}\})$$

TABLE 3.2. Computed values of posterior means for different values of $K$

| $K$ | 0 | 100 | 300 | 500 | 1000 | 5000 | 10000 | $\infty$ |
|---|---|---|---|---|---|---|---|---|
| $\hat{\theta}_{111}$ | .0000 | .0050 | .0129 | .0189 | .0290 | .0508 | .0560 | .0618 |
| $\hat{\theta}_{112}$ | .0009 | .0058 | .0136 | .0195 | .0295 | .0509 | .0561 | .0618 |
| $\hat{\theta}_{121}$ | .0026 | .0074 | .0150 | .0207 | .0304 | .0513 | .0563 | .0618 |
| $\hat{\theta}_{122}$ | .0208 | .0241 | .0294 | .0334 | .0402 | .0547 | .0582 | .0620 |
| $\hat{\theta}_{211}$ | .0009 | .0058 | .0136 | .0195 | .0295 | .0509 | .0561 | .0618 |
| $\hat{\theta}_{212}$ | .0035 | .0082 | .0156 | .0213 | .0309 | .0514 | .0564 | .0618 |
| $\hat{\theta}_{221}$ | .0026 | .0074 | .0150 | .0207 | .0304 | .0513 | .0563 | .0618 |
| $\hat{\theta}_{222}$ | .0165 | .0201 | .0260 | .0304 | .0378 | .0539 | .0577 | .0620 |
| $\hat{\theta}_{311}$ | .0407 | .0425 | .0452 | .0473 | .0508 | .0584 | .0602 | .0623 |
| $\hat{\theta}_{312}$ | .0650 | .0648 | .0645 | .0642 | .0638 | .0630 | .0628 | .0625 |
| $\hat{\theta}_{321}$ | .0919 | .0895 | .0858 | .0830 | .0782 | .0680 | .0655 | .0628 |
| $\hat{\theta}_{322}$ | .3120 | .2921 | .2605 | .2365 | .1961 | .1093 | .0883 | .0653 |
| $\hat{\theta}_{411}$ | .0295 | .0321 | .0363 | .0394 | .0448 | .0563 | .0591 | .0621 |
| $\hat{\theta}_{412}$ | .0685 | .0680 | .0672 | .0667 | .0657 | .0636 | .0631 | .0626 |
| $\hat{\theta}_{421}$ | .0338 | .0361 | .0397 | .0425 | .0471 | .0571 | .0595 | .0622 |
| $\hat{\theta}_{422}$ | .3111 | .2913 | .2598 | .2359 | .1957 | .1091 | .0882 | .0653 |

and $\pi_{2,1}$ and $\pi_{2,2}$ are the joint densities of $\{\eta_{i\cdot k}\}$, and $\{\eta_{\cdot jk}\}$ respectively. To estimate the integrals in (3.22), we randomly generated $n_0$ pairs of values $(\eta_{i\cdot k}, \eta_{\cdot jk})$ using

$$\eta_{i\cdot k} = \frac{\exp(u_{1(i)} + u_{3(k)} + u_{13(ik)}) \sum_j \exp(u_{2(j)} + u_{23(jk)})}{\exp(D(\{\gamma_{ijk}\}))}$$

$$\eta_{i\cdot k} = \frac{\exp(u_{2(j)} + u_{3(k)} + u_{23(jk)}) \sum_i \exp(u_{1(i)} + u_{13(ik)})}{\exp(D(\{\gamma_{ijk}\}))}.$$

Calling these pairs $(e^l_{13(ik)}, e^l_{23(jk)})$, $l = 1, \ldots, n_0$, and defining $e^l_{3(k)} = \sum_i e^l_{13(ik)}$ for $l = 1, \ldots, n_0$, and $k = 1, 2$, we see that the summation

$$\sum_{l=1}^{n_0} \frac{e^l_{13(ik)} e^l_{23(jk)}}{e^l_{3(k)}} \Pi_{i,j,k} \left\{ \frac{\Gamma(x_{ijk} + K \frac{e^l_{13(ik)} e^l_{23(jk)}}{e^l_{3(k)}})}{\Gamma(K \frac{e^l_{13(ik)} e^l_{23(jk)}}{e^l_{3(k)}})} \right\}$$

estimates the integral in the numerator of (3.22) and

$$\sum_{l=1}^{n_0} \Pi_{i,j,k} \left\{ \frac{\Gamma(x_{ijk} + K \frac{e^l_{13(ik)} e^l_{23(jk)}}{e^l_{3(k)}})}{\Gamma(K \frac{e^l_{13(ik)} e^l_{23(jk)}}{e^l_{3(k)}})} \right\}$$

estimates that of the denominator of (3.22). Note that the numbers $e^l_{3(k)}$ should be the same if the are computed in terms of the $e^l_{13(ik)}$ or in terms of the $e^l_{23(jk)}$. Table 2 contains the computed values using $n_0 = 5000$. We recomputed the estimates several times using new sets of 5000 pairs $(e^l_{13(ik)}, e^l_{23(jk)})$. The new estimates deviated no more than .0001 (in each cell) each time from the estimates presented in table 2.

## 5. CONCLUDING REMARKS

We have put forward a new class of prior distributions for use in the Bayesian analysis of multidimensional contingency tables and we have presented some properties of the resulting estimators that lend support to our approach. Our priors use a loglinear parametrization on the means of the Dirichlet stage of the prior, inducing a specific multiplicative structure. This approach is particularly appealing because it brings to bear the richness of loglinear models and the independence structures they induce into the realm of Bayesian statistics. Bayesian methods are particularly relevant for the treatment of sparse high-dimensional tables because the traditional maximum likelihood estimates are often either inadequate or nonexistent.

The class of priors we suggest is appropriate when the user feels uncertain about the presence of a particular loglinear structure in the cell probabilities. The resulting estimates represent a compromise between maximum likelihood estimates under a saturated model and estimates reflecting uncertainty about the values of the Dirichlet means. The flattening constant, $K$, permits the user to modulate in the prior the strength of his prior belief about the structure. On the other hand, if the user is certain about a structure he may use the loglinear parametrization directly on the cell probabilities.

Nazaret's (1987) approach, as did those of Laird (1978) and Leonard (1975), requires exchangeability assumptions. Our approach does not, but it can be easily modified to accommodate (vague) prior knowledge of exchangeability of effects or exchangeability of a suitable set of marginal means.

We expect to report separately on at least two other aspects of the work presented here. The first relates to facilitating the specification of the prior in two-way tables, in the approach of section 2. Johnson (1949) furnishes approximations for the prior moments of $\boldsymbol{\theta}$, which allow us to regard the parameters $\mu_{1(i)}$ and $\mu_{2(j)}$ as prior guesses of the main effects in the logit scale, of the table of probabilities $\{\theta_{ij}\}$, and the parameters $\sigma_{1(i)}$ and $\sigma_{2(j)}$ as reflecting the uncertainty about the guesses at the effects.

The second, relates to estimation under a non-saturated model. We may know a priori that there exists a structure underlying the cell probabilities and at the same time feel uncertain whether there is a further substructure. Consider for example the table of section 4, where it is possible that we believe a priori that type of cancer and sex are independent factors, i.e.,

$$\log \theta_{ijk} = u_{1(i)} + u_{2(j)} + u_{3(k)} + u_{12(ij)} + u_{13(ij)}$$

but felt uncertain whether the terms $\{u_{12(ij)}\}$ are zero. A variant of this paper's approach applicable to such a circumstance uses one loglinear model to parametrize the cell probabilities and another to parametrize the Dirichlet means. The cells posterior means appear to be a compromise between estimates corresponding to the two models.

## REFERENCES

[1] Albert, J.H. & Gupta, A.K. (1982). Mixtures of Dirichlet distributions and estimation in contingency tables. *Ann. Statist.* **10** *,61-68.*

[2] Albert, J.H. & Gupta, A.K.(1983). Estimation in contingency tables using Prior information. *J. Roy. Statist.Soc. B,* **45** *, No.1,60-69.*

[3] Andersen, A.H. (1974). Multidimensional contingency tables. *Scand. J. Statist,* **1**, *115-127.*

[4] Bishop, Y.M.M., Fienberg, S.E. & Holland, P.W.(1975). *Discrete Multivariate Analysis: Theory and Practice.* Cambridge, Mass: M.I.T. Press.

[5] Epstein, L.D. (1989). Bayesian Estimation in Multidimensional Contingency Tables. Ph.D. Thesis. Department of Statistics, Carnegie-Mellon University.

[6] Fienberg, S.E., & Holland, P.W. (1972). On the choice of flattening constants for estimating multinomial probabilities. *J. Mult. Analysis,* **2**, *127-134.*

[7] Fienberg, S.E., & Holland, P.W. (1973). Simultaneous estimation of multinomial cell probabilities. *J. Amer. Statist. Assoc.,* **68**, *683-691.*

[8] Johnson, N.L. (1949). Systems of frequency distributions curves generated by methods of translation. *Biometrika* **36**, *149-76.*

[9] Good, I.J. (1965). *The Estimation of Probabilities: An Essay in Modern Bayesian Methods.* Cambridge, Mass: M.I.T. Press.

[10] Good, I.J. (1967). A Bayesian significance test for multinomial distributions (with discussion). *J. Roy. Statist. Soc. B,* **29**, *399-431.* (Corrigendum(1974), **36**, 109.)

[11] Good, I.J. & Crook, J.F. (1980). On the application of symmetric Dirichlet distributions and their mixtures to contingency tables: II. *Ann. Statist.*, **8**, *1198-218.*

[12] Haberman, S.J. (1973). Loglinear models for frequency data: sufficient statistics and likelihood equations. *Ann. Statist.*, **1**, *617-632* .

[13] Laird, N.M. (1978). Empirical Bayes methods for two-way contingency tables. *Biometrika,* **65**, *581-90.*

[14] Lindley, D.V. & Smith, A.F.M. (1972). Bayes estimates for the linear model (with discussion). *J. Roy. Statistic. Soc. B,* **34**, *1-42* .

[15] Leonard, T. (1973). A Bayesian method for histograms. *Biometrika,* **60**, *297-308.*

[16] Leonard, T. (1975). Bayesian estimation methods for two-way contingency tables. *J. Roy. Statist. Soc. B,* **37**, *23-37.*

[17] Leonard, T. & Novick, M.R. (1986). Bayesian full rank marginalization for two-way contingency tables. *Journal of Educational Statistics,* **11**, *33-56* .

[18] Nazaret, A. (1987). Bayesian loglinear estimates for three way contingency tables. *Biometrika* **74**, *401-10.*

[19] Wang, C.R. (1985). Bayesian modeling for cross classified data. Ph.D. Thesis. Department of Mathematics and Statistics, State University of New York at Albany.

4

# Bayesian Nonparametric Prediction and Statistical Inference


## Bruce M. Hill [1]



**Abstract**

The problem of Bayesian nonparametric prediction and statistical inference is formulated and discussed. A solution is proposed based upon $A_n$ and $H_n$ as in Hill (1968, 1988). This solution gives rise to the posterior distribution of the number of species (or more generally, the number of distinct values) in a finite population of Hill (1968, 1979), and to the posterior distribution of percentiles of the population of Hill (1968). Next, the meaning of parameters in the subjective Bayesian theory of Bruno de Finetti is discussed in connection both with $H_n$ and with conventional parametric models. It is argued that the usual sharp distinction between prediction and parametric inference is largely illusory. The finite version of de Finetti's theorem is emphasized for the practice of statistics, with the infinite case used only to obtain approximations and insight.



*Key words and Phrases:* Nonparametric Bayes; Prediction; Finitely additive priors; Finite exchangeability; de Finetti; Bose-Einstein statistics; Robustness.


## 1. INTRODUCTION

Bayesian nonparametric statistics consists of methods for statistical inference and prediction based upon weak a priori knowledge as to the form of the underlying population. In real world problems one often does not have the type of sharp a priori knowledge usually assumed about models. Indeed, it is well known that in the practice of statistics the most difficult and important phase consists of the specification of such models. In this article we wish to discuss the case in which it is difficult, or impossible, to model the data in terms of conventional parametric models such as the Gaussian, exponential, or even exponential family, at least without resorting to complex mixtures of such distributions. Our inference will instead be based upon $H_n$, the nonparametric Bayesian approach of Hill (1968, 1980a, 1987b, 1988). A version of this approach was originally suggested from a

---


[1] University of Michigan

fiducial point of view by R. A. Fisher (1939, 1948). See also Dempster (1963).

In his celebrated article "La Prévision" (1937), Bruno de Finetti proposed a subjective Bayesian solution to the problem of scientific induction, as formulated, for example, by the Scottish philosopher, David Hume (1748). De Finetti did so in terms of the concept of exchangeability, which is a special form of dependence that he introduced and studied extensively (1937, 1974). Other key references are Hewitt and Savage (1955), Savage (1972), Heath and Sudderth (1976), and Diaconis and Freedman (1980, 1981). In this article I shall first give a somewhat personal review of the history and substance of the connection between induction and subjectivistic perceptions of symmetry, with particular attention to $A_n$ and $H_n$, which I developed for the case of vague or diffuse prior knowledge as to the shape of the underlying distribution of the observables.

The problem of induction is the problem of drawing inference about the future based upon the past. This problem has long plagued philosophers and others, partly because there is no way to prove that induction works (apart from induction itself), and also because in the real world it can be extremely difficult to formulate inferential or decision procedures, i.e., inductive techniques, that are appropriate in a given situation. The problem is best thought of in terms of the probabilistic prediction of potentially observable random quantities (not necessarily exchangeable), say $X_1, \ldots, X_n$. Given the values of the first n observations, $X_1 = x_1, \ldots, X_n = x_n$, what can we say about $X_{n+1}$ or any other future observations? In the Bayesian approach this is done in terms of the evaluation of a probability distribution for the future observables, given the data $X_1 = x_1, \ldots, X_n = x_n$. Conventional Bayesian methods, using a prior distribution for a 'parameter,' such as the parameter of a Bernoulli sequence, yield such a predictive posterior distribution, in addition to the more customary posterior distribution for the parameter. In such situations, once a statistical model and prior distribution have been formulated and specified, the posterior distribution of the future observations, given the past, is thereby completely determined. Such a scheme may be called inductive, since it prescribes a (coherent) mode of inference and behavior with respect to the future observables, given any set of data. See Hill (1990b) for discussion of Bayesian statistics and induction.

This scheme, as usually interpreted, requires that there exist 'true' known probabilities that represent the conditional distribution of the data, given the parameter, i.e., a conventional statistical model. However, at the deepest level, where 'true' probabilities either do not exist, or even if in some sense they do exist they are unknown, the conventional model-based Bayes theory is incomplete, since it is difficult even to give operational meaning to the assertion that a particular model is 'true,' much less to find such a model.[2]

The problem that de Finetti clearly formulated and largely solved was the problem of giving meaning to Bayesian inferential procedures without relying upon the usual crutch of an assumed statistical model. Before the fundamental work of de Finetti, the assumption of such a true model was simply an unjustified act of faith. One could, of course, refer to some underlying physical theory, or to the central limit theorem, or some previous analogous data, to support belief in such a model. But deep down this remained at best a matter of delicate subjective judgment, and it was not even clear how to express what such subjective judgments concerned. For example, consider the use of the normal or Gaussian distribution. Poincaré (1912, p. 171) states in connection with this distribution, "Tout le monde y croit cependant, me disait un jour M. Lippmann, car les expérimentateurs s'imaginent que c'est un théoréme de mathématiques, et les mathématiciens que c'est un fait expérimental," or "everybody believes in the law of errors, the experimenters because they think it is a mathematical theorem, and the mathematicians because they think it is an experimental fact." In the real world it is justified, in fact, by neither. In Hill (1969) it is shown that the use of the normal distribution can instead be based simply upon a subjective judgment of spherical symmetry for the 'actual' errors in the observations. See also Borel (1914, p. 66, 90-93) and Borel (1906). Hill (1969, p. 95) gives the exact density for the marginal distribution of n coordinates based upon spherical symmetry, or conditional uniformity, on the N-dimensional sphere. By Scheffé's lemma that convergence of densities to a proper density implies convergence in distribution, it immediately follows that each fixed r-dimensional marginal distribution of the joint distribution of the n coordinates converges to the Gaussian, even as n goes to infinity as well as N. In this sense spherical symmetry implies approximate normality. It should be noted that my statement of the result, which is for the case of spherical symmetry without a constraint on the average of all N coordinates, agrees with that of Borel, who discovered and stated the result for the case of one coordinate, and appears to have understood the general case. When there is also a constraint on the average of the N coordinates, then my exponent N - n - 2 should be changed to N - n - 3.

In the theory that I proposed spherical symmetry, or more generally, conditional uniformity on surfaces, is itself only an approximation based upon

[2]The only situation known to me where it may be appropriate to regard the probabilities as existing but unknown is where there is a referee or mechanism as in Hill (1974b, 1990c), and one makes the subjective judgment that the referee generates the data according to some algorithm or method of randomization. In this case one must take into account the various constraints, such as finite computer memory, used in generating the data. Probability and utility can be developed together, as in Hill (1978), where the axioms are such as to include both subjective probability and 'objective' probability in the sense of such a referee or mechanism.

the available knowledge, and does not purport to be more than this, or to have any other objective meaning. For example, in the case of errors of measurement, one may view the usual orthogonal axes of a coordinate system as arbitrary, and therefore introduce rotational symmetry. Ultimately, it is simply a matter of judging that spherical symmetry represents a sufficiently good approximation to one's opinions in order to be useful for inference, prediction, and decision-making. In my opinion there is no hope to demonstrate that such a judgment is either 'correct' or 'incorrect,' other than empirically, for example, by seeing how well it works predictively.

At an even more basic level, as de Finetti first realized, induction can be based upon a direct subjective judgment of exchangeability for the sequence of observables. Once this subjective judgment is made, it is a mathematical fact that one will be acting (nearly) as though some statistical model were true. Conditional upon the parameters of such a model, the data will be regarded as independent and identically distributed if the exchangeable sequence is infinite, and approximately so if it is a sufficiently long finite sequence. See Diaconis and Freedman (1980). [3] This is de Finetti's theorem, and its significance is that it provides a subjective justification for the use of statistical models and for conventional model-based Bayesian inference, provided that care is taken in the interpretation of such techniques. See Savage (1972) for a treatment of exchangeability from this point of view. De Finetti stressed the fact that the judgment of exchangeability is itself only a subjective judgment, and perhaps only an approximation to one's actual opinions. To the extent that one judges the sequence as exchangeable, then one is led to conventional Bayesian inferential techniques. Often, in fact, approximate or even partial exchangeability will suffice to justify such techniques. Furthermore, as will be discussed in Section 3, I believe that it is necessary to integrate the conventional Bayesian theory with data-analytic methods for the selection of models, parameters, and hypotheses. See Hill (1987a) for discussion of the deeper structures that may underlie conventional statistical models, Hill (1985-86) for Bayesian selection of models, and Hill (1990a, 1990d) for a theory of Bayesian data analysis.

In a practical sense, based upon the subjective judgment of exchangeability, de Finetti had completely solved the problem of inductive inference for the case of Bernoulli data, or, more generally, for multinomial data with a known finite number of categories. Combined with the beautiful result of W. E. Johnson (1932), as discussed in Zabell (1982), there was little more to be said at the foundational or even practical level for these simple cases, other than to elaborate on the choice of prior distribution for the Bernoulli parameter p or for the parameter $\theta$ of a multinomial distribution. Thus

[3] It should be understood that the conventional notion of independent, identically distributed observations, with an unknown distribution, corresponds roughly to the subjective Bayesian assumption of exchangeability.

the precise measurement (or stable estimation) argument of L. J. Savage (1961, p. Ch. 4; 1962, p.20), or as presented in Degroot (1970, p. 199), deals with the case in which the prior distribution is diffuse relative to the likelihood function, so that it is of relatively little consequence, and the posterior density for the parameter can be approximated by the likelihood function. On the other hand, the Jeffreys-Savage theory of hypothesis testing covers the most important situations in which the prior is not diffuse. See Jeffreys (1961), Savage (1962), Edwards, Lindman and Savage (1963), and Hill (1974b, 1982, 1990d) for discussions. The problem of so-called 'uninformative' priors has also been dealt with very effectively by a number of people for the case of multinomial data. See Good (1965, Ch. 4) for a review and discussion. Furthermore, it has long been recognized that many real world problems can be adequately modelled by such finite partitions, Fisher (1959, p. 111 ), Savage (1961, p. 4.23), so when this can be done effectively there is available a more or less complete system (apart from details and various complications that arise in practice) for inductive inference and decision-making, including the prediction of future observations. In de Finetti's own words (1937, p. 147):

> It is thus that when the subjectivistic point of view is adopted, the problem of induction receives an answer which is naturally subjective but in itself perfectly logical, while on the other hand, when one pretends to *eliminate* the subjective factors one succeeds only in *hiding* them (that is, at least, in my opinion), more or less skillfully, but never in avoiding a gap in logic. It is true that in many cases–as for example on the hypothesis of exchangeability–these subjective factors never have too pronounced an influence, provided that the experience be rich enough; this circumstance is very important, for it explains how in certain conditions more or less close agreement between the predictions of different individuals is produced, but it also shows that discordant opinions are always legitimate. This does not make any change in the purely subjective character of the whole theory of probability.

See also de Finetti (1974, Ch. 11).

Thus in the exchangeable case the only situation that had not been essentially resolved is that in which no fixed finite partition model is thought to be appropriate, or more generally, when the number of parameters requisite realistically to model the data is large relative to the number of observations. One can speak of this either in terms of multinomial data with an infinite number of categories, or alternatively, in terms of an unknown and possibly quite large finite number of categories. [4] Still again, to

[4] All real world problems must eventually be 'finitized' in order for statisti-

suggest the general type of problem, one can speak of Bayesian nonparametric statistics, or of Bayesian inference about an 'unknown' distribution function. Whatever words we may use, what we are trying to describe is the situation in which no known conventional parametric statistical model is thought to be appropriate for the exchangeable sequence of observations. This situation seems often to arise in the practice of statistics.

What then can be said about the nonparametric case from a subjective Bayesian point of view? The first thing to observe is that de Finetti's theorem still holds, so that in the case of an infinite exchangeable sequence of observables, one will be mixing over a dummy variable that represents the 'unknown' distribution, say F, in the population. De Finetti (1937, Ch. 4), had already given an insightful development of the mathematics of this situation for exchangeable random quantities, while Hewitt and Savage (1955) greatly generalized the ideas of de Finetti and presented them in the modern style of mathematics. Diaconis and Freedman (1980, 1981) present easily accessible proofs. Just as in the case of exchangeable events, what must be specified in order to implement the Bayesian approach is the mixing function, or a priori distribution for F. In the case of exchangeable events, F is concentrated at only two known values, 0 and 1, while in the present case the distribution F can in principle be any distribution function on the real line. Special mixing distributions will correspond to the subset of F appropriate for a finite multinomial situation, in which the categories are coded numerically, or for knowledge that F is Gaussian, etc. The fully nonparametric case is that in which F cannot be restricted to such special subsets of the space of all distribution functions. In principle, what the Bayesian must do is to specify a prior distribution, $\pi$, on the space of all distribution functions; and then, given the data, update this prior distribution to become a posterior distribution, $\pi^\star$, in accord with Bayes's theorem.

How then can a subjective Bayesian specify a prior distribution that expresses a realistic degree of vagueness about F? Since we are dealing with possibly infinitely many parameters, it is clear that the problem is formidable even from the point of view of the mathematics involved, and of course even much more so conceptually. Furthermore, after observing a sample from the population, and obtaining $\pi^\star$, in order to obtain the posterior predictive distribution of future observations, one would have to integrate the conditional distribution of the future observations, given F, with respect to this posterior distribution of F. Here the 'unknown' F plays the same role as the 'unknown' $\theta$ of a conventional multinomial model, but the mathematics is again enormously more complicated. In

---

cal procedures to be implemented on a computer with finite memory in finite time. There is no harm in visualizing matters in terms of infinite partitions, provided that these are eventually used to obtain some procedure that can be so implemented. See Hill (1990c).

addition, a basic difficulty arises here that did not appear in the case of finite multinomial models. It is no longer the case that one can usually rely on some form of Savage's precise measurement argument. Thus the distribution function F may have infinitely many parameters, and no matter how large a finite sample is taken from the population, the prior distribution for F may still play a crucial role. Even if, more realistically, we regard F as having a large finite number of parameters, in a practical sense the same phenomenon occurs, since realistic sample sizes will be small relative to the total number of parameters. See Hill (1975b) for a discussion of this phenomenon. Typically there is no such thing as global robustness (for all possible distributions of F), or in other words, the posterior distribution may be extremely sensitive to the prior distribution for F.

The problem is not, however, so hopeless of solution as may first appear. The first modern day hint or suggestion as to the nature of a possible solution occurs in the work of R. A. Fisher (1939, 1948), who proposed a fiducial interpretation for what I later called $A_n$, and who gives credit to 'Student' for the underlying idea.

Consider a conventional formulation of statistical inference, in which the observations are conditionally independent with cumulative distribution function $F(x;\phi)$, where $\phi$ is a conventional unknown parameter. Assume that the distribution function is continuous in x for each $\phi$. Let $X_{(i)}$ denote the ascending order statistics of the data, for $i = 1, \ldots, n$. Then let $\theta_i = F(X_{(i+1)};\phi) - F(X_{(i)};\phi)$, for $i = 0, \ldots, n$, where by definition $X_{(0)} = -\infty$ and $X_{(n+1)} = \infty$. *Before* the data are drawn, clearly the distribution of the $\theta_i$ is a uniform distribution on the (n+1) dimensional simplex, i.e., a special Dirichlet distribution in which all the parameters are equal to unity. This is the fundamental frequentistic intuition with regard to $A_n$, and which Fisher presumably used to put forth his proposed fiducial solution. Thus Fisher suggested (or implied) that even when the random variables $X_{(i)}$ are replaced by their observed values $x_{(i)}$, that the uniform distribution for the $\theta_i$ would still be appropriate. It should be mentioned that the articles by Fisher (1939, 1948) only briefly and cryptically discuss the nonparametric fiducial case that we are concerned with. The first clear formal statement of something like $A_n$ (although not in terms of exchangeable sequences) is by Dempster (1963), who stated the Fisherian argument precisely, changed the name from fiducial to 'direct' probability, and applied the argument for prediction of future observables. Dempster also asserted that what I later called $A_n$ "does not appear to have a Bayesian interpretation."

In my 1968 article I showed that in fact $A_n$ does have a Bayesian interpretation. Before discussing this, however, let me note that Fisher's proposed fiducial distribution is an example of a posterior predictive distribution, since whatever the rationale, it is posterior to the data, and does partially specify a probability distribution for the future data. This predictive distribution is not completely specified, since what it does is to attach a probability of $\frac{1}{n+1}$ to each of the n + 1 open intervals formed by the consecutive order statistics of the given sample, assuming that there are no ties, and goes no further. The fiducial argument that Fisher gave for this evaluation depends upon one's willingness to persist with the pre-data evaluation of the distribution of, say, $F(X_{(i+1)};\phi) - F(X_{(i)};\phi)$, *after* the $X_{(i)}$ are replaced by their observed numerical values. Such a fiducial argument, although intriguing, was logically suspect. See Edwards (1972, p. 207) for a totally devastating example against the logic of the fiducial argument. Also, Lindley (1958) had already shown, *under certain very special conditions,* that the fiducial argument leads to a genuine posterior distribution only in simple cases reducible to that of a location parameter. However, Fisher, in a footnote (1959, p.51), asserted:

> Probability statements derived by arguments of the fiducial type have often been called statements of 'fiducial probability'. This usage is a convenient one so long as it is recognized that the concept of probability involved is entirely identical with the classical probability of the early writers, such as Bayes. It is only the mode of derivation which was unknown to them.

In short, the situation with regard to $A_n$ was anything but clear, and at the time it was not even known whether $A_n$ was a coherent evaluation in the sense of de Finetti. If it was, then presumably it could have been derived by means of a prior distribution for F, Bayes's theorem, and an integration with respect to the posterior distribution of F, as discussed above. The problem that I addressed in my 1968 article was that of giving such a derivation of $A_n$. The first step in my formulation was to consider the case of arbitrary finite populations, in which the size of the population may be unknown, rather than infinite populations, or in other words, to deal with finite exchangeable sequences.

This step is not only more realistic, since in the real world we are not ordinarily called upon to deal with more than finite populations or sequences of observables, but it also greatly simplifies the mathematics, and has clear operational meaning. [5] The proofs of de Finetti's theorem for the infinite case by Heath and Sudderth (1976) and by Diaconis and Freedman (1980, 1981), proceed by taking limits for the finite case. Thus for the case of events, one can condition on the sum of the indicators after N observations, and note that because of exchangeability, all paths to a given sum are equally likely. The exchangeable distribution of any N indicators (for example, for red and white balls in an urn) is then identical with one that

[5] A number of distinguished mathematicians have questioned whether some of the usual laws of logic, such as the law of the excluded middle, are valid with respect to infinite collections. See Hill (1990c) for a discussion of the operational meaning of the admissibility principle in connection with infinite partitions for a parameter, and Hill (1987b) for discussion of finite additivity in connection with $A_n$.

arises from sampling without replacement from a 'randomly selected' urn with N balls, i.e., the draws are made from an urn with composition (R,W), R + W = N, with a probability equal to the original subjective probability that the sum of the N indicators is R. More generally, one can consider the empirical distribution function that arises from 'sampling' the first N coordinates from an exchangeable sequence. Conditional on this empirical distribution function, the individual coordinates are distributed uniformly over the collection of N-tuples having this empirical distribution. Because sampling without replacement is, for large N, close to sampling with replacement from this empirical distribution, and because of the convergence of this empirical distribution to some limiting distribution (with probability one, under exchangeability), one obtains de Finetti's theorem for the infinite sequence in this way. See Diaconis and Freedman (1980, p. 749; 1981, p. 209) for details. These authors have emphasized the importance of the finite case even for the underlying mathematics, and as I argue in Hill (1990c), it is also the appropriate formulation for inferential purposes. The model for $A_n$ in Hill (1968, p. 679) is actually equivalent to the specification of a diffuse prior distribution for the empirical distribution of a finite population. In the context of Heath and Sudderth, or of Diaconis and Freedman, it is equivalent to specifying the subjective distribution for the sufficient statistic based upon N trials within their models. Thus instead of specifying a diffuse prior distribution on the space of all distribution functions F, what I have done is to specify such a prior distribution for the empirical distribution function of the entire finite population from which a simple random sample has been drawn. Here the number of units in the finite populations can be unknown, as well as the number of jump-points of the empirical distribution of the population, and the points at which the jumps occur and the sizes of the jumps. The details concerning this diffuse prior distribution will be given in Section 2. Here what I want to discuss is the underlying sampling model.

The statistical model for this problem can be thought of in terms of sampling with or without replacement from a finite population of units. Imagine that each unit carries an attached tag or label, for example giving the color of the unit, or the name of the species to which the unit belongs, or a numerical value such as the mass of the unit. It does no harm to visualize the population of units, with their attached labels, as sitting in an urn. A simple random sample, with or without replacement, is then drawn, and we observe the value of the label for each unit in the sample. It is assumed for simplicity here that the label or numerical value is observed without error, although the theory can easily be extended to deal with errors of measurement.

The population of labels or numerical values can be described in terms of the empirical distribution of such labels or values. Indeed, because we are dealing with only finite populations, the case of colors or names can be viewed as a special case of the case of numerical values. Thus we can imagine, without loss of generality, that the finite collection of colors or names in the entire population have been encoded numerically, thus yielding numerical values. For simplicity, we shall describe the situation for the case of such numerical values. When we return to the case of 'colors,' as in the species sampling problem, we shall point out the special features that arise in this case. For the time being, visualize the urn population as consisting of the numerical values attached to the units, such as their masses. This population can then be described in terms of the number of units in the population, say N, and the empirical distribution of the values in the population. Note, for example, that if sampling is with replacement from this population, and if the number of distinct values in the population is known to be say, M, then this model is a special case of a conventional multinomial model with exactly M non-empty categories. In general, of course, M need not be known, except that $1 \leq M \leq N$, and sampling can be without replacement. In any case, the number of units in the population, N, and the empirical distribution of population values, completely characterize the finite population of values. In fact here the empirical distribution for the entire finite population of values plays a similar role to that of the 'unknown' probabilities in a conventional statistical model. Following the spirit of de Finetti (1937), I regard the fundamental problem of induction to be reducible to that which arises in sampling without replacement from an urn consisting of units that are labelled with numerical values.

The solution that I proposed for this problem consists of a model for a generalized version of $A_n$, called $H_n$, in which ties can occur, and in which smoothing is possible, and will be discussed in Section 2. Historically, the sequence of events concerning $A_n$ after Dempster (1963) was as follows. I proved in my 1968 article that $A_n$ cannot hold exactly for *countably additive proper prior distributions,* in the case of exchangeable sequences in which ties have probability 0. At the same time I recommended it as an approximation for a variety of situations that can be roughly described as situations in which the data is measured on a "rubbery scale," as for example when the finite population values are distorted by some measuring instrument that is equivalent to making an unknown monotonic increasing transformation of the original values. It was also proved in Hill (1968, p. 686) that $A_n$ for all n implies that the posterior distribution of the $\theta_i$ defined earlier is the uniform Dirichlet distribution on the (n + 1) dimensional simplex, thus giving support to Fisher's fiducial argument. The next historically significant development regarding $A_n$ was the proof by Lane and Sudderth (1978) that $A_n$ for all n is coherent in the sense of de Finetti, i.e., it is impossible to be made a sure loser, and related results by the same authors (1984). The robustness and invariance properties of $A_n$ were investigated by myself in Hill (1980a) with the general result that it is robust in the modern Bayesian sense of Berger (1985), Hill (1980b). Then in his doctoral dissertation, Peter Lenk (1984) showed, along with many other things, that $A_n$ can arise as a limit of proper priors, using a log gaussian model for the prior distribution of the unknown density function of the population. Next, Berliner and Hill (1988) used $A_n$ to obtain the predictive distribution for future observations in the case of censored data, as for example in survival analysis. Finally, in Hill (1987b) I have constructed a class of simple parametric models, called splitting processes, such that $A_n$ holds for all n. A modification of this construction also yields $H_n$ for all n, and the coherency of $H_n$ is proved. $A_n$ and the Dirichlet process of Ferguson (1973) are the finitely additive and countably additive versions, respectively, of the same underlying splitting process.

In the next Section I will restate $A_n$, give my model for inference and prediction, and suggest a subjectivistic argument for $A_n$.

## 2. $A_n$ AND $H_n$

In Hill (1968) a direct specification, denoted $A_n$, for the posterior predictive distribution of future observations was proposed. $A_n$ was meant to express extremely vague subjective prior knowledge as to the form of the underlying population distribution. For the case of n = 1 and 2, $A_n$ follows from conventional parametric models (Gaussian, for example) with a uniform prior distribution on the location parameter, or on the location parameter and logarithm of the scale parameter, respectively, Jeffreys (1961, p. 171), Hill (1968, p. 688). For example, when n = 1, suppose that the parameter $\theta$ is the mean of a normal population with known standard deviation of unity. Given an observation $X_1 = x_1$ from this population, the posterior distribution of $\theta$ is $N(x_1, 1)$. The predictive distribution of the next observation, $X_2$, is then easily seen to be $N(x_1, 2)$, which of course agrees with the standard tolerance interval analysis. Hence the posterior probability that $X_2 \leq x_1$ is .5. Note that for any prior distribution of $\theta$ which is diffuse relative to the likelihood function, $A_1$ will hold to a good approximation, since the posterior distribution of $\theta$ will still be approximately $N(x_1, 1)$. A similar analysis applies in the case n = 2. At the time of Hill (1968), it was not known whether $A_n$ could be obtained for conventional parametric models when $n \geq 3$. However, Hill (1987b) shows that this is the case for both $A_n$ and $H_n$.

$A_n$ for untied data, or $H_n$ for the case of ties, are exactly appropriate for data measured on a merely ordinal scale, or with a trivial modification, for data that consists of labels (such as the names of species, as in the species sampling problem), and can yield an extremely good approximation for data on a ratio or interval scale, such as the weights in a population of penguins, as will be discussed at the end of this section. Just as with other nonparametric models, it is hardly necessary for the assumptions to hold literally, in order that the conclusions be appropriate to a very good approximation. Indeed, $A_n$ is both a fiducial and Bayesian procedure, and as with other procedures that combine these two properties, it is robust to variation in the specification of the a priori distribution. See Hill (1990e).

The condition $A_n$ is defined as follows. $A_n$ asserts that conditional upon $X_1, \ldots, X_n$, the next observation $X_{n+1}$ is equally likely to fall in any of the open intervals between successive order statistics of the given sample (Hill, 1968, p. 677). Note that in our original definition of $A_n$ we did not assume that the sequence is necessarily exchangeable or that ties have probability 0. Thus we also included cases where there is a positive probability that the next observation ties one of the previous observations, and also partially exchangeable situations that satisfy the above definition. For the purposes of this article I wish to modify the notation, use $H_n$ to denote the situation in which ties can occur, and reserve $A_n$ for the special case of $H_n$ in which there are no ties (or ties have probability 0). In this article I will also assume that the observations are exchangeable, although this will not be included in the definition of $A_n$ or $H_n$.

$A_n$ specifies a predictive distribution for one future observation. If also $A_{n+1}$ holds, then by conditioning upon which interval the first new observation falls in, we can obtain a predictive distribution for two new observations, and by extension for an arbitrary number of new observations. See Hill (1968, p.684) for such predictive schemes. [6] Furthermore, we can use this same idea to deal with censored data, again by conditioning upon which intervals the censored observations will fall in. Berliner and Hill (1988) carry through such an analysis for the case of survival data, present upper and lower bounds for the survival function, and simple algorithms with which to make the analysis. In the survival problem, for example, we assess the predictive probability distribution for the times of death of new patients given a treatment, using as data the death times, and the intervals in which censoring occurred, i.e., the partial censoring information, for a previous group of patients who were given the treatment and with whom the new patients are regarded as exchangeable. Chang (1989) provides additional computational algorithms, extends the results to the two sample case, and compares our procedure with the Kaplan-Meier procedure by means of Monte Carlo studies.

The definition of exchangeability that we shall use is that motivated by the subjective Bayesian viewpoint, namely, in terms of a subjective judgment that the order is irrelevant. [7] To be precise, let $X_1, \ldots, X_k$, be k random variables that are (finitely) exchangeable in the sense that the joint distribution function of any r distinct variables is the same as that for any

---

[6] Note that the equation on the top of page 684 is only valid if $i \neq j$. When i = j, it is necessary to add another term which corresponds to the possibility that the second new observation ties the first. A similar correction is necessary in the formula for $E(\theta_i \times \theta_j)$.

[7] Mathematically, this judgment is equivalent to the usual definition of exchangeability, but psychologically it is different, in that we do not assume the sequence is 'truly' exchangeable, but merely that one regards it as exchangeable, perhaps only as an approximation.

other r such variables, r = $1,\ldots,k$. An infinite sequence of such variables is then said to be exchangeable if the above condition is true for each k. [8] Such models arise from the following Bayesian formulation: Assume that, given some distribution, say F, $X_1,\ldots,X_n$, are independent and identically distributed according to F. For F unknown it is natural for the Bayesian to model F itself as 'random' with some a priori probability specification. [9] This can be done either parametrically or nonparametrically. In either case, 'integrating out' F leads to an exchangeable unconditional joint distribution for the X's. Conversely, de Finetti's theorem implies that if the exchangeable sequence is infinite, then there exists a distribution for F, called the prior distribution of F, for which the joint distribution of the observations obtained by 'integrating out' F is the original exchangeable distribution. See Hewitt and Savage (1955), Heath and Sudderth (1976), and Diaconis and Freedman (1980, 1981) for proofs. The 1980 article, which emphasizes the finite exchangeable case, is particularly appropriate for my purposes. The authors show that the most general exchangeable sequences arise by taking limits of the finite exchangeable sequences that arise in sampling without replacement from urns.

I will now present my model for $A_n$, or more precisely, for my generalization of $A_n$, called $H_n$, which allows for ties and for the possibility of smoothing, and of which $A_n$ is a special case. We assume that there exists a finite population of units, with each unit having an attached value or label. For example the value might be the mass of the unit, or the label might be the name of the species to which the unit belongs, or the color of the unit. We assume that the set of values is simply ordered, or at least can be simply ordered. By a simple ordering we mean a relationship, say $\leq$, which for any two elements x, y, of the set of values, is such that either $x \leq y$ or $y \leq x$, and which is transitive. See Jeffreys (1957, Chs. 5-6), Luce and Narens (1987), and Whitrow (1980, Sec. 4.7) for discussions of the concept of measurement.

Thus masses would be on a ratio scale, and are certainly simply or-

---

[8] In this article we ignore the issue raised in Hill (1987b) as to whether the definition is to be made in terms of equality of joint distributions or equality of joint distribution functions. Such a distinction is of theoretical importance in the finitely additive theory of probability, but is not of practical importance.

[9] Such randomness, however, is merely a way to represent uncertainty about F, and does not imply that F is a conventional stochastic process, with the usual structure for such processes. For example, F can be fixed, as when it is the complete empirical distribution for some finite population, and not in any sense 'truly' random, but merely unknown. In the finitely additive theory, it is unnecessary to introduce the usual measure-theoretic structure, since one need not restrict attention to sigma fields of events. In practice a statistical problem must in any case be finitized to be representable on a computer with finite memory, so the real-world distinction between finite and countable additivity concerns only the class of approximations that is allowed. The finitely additive theory allows a much richer class of such approximations.

dered; while labels can be simply ordered for a finite population simply by designating an ordering. (Some think that this can be done for infinite populations as well, using the well-ordering theorem, but there is no need to go into such things here. See Hill (1990c, 1990d) for discussion of the issues and paradoxes that arise in connection with the use of infinite partitions.) Suppose there are N units in the population, and that the set of attached values or labels is $\{Z_i, i = 1, \ldots, N\}$. We shall now refer only to values, with it being understood that we include labels as a special case after the finite population has been simply ordered. Some of the values $Z_i$ may be equal to one another. Suppose that in fact there are only M distinct values amongst the $Z_i$, and denote these in ascending order of magnitude as $X_{(1)} < X_{(2)} < \ldots < X_{(M)}$, where of course $M \leq N$. Finally, suppose that the value $X_{(i)}$ occurs in $L_i$ units, where $L_i \geq 1$ (since by assumption the value $X_{(i)}$ does in fact occur), and note that $\sum_{i=1}^{M} L_i = N$.

The above model constitutes our description of the finite population of values $Z_i$. Note that this determines the empirical distribution of values in the finite population, i.e., the empirical distribution has jumps occurring at $X_{(1)}, \ldots, X_{(M)}$, and the jump that occurs at $X_{(i)}$ has height $L_i/N$. Of course in general all of these quantities are unknown, i.e., N, M, the $X_{(i)}$, and the $L_i$. From the subjective Bayesian point of view one must then specify a probability distribution for all of these quantities. It should be noted that this point of view corresponds exactly to the recent probabilistic treatments of exchangeability for finite sequences, as in Diaconis and Freedman (1980), where the finite exchangeable sequence of length N is the vector $Y_1, \ldots, Y_N$, which would be generated by sampling without replacement all N elements of the finite population, so that these $Y_i$ are some permutation of the $Z_j$.

The case of an infinite exchangeable sequence may be viewed as an idealization of this scheme, and gives rise to de Finetti's theorem. But the model in terms of sampling from a finite population is simpler, avoids difficulties and paradoxes of infinity, and loses no real generality since real world populations are in fact finite. In my model we then require only a prior distribution for the composition of the finite population, i.e., for (M, N, $\underline{X}$, $\underline{L}$), rather than a prior distribution for F, the theoretical distribution for an infinite exchangeable population. It is far simpler to specify such a prior distribution for the finite number of parameters (at most 2N +1) needed to describe this finite population, than to do so on the infinite dimensional space of distribution functions F. Furthermore, we shall argue that there is a natural way to represent vagueness for the finite population, which would be much more difficult to achieve for an infinite population (for example, one would have to confront some basic issues concerning the difference between countable and finite additivity).

Now let us consider the data that we shall be analyzing. It is assumed that a simple random sample is drawn without replacement from the finite population that we have described above. Let the sample size be n. The data will consist of the numerical values attached to the n units that are thus selected from the finite population. Let $x_{(1)} < x_{(2)} < \ldots < x_{(m)}$, be the ascending order statistics of the sample, with m distinct values, $1 \leq m \leq n$, with $n_i$ sample units having the value $x_{(i)}$, and define $\underline{x} = (x_{(1)}, \ldots, x_{(m)})$. Also let $I_i$ be the open interval between $x_{(i)}$ and $x_{(i+1)}$. Then $n_i \geq 1$ , and $\sum_{i=1}^{m} n_i = n$. It is assumed here that the values are measured without error, so that each $x_{(i)}$ is necessarily some $X_{(j)}$ in the population, but of course we do not know with certainty which. By the data we mean the m distinct $x_{(i)}$ values, and the $n_i$. Thus the data determines the empirical distribution of the sample. We now require only one further bit of notation. Given the data, define $J_i$ to be the rank, in the population, of the value $x_{(i)}$ in the sample, for $i = 1, \ldots, m$. The vector $\underline{J} = (J_1, \ldots, J_m)$ then gives the true ranks, in the population, corresponding to the sample values $x_{(i)}$. Thus $1 \leq J_1 < J_2 < J_3 \ldots < J_m \leq M$, because of the fact that the $x_{(i)}$ and $X_{(j)}$ are strictly ordered. We now are ready for the basic equations of the model for $H_n$.

In the first equation, we condition on the true composition of the finite population, by which we mean the unknown quantities $\underline{X}$, $\underline{L}$, M, and N. This equation gives the probability for observing the data together with $\underline{J} = \underline{j}$, for each possible vector of ranks $\underline{j}$:

$$Pr\{\ data,\ \underline{J} = \underline{j} \mid \underline{X},\ \underline{L} = \underline{l},\ M,\ N\ \} \quad = \quad \binom{N}{n}^{-1} \times \prod_{i=1}^{m} \binom{l_{j_i}}{n_i}, \qquad (1)$$

if $X_{(j_i)} = x_{(i)}, i = 1, \ldots, m$, and is otherwise 0.

Note that this would be the likelihood function for the finite population parameters, except that we have included $\underline{J} = \underline{j}$ together with the data, because this is the key to making an effective evaluation; the ordinary likelihood function would involve a mixture of (1) with respect to $\underline{j}$. For sampling with replacement, it is necessary to replace the factor $\binom{l_{j_i}}{n_i}$ by $(l_{j_i})^{n_i}$, etc. We shall not further deal with the case of sampling with replacement, since sampling without replacement is the more common, more difficult to analyze, and more important form of sampling. In fact, even if sampling is with replacement, then one could throw out repetitions of the same unit, without loss of information.

The next step is to integrate out over the unknown $\underline{X}$ values in the population. In general such an integration requires the assumption of countable additivity, or equivalently in countable spaces, conglomerability in the finitely additive theory, as in Schervish et al (1984), Hill and Lane (1985). However, in the present case with only two values for the probability in question, i.e., that given by (1), or else the value 0, such an assumption is not necessary.[10] Define

[10]There is a subtlety, however, underlying our use of notation for probability

$$h(\underline{x}, \underline{j}, \underline{l}, M, N) = \prod_{i=1}^{m} \binom{l_{j_i}}{n_i} \times Pr\{\, X_{(j_i)} = x_{(i)}, i = 1, \ldots, m \mid \underline{L} = \underline{l},\ M,\ N\},$$

so that

$$Pr\{\ data,\ \underline{J} = \underline{j} \mid \underline{L} = \underline{l},\ M,\ N\ \} \quad = \quad \binom{N}{n}^{-1} \times h(\underline{x}, \underline{j}, \underline{l}, M, N).$$

We thus obtain the basic result:

$$\begin{aligned} Pr\{data,\ \underline{J} = \underline{j},\ \underline{L} = \underline{l} \mid M,\ N\} &= Pr\{\ data,\ \underline{J} = \underline{j} \mid \underline{L} = \underline{l},\ M,\ N\} \\ &\times\ Pr\{\ \underline{L} = \underline{l} \mid M, N\ \} \\ &= \binom{N}{n}^{-1} \times\ h(\underline{x}, \underline{j}, \underline{l}, M, N) \\ &\times\ Pr\{\ \underline{L} = \underline{l} \mid M, N\}. \qquad (2) \end{aligned}$$

Clearly, for given N, all that must be specified in order to make further evaluations are the three components of the prior distribution on the composition of the population, namely

$$Pr\{\ \underline{L} = \underline{l} \mid M,\ N\ \}, \qquad (3)$$

$$Pr\{\ X_{(j_1)} = x_{(1)}, \ldots, X_{(j_m)} = x_{(m)} \mid \underline{L} = \underline{l},\ M,\ N\}, \qquad (4)$$

$$\text{and } Pr\{\ M \mid N\}. \qquad (5)$$

Although our primary interest in this article is the specification of (4) in such a way as to express diffuse or vague knowledge about the underlying population of values, we note that our formulation is sufficiently general so as to include conventional parametric specifications as well. For example, we may be of the opinion that the population distribution is approximately normal, in which case the distribution of $\underline{X}$ can be chosen so that the $X_{(i)}$ are distributed like the order statistics of a sample from a normal population, and similarly for any other parametric distribution. In general, specification of (4) can stem from a desire to smooth the data, which is appropriate when one has some strong a priori knowledge as to the shape of the finite population, or can arise through data analysis as in Hill (1990a). In such cases one can let (4) depend upon $\underline{j}$, in some way or other. We shall not pursue this idea at the present time, however, since the most basic case

rather than the more conventional notation in terms of probability density functions. We have not space to go into this subject here, but it is related to the previously mentioned considerations concerning finitization of statistical procedures. Thus when actually implemented, the $x_{(i)}$ must be values represented in the finite memory of some computer.

is the nonparametric diffuse case in which very little is known about the shape of the underlying population.

In this article, as in my previous work, we shall specify (3) and (4) as follows:

$$Pr\{\ \underline{L} = \underline{l} \mid M, N\} = \binom{N-1}{M-1}^{-1}, \tag{6}$$

while for each possible $\underline{x}$,

$$Pr\{\ X_{(j_1)} = x_{(1)}, \ldots, X_{(j_m)} = x_{(m)} \mid \underline{L} = \underline{l},\ M,\ N\} \tag{7}$$

depends only upon $\underline{x}$. [11]

Any specification of (3) and (4) is equivalent to a specification of the prior distribution for the empirical distribution of the population, given M and N. Obviously this can be done in infinitely many ways, any one of which might be appropriate in a specific real world situation. But it is of value to single out those specifications that are of special significance, such as for example correspond to a diffuse prior distribution (as is commonly done with improper prior distributions for conventional parameters), and also those that are known to be compatible with much real world data. The specification (6) is the Bose-Einstein distribution for non-empty cells, as in Feller (1968, p. 40). Thus the results in Hill (1968) are based upon this choice, while those in Hill (1980a) discuss the robustness of this choice within the class of exchangeable distributions for $\underline{L}$. Wen-Chen Chen, in his Ph. D. dissertation (1978) and Chen (1980) generalized this choice to include arbitrary symmetrical Dirichlet-multinomial distributions, and argued that for some data it is desirable to choose a Dirichlet-multinomial prior distribution other than the Bose-Einstein, which of course is a Dirichlet-multinomial corresponding to a uniform Dirichlet distribution. See also Lewins and Joanes (1984) and Boender and Kan (1987) who use the same basic Dirichlet-multinomial model, and obtain the same results as myself and Chen. My primary motivation for the Bose-Einstein distribution (which I still regard as the single most appropriate choice) is the connection with Zipf's Law. This law represents more real world data than any other known law, including the Gaussian. It is shown in my articles Hill (1970, 1974a, 1975a, 1979, 1980a, 1981), and in Hill and Woodroofe (1975) and Woodroofe and Hill (1975), that the Bose-Einstein choice yields Zipf's Law. This is why I singled it out as of special significance within the class of exchangeable prior distributions for $\underline{L}$. See also Ijiri and Simon (1975) for discussion of the Bose-Einstein distribution in connection with Gibrat's law. Of course it is mathematically straightforward to replace the Bose-Einstein distribution by any other Dirichlet-multinomial distribution, and sometimes this may

[11] This last requirement can be relaxed to allow dependence upon M and N, in which case only the posterior distribution of M and N will be altered.

be of value in modelling the data. The *logic* underlying my model would at best only suggest that the distribution of $\underline{L}$ should be chosen to be exchangeable, and even this is not really necessary. For example, one might know that there were fewer units with small values than with large values, and use a non-symmetrical Dirichlet-multinomial prior distribution.

Next, one must also make some specification for the prior distribution of M and N. The most basic case for inference is simply where N is known to be large, and M has a uniform distribution, given N. This was the case considered in Hill (1968). Hill (1979) then considered the case where M has a truncated negative binomial distribution, of which the uniform is a special case. Although the specification of the distribution for M, given N, is of lesser importance here than the specification of (3) and (4), it does play a crucial role in obtaining Zipf's Law, as in the cited articles by myself and by Chen.

Even more important than the choice of the Bose-Einstein distribution for $\underline{L}$ is the choice of (7). Here we directly confront the problem of formulating a diffuse prior distribution on the empirical distribution of the population. Note that if M = N, so that all $L_i = 1$, then we obtain the case where ties have probability 0, i.e., $A_n$, and must then only express vagueness of opinion about the jump-points $X_{(i)}$ in the population. The problem of expressing a diffuse distribution for the jump-points is logically independent of that of expressing one for $\underline{L}$. It was shown by Lane and Sudderth (1978, Theorem 1), defining $A_n$ for the case where ties have probability 0 and where the sequence is exchangeable, that (7) is equivalent to $A_n$.

Consider then the specification (7). What it says is that no matter what the distinct values $x_{(i)}$ may be, they contain no information whatsoever about the ranks $J_i$ of these values in the population. Clearly this is not always appropriate. For example, if one believed that the population was approximately Gaussian in form, then one would favor some $\underline{j}$ vectors over others. Or if one knew sufficiently much about the set of values in the population, then one might know, for example, that $x_{(m)}$ was in fact the largest value in the population. Or again, if the $X_{(j)}$ are necessarily integers, and if two observations are consecutive, then one knows that the corresponding $J_i$ are also consecutive. It is easy to make adjustments for such special cases. But note that the specification (7) is the only general specification that commands any credibility for the case of vague a priori knowledge as to the shape of the finite population. Any other specification requires substantive knowledge as to the shape. And because the formula for the conditional posterior distribution of $\underline{J}$ and $\underline{L}$, as given by equation (2), is mathematically valid, any Bayesian procedure for smoothing must imply *some* specification for

$$Pr\{ X_{(j_1)} = x_{(1)}, \ldots, X_{(j_m)} = x_{(m)} \mid \underline{L} = \underline{l},\ M,\ N\}.$$

Such a specification for smoothing would then have to be justified on the basis of special a priori or post-data knowledge.

If one visualizes the $X_{(i)}$ in the population as being physically meaningful numbers such as masses, and then as being distorted by measurement on an unknown rubbery scale, i. e., distorted by an unknown monotonic transformation, one easily comprehends that in this case only order properties remain meaningful, and these are precisely what $A_n$ and $H_n$ specify. It is clear that in such a situation any a priori knowledge about the distribution of the values in the finite population would have been lost because of the distortion. The real world situation is usually somewhere between this case where only order properties are meaningful, and the case where conventional quantitative models, such as the Gaussian, are appropriate. When the true situation is that the numbers do have a physical meaning, and have not been distorted, but there is little evidence as to the shape of the population, then the question as to whether to use $H_n$ is more subtle. On the one hand, possibly via data analysis, one may find a parametric model that one regards as appropriate. On the other hand, one may not find such a model, and therefore have no rational way to smooth the data. Clearly one does not want merely to do some form of 'random smoothing,' as for example by picking a model out of a hat, when that model has no real credibility. There is much that needs to be learned about such issues. To understand the force of the argument for (7), even in the case of a genuine quantitative measurement, consider the following example.

Suppose for the sake of argument that there are 100,000 adult male emperor penguins, and that their weights can be measured sufficiently precisely so that no two agree exactly. (This is assumed only to make the essential point clear. My model $H_n$, with ties, can deal with any degree of rounding.) Consider your a priori subjective opinions about the population of weights of these penguins. Suppose now that I were to give you all but one of these weights as the data $x_{(i)}$, i.e., 99999 positive numbers, no two of which are equal. The question I wish you to think about concerns your opinions about the vector $\underline{J}$, which specifies the ranks of these 99999 numbers in the population of all 100000 numbers. Condition (7) here would require that you be indifferent a posteriori as to which ranks these observations have in the population with M = N = 100,000. Note that this is meant to apply no matter what the $x_{(i)}$ values are, provided that only possible values are included, so that negative values are excluded, as well as weights that are known to be impossibly large or impossibly small. For example, if (7) holds, then you are indifferent as to which of the 100,000 possible values is missing in the data. It could just as well be the largest as the smallest, or any other member of the population. Thus it would be the largest that is missing if it were the case that $\underline{J}$ consists of the ranks $1, \ldots, 99999$ in the population, and it would be the smallest that is missing if $\underline{J}$ consists of the ranks $2, \ldots, 100,000$ in the population. Are you so indifferent?

A fairly natural first reaction is to say that you might or might not be indifferent, depending upon what numbers $x_{(i)}$ I give you. And you might think that for lots of such sets of 99999 numbers you would be, and for others you would not be. But think again. Suppose, to take an extreme case that might seem to speak against $A_{(99999)}$, that the $x_{(i)}$ that I give you are such that there is an enormous gap between the largest, $x_{(99999)}$, and all the others. In fact, suppose that $x_{(99999)}$ is an extremely large value, say 200 pounds, one that (although perhaps not impossible), seems highly improbable, while the other sample weights are all less than 100 pounds. You do not appreciate the full force of $A_n$ until you realize that if the largest weight in the sample were in fact 200 pounds, then there might well be another penguin that weighs even more than this! Thus the naive reaction, which would be that no penguin weighs anything like 200 pounds, is immediately dispelled once one fully appreciates the fact that you have already seen one such (in the scenario of the problem, where it is assumed that no cheating occurs) and may therefore well see another one. Still another example of this type concerns human age. One might regard it as extraordinarily improbable that any human being has lived to the age 500 years. But if one such could be demonstrated, then you might well think that another might also, and even find that your opinions were roughly in accord with $A_n$.[12]

What condition (7) is expressing is a completely pragmatic attitude towards the population. Such an attitude is not only a subjectively coherent Bayesian attitude but in the case at hand even seems quite compelling; and this is for the case of weights on a ratio scale, which is the worst type of example for $A_n$, as opposed to data on a merely ordinal scale, such as the Mohs scale for hardness of rocks, where the hardness values are more or less meaningless. See, for example, Whitrow (1980, p. 216). And yet I think, after reflection, you may find it compelling even in the extreme example I have given. It would of course be even more compelling if the largest weight in the data were say, 120 pounds, rather than 200 pounds. The general argument that I would give is that (7) with n = m = M - 1, and N = M (so there are no ties), is a highly compelling subjective evaluation, and this implies $A_{N-1}$. Note that there is no possibility of a mathematical proof that (7) is 'correct,' just as there is never any way of 'proving' that one ordinary prior distribution is more 'correct' than any other. All prior distributions are possible, and each is to be given 'equal rights,' as de Finetti says. But just as some prior distributions are sometimes regarded as more appropriate than others, for example, a uniform prior distribution on the parameter of a Bernoulli process is sometimes regarded as particularly appropriate, so too I claim that (7) is quite compelling, and I personally regard it is the most generally appropriate specification. My reasons are perhaps not

[12]The Encyclopedia Americana, 1981, referring to penguins, states "In size they range from the gigantic emperor penguins, standing about 40 inches high, and weighing up to 90 pounds, to the diminutive fairy penguin of the Australian region that attains a length of just over a foot.

entirely unrelated to those of Bayes, and the fiducial intuitions by 'Student' and Fisher.

That $A_n$ for large n should be highly compelling also agrees with certain frequentistic ideas in conventional nonparametric statistics. Very few statisticians use parametric models when dealing with large samples from some underlying population F. The reason is that one is nearly certain that the true distribution is not of any specific parametric form, for example, Gaussian, and that with a sufficiently large sample the discrepancies will almost certainly appear and be serious. This is part of the approach to hypothesis testing of J. Berkson (1938) who pointed out that with a sufficiently large sample you will reject most conventional null hypotheses. Thus for a sufficiently large sample one might be nearly certain that the data will allow rejection of any prespecified fixed dimensional parametric model, even using a subjective Bayesian test of the hypothesis, for which it is more difficult to reject the null hypothesis. On the other hand, if the sample size from the very same population F were sufficiently small, then one might well use the Gaussian or some other parametric model. Because of the relationship of $A_n$ to the empirical distribution function, as in Berliner and Hill (1988, p. 773), it is clear that the same considerations that make conventional statisticians prefer the empirical distribution function when dealing with large samples should also apply to $A_n$.

Now we come to a rather strange and interesting fact. Suppose I have managed to convince you of the appropriateness of $A_{N-1}$. But it is a mathematical fact, proved in Hill (1968, p.688), that $A_k$ implies $A_j$ for $j \leq k$. Thus if you accept $A_{N-1}$ as appropriate exactly, then you are forced into $A_1$ as well. Of course, both $A_1$ and $A_2$ correspond to conventional Bayesian and frequentistic procedures, with a diffuse prior on location, or on location and scale parameters, respectively, and they are certainly sometimes appropriate as an approximation. But it is equally clear that they are not always appropriate. How are we to explain this? My argument for $A_{N-1}$, which I regard as extremely compelling when N is large, if accepted, then implies $A_1$ as well, which is not always compelling. I believe that the explanation is as follows. In my proof that $A_k$ implies $A_j$ for $j \leq k$ there is a backwards induction. In carrying the argument backwards, it is possible that slight discrepancies from $A_{N-1}$ may build up, yielding a possibly much larger discrepancy for $A_1$. I should also point out that even $A_{N-1}$ need not hold literally. For example, suppose that you knew a great deal about the average weight in the population of penguins. Then if I gave you all but one of these weights, you would have a good idea about the missing weight. Indeed, you would know it exactly if you knew the average weight exactly. Similarly, one might observe that a particular weight that one knows occurs in the population is missing in the sample. Thus one might have a discrepancy even from $A_{N-1}$, and this could build up even more in reaching down to $A_1$. It is considerations such as these which point out the importance of recognizing, once and for all, that we are at best only dealing with approximations. See Hill (1990c). These approximations can nonetheless be very useful. It is my opinion that the nonparametric formulation, as in $H_n$, although itself only an approximation, is ordinarily the most important way to perform predictive inference, with parametric representations, such as the Gaussian, being useful primarily for inference and prediction when the sample size is small.

I remarked earlier that $A_n$ is exactly appropriate for merely ordinal data, such as hardness of rocks, in the absence of ties. The argument is as follows. Suppose one draws a simple random sample of n rocks from a population of N rocks in which no two are of the same composition or of the same hardness. (Here, as is usual, one rock is said to be harder than another if it scratches the other rock.) Before the data is taken you are surely of the opinion that $\underline{J}$ is equally likely to be any of the $\binom{N}{n}$ possible $\underline{j}$ vectors, since this is precisely what sampling without replacement means. In the present case, however, even after the sample is drawn you must still be of the same opinion, since no 'data' becomes available other than the relative orderings of the rocks in hardness, i.e., there are no values. (See Whitrow (1980, p. 216) for a discussion of the uninformativeness of the Mohs scale.) Thus in this situation one is forced to make the evaluation (7). Furthermore, this provides a justification for the original fiducial intuition, which also ignores the 'values' of the observations. Finally, if we now consider the case of ties, as for example if we draw a simple random sample from the rocks on some mountain, then it can easily be seen that $H_n$ rather than $A_n$ applies, using an appropriate distribution for $\underline{L}$. In general, $H_n$ is more flexible and realistic than $A_n$, because although ties may be rare in some examples, it would seem foolish to pretend that the probability of a tie is exactly 0.

Finally, in the case of 'colors' or 'species,' the natural way to proceed is as follows. Suppose that we go to a new region and, taking samples, find n living creatures that we decide belong to m distinct species. We can number these species in any way we like, for example, we can take species 1 to be the first type caught, etc., with species m the last type caught in our sample. Define the quantity $\gamma_i$, i $= 1, \ldots, m$, to be the proportion of the unsampled population belonging to the same species as the $i^{th}$ sample species, and $\theta_i$ to be the proportion of the unsampled population with value strictly between $x_{(i)}$ and $x_{(i+1)}$, as in Hill (1968, p. 682). In this case the $\theta_i$ are necessarily 0 for $i \leq m - 1$, since any creature belonging to a new species must then be assigned a number larger than m. Although neither $A_n$ nor $H_n$ is exactly appropriate, it is shown in Hill (1980a) that the posterior distribution of the quantities $\gamma_i$, $i = 1, \ldots, m$, is exactly as under $H_n$, and that the posterior distribution of M and the posterior probability of catching a new species are as in Hill (1968, p. 681, p. 691; 1979)

## 3. ON THE MEANING OF PARAMETERS

The role and meaning of parameters in the subjective Bayesian theory is quite different from that in conventional statistics. Consider a finite exchangeable sequence of 0-1 valued observations, $X_i$, for $1 = 1, \ldots, N$. In the de Finetti approach, there need not be any pre-existing 'true' probability, p, for a success, i.e., for $X_i = 1$. However, according to de Finetti's theorem, if the sequence were infinite, then one would implicitly be acting exactly as though there were such a p; and the implicit prior distribution for this p would be simply the de Finetti measure, $\pi$, for the sequence. In other words, it is as though there exists a parameter p with a priori distribution $\pi$, and conditional upon p, the observations form a Bernoulli sequence. If the sequence is only finite, but N is sufficiently large, to a good approximation the same thing occurs. In this case, p is simply the average of the N random quantities, $p = \bar{X} = (\sum_{i=1}^{N} X_i)/N$. Conditional upon this p, one no longer has exact independence of the $X_i$, but some degree of dependence. The difference between the infinite case and the finite case amounts to the difference between sampling with replacement versus sampling without replacement from an urn. See Heath and Sudderth (1976) and Diaconis and Freedman (1980). Of course all real world sequences are necessarily finite, but provided one is careful the infinite case can be used as a convenient approximation to the finite case. This is also the spirit in which I originally proposed $A_n$.

In this formulation note that before the sequence is actually determined, for example, before the coin is flipped, there is no pre-existing p, and what p actually represents is the random average $\bar{X}$ of the N observables, which is as yet to be determined. The a priori distribution for p is merely the prior distribution for $\bar{X}$, and this is in fact a useful way to elicit opinions about the conventional Bernoulli parameter p. Although p is usually thought of as a quantity with an objective existence even before the coin is tossed, this would be very difficult to demonstrate. Of course, if one likes, one can imagine that the coin has already been tossed N times, so that the $X_i$ have already been determined, but that one has not yet observed them. In this case there would be an existing quantity, which is as yet unknown, and is simply $\bar{X}$ for the realized sequence $X_i$. Provided there is no additional information, in the subjective Bayesian framework it is then much the same as if the tosses had not yet been made. See Hill (1990a, Sec. 3) for futher discussion. It is then largely immaterial, for practical purposes, which point of view one takes as to the 'objective' existence of p. (Note, however, that even in the case where $\bar{X}$ has already been realized, the interpretation of this quantity as the 'true' probability cannot be made without assumptions as to the sampling mechanism.) In this framework the distinction between 'inference' and 'prediction' becomes blurred. On the one hand, if the sequence has not yet been determined, one would view p as a random quantity which one might want to predict, i.e., it is the future proportion of heads. On the other hand, if the sequence has been determined, but is as yet unobserved, then p might be thought of as a parameter in the conventional sense. The upshot of this discussion is that the usual sharp distinction between prediction and parametric inference is largely illusory.

The situation with regard to parameters in $A_n$ is more subtle. Given the data, $X_i$, for $i = 1, \ldots, n$, I have defined the 'parameters' $\theta_i$ and $\gamma_i$ to be the proportions of observations in the unsampled population, between and at the order statistics of the data. Such parameters are defined in terms of the data, and so are not the usual kinds of parameters. Nonetheless, they are unknown quantities, and so in the de Finetti theory one can deal with them just as with any other unknown or 'random' quantities. Because the sequence of observable random quantities, $X_i$, is viewed as exchangeable, it follows from the general form of the theorem of de Finetti, that one is acting as though one had a distribution $\pi$ on the space of all possible distribution functions, F. In principle the situation is as follows. Given the data, the prior distribution $\pi$ is updated, as usual, to become a posterior distribution $\pi^\star$, and posterior predictive probabilities for future observables can be obtained by taking expectations with respect to $\pi^\star$. Then we have

$$\begin{aligned} Pr\{X_{n+1} \in I_i \mid data\} &= E[\theta_i | data] && (8)\\ &= E[\hat{F}(x^-_{(i+1)}) - \hat{F}(x_{(i)}) \mid data] && (9)\\ &= N/(N-n) \times E[F(x^-_{(i+1)}) - F(x_{(i)}) | data]. && (10)\end{aligned}$$

In this equation $\hat{F}$ is the empirical distribution function for the unsampled population, F is the empirical distribution for the entire population, and expectations are taken with respect to the posterior distributions of $\hat{F}$ and $F$. For small n/N, $\hat{F}$ is approximately equal to $F$, which plays much the same role as does $\bar{X}$ for a Bernoulli sequence. Thus despite the fact that the $\theta_i$ depend upon the data for their definition, in principle their posterior expectation can be defined in terms of the 'parameter' $F$, just as in conventional parametric statistics.

The two cases we have considered here, namely the 0-1 case, and the fully nonparametric case, are in fact the extreme cases with respect to complexity of the underlying model. Much of statistical inference and prediction takes place in an intermediate case, namely of a conventional parametric model for real-valued observations. However, such intermediate cases can be considered in much the same way. Consider, for example, the case of an exponential model for data, which is to be used in connection with a real or conceptual finite population whose empirical distribution is known to be approximately of exponential form. Let the parameter be $\alpha$, the expectation of the distribution. Imagine the finite population of values, $X_i$, for $i = 1, \ldots, N$, and consider $\bar{X}$ for this population. Then one's a priori distribution for $\alpha$ is approximately one's prior distribution for $\bar{X}$, and conditional upon $\bar{X}$, the observations are approximately independently distributed according to an exponential distribution with 'parameter' $\bar{X}$.

The final point I wish to discuss concerns the role of Bayesian data analysis with respect to $A_n$. In Hill (1987b) two theorems are proved. The first gives a simple parametric model, called a nested splitting process, that gives rise to $A_n$ exactly. The second shows that $A_m$ holds in sampling from complex mixtures of distributions, where here m represents the number of groups or types formed from the n observations via data analysis. For example, in sampling from the population of cetaceans (whales, porpoises, dolphins) the sample animals may be classified according to species (or other variables) into m groups. From my point of view, such classification should be done by a form of data analysis. After performing such classification, the statistical problem can be reduced to one concerning the random effects model in the analysis of variance. See Box and Tiao (1973), Hill (1965, 1967, 1977, 1980b), and Lindley and Smith (1972) for Bayesian analysis of such models.

In Hill (1990a) a theory of Bayesian data analysis is put forth in which, because of computational complexity, or because of thoughts that are triggered off during the analysis of the data, a departure is made from the classical Bayesian theory in which models and prior distributions are all specified before seeing the data. I believe that this modification is essential in order to make the classical Bayesian approach more realistic in applications. Any scientist worth his salt would play with his data, analysing it in a variety of ways, and giving free rein to his imagination and creativity. As argued in Hill (1985-86), classical non-Bayesian theory breaks down completely in connection with such data analysis, since all probabilities would have to be conditional on the exact procedures employed, including their order, and even the thoughts that cross one's mind. This also poses a challenge for the Bayesian approach. However, one can view the data-analytic procedures as occurring prior to the point at which the Bayesian analysis, proper, begins; and it should be observed that the Bayesian theory has always had an arbitrary element in it as to the time point at which one proceeds to make a formal Bayesian analysis. I have suggested that often the appropriate point is following the process of data analysis. After one has reformulated old models, or formulated entirely new models, by means of data analysis, one can proceed with the classical form of Bayesian reasoning, including robustness and sensitivity analysis, as in Berger (1984), Hill (1980b). See also Hacking (1967) and Smith (1986).

In the context of $A_n$ and $H_n$ this means that the 'parameters' $\theta_i$ and $\gamma_i$ are actually the results of such Bayesian data analysis, as discussed in Hill (1987b). This need not, however, change the basic nature of the interpretation of these quantities. As shown in Hill (1990a), a generalization of the restricted likelihood principle of Hill (1987a) remains valid in the context of data analysis. Of course classical non-Bayesian reasoning, for example conventional asymptotic theory, becomes almost entirely irrelevant. However, in low dimensional problems one can still plot likelihood functions, and these may turn out to be sharp relative to 'a priori' distributions for such parameters as are introduced by means of the data analysis. The force of such a Bayesian analysis of data must depend upon an agreement amongst scientists that specific prior distributions and likelihood functions are pertinent to the problem, and can be considered on their own merits, even after the data has been observed. In high dimensional problems one must learn new techniques for the analysis and display of likelihood functions, as in Hill (1975a) with an example concerning the tails of distributions. See Mosteller and Wallace (1964), Chen et al (1983), and Hill (1987b, 1990a, 1990d) for examples of Bayesian data analysis.

## 4. CONCLUDING REMARKS

The initial intuition as regards $A_n$ seems to be due to 'Student,' or at least Fisher (1939) implies that this is the case. Fisher then generalized the idea and interpreted it in a fiducial spirit, which Dempster (1963) crystallized and called 'direct probability.' Note that for all three of these authors the justification for $A_n$ seems to be purely intuitive. Thus none give anything vaguely representing a 'proof' for $A_n$, or suggest a way in which its coherency or rationale can be discussed, or even indicate in what circumstances it might or might not be appropriate. While I believe that sound intuition (or inspiration) is what all scientific progress ultimately comes from, it is nonetheless the case that a critical attitude is necessary, and that one must ask when and why $A_n$ is sensible, and whether there are any qualifications and pitfalls associated with it. For example, one must ask immediately, when, if ever, should one be using conventional parametric models as opposed to $A_n$. It is in helping to understand such questions that I believe the subjective Bayesian approach plays a fundamental role.

Consider, for example, the case of a normal model with known standard deviation of 1 and unknown mean, $\theta$. The fiducial argument of Fisher suggests that the pivotal quantity $(X - \theta)$ should continue to have the N(0,1) distribution even after X is replaced by its observed value x, which is a number. The confidence argument of Neyman does not assume this, but provides no way of telling whether there is anything peculiar about the particular x for which the confidence is quoted. As is well known, there are many examples, such as the Fieller-Creasy example, in which such a procedure is patently absurd, in that the whole real line may have confidence 95 percent. More generally, such confidence procedures do not provide a way to allow one to deal with data for which the conventional confidence level is obviously inappropriate based upon prior knowledge that is generally accepted. Thus the confidence argument, as applied in practice by sensible statisticians, is instead a conditional argument, i.e., it is conditional upon not getting data that is wildly contrary to prior knowledge. As shown in Hill (1985-86), the 'true' confidence coefficients, when adjusted to be conditional upon not getting such data, are necessarily both unknown and unknowable.

The Bayesian argument goes far beyond this. It first shows that if one has a prior distribution for $\theta$ which is sufficiently diffuse relative to the likelihood function, then in fact Fisher's fiducial conclusion is often justified. (This fact, which Harold Jeffreys had been telling Fisher for years, seems finally to have been accepted by Fisher, as the previously quoted footnote of Fisher (1959, p. 51) seems to indicate.) Next, there are many situations where the prior distribution is sharp relative to the likelihood function, in which case the appropriate conclusions are quite different, as in Hill (1965, 1967) concerning variance components; and still again, there are important cases in which the prior distribution and the likelihood function are of comparable magnitude. Thus one sees clearly in what situations the fiducial argument is relevant, and what the nature of its limitations are. See Hill (1974b, p.570) for a mathematical discussion of the behavior of a posterior distribution for various kinds of extreme data.

The situation with regard to $A_n$ is of the same general nature as that for a normal mean, except it is much more complicated. Thus the primary basis for $A_n$ or $H_n$ is (7), which taken literally says that the observed values $x_i$ are totally uninformative about $\underline{J}$. Once again the initial intuition comes from a form of Fisher's fiducial argument. Even if one finds his argument compelling, however, one would presumably want to put it into a broader context, including at least sampling without replacement and the case of ties. Thus the generalization from $A_n$ to $H_n$, and from sampling with replacement to sampling from a finite population without replacement, is important, since the case of ties and of sampling without replacement is both more fundamental and more realistic. A generalization of $H_n$ could also allow for the possibility of smoothing using (4) without assuming (7), with such smoothing perhaps stemming from data analysis. The question of choice between $A_n$ and $H_n$ comes down to a very refined question as to the appropriate a priori information to utilize, and the quality of approximation that is desired. The real world situation is quite subtle and complex. As Einstein (1921, p. 3) says:

> As far as the propositions of mathematics refer to reality they are not certain, and in so far as they are certain they do not refer to reality.

Finally, because $A_n$ is a de Finetti coherent procedure, one knows that there are no operationally meaningful ways in which one can be made a sure loser by using $A_n$ or $H_n$. In Hill (1974b, 1990c) an operationally meaningful argument for admissibility is given, in much the same spirit as de Finetti had earlier supplied operational meaning to the evaluation of probabilities by showing that violation of the laws of finitely additive probability leads to a Dutch book. See Hill (1990d) for futher discussion of this argument. However, it is argued in Hill (1990c), that Hill (1974b) notwithstanding, this argument is only valid in the context of finite partitions. It is suggested that any supposedly bad consequences to be associated with the use of a statistical procedure should be required to be demonstrable in the context of a game with pre-specified rules, that can with certainty be played and completed by two players in a finite time, with a referee in charge to keep order, and on a computer with a known finite bound on its memory. With respect to the use of $A_n$ and $H_n$ it is not possible, because of their established coherency, theoretically to prove any such bad consequences. Of course in practice these matters must be discussed in the context of implementations of $A_n$ and $H_n$ on a computer with finite memory, as in Hill (1990c), rather than in their idealized versions. The articles Hill (1990a, 1990c, 1990d) also emphasize the distinction between pre-data distributions, which are relevant in the design stage, or when the statistical analysis and decision-making are constrained to be mechanically implemented on a computer, without thought; versus post-data statistical analysis and decision-making. The former is characterized as the policy-maker or bureaucratic problem, while the latter is the creative scientist or creative decision-maker problem, and requires Bayesian data analysis. These are quite different problems, and problems of the one type should not be forced into the mold for the other.

Although the property of coherency is desirable, it is of itself not sufficient to justify use of procedures such as $A_n$ and $H_n$. Thus in addition to the internal coherency property, one wants also to know whether the procedure is 'reasonable,' that is to say, whether it corresponds to prior knowledge that is generally considered to be appropriate for the situation at hand, or for which a reasonable case can be made. In my opinion this is ordinarily true, but with a few minor qualifications that there is not space to go into at this time.

Let me conclude by observing that $A_n$ is supported by all of the major approaches to statistical inference. It is Bayesian, fiducial, and a confidence/tolerance procedure. It is simple, coherent, and plausible. It can even be argued, I believe, that when viewed in the context of Bayesian data analysis, $A_n$ and $H_n$ constitute the best solution we now have to the problem of induction as formulated by Hume.

## ACKNOWLEDGEMENTS

This work was supported by the U. S. Air Force under grant AFOSR-87-0192, and by the National Science Foundation under grant DMS-8901234.

## REFERENCES

Aitchison, J., and Dunsmore, I. R. (1975), *Statistical Prediction Analysis,* Cambridge University Press.

Berger, J. (1984), The robust Bayesian viewpoint," in *Robustness of Bayesian Analysis,* J. Kadane, ed., North-Holland: Amsterdam, (with discussion), 321-372.

Berkson, J. (1938), Some difficulties of interpretation encountered in the application of the chi-square test." *Journal of the American Statistical Association,* **33,** 526-542.

Berliner, L. M., and Hill, B. M. (1988), Bayesian nonparametric survival analysis, *Journal of the American Statistical Association*, **83**, 772-784 (with discussion).

Blackwell, D., and MacQueen, J. B. (1973), Ferguson distributions via Pólya urn schemes, *The Annals of Statistics,* **1**, 353-355.

Boender, C. G. E., and Kan, A. H. G. Rinnooy (1987), A multinomial Bayesian approach to the estimation of population and vocabulary size, *Biometrika,* **74,** 849-856.

Borel, E. (1906), Sur les principes de le théorie cinétique des gaz, *Annales de lÉcole Normale Supérieure,* 9-32. Reprinted in *Selecta: Jubilé Scient. de M. Émile Borel,* Gauthier-Villars, Paris, 1940, 243-265.

Borel, E. (1914). *Introduction Géométrique à Quelques Théories Physiques,* Gauthier-Villars, Paris.

Box, G. E. P., and Tiao, G. C. (1973), *Bayesian Inference in Statistical Analysis,* Addison-Wesley, Reading.

Chang, C. (1988), *Bayesian Nonparametric Prediction Based on Censored Data,* University of Michigan Doctoral Dissertation.

Chen, Wen-Chen (1978), *On Zipf's Law*, University of Michigan Doctoral Dissertation.

Chen, Wen-Chen (1980), On the weak form of Zipf's law, *Journal of Applied Probability,* **17**, 611-622.

Chen, Wen-Chen, Hill, B. M., Greenhouse, J., and Fayos, J. (1985). Bayesian analysis of survival curves for cancer patients following treatment (with discussion), in *Bayesian Statistics* **2,** J. M. Bernardo, M. H. DeGroot, D. V. Lindley, A. F. M. Smith eds. North-Holland

de Finetti, B. (1937), La prévision: ses lois logiques, ses sources subjectives, *Annales de l'Institut Henri Poincaré,* **7**, 1-68.

de Finetti, B. (1974). *Theory of Probability, Vol. 1*, London: John Wiley & Sons, Inc.

Dempster, A. P. (1963), On Direct Probabilities, *Journal of the Royal Statistical Society* **B, 25,** 100-114.

Diaconis, P., and Freedman, D. (1980), Finite exchangeable sequences, *The Annals of Probability*, **8**, 745-764.

Diaconis, P., and Freedman, D. (1981), Partial exchangeability and sufficiency, *Proceedings of the Indian Statistical Institute Golden Jubilee International Conference on Statistics: Applications and New Directions,* 205-236.

Dickey, J., and Kadane, J. (1980), Bayesian decision theory and the simplification of models, in *Evaluation of Econometric Models,* J. Kmenta and J. Ramsey, eds., Academic Press, 245-268.

Einstein, A. (1921). *Geometrie und Erfahrung.* J. Springer, Berlin.

Feller, W. (1968), *An Introduction to Probability Theory and its Applications, Third Edition, Revised Printing,* New York: John Wiley & Sons.

Feller, W. (1971), *An Introduction to Probability Theory and its Applications, Vol. 2, Second Edition,* New York: John Wiley & Sons.

Ferguson, T. (1973), A Bayesian analysis of some nonparametric problems, *The Annals of Statistics,* **1**, 209-230.

Fisher, R. A. (1939), Student,*Annals of Eugenics,* **9**, 1-9.

Fisher, R. A. (1948), Conclusions Fiduciare,*Annales de l'Institut Henri Poincaré,* **10**, 191-213.

Fisher, R. A. (1959). *Statistical Methods and Scientific Inference,* Second Edition, New York: Hafner Publishing Co.

Good, I. J. (1965), *The Estimation of Probabilities,* MIT Research Monograph No. 30.

Hacking, I. (1967), Slightly more realistic personal probability, *Philosophy of Science,* **34,** 311-325.

Hartigan, J. (1983), *Bayes Theory,*, New York: Springer-Verlag.

Heath, D., and Sudderth, W. (1976), de Finetti's theorem for exchangeable random variables, *The American Statistician,* **30,** 188-189.

Hewitt, E., and Savage, L. J. (1955), Symmetric measures on cartesian products, in *The Writings of Leonard Jimmie Savage-A Memorial Selection,* Published by The American Statistical Association and The Institute of Mathematical Statistics, 1981, 244-275.

Hill, B. M. (1965), Inference about variance components in the one-way model, *Jour. Amer. Statist. Assoc.,* **58**, 918-932.

Hill, B. M. (1967), Correlated errors in the random model, *Journal of the Americann Statistical Association,* **62**, 1387-1400.

Hill, B. M. (1968), Posterior distribution of percentiles: Bayes theorem for sampling from a finite population, *Journal of the American Statistical Association,* **63**, 677-691.

Hill, B. M. (1969), Foundations for the theory of least squares, *Journal of the Royal Statistical Society,* **Series B, 31,** 89-97.

Hill, B. M. (1970), Zipf's law and prior distributions for the composition of a population, *Journal of the American Statistical Association,* **65**, 1220- 1232.

Hill, B. M. (1974a), The rank frequency form of Zipf's law, *Journal of the American Statistical Association,* **69**, 1017- 1026.

Hill, B. M. (1974b), On coherence, inadmissibility and inference about many parameters in the theory of least squares, in *Studies in Bayesian Econometrics and Statistics in Honor of L. J. Savage,* S. Fienberg and A. Zellner, eds., North Holland, 555-584.

Hill, B. M. (1975a), A simple general approach to inference about the tail of a distribution, *Annals of Statistics,* **3**, 1163-1174.

Hill, B. M. (1975b), Aberrant behavior of the likelihood function in discrete cases, *Journal of the American Statistical Association*, **70**, 717-719.

Hill, B. M. (1977), Exact and approximate Bayesian solutions for inference about variance components and multivariate inadmissibility, in *New Developments in the Application of Bayesian Methods*, A. Aykac and C. Brumat, eds., North Holland, 129-152.

Hill, B. M., (1978), Decision theory. In *Studies in Statistics*, **Vol. 19**, R.V. Hogg, ed., The Mathematical Association of America, 168-209.

Hill, B. M. (1979), Posterior moments of the number of species in a finite population, and the posterior probability of finding a new species, *Journal of the American Statistical Association*, **74,** 668-673.

Hill, B. M. (1980a), Invariance and robustness of the posterior distribution of characteristics of a finite population, with reference to contingency tables and the sampling of species. In *Bayesian Analysis in Econometrics and Statistics: Essays in Honor of Harold Jeffreys,* ed A. Zellner, North-Holland, 383-395.

Hill, B. M. (1980b), Robust analysis of the random model and weighted least squares regression, in *Evaluation of Econometric Models,* ed. by J. Kmenta and J. Ramsey, Academic Press, 197-217.

Hill, B. M. (1980c), On finite additivity, non-conglomerability, and statistical paradoxes, (with discussion) in *Bayesian Statistics,* J. M. Bernardo, M. H. Degroot, D. V. Lindley, A. F. M. Smith, eds., University Press: Valencia, Spain, 39-66.

Hill, B. M. (1981), A theoretical development of the Zipf (Pareto) law, in *Studies on Zipf's Laws,* ed. by H. Guiter, Sprachwissenschaftliches Institute, Ruhr-Universitat, Bochun.

Hill, B. M. (1985-86), Some subjective Bayesian considerations in the selection of models, *Econometric Reviews* **4, No. 2,** 191-288 (with discussion).

Hill, B. M. (1987a), The validity of the likelihood principle, *The American Statistician*, **41,** 95-100.

Hill, B. M. (1987b), Parametric models for $A_n$: Splitting Processes and Mixtures, Unpublished, Department of Statistics, The University of Michigan.

Hill, B. M. (1988), De Finetti's theorem, induction, and $A_n$, or Bayesian nonparametric predictive inference, in *Bayesian Statistics* **3,** J. M. Bernardo, M. H. Degroot, D. V. Lindley, and A. F. M. Smith, eds., Oxford University Press, 211-241 (with discussion).

Hill, B. M., (1990a), A theory of Bayesian data analysis, in *Bayesian and Likelihood Methods in Statistics and Econometrics: Essays in Honor of George A. Barnard,* S. Geisser, J. Hodges, , S. J. Press, A. Zellner (Eds.), North-Holland, 1990, 49-73.

Hill, B. M., (1990b), Bayesian Statistics, in *1990 Yearbook of Encyclopedia of Physical Science and Technology,* Academic Press.

Hill, B. M. (1990c), Discussion of An ancillarity paradox that appears in multiple linear regression, by L. D. Brown, in *The Annals of Statistics,* 1990.

Hill, B. M. (1990d), Dutch books, the Jeffreys-Savage theory of hypothesis testing, and Bayesian reliability. Unpublished technical report, the University of Michigan. Presented at the course-congress on *Reliability and Decision Making,* Siena, Italy.

Hill, B. M. (1990e), Robustness in Statistics. Unpublished technical report, the University of Michigan.

Hill, B. M. and Lane, David (1985), Conglomerability and countable additivity, *Sankhyá,* **47, Series A,** 366-379.

Hill, B. M., Lane, David, and Sudderth, William (1980), A strong law for some generalized urn processes, *Annals of Probability,* **8**, 214-226.

Hill, B. M., Lane, David, and Sudderth, William (1987), Exchangeable urn processes, *Annals of Probability,* **15**, 1586-1592.

Hill, B. M. and Woodroofe, M. (1975), Stronger forms of Zipf's law, *Journal of the American Statistical Association,* **70**, 212-219.

Hoppe, F. (1987), The sampling theory of neutral alleles and an urn model in population genetics, *Jour. of Mathematical Biology,* **25**, 123-159.

Hume, David (1748), *An Enquiry Concerning Human Understanding,* London.

Ijiri, Y., and Simon, H. A. (1975), Some distributions associated with Bose-Einstein statistics, Proceeding of the National Academy of Science, USA, **72**, 1654-1657.

Jeffreys, H. (1957), *Scientific Inference,* Second Edition, Cambridge University Press.

Jeffreys, H. (1961), *Theory of Probability,* Third Edition, Oxford at the Clarendon Press.

Johnson, W. E. (1932), Probability: the deductive and inductive problems, *Mind*, **49**, 409-423. [Appendix on pages 421-423 edited by R. B. Braithwaite].

Kingman, J. F. C. (1975), Random discrete distributions, *Journal of the Royal Statistical Society*, **Series B, 37,** 1-22 (with discussion).

Lane, D., and Sudderth, W. (1978), Diffuse models for sampling and predictive inference, *Annals of Statistics*, **6,** 1318-1336.

Lane, D., and Sudderth, W. (1984), Coherent predictive inference, *Sankhyá* **Ser. A, 46,** 166-185.

Lenk, P. (1984), *Bayesian Nonparametric Predictive Distributions,* Doctoral Dissertation, The University of Michigan.

Lewins, W. A., and Joanes, D. N. (1984), Bayesian estimation of the number of species, *Biometrics*, **40,** 323-328.

Lindley, D., and Smith, A. F. M. (1972), Bayes estimates for the linear model, *Journal of the Royal Statistical Society, Series B,* **34**, 1-41.

Luce, R. D., Narens, L. (1987), Measurement scales on the continuum, *Science,* **236,** 1527-1531.

Mandelbrot, B. B. (1982), *The Fractal Geometry of Nature,* W. H. Freeman and Co., New York.

Poincaré, H. (1912), *Calcul des Probabilités,* Deuxiéme Edition, Gauthier-Villars.

Ramakrishnan, S. and Sudderth, W. (1988), A sequence of coin- toss variables for which the strong law fails, *American Mathematical Monthly,* **95,** 939-941.

Rényi, A. (1970), *Probability Theory,* New York : American Elsevier.

Savage, L. J., 1961. *The Subjective Basis of Statistical Practice.* Unpublished technical report, Department of Mathematics, University of Michigan.

Savage, L. J. et al., 1962. *The Foundations of Statistical Inference.* Methuen, London.

Savage, L. J. (1972), *The Foundations of Statistics*, Second Revised Edition, New York: Dover.

Schervish, M., Seidenfeld, T., and Kadane, J., (1984), The extent of non-conglomerability, *Z. f. Wahrscheinlichkeitstheorie*, **66**, 205-226.

Smith, A. F. M., (1986), Some Bayesian thoughts on modelling and model choice, *The Statistician,* **35**, 97-102.

Whitrow, G. J. (1980), *The Natural Philosophy of Time, Second Edition,* Oxford University Press.

Woodroofe, M. and Hill, B. M. (1975), On Zipf's law, *Journal of Applied Probability*, **12,** 425-434.

Zabell, S. L. (1982), W. E. Johnson's sufficientness postulate, *The Annals of Statistics*, **10,** 1091-1099.

## COMMENTS

James Berger
*Purdue University*

I very much enjoyed this paper, both for its philosophy and its enlightening review of the author's (and previous) work on $A_n$ and $H_n$. In this discussion I will consider the issues of robustness, coherence, and general applicability.

The assumptions under which Hill derives $A_n$ and $H_n$ are quite believable and weak, at least for large $n$. The most troubling assumption is, of course, the assumption that the actual values of the data carry no information beyond that in their ordering. By concentrating only on the ordering, a great deal of robustness is potentially achieved, but this robustness is "real" only as long as the assumption is believable. While I am happy with the assumption for large $n$, my unease at the assumption for small $n$ can be best illustrated with an example.

Example 1. Suppose the data consists of the three failure times 3, 4 and 6. The conclusion of $A_3$ is that the probabilities that a future failure time falls in the intervals (0,3), (3,4), (4,6) and (6,$\infty$) are each $\frac{1}{4}$.

For comparison, consider the common exponential model $f(x|\theta) = \theta^{-1} \exp(-x/\theta)$ for failure times, and assign $\theta$ the usual noninformative prior $\pi(\theta) = \theta^{-1}$. Then the posterior distribution of $\theta$ for the above data is $\pi(\theta|\text{data}) \propto \theta^{-4} \exp(-13/\theta)$, and the predictive probability that a future observation will fall in (a,b) is

$$\begin{aligned} P(a < x < b) &= \int \left[ \int_a^b f(x|\theta) d\theta \right] \pi(\theta|\text{data}) d\theta \\ &= \left(1 + \frac{a}{13}\right)^{-3} - \left(1 + \frac{b}{13}\right)^{-3}. \end{aligned}$$

For the four intervals, (0,3), (3,4), (4,6) and (6,$\infty$), the predictive probabilities are 0.46, 0.09, 0.13 and 0.32, respectively. These numbers differ substantially from those obtained for $A_3$, but this is not that surprising. One cannot really hope to have robustness with only three observations.

Which probabilities are best? Note that the question is not simply whether or not one believes the exponential assumption; for a very wide range of models and priors the probabilities arising from the exponential model will be closer to the actual predictive probabilities than will those arising from $A_3$. Indeed, I would guess that for most model–prior pairs that one would entertain for this data, the exponential–based probabilities will be closer.

The basic point of the above example is that it is wrong to think of $A_n$ as "assumption–free" for small $n$, in the sense that the resulting probabilities would necessarily be closer to the true predictive probabilities than would those obtained by using even a substantially wrong model and prior. For larger $n$ there are reasons to think that this might be the case, however, supporting $A_n$'s claim of robustness when $n$ is large.

Hill reports that $A_n$ is coherent in the sense of deFinetti. This type of coherence is not well understood, and is a weaker property than many would like. Consider the following result, which is related to the parametric example discussed in Berger (1985, p. 260).

Lemma. Define a betting strategy against $A_1$ as follows:

$$\begin{aligned}&\text{If } X_1 > K, \text{ bet at even odds that } X_2 \in (-\infty, X_1);\\&\text{If } X_2 < K, \text{ bet at even odds that } X_2 \in (X_1, \infty).\end{aligned} \qquad (1)$$

If $X$ has density $f(x)$ with respect to Lebesgue measure and if $P_f(X < K) > 0$ and $P_f(X > K) > 0$, then $E_f$ (Gain) $> 0$.

Proof. It suffices to show that

$$\begin{aligned}\tfrac{1}{2} < P_f(\text{Winning}) &= P_f(X_1 > K,\ X_2 < X_1)\\&\quad + P_f(X_1 < K,\ X_2 > X_1)\end{aligned}$$

(since $\{X_1 = K\}$ and $\{X_1 = X_2\}$ have probability zero). Defining $F$ as the c.d.f. of $X$, it is clear that

$$\begin{aligned}P_f(\text{Winning}) &= \int_K^{\infty} F(x)f(x)dx + \int_{-\infty}^{K} (1 - F(x))f(x)dx\\&= \int_{-\infty}^{\infty} F(x)f(x)dx + \int_{-\infty}^{K} (1 - 2F(x))f(x)dx. \qquad (2)\end{aligned}$$

Observe that

$$\int_{-\infty}^{\infty} F(x)f(x)dx = \tfrac{1}{2}, \qquad (3)$$

so that

$$\int_{-\infty}^{\infty} (1 - 2F(x))f(x)dx = 0.$$

But $g(x) = 1-2F(x)$ is a nonincreasing function which is negative for large enough $x$. Hence it must be the case that

$$\int_{-\infty}^{K} (1 - 2F(x))f(x)dx > 0 \tag{4}$$

(using also $P_f(X < K) > 0$ and $P_f(X > K) > 0$). Combining (2), (3) and (4) yields the

Example 2. To see that substantial expected gains are possible, consider the exponential distribution in Example 1. Calculation yields

$$P_\theta(\text{Winning}) = \exp(-k/\theta) - \exp(-2K/\theta) + \tfrac{1}{2},$$

which can be as large as 3/4 (when $\theta = K/\ln 2$). r Coherency is an artificial game that one uses to "test" an inferential method. Thus there is considerable scope in deciding what constitutes a fair test. My own view is that the scenario of the above lemma is a fair test from the coherency perspective. My reasoning is that $A_1$ states that it will allow bets on $(-\infty, X_1)$ or $(X_1, \infty)$ at even odds. I then propose the allowable scheme (1), and give over to my opponent (the supporter of $A_1$) the freedom to choose any $f$ he or she wants, modulo the restrictions placed on $f$ in the lemma. (These restrictions had the sole purpose of eliminating trivial cases, where the expected gain is zero because only half of the betting scheme can occur. Note that, for any given $f$, one could certainly choose $K$ to satisfy the conditions.) If we now repeatedly play the game with independent data generated from $f$, the law of large numbers assures me of eventual large gains. This can also be generalized to allow a "random" choice of $f$ at each play, as long as $f$ is chosen in a realizable way (i.e., according to a countably additive distribution).

Why then is $A_1$ coherent in the sense of deFinetti? It is because this version of coherence allows $f$ to be chosen according to a finitely additive prior distribution, and there exist finitely additive distributions that essentially concentrate on $f$ for which $E_f$ (gain) is arbitrarily close to zero, so that the overall expectation is zero. Such prior distributions are not implementable, however, in the sense of being used to actually play the game, so I would argue against their use.

In spite of the above, I do not think $A_1$ is actually bad. The reason is that I feel coherency itself to be an inadequate method of judging an inference. To me, $A_1$ is a "confidence statement," to the effect that one chooses to measure the confidence to be placed in $(-\infty, X_1)$ and $(X_1, \infty)$ by the probability 1/2. It is not apriori obvious that one should be forced to bet based on these probabilities. Among the serious alternatives is to "score" reported confidences, $\alpha(X_1)$, using a scoring rule such as

$$(\alpha(X_1) - I_{C(X_1)}(X_2))^2,$$

where $I_{C(X_1)}(X_2)$ is 1 if $X_2 \in C(X_1)$ (the proposed set) and 0 otherwise. In the parametric version of this example (see Berger, 1985, p. 260) the confidence statement analogous to $C(X_1)$ is admissible in the sense that

$$E_f(\tfrac{1}{2} - I_{C(X_1)}(X_2))^2$$

cannot be improved upon for all $f$ by an alternative choice of $\alpha(X_1)$, and I suspect that the same is true here. Thus I feel that $A_1$ (and indeed $A_n$) will be "admissible" inferences according to reasonable "inference losses."

Another way of saying this, is that countably additive coherence may be too stringent a test, while finitely additive coherency is too lax a test. Admissibility criteria seem to fall midway between the two (see also Robinson, 1979) and can be argued to be at the sensible level of discrimination.

I can readily perceive the usefulness of $A_n$ when there are a large number of exchangeable univariate observations, or when one can "visually" separate the observations, on the basis of clusters or covariates, into exchangeable groups, each having a large number of observations. When one is dealing with multivariate problems or problems having multilevel structure, however, it is harder to judge whether or not versions of $A_n$ can be useful. To put this another way, when in situations in which models are used to help discover or measure non–obvious structures in the data, the usefulness of $A_n$ is certainly not apparent.

My guess is that $A_n$ and $H_n$ (and modifications) will become part of our collection of useful Bayesian techniques. They will be used in probably three circumstances. First, when a predictive distribution is desired, the "structure" of the data is clear, and the number of observations is large, $A_n$ and $H_n$ can provide an excellent and quick Bayesian answer. Indeed, in such a situation I would be hard put to imagine suggesting use of anything else. The second situation in which they can clearly be useful is in performing a robustness check. For instance, in Example 1 the substantial difference between $A_3$ and the exponential–based predictive probabilities provides a warning that the exponential assumption requires a careful look. Finally, I can see $A_n$ and $H_n$ being used in a noninformative prior sense, when one must avoid (for sociological reasons, say) a Bayesian model–prior approach. In such situations the proper reference for evaluating $A_n$ and $H_n$ is not careful, subjective Bayesian analysis, but is instead nonparametric classical statistics. Given the Bayesian justifications and clearly conditional nature of $A_n$ and $H_n$, I would certainly tend to bet on them in any such confrontation.

Finally, let me heartily endorse Professor Hill's comments concerning Bayesian data analysis. It is an all–too–common misconception about Bayesian analysis that one must fix the model and prior ahead of time, and then blindly update (via the posterior distribution) once the data is at hand. This has never been the practicing Bayesian position, as has been eloquently argued by many Bayesians.

The way I like to think of the situation is to imagine the reader of a Bayesian statistical analysis. All he cares about is whether the model and prior considered are reasonable, and whether or not the answers are robust to reasonable departures from the model or prior. (Of course, Bayesian answers could be given for a variety of models and priors.) Knowing where the models and priors came from is relevant only to the extent that such knowledge might bear on the determination of what other models and priors might be reasonable and should be a part of the robustness study. Indeed, it is probably only in the worst–case scenario, where robustness is lacking yet a conclusion must be reached, that one needs knowledge of the process by which the models and priors were obtained by the analyst; then, a determination (possibly retrospective) of the prior probabilities of the various models is needed. (Yes, it may be difficult to forget about the data in such a retrospective assessment, but there is no other ball game in town.)

## REJOINDER

I wish to thank Professor Berger for his interesting comments. He raises a number of subtle points, for which I am especially grateful, since it gives me the opportunity to respond to certain often unvoiced criticisms (not necessarily his own) that I have sometimes heard. It is the subconscious, unvoiced assumptions that people make that I believe have most hampered the progress of Bayesian statistics, and therefore all serious statistics, in the past. To illustrate, the second time that I spoke about $A_n$ was in 1966, and an eminent statistician and probabilist in the audience objected with respect to $A_1$ that it was not a possible evaluation, since he claimed that under my assumptions the second observation must be independent of the first. This is of course not true, and nowadays more probabilists and statisticians understand the difference between exchangeability (which implies conditional independence) and independence. [13] To put it another way, even if there were a 'true' F, this F would be unknown, and there would be mixing induced by the a priori distribution for such an F, thus introducing dependency. See for example our rejoinder to Genest and Kalbfleisch in Berliner and Hill (1988, p. 783). In fact $A_n$ does not even require that there exists such a true F, and only concerns one's opinions about the next observable value, given what has been observed. I will return to this aspect later, since in a slightly different form it is also relevant to Berger's discussion. Another sometimes unvoiced criticism of $A_n$ is that it is absurd because it gives equal probability to intervals of vastly different length.

---

[13] For some strange reason there has been an extraordinary degree of confusion amongst conventional probabilists, mathematicians, and statisticians, regarding Bayes's theorem, de Finetti's theorem, and even the concept of *dependency*. See Hill (1990b) for a brief history and review.

This way of thinking stems from a naive interpretation of Lebesgue measure, and fails to recognize that the reason that an interval is long may be because F (in the case, if one likes, of a 'true unknown' F) may have very little mass in that region.

Still another question that arises concerns the distinction between finite and countable additivity. For example, Dennis Lindley in his SIAM review (1971, p. 67) argued that because Hill had shown that $A_n$ cannot hold literally for countably additive distributions, that therefore (despite Hill's recommendations to the contrary) it should be abandoned. This criticism ignored the fact that there has never been a serious attempt to justify countable additivity other than by convenience or expediency. For example, Kolmogorov (1950, p. 15), in discussing the axiom of continuity, which is equivalent to the axiom of countable additivity, says:

> For infinite fields, on the other hand, the Axiom of Continuity, VI, proved to be independent of Axioms I-V. Since the new axiom is essential for infinite fields of probability only, it is almost impossible to elucidate its empirical meaning, as has been done, for example, in the case of Axioms I-V in 2 of the first chapter. For, in describing any observable random process we can obtain only finite fields of probability. Infinite fields of probability occur only as idealized models of real random processes. *We limit ourselves, arbitrarily, to only those models which satisfy Axiom VI.* [14] This limitation has been found expedient in researches of the most diverse sort.

The distinguished probabilists such as Kolmogorov who founded the modern measure-theoretic approach to probability were thus under no illusions as to the 'truth' or validity of the axiom of countable additivity. On the other hand, the other axioms of probability can be given serious justification whether from a Bayesian point of view, as by de Finetti and Savage, or even from a frequentist point of view. But these other axioms are precisely the axioms for finite additivity. [15] Also note that standard Bayesian procedures, based upon improper prior distributions that we all use as convenient approximations, have only a finitely additive interpretation in the rigorous theories of de Finetti and Savage. The unfortunate confusion about foundational issues in probability and statistics has led to a 20 year period in which$A_n$ was largely neglected. $A_n$ is a de Finetti

[14] Author's italics.

[15] The controversy regarding finite additivity is related to the older controversy in mathematics and logic concerning the validity of some operations that involve infinite collections, for example, formalism versus 'intuitionism' and 'constructivism'. The finitely additive approach has operational meaning for all finite partitions, and along with Bayesian statistics, can be viewed as part of the constructive approach. See Hill (1987b, 1990c, 1990d) for discussion.

coherent Bayesian procedure, and as shown in Hill (1987b) it arises as a posterior distribution for a class of models called splitting processes, of which the Dirichlet process is a very special case. [16] As such it inherits all the desirable properties of Bayesian procedures, both for small samples and large samples, and in addition represents a diffuse form of a priori knowledge that often seems appropriate when theoretical and/or empirical knowledge about conventional models is not available, as so often happens in practice.

Now let me turn to the specific questions raised by Berger.

(1) Berger comments that "the robustness is 'real' only as long as the assumption is believable," where the assumption is that the actual values of the data carry no information beyond that in their ordering.

I think that both are true, namely $A_n$ is robust, and the assumption is believable in many real world problems, provided that the words 'no information' are replaced by 'little information'. I have shown in Hill (1968) that $A_n$ holds exactly in the case of merely simply ordered data, and in Hill (1987b) that this is also true for the quantitative data arising from splitting processes, so it is indeed logically possible for the assumption to hold exactly. But for typical real-world applications, I do not think that $A_n$ will *exactly* represent anyone's opinions, and I have instead repeatedly advocated it as a very realistic approximation. The situation in which $A_n$ or $H_n$ would be most appropriate is that in which a referee selects a distribution F from a huge collection of distributions, using a computer with a large (but finite) memory, and then given F, generates the $X_i$ as being conditionally independent, with common distribution F. Imagine that you are given the value of $X_1$, and must then put forth your posterior predictive probability that $X_2$ will be larger than the observed value of $X_1$. If you are told nothing further about the way in which the referee chooses F, so that F could be any distribution whatsoever, provided only that it can be represented on the computer, then upon reflection you may find that $H_1$ reflects your opinions to a very good approximation. (Recall that $A_n$ is the special case of $H_n$ in which M = N so that ties cannot occur, and therefore $H_n$ offers more flexibility. However, even $A_n$ will often provide a satisfactory approximation.) To turn Berger's question around, suppose that I were to object to the normal distribution and the exponential distribution on the grounds that neither of these could possibly represent anyone's exact opinions about the *observed data,* for example, because it would take eternity even to record the digits of a single irrational number. I am sure that he would respond that such distributions are only meant as useful approximations, and that is precisely my response to him. Furthermore, it would appear to me that in real-world situations we are ordinarily much 'closer'

---

[16] The conventional Dirichlet process is the countably additive version, while $A_n$ is the finitely additive version, of a nested splitting process.

(in some appropriate sense) to the view represented by $A_n$ than to that represented by conventional models, since models are often little more than convenient fictions that in principle require extensive justification for their use, while $A_n$ requires only a quite common form of ignorance. This is of course not to imply that models cannot be extremely useful, as discussed in Hill (1985-86) and in my present article in connection with predictivism.

(2) In his Example 1, Berger states "...the question is not simply whether or not one believes the exponential assumption; for a very wide range of models and priors the probabilities arising from the exponential model will be closer to the actual predictive probabilities than will those arising from $A_3$."

I don't understand this comment. All that he has assumed is that we are dealing with failure times, and therefore nonnegative variables. How can this imply the conclusion he draws? There are infinitely many possible models. For example, it may be that $\log(X)$ has a normal distribution, or that some other function of X has some other specified distribution, i.e., the Makeham, Weibull, or Zipf distribution, etc. The 'true' distribution may also be a mixture of these and infinitely many other distributions. It is plain that one can choose the model in such a way as to produce essentially any probabilities that one desires for the intervals, provided only that they are coherent, i.e., are nonnegative and sum to unity. Hence in drawing his conclusion as to closeness, he must implicitly be excluding, among other things, the situation in which one simply does not know which of these infinitely many possible models, if any, is true. But such an exclusion would have to be justified either theoretically or empirically, and he has made no attempt to do so. In this context it is not at all clear in what sense the class of models and priors which yield results closer to those for the exponential is 'large'. Indeed, it would instead appear to be quite 'small' and special within the class of all mixtures. Furthermore, with regard to $A_n$, because it implicitly mixes over all such unknown models, and corresponds to the state of knowledge in situations where one does not have the type of sharp a priori model knowledge that Berger is implicitly assuming, it would appear to me to be more generally useful in typical real-world situations. Finally, since the data he brings up are fictitious it is difficult to understand on what basis he would guess that "for most model-prior pairs that one would entertain for this data, the exponential-based probabilities will be closer." I would have thought that this question could at best only be discussable in the context of specific *real-world* data, not made up data.

(3) The question as to the size of sample for which $A_n$ might be appropriate is an extremely subtle one. Hill (1968, p. 688) showed that $A_k$ logically implies $A_j$ for $j \leq k$. (This same argument is also used by Lane and Sudderth (1978, p. 1322) in their demonstration of coherency of $A_n$.) This means that if you contemplate a large sample size, n, and believe that $A_n$ would be exactly appropriate for that n, then in fact $A_1$ automatically

holds for you, even if the actual sample size is only 1. I agree with Berger that $A_n$ for large n is extremely compelling, but it therefore follows that the same is true even of $A_1$! In my article I have given a subjectivistic argument for $A_{N-1}$, where N is the population size. It seems to me that the situations where one would want seriously to violate $A_{N-1}$ (apart from allowing for ties as in $H_{N-1}$ ) are quite rare, and require extremely strong prior knowledge about some aspect of the population. In some respects this analysis disturbs me, since like Berger, I have an intuitive reaction that $A_1$ is often not so compelling as $A_{N-1}$. One might wish, for example, to utilize strong a priori information about the mean and the variance of the finite population, as one does with the conventional normal model for the data. In part this arises through the *utility* of the model. Thus in the small sample case it is relatively more useful to smooth according to some parametric model, since in this case the predictive distribution given by $A_n$ may not be sufficiently informative for decision-making purposes; while when n is large, the predictive distribution given by $A_n$ can be quite informative and useful for such purposes. Also one must bear in mind that even $A_{N-1}$ would only be an approximation, and minor departures from $A_{N-1}$ could build up (in the backwards induction) to yield a serious departure from $A_1$.

(4) Berger's and my own view are close on some aspects of statistics. The questions on which we seem most to differ concern coherency and finite additivity. These are the most subtle questions in all of statistical theory, and I am hopeful that we can resolve them, since I think that like the other typically unvoiced criticisms, failure to reach a consensus on these issues has greatly harmed the profession. Berger shows that given the 'true' f, if I were to play his game then there is a sense in which I would have a positive expected loss for each such f, relative to someone who used his strategy. This is a nice result, which contributes to our understanding of $A_1$. It is a special case of a result attributed to T. Cover by S. Samuels (1989, p. 290) in connection with the secretary problem. Cover's argument relates even more directly to $A_1$ than does Berger's, since it does not require assumptions as to the form of the underlying distribution F, if there should happen to be such. Samuels's version of Cover's result is as follows. Pick any number x. If $X_1 > x$, then select $X_1$ as the larger; if $X_1 < x$, select $X_2$ as the larger. [17] If both $X_1$ and $X_2$ turn out to be bigger than x, or if both are smaller than x, then Samuels argues that by exchangeability this rule selects the larger of the two with probability 1/2; while, if one random variable is larger than x while the other is smaller, the larger one is sure to be chosen. Thus, the probability of selecting the larger with this strategy is

$$\frac{1}{2}[1 + P\{min(X_1, X_2) < x < max(X_1, X_2)\}].$$

[17] If $X_1 = x$, then one could presumably select either. In the context of $A_1$, however, $P(X_1 = x)$ would be negligible.

This probability is strictly greater than 1/2 so long as x is between the essential inf and the essential sup of the $X_i$. Samuels also notes that Cover's argument is related to that of Hill (1968) in the proof that $A_1$ cannot hold exactly for countably additive distributions.

There are a number of comments to be made, some general, some specific. First of all, although the 'true' F can be interpreted by means of de Finetti's exchangeability theorem, as in my article, strictly speaking $A_1$ does not require the existence of such a 'true' F. The situation is similar to that in contrasting the Pólya process for generating a sequence of 0-1 valued random variables, with the Bayes-Laplace model in which first a 'true' p is generated according to a uniform distribution on the unit interval, and then the observable variables are generated as a Bernoulli sequence with that value of p. See de Finetti (1975, p. 220) and Hill, Lane, Sudderth (1987). For the Pólya process there is no fixed true p, just a sequence of ever-changing urns. Yet these two processes are probabilistically identical. In the same way as with the Pólya process, $A_1$ can be viewed as specifying a predictive distribution, even when there is no fixed true distribution. In the absence of such a true f, what would Berger's result mean? Note that in the subjectivistic theory, excluding the case of a random sample from a fixed finite population, there is no hope ever to prove the existence of such a true f, so its existence would represent an unverifiable assumption, just as with countable additivity in the quotation by Kolmogorov. Even when in a certain sense a true F exists, as for example in Section 3 of my article, when the data have been realized but not observed, it is still different in a very critical way not to know F, and therefore to represent one's opinions as a mixture of distributions, than to view things from the perspective of the true (but unknown) F. For example, a property may be true of every such F, without being true of the mixture. Independence is an example. Also non-conglomerability may raise its ugly head.

Note that a basic difference in perspective between Berger and myself may occur here insofar as I believe that all probabilities must be conditional upon the actual state of knowledge, and therefore do not *necessarily* accept the relevance of looking at matters from the perspective of the 'true' distribution, which in fact may not even exist. In part this concerns the difference between the 'creative scientist' and 'policymaker' problems of Hill (1990c, 1990d). I do agree with Berger that from a pre-data perspective, or when data analysis is not possible, it can be useful to view matters from the perspective of a referee or mechanism, as in my argument for the admissibility principle in Hill (1974, 1990c). Indeed, it was for this reason that the derivation of probability and utility in Hill (1978) included not only subjective probability, but also 'objective' probability, in the sense of such a distinguished referee or mechanism. In the context of Bayesian data analysis of Hill (1990a), however, or in the creative scientist problem, this mode of pre-data reasoning is often not relevant.

Now let us consider the game that Berger wishes to play. He would let me choose the f, and he will bet in accord with his strategy as given by the lemma. To avoid an asymmetry in his knowledge and mine, let us first modify the game so that instead of myself picking f, an impartial referee chooses f, selects a value $x_1$ from f, and reports its value to Berger. Assume that we are told nothing about the method of selection other than that the referee will use a computer to generate the data. This is then a special case of the gambling scenario of Hill (1974b, 1990c), Heath and Sudderth (1978). However, even this knowledge dramatically changes things. Computers have finite memories, and it would not be difficult for me to obtain an upper bound on the memory of the computer. I may know, for example, that $X_1$ must fall between 0 and C + 1, where C is known to me, and that the observable value $x_1$ must be an integer in that domain. There is no loss in generality in assuming this to be the case, since realizable fractions can be so converted to integers. We have thus reduced the problem to a finite problem. See the example below for a concrete illustration in connection with Berger's lemma. As argued in Hill (1990c, 1990d), it is only the specific *implementation* on a computer with finite memory, of any procedure, such as $A_n$, or shrinkage estimators, that should in principle be admissible. One can use infinite models to gain insight, and for communication, but eventually these models must be implemented on such a computer. For the case of vague a priori knowledge I might still wish to use $A_1$ as an *approximation* to my posterior predictive probability that the second observation is larger than the first (although not for all realizable $x_1$ values, such as 0 or C + 1), especially in preference to any other known way of evaluating this probability. In the example below it is shown that there are no non-trivial shortcomings for $A_1$ in making such an evaluation. This evaluation in fact agrees with the conventional inference based upon diffuse a priori knowledge. Thus for n = 1, if there were anything to worry about due to Berger's argument, it would imply that standard Bayesian and non-Bayesian inference about the mean of a normal distribution was in some sense objectively defective, which has never been demonstrated, and in my opinion is simply not so. The example discussed below for the case of a finite space shows that the 'improvement' suggested by Berger is negligible under the circumstances in which anyone would contemplate using $A_n$. When n is larger than 1, the motivation for use of $A_n$ is even stronger, since one must be very uncertain about the nature of the f chosen by the referee, and could not presume it to be any of the standard parametric forms. Berger seems to agree with this, insofar as he finds $A_n$ acceptable for large n. See also Hill (1980c), and my rejoinder to Professor Zabell with regard to the exchange paradox, in Hill (1988b), for discussion of the behavior of $A_n$ in the finite case.

Such a reduction to a finite model is a special case of the argument I gave in my discussion of the Monette-Fraser example in the monograph by Berger and Wolpert (1988, p. 167). In that example there is a procedure that is known to be inadmissible in the extended sense, i.e., that can be beaten uniformly by some $\epsilon > 0$. See Heath and Sudderth (1978). I argued that even if the 'true' parameter could not be bounded, it was necessarily the case that only a bounded integer could be reported by the referee in a finite time, using a pre-specified method of reporting, as would be required in order for the gamble to be operationally meaningful. I argued that even such strong inadmissibility was operationally meaningless, because the gambles become conditional gambles, when one allows for finite computers, finite time, etc. The aspect that I would most like here to emphasize concerns the need that procedures be operationally meaningful. This was one of de Finetti's greatest contributions to statistical theory. Many notions that are current in statistics have never been given operational meaning, while others have operational meaning only in very special circumstances.

For example, consider the concept of admissibility. If two risk functions are identical, except that one is smaller at a single totally undistinguished irrational value of the parameter (one that no one has bothered to name or even thought of, and that it would take forever to record), can it be seriously argued from an operational point of view that the one is better than the other? The first attempt that I know of to provide some operational justification for the concept of admissibility was my own in Hill (1974b, Sec. 2). See also Berger (1984, p. 88; 1985, p. 257), who discusses this type of evaluation game. I know of no other attempts to address the question. My own justification, as in the footnote of Hill (1974b, p. 559) referred to very specific concrete knowledge, with again a referee using a computer to choose the parameter at random according to some a priori distribution. I argued that from the point of view of the referee, the person who uses the inadmissible estimator would be expected to do less well than if he used an estimator that dominates the inadmissible one (in the Waldian sense) on a set of positive a priori measure in the parameter space. This would especially be the case if the inadmissible estimator were actually inadmissible in the extended sense. However, even this argument requires that the parameter be generated on a computer or something equivalent, and also that one knows of an estimator whose risk function dominates that of the inadmissible one. Futhermore, as pointed out in Hill (1990c) in discussion of an article by L. Brown, even my own argument for admissibility has no known operational meaning except in the case of a finite partition for the parameter. In short, I believe that apart from finite or other special situations, where the admissibility principle is essentially equivalent to the use of Bayesian procedures, the argument for the *strong condition* of admissibility is very weak. The argument for extended admissibility is stronger, but still only compelling in the case of a finite partition for the parameter. The argument for the *weak condition* of coherency is I believe a very strong one. It is based upon the avoidance of a Dutch book, or sure loss, on finite collections of gambles. See Hill (1990d).

No one wishes to use a procedure that would make him a sure loser, such as if one violates the laws of finitely additive probability. Plainly, such a procedure is also both inadmissible and extended inadmissible. Countable additivity implies finite additivity, but makes unverified and unverifiable presumptions about infinity. The realism of the finitely additive approach stems from the fact that it makes no such presumptions. It is based squarely on the logic of uncertainty for finite partitions, but need not exclude the use of infinite parameter spaces to provide insight and approximations. My own version of the finitely additive approach may differ from that of de Finetti in that while I would allow infinite parameter spaces to be used in the conceptualization of a problem, I require that after this is over with, any suggested procedure must be capable of an implementation on a computer with finite memory. According to Theorem 1 of Hill (1990c), the implementation of any Bayesian procedure on such a computer is necessarily admissible with respect to the finite partition of parameter values represented on the computer, provided that all prior probabilities are positive. This argument applies for any coherent procedure whatsoever, including $A_n$, as well as procedures based upon Lebesgue measure as prior distribution, such as the Bayesian truncated version of least squares estimators. In my opinion the various paradoxes and theorems concerning inadmissibility are only pale reflections of more fundamental paradoxes concerning infinity, and have virtually nothing to do with real world statistical problems. It would seem plain that once it is agreed that the implementation of any procedure on a computer necessarily reduces the problem to a finite one, that non-Bayesian procedures should be ruled out whenever possible, at least in the policymaker problem of Hill (1990c). [18] Hence it is ultimately a question of choosing a prior distribution on a finite set. If one prefers, the choice can be made in terms of the risk function as well. Of course, in typical real-world applications, small finite partitions are quite adequate for decision-making purposes. Note the important fact that any result that can be proved in the finitely additive framework cannot be disproved without disproving finitary mathematics and logic themselves, while this is not the case for results based upon countable additivity. See also Scozzafava (1984).

In discussing finitely additive priors, Berger asserts that "Such prior distributions are not implementable, however, in the sense of being used to actually play the game..." In fact no distribution that gives positive probability to more than a finite number of distinct points is implementable by human means, so one wonders why he singles out the merely finitely additive ones in this regard. In Hill (1980c), and again in Hill (1990c), I put forth a challenge for anyone to give an operationally meaningful argument against the finitely additive approach. By 'operationally meaningful', I mean an explicit scenario, a game that can actually be played out in gory

[18] This is also true of the creative scientist problem, but for different and more complicated reasons that arise in data analysis.

detail. Such a game is necessarily finitistic. Berger attempts to meet this challenge, but for a variety of reasons, I do not think he succeeds.

Berger (like Cover) shows that his expected winnings are larger than 1/2 for all f, subject to a minimal condition. The relevance of the 1/2 is due both to the fact that under $A_1$ the predictive probability for the event $X_2 > x_1$, given $X_1 = x_1$, is 1/2 for all possible $x_1$; and also that under exchangeability, if one picks $X_2$ as larger, then one expects to win about half of the time in repeated plays of the game. He shows that for each fixed 'true' f he can achieve an unconditional expected winnings larger than 1/2, while for the person who uses $A_1$ the unconditional and conditional expected winnings are 1/2, assuming a unit stake. Hence he argues that over a long sequence of repetitions the person who uses $A_1$ can be expected to do less well than himself. However, I think that Berger fails properly to distinguish between a strategy such as he proposes to win a particular betting game, namely to pick the larger value, and a genuine (if only approximate) evaluation of a predictive probability under realistic conditions of uncertainty. These serve quite different purposes. If my primary objective were to win the game that Berger suggests playing, under repeated trials, then I would choose an appropriate strategy to do so, as in my discussion of the policymaker problem. But that is not the purpose of $A_n$. There is no conflict that I know of between making the probability evaluation given by $A_1$, on the one hand, and using a strategy such as that proposed by Berger to play his game, on the other hand. It is one thing to show that there exist strategies that may be better for achieving certain purposes, such as winning a particular gamble, and another thing to suggest that a procedure such as $A_n$, which has no such purpose, is in some sense 'objectively' inadequate. For example, it would be entirely compatible with $A_1$, under which one has posterior probability 1/2 that $X_2$ will be the larger, to also use the strategy of Berger or Cover to pick the larger, since neither $A_1$ nor Bayesian decision theory dictates how to choose in this case.

The optimal exact Bayes rule for a completely specified joint distribution of the $X_i$, and with 0-1 loss, would of course be to choose $X_2$ if the posterior probability that $X_2$ is the larger is more than 1/2, choose $X_1$ if it is less than 1/2, and to be indifferent if this probability is exactly 1/2. However, this optimal rule can only be used if one has completely specified the joint distribution of the $X_i$, and neither Cover nor Berger do so, while $A_n$ does implicitly give at least a partial specification. It is because $A_1$ is only an approximation that Cover and Berger can appear to do better in repeated plays of the game. To have the *approximate* value of 1/2 does not imply that there are not procedures with which one can select the larger value in more than half of the cases under repeated trials. [19]

[19]To implement the Cover or Berger procedure in such a way as to actually achieve a non-trivial improvement, even in repeated trials, is quite another matter. See the example below.

The distinction is closely related to that between the creative scientist and policymaker problems of Hill (1990c), and is also fundamental to the distinction between the viewpoints of R. A. Fisher and J. Neyman. There are quite different objectives involved in these two problems, with the policymaker version being appropriate for picking an optimal strategy to select the largest value under repeated trials, and the creative scientist version being appropriate for putting forth an approximate posterior predictive probability that expresses one's best judgment given the conditions of the problem. Both problems have Bayesian solutions, but naturally these can be different, since a different objective (or utility function) is involved.

The critical point to understand is that the procedures under consideration are only approximations, so that improvements can be possible. For example, in principle Cover could evaluate his constant x, so as to yield the largest possible probability of picking the larger observation. But in order to do so he would have to specify explicitly his joint a priori distribution for the $X_i$. In the same vein, observe that Berger does not suggest how his constant K be chosen, nor does he suggest a way actually to evaluate the posterior predictive probability that the next observation is larger than the first, given the value of the first. Such an evaluation is what would be required in order to make the argument of Hill (1974b, 1990c) for admissibility be operationally meaningful. To actually make such an evaluation Berger too would have to specify his marginal a priori distribution for the $X_i$, and then using his argument or that of Cover he could choose the value x for which

$$P\{min(X_1, X_2) < x < max(X_1, X_2)\}$$

is maximal. Berger does not do so, and his argument seems to prove too much, in that it suggests that any positive K will work, subject to his positivity condition. But plainly this is not true in any practically meaningful sense. For example, suppose the observations are the weights of whales in pounds, when two whales are drawn at random from some population of whales. Then his argument would apply with a choice of K = 50 pounds! The following simple finite example should make crystal clear the reason why the arguments of Cover and Berger are misleading.

**EXAMPLE:** Suppose that the possible values for the parameter $\theta$ are the integers from 1 to C. Suppose that given the value of $\theta$, that an observation $X_i$ is either $\theta - 1$, $\theta$, or $\theta + 1$, with probability $\delta$ for the value $\theta$, and equal probability $(1-\delta)/2$ for both $\theta - 1$ and $\theta + 1$, where the $X_i$ are conditionally independent, given $\theta$, for i = 1, 2. Take the a priori distribution $\pi$ for $\theta$ to be uniform on the integers from 1 to C, and note that the prior distribution for each $X_i$ also has the constant value 1/C, except at the four endpoints 0, 1, C, C + 1, so that the observable variables are very nearly uniformly distributed. This specifies a joint distribution for $X_1$ and $X_2$, which is exchangeable. For any value $2 \leq x_1 \leq C - 1$, the posterior distribution of $\theta$ is symmetrical about $x_1$, so a Bayesian would be indifferent as to which $X_i$ he chose as larger. For $x_1 = C + 1$, however, it is a posteriori certain that $\theta = C$, and similarly if $x_1 = 0$, it is certain that $\theta = 1$. For $x_1 = C$, it is certain that $\theta$ is either C or C - 1, and for $x_1 = 1$, it is certain that $\theta$ is either 1 or 2. It follows that there is no conditional improvement possible over picking at random between the first and second observation as larger, unless $x_1$ is one of the four distinguished endpoints. The unconditional probability of the union of these four point is 2/C, so if C is large, then there is a negligible probability that an $x_1$ will be observed for which one can do better than the prescription given by $A_1$. It then follows immediately from the connection between conditional optimality and unconditional optimality that there is also only a negligible improvement possible even from an unconditional perspective, such as in the arguments of Cover and Berger. See DeGroot (1970, p. 139). Thus one can easily verify that the improvement over 1/2 in Cover's argument is

$$P\{min(X_1, X_2) < x < max(X_1, X_2)\} = \frac{(1-\delta)^2}{2C} \leq \frac{1}{2C}$$

for $1 \leq x \leq C$, and otherwise is 0. If C is large, say a googol ($10^{100}$,) would anyone think that such an improvement would be worth having on a single play of the game? Note also that even to the extent that such an improvement would be worth having in repeated plays of the game, this improvement would already be available to a Bayesian who modified $A_1$ by using the optimal Bayes decision rule when the observed $x_1$ was one of the four distinguished endpoints. Such a Bayesian would, for example, choose $X_1$ as the larger, if the observed value $x_1$ was either C or C + 1, and choose $X_2$ as larger if the observed value was 0 or 1.

It can be verified that the unconditional probability of a tie is

$$\eta \equiv P(X_1 = X_2) = \delta^2 + \frac{1}{2}(1-\delta)^2,$$

and also that

$$P(X_2 = x_1 \mid X_1 = x_1) = \eta,$$

except when $x_1$ is one of the four distinguished endpoints. The quantity $\eta$ has a minimum value of 1/3, which occurs when $\delta = 1/3$. Thus the probability of a tie is not negligible in this model. One can easily alter the example, however, so that there are more than three possible values of $\theta$ compatible with the observed $x_1$, for all $x_1$ except for values near the endpoints, in which case the probability of a tie can be made small. It may be noted that the original definition of $A_n$ of Hill (1968, p. 677) implies only that there be equal posterior probability that the next observation is smaller or larger than $x_1$, not necessarily a probability of 1/2, which thus allows for ties. In real-world problems, although ties may be unlikely, it would be unusual for ties to have exactly a probability of 0. This is one of the reasons why $H_n$ can provide a real improvement over $A_n$ in terms of representing one's opinions. Although $A_n$ is the starting point for evaluation of the nonparametric posterior predictive distribution, since the time of my 1968 article I have become increasingly convinced that the ability to draw inference about M and the $\gamma_i$, which only $H_n$ allows, is of great importance.

This example illustrates how negligible the 'improvement' given by the method of Cover and Berger can be. It should not be thought that the conclusion that the improvement is negligible holds only for the above exact model. There are many other models for the data where the improvement is negligible. For example, $\pi$ need not be exactly uniform, there can be more than two points compatible with the value of $\theta$, and also there can be uncertainty as to the value of the upper and lower bounds for $\theta$. Whether or not such a priori distributions for the data are appropriate in a particular example is, as always, a question of empirical evidence, theoretical knowledge, and subjective judgment. My basic thesis is that there is no 'objective' sense in which one can do non-negligibly better than $A_1$ on a single play of the game, and that if one appropriately modifies $A_1$ near the endpoints, then this is true even on a number of plays of the game. In any case, to provide an improvement over $A_1$ on a number of plays of the game it would be necessary actually to specify the x of Cover or the K of Berger. Unless one has chosen K optimally, which requires at least partial specification of the marginal a priori distribution for the $X_i$, one might find oneself vacillating between various choices for K, even if the goal were merely to pick the larger.

Even in circumstances where there exists a 'better' strategy for betting on the size of the next observation in repeated plays of the game, how could this in anyway impugn the validity of $A_1$ ? The evaluation given by $A_1$ is proposed as a realistic evaluation of a probability under certain conditions, not as a strategy to win betting games having only a tenuous connection with the underlying scientific problem. It is coherent (so that one cannot be made a sure loser even in such games), in accord with both conventional non-Bayesian methods and Bayesian approximations based upon diffuse (or improper) prior distributions, and I believe quite sensible in certain situations. By contrast, it would hardly be sensible to choose $K = 50$ in the whale example, despite the lemma of Berger, and I doubt that anyone would do so. The situation is reminiscent of the approach of C. Stein, who showed that shrinking towards any value 'beats' the classical MLE for the mean of a multivariate normal distribution, with respect to squared error loss, when the dimension is more than 2. So one might shrink towards a totally absurd value, and still beat the MLE. But some of us think that this argument really shows, by *reductio ad absurdum,* that admissibility is not a sensible criterion (except of course in the finite case with all probabilities positive, when it is essentially equivalent to Bayesianity). See also Heath and Sudderth (1978) and Hill (1990c).

It should also be mentioned that replaying of the game introduces other game theoretic elements as to the method of selecting f, learning about the method of selecting f, etc., as discussed in Hill (1974b, p. 560; 1990c, 1990d). These are very subtle questions, and the arguments of Hill (1974b, p. 559) and Berger (1984, p. 88; 1985, p. 287) for admissibility, are not necessarily applicable, even apart from finiteness considerations. See for example Zabell's (1988) resolution of the exchange paradox, which involves the complex flow of information between the participants in the game. There is not space to discuss these delicate questions here, other than to observe that my approach is basically a finitistic one, in the spirit of traditional science, and with conventional models being used only as approximations in the special circumstances in which such approximations can be justified. For example, the 'true' F, when it can be said to exist at all, as in my article, is necessarily only a finite population empirical distribution function. I would especially question theoretical results that depend for their validity upon the existence of densities, such as the exponential, which model an uncountable infinity of possible values, with infinite precision for both data and parameter values. In my opinion such densities are primarily useful only in obtaining carefully thought out convenient approximations to the case of real interest, the finite case. The conventional use of such distributions in statistical theory unfortunately opens the door to infinitely many paradoxes of the infinite, most of which are not of much relevance to the practice of statistics. Also, in the finitely additive context, notions based upon countable additivity are often false. For example, Ramakrishnan and Sudderth (1989) show that the strong law of large numbers does not hold even for tossing a fair coin. Given that we have so little real knowledge of matters involving infinity, it would seem prudent to withhold judgment about questions that depend crucially upon such matters. It is my opinion that $A_n$ and $H_n$ find their real justification in the practice of statistics not in terms of infinite models, but rather because when used with judgment they do yield sensible approximations for finite models. Infinite models, such as my own splitting processes, are mathematical idealizations of the finite reality. They may provide insight, but such models should never be taken literally.

(5) I very much agree with the last two paragraphs of Berger regarding data-analysis and the only game in town. Berger (1984) and Smith (1986) present realistic versions of the Bayesian argument. So have others. My own views may be slightly different, in that in Hill (1980d), Hill (1985-86), and especially Hill (1990a), I give specific arguments for the validity of a generalized form of the (restricted) likelihood principle. This allows for Bayesian data analysis, and attempts to answer the important criticisms of Hacking (1967). See Hill (1990d) for an application of my approach to Bayesian data analysis in the context of hypothesis testing.

## ADDITIONAL REFERENCES

Berger, J. (1985), *Statistical Decision Theory and Bayesian Analysis, Second Edition*, Springer-Verlag.

Berger, J. and Wolpert, R. (1988), *The Likelihood Principle, Second Edition,* IMS Lecture Notes-Monograph Series.

de Finetti, B. (1975) *Theory of Probability*, Vol. 1, London: John Wiley & Sons, Inc.

Heath, D., and Sudderth, W. (1978), "On finitely additive priors, coherence, and extended admissibility," *Annals Statist.*, **6**, 333-345.

Hill, B. M. (1980d), Review of *Specification Searches,* by Edward Leamer, *Journal of the American Statistical Association,* **75**, 252-253.

Kolmogorov, A. N. (1950), *Foundations of Probability,* New York: Chelsea Publishing Co.

Lindley, D. (1971), *Bayesian Statistics, A Review,* Philadelphia: SIAM.

Robinson, G. K. (1979). Conditional properties of statistical procedures. *Ann. Statist.* **7**, 742–755.

Samuels, S. (1989), "Comment: Who will solve the secretary problem," *Statistical Science*, **4**, 289-29

Scozzafava, R. (1984), "A survey of some common misunderstandings concerning the role and meaning of finitely additive probabilities in statistical inference," *Statistica, anno XLIV,* 21-45.

Zabell,S. (1988), Comment on "De Finetti's theorem, induction, and $A_n$, or Bayesian nonparametric predictive inference." In *Bayesian Statistics* **3,** J. M. Bernardo, M. H. DeGroot, D. V. Lindley, and A. F. M. Smith, eds., Oxford University Press, 233-236.

# 5

# Homogeneity of Subpopulations for Two-By-Two Contingency Tables

## Yashaswini Mittal [1]

**Abstract**

Pooling of information from subpopulations achieves data compactification but some characteristics of the data are lost in the process. The greatest danger in amalgamating contingency tables is the possibility of a resulting paradox. Many such paradoxes have been noted since 1903 when Yule first noticed this phenomenon. We note three types of paradoxes. The definition of homogeneity proposed here is a sufficient condition to avoid Yule's Reversal Paradox. We look at interpretations of this definition and some examples.

*Key Words and Phrases:.* Contingency tables, paradoxes, pooling, homogeneity of subpopulations, collapsibility.

## 1. INTRODUCTION

Homogeneity of subpopulations is a very hard concept to define and yet, appropriate judgments of homogeneity are necessary in statistical studies at every stage of the game starting with the design of the experiment to reporting of the data. These perceptions of homogeneity are and should be different in different situations according to the different aims of the study. Suitable definitions of homogeneity should highlight the similarities for the current interest and downplay the remaining heterogeneity of the subpopulations. In this paper we study the concept of homogeneity of subpopulations from the point of view of amalgamating the corresponding $2 \times 2$ contingency tables. We propose a new definition of homogeneity which is different from the one given in Good and Mittal (1987) [abbreviated as G&M from now on]. There we defined homogeneity with respect to various measures of association. Stephen Fienberg has pointed out that this definition is similar to the notion of "collapsibility" defined in Bishop, Fienberg and Holland (1975). G&M found necessary and sufficient conditions for

[1] Virginia Polytechnic Institute and State University

two subpopulations to be homogeneous with respect to various measures of association. We use the following conventions for the attributes and the cell entries in a 2×2 contingency table.

**Table 1**

$$\begin{matrix} S & \bar{S} \end{matrix}$$
$$\begin{bmatrix} a & b \\ c & d \end{bmatrix}$$

with $a + b + c + d = N$. We represent the subpopulations by $M$ and $\overline{M}$, respectively. The following definition was proposed by G&M.

**Definition 1.1**: Two subpopulations represented by their 2×2 contingency tables $\boldsymbol{a}_1 = (a_1, b_1; c_1, d_1)$ and $\boldsymbol{a}_2 = (a_2, b_2; c_2, d_2)$ are *homogeneous* with respect to $\alpha$ if

$$\alpha(\boldsymbol{A}) = \alpha(\boldsymbol{a}_1) = \alpha(\boldsymbol{a}_2) \tag{1.1}$$

where $\mathbf{A}$ is the amalgamated table.

The notion of collapsibility relates to the invariance of factor interactions with respect to the collapsed factor. G&M show in Section 5 that one or the other of the following two conditions is sufficient for two subpopulations to be homogeneous with respect to a number of different measures of association.

$$\begin{gathered}\text{(i) Conditional on } S, T\&M \text{ are independent} \\ \\ \text{(ii) Conditional on } T, S\&M \text{ are independent.}\end{gathered} \tag{1.2}$$

Whittemore (1978) has shown that conditions in (1.2) are not necessary for collapsibility with respect to one of the factors. Shapiro (1982) has a geometric proof of the fact that collapsing preserves the odds ratio for two contingency tables if and only if one of the two conditions (1.2 i), (1.2 ii) is satisfied. (This is the same as part of the results of G&M which were obtained independently.) Shapiro also shows that (1.2) is not necessary for collapsibility (with respect to "odds ratio") of three or more tables. (In terms of Definition 1.1, this means that (1.2) is not necessary for three or more tables to be homogeneous with respect to "odds ratio". This can be thought of as a paradox of some sort since one expects stricter conditions for homogeneity of larger numbers of subpopulations).

Since the conditions in (1.2) are so strong, they provide little guidance in practical situations when data compactification is necessary. One of the major problems in amalgamating contingency tables from various subpopulations is the possibility of a resulting paradox. In Section 2 we review three types of paradoxes and their history. We identify one out of these, viz., Yule's Reversal Paradox, as potentially the most serious due to the inference problems it generates. The new definition of "homogeneity" in Section 3 will be shown to be sufficient to avoid this paradox. We look at one interpretation of this definition and propose some practical guidelines for amalgamation of $2 \times 2$ contingency tables. We illustrate our point with the help of some data sets.

What would be best is to find a "statistical explanation" of the paradox itself viz., a necessary and sufficient condition for the paradox that can be described in statistical terms. Unfortunately, none seems to have been discovered so far. Since the paradox is just a mathematical oddity, it is not difficult to find necessary and sufficient conditions for it but they are generally technical in nature (See eg. Thornton and Innes (1985).) Many attempts have been made at the "statistical explanations" as well. G&M (1987) found sufficient conditions for the avoidance of the Amalgamation Paradox which often, but not always, could be incorporated into the design of the experiment. Lindley and Novick (1981) state that the paradox can only arise if, $S$ and $T$ being positively associated, $M$ is positively associated both with $S$ and with $T$. This suggests that a specified order in the confounding of the attributes can be seen as the statistical explanation of the paradox.

## 2. VARIOUS PARADOXES AND THEIR HISTORY

Practitioners have been warned of the dangers of pooling information for a long time. The phenomenon seems to have been first noted by Yule (1903). For the study of contingency tables, the paradoxical behavior can be classified in terms of three different types of paradoxes. Yule notes that a spurious association between attributes may arise in a population as a result of amalgamation even though the attributes are independent in the subpopulations. In the above notation this implies that $\alpha(\boldsymbol{a_1}) = \alpha(\boldsymbol{a_2}) = 0$ but $\alpha(\boldsymbol{A}) \neq \mathbf{0}$. (Note that here $\alpha = \mathbf{0}$ implies independence of the attributes). There are many measures of association, such as the Odds Ratio, where independence is implied by $\alpha = 1$. A corresponding statement for these measures is apparent. However, for all measures considered in G&M (1987), viz., Peirce's, Yules, Odds Ratio, Weight of Evidence and Causal Propensity, this paradox occurs if and only if $a_i d_i = b_i c_i, i = 1, 2$ but $(a_1 + a_2)(d_1 + d_2) \neq (b_1 + b_2)(c_1 + c_2)$). We will call this "Yule's Association Paradox" (YAP). Cohen and Nagel (1934) notice the paradoxical behaviour in the form that is now popularly called "Simpson's Paradox", $\alpha(\boldsymbol{a_i}) \geq \mathbf{0}$ $(\leq 0)$ $i = 1, 2$ but $\alpha(\boldsymbol{A}) \leq \mathbf{0} (\geq \mathbf{0})$. Since Nagel suspects that he learned of the paradox from Yule's work, it seems fair to call it "Yule's Reversal Paradox" (YRP) (Messick and van de Geer (1981) call this a reversal paradox). G&M state that YRP is a stronger paradox than YAP because of its perceived paradoxical content and the fact that YAP implies

YRP. The amalgamation paradox (AMP) noted in G&M occurs if

$$\alpha(\boldsymbol{A}) < \min_i(a_i) \text{ or } \alpha(\boldsymbol{A}) < \max_i(a_i).$$

We notice that AMP is implied by both YAP and YRP. We hesitate to call AMP a stronger paradox than either YAP or YRP because the paradoxical content in AMP is not as disturbing as YRP or YAP. Whatever adjectives may be used, it is clear that

$$\text{YAP} \Rightarrow \text{YRP} \Rightarrow \text{AMP}.$$

Hence a sufficient condition for the avoidance of AMP must be sufficient to avoid YAP as well as YRP. In terms of the frequency of occurrence in practice, AMP is the most frequent and YAP least frequent. However, it is YRP that when it occurs, it poses critical problems of interpretation and inference. Thus it seems highly desirable to avoid YRP in pooling of information from subpopulations.

G&M proved that one or the other of the following conditions is sufficient to avoid AMP for various different association measures.

(a) Row Fair Design viz., $(a_i + b_i)/\ (c_i + d_i) = \lambda$, i = 1, 2.......n.

(2.1)

(b) Column Fair Design viz., $(a_i + c_i)/(b_i + d_i) = \delta$, i = 1,2.......n.

A little reflection shows that the weaker paradox YRP occurs for all the measures of association mentioned above, if and only if

$$a_i d_i - b_i c_i \geq 0 (\leq 0)$$

but

$$(\sum a_i)(\sum d_i) - (\sum b_i)(\sum c_i) \leq 0 (\geq 0). \tag{2.2}$$

Since the phenomenon described in (2.2) is paradoxical from the point of view of so many measures of association (including Yule's measure $y(\boldsymbol{a}) = a/N - \{(a+b)(a+c)/N^2\} = (ad - bc)/N^2)$, we call it "The YRP in $2 \times 2$ contingency tables". It is easy to see from the proofs in G&M that either of the two conditions in (2.1) is sufficient to avoid the paradox. Conditions (2.1), being the "design" conditions, provide some practical guidance for the practitioner to avoid the paradox viz., to employ either a row fair or a column fair design whenever possible. However, it is not always possible to do so. Also the conditions are too strong to expect to be satisfied in an observational data. Definition 3.1 of homogeneity is based on the conditional probabilities of the two attributes under study. It is clear that the definition is very closely linked with the particular relationship of the attributes of interest. Our effort here is to find practical conditions which have a statistical interpretation. The attempt in the following is to link the occurrence of the paradox to statistically unacceptable behavior viz., pooling of non-homogeneous subpopulations. The condition of homogeneity as defined in (3.1) below is sufficient to avoid YRP but not necessary. It is both necessary and sufficient for YAP if only two tables are amalgamated and neither necessary nor sufficient for AMP. Thus we do not recommend that the subpopulations be pooled if and only if (3.1) is satisfied, but we do recommend that a closer look at the data is warranted before amalgamation if (3.1) is not satisfied.

## 3. HOMOGENEITY AND THE PARADOXES

We represent the $n$ subpopulations by the corresponding $2 \times 2$ tables $\boldsymbol{a}_i = (a_i, b_i; c_i, d_i), i = 1, 2......n$.

**Definition 3.1**: The subpopulations are called *homogeneous* if

$$\max_i \ r_i \leq \min_i \ s_i \tag{3.1}$$

where $(r_i, s_i)$ is one of the four pairs $(\frac{b_i}{a_i}, \frac{d_i}{c_i}), (\frac{d_i}{c_i}, \frac{b_i}{a_i}), (\frac{c_i}{a_i}, \frac{d_i}{b_i})$ or $(\frac{d_i}{b_i}, \frac{c_i}{a_i})$.

Note: We can make a sub-classification into *row-homogeneous* and *column-homogeneous* according to whether the first two pairs are used for $(r_i, s_i)$ or the last two and call the sub-populations *homogeneous* if they are either *row* or *column homogeneous*. These two conditions are distinct since the sub-populations denoted by the tables (15,12;5,8) and (18,25;2,5) are row-homogeneous but not column-homogeneous. What follows is an interpretation of homogeneity in terms of the conditional probabilities.

If we denote the subpopulations by $M_i$, the condition can be interpreted as follows. Suppose that the treatment T is found to be better than the non-treatment $\overline{T}$ in all of the $n$ subpopulations i.e., $P(S|T\&M_i) \geq P(S|\overline{T}\&M_i)$ for all $i$, and the subpopulations are not (row-) homogeneous according to the above definition, then we must find two subpopulations $M_i$ and $M_j$ so that both the treatment and non-treatment in one subpopulation are more effective than either the treatment or the non-treatment in the second population i.e., max $(b_i/a_i,\ d_i/c_i) \leq$ min $(b_j/a_j\ ,\ d_j/c_j)$ indicating that the treatment acts quite differently in the two subpopulations $M_i$ and $M_j$ rendering them to be non-homogeneous with regards to this particular treatment. If the sub-populations are non-column-homogeneous, then we must find two such that both the conditional probabilities of treatment given success and non-success in one sub-population must exceed those of the second sub-population, indicating an uneven application of treatment. However, in cases where the rows are perceived as possible causes of the columns, row-homogeneity in the data will be more meaningful than column-homogeneity.

The geometrical representation of Table 1 by two vectors $(a, b)$ and $(c, d)$ illuminates both the import of Definition 3.1 and why it is sufficient to avoid YRP. In Figure 1 we represent a collection of $n$ homogeneous populations. Note that the area of the parallelogram made by the two vectors $(a, b)$ and

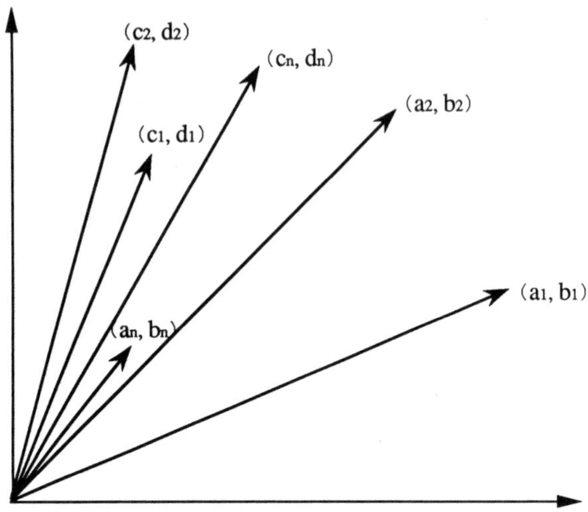


Figure 1
Representation of Homogeneous Populations

$(c, d)$ is $ad - bc$ if the slope of the vector $(c, d)$ is bigger than that of $(a, b)$. Also note that the treatment $T$ is more effective than the non-treatment $\overline{T}$ if and only if $a/(a + b) \geq c/(c + d)$ i.e., if $b/a \leq c/d$ i.e., if the $(a, b)$ vector has a smaller slope than the $(c, d)$ vector. The definition says that the subpopulations will be considered homogeneous if $\min_i P(S|T\&M_i) \geq \max_i P(S|\overline{T}\&M_i)$. This is true if the maximum slope of the $(a, b)$ vectors is less than the minimum slope of the $(c, d)$ vectors. Adding all of the $(a, b)$ vectors, the slope still remains smaller than that of the sum of all $(c, d)$ vectors. Thus showing that (3.1) is sufficient to avoid YRP.

Condition (3.1) is not necessary for avoidance of YRP and hence a paradox will not necessarily occur if non-homogeneous subpopulations are amalgamated. However, the following lemma shows that given any two non-homogeneous subpopulations, we can change the sample size without changing the probabilities of success under either the treatment or the non-treatment, but so that YRP will result after pooling the tables.

Suppose that the two subpopulations $\boldsymbol{a}_i = (a_i, b_i; c_i, d_i), i = 1, 2$ are not homogeneous due to

$$r_1 < s_1 < r_2 < s_2 \tag{3.2}$$

where $r_i = b_i/a_i$ and $s_i = d_i/c_i$ , $i = 1, 2$. Then $a_i d_i$ - $b_i c_i > 0$ for $i = 1, 2$.

If YRP does not occur when the tables are pooled then

$$(d_1 + d_2)/(c_1 + c_2) > (b_1 + b_2)/(a_1 + a_2). \tag{3.3}$$

We can produce a table $\boldsymbol{a}_2$ with same ratios $r_2$ and $s_2$ (i.e., by keeping $P(S|T\&\overline{M})$ and $P(S|\overline{T}\&\overline{M})$ constant) such that YRP will occur in the pooled table.

**Lemma 3.1**: For a given table $\boldsymbol{a}_1$ satisfying (3.1), let $q$ be any number $s_1 \leq \text{q} \leq r_2$ and let

$$a_2 = a_1(q - r_1)/(r_2 - q); \quad b_2 = r_2 a_2,$$

$$0 \leq c_2 \leq c_1(q - s_1)/(s_2 - q); \quad d_2 = s_2 c_2. \tag{3.4}$$

Then

$$(a_1 + a_2)(d_1 + d_2) - (b_1 + b_2)(c_1 + c_2) \leq 0.$$

**Proof**: We can write (3.3) as

$$\alpha r_1 + (1 - \alpha) r_2 > \beta s_1 + (1 - \beta) s_2. \tag{3.5}$$

for $\alpha = a_1/(a_1 + a_2)$ and $\beta = c_1/(c_1 + c_2)$. In view of (3.2), (3.5) can hold only if both sides of (3.5) take values between $s_1$ and $r_2$. Suppose for $q$, $s_1 \leq q \leq r_2$, $\alpha r_1$ + (1-$\alpha$)$r_2 = q$ then $\alpha = (r_2 - q)/(r_2 - r_1) = a_1/(a_1 + a_2)$ and hence $a_2 = a_1(q - r_1)/(r_2 - q)$. If $\beta_0$ is such that $\beta_0 s_1$ + (1-$\beta_0$)$s_2 = q$, then for all $1 > \beta > \beta_0$, (3.5) will hold with the above choice of $\alpha$. But $\beta_0 = (s_2 - q)/(s_2 - s_1)$. Hence $\beta = c_1/(c_1 + c_2) > (s_2 - q)/(s_2 - s_1) \Rightarrow c_1(q - s_1)/(s_2 - q) > c_2 \geq 0$. We note that $P(S|T\&\overline{M})$ and $P(S|\overline{T}\&\overline{M})$ are preserved in this table but the cell entries and $N_2 = a_2 + b_2 + c_2 + d_2$ are changed. Substituting values for the cell entries, we see that,

$$\frac{(1 + s)(q - r)}{(r - q)} a_1 \leq N_2 \leq \frac{(1 + s_2)(q - r_1)}{(r_1 - q)} a_1 + \frac{(1 + r_2)(q - s_1)}{(s_2 - q)} c_1. \tag{3.6}$$

Since the L.H.S. of (3.6) is increasing in $q$, the smallest possible value of $N_2$ is $(1 + s_2)(s_1 - r_1)/(r_1 - s_1)$.

We see that homogeneity, viz., (3.1), is not sufficient for avoidance of AMP (for say the "Odds Ratio" or "Peirce's Measure" $\Pi_R$), since the two subpopulations represented by the tables (15,12;5,8) and (18,25;2,5) are row-homogeneous but the "Odds Ratios" for the two tables and the amalgamated table are 2, 1.8, and 1.6 respectively. The corresponding values of the Peirce's measure are 0.1709, 0.1329 and 0.1214. Condition (3.1) is not necessary for avoidance of AMP (and hence of YAP and YRP) either since the tables (13,17;8,10) and (15,23;5,7) (with the pooled table (28,40;13,17)) are neither row nor column-homogeneous and AMP does not occur for either the Peirce's measure or the Odds Ratio since the values of the Peirce's measures are -0.0111, -0.0219 and -0.0216 respectively. The "Odds Ratios" are respectively 0.9559, 0,9130 and 0.9154.

The following lemma shows that homogeneity is sufficient to avoid YAP. This is implied by the fact that homogeneity is sufficient to avoid YRP, but we repeat the argument for clarity. It also shows that homogeneity is necessary for avoidance of YAP when two subpopulations are pooled but is not necessary when more than two subpopulations are involved.

**Lemma 3.2**: Suppose that the attributes are independent in each of the $n$ sub-populations viz.,

$$a_i d_i = b_i c_i \text{ for } i = 1, 2, \cdots, n. \tag{3.7}$$

If the sub-populations are homogeneous (row or column), then YAP is avoided viz.,

$$(\sum a_i)/(\sum d_i) = (\sum b_i)(\sum c_i). \tag{3.8}$$

On the other hand, if YAP is avoided and $n = 2$, then the subpopulations must be both row and column homogeneous.

**Proof**: Without loss of generality, suppose $\max_i b_i/a_i \leq \min_i d_i/c_i$ viz., the subpopulations are row-homogeneous, and (3.7) holds then we must have

$$b_1/a_1 = b_2/a_2 = \cdots b_n/a_n = d_1/c_1 = d_2/c_2 = \cdots d_n/c_n. \tag{3.9}$$

which easily implies (3.8). On the other hand, suppose (3.7) and (3.8) hold then $(\sum c_i)/(\sum a_i) = (\sum d_i)/(\sum b_i)$ gives convex combination of $n$ numbers on each side and the set of these $n$ numbers is the same for each side due to (3.7). For $n = 2$ viz., for two subpopulations, this will be true if and only if the corresponding weights on each side are the same, which implies (3.9) and hence the subpopulations are row-homogeneous. Due to the symmetry of the argument, they must be column-homogeneous as well. If $n > 2$, then the weights do not have to be the same.

Since homogeneity is not necessary for avoidance of YAP, we do not recommend that no subpopulations ever be pooled unless they are homogeneous. However, we do suggest that in case the subpopulations are found to be non-homogeneous, a second look at the data is warranted. As an example, we look at two data sets.

## 4. TWO EXAMPLES

The first example is taken from Bishop, Fienberg and Holland (1975, pp. 41), who say that the original source of the data is unknown.

Table 1. Three-Dimensional Array Relating Survival of Infants to Amount of Prenatal Care Received in Two Clinics

| Place Where Care Received | Amount of Prenatal Care | Infants' Survival Died | Survived | Mortality Rate(%) |
|---|---|---|---|---|
| Clinic A | Less | 3 | 176 | 1.7 |
| | More | 4 | 293 | 1.4 |
| Clinic B | Less | 17 | 197 | 7.9 |
| | More | 2 | 23 | 8.0 |

The Odds Ratios of the tables for Clinics A and B and the amalgamated table are, respectively, 1.2, 1.0 and 2.8. Bishop etal conclude that one looking at the pooled table alone will erroneously conclude that survival of the infants was related to the amount of prenatal care received by the mother. Since the tables are non-row-homogeneous (they are non-column-homogeneous as well, but due to the causal relationship under study, it is the row-non-homogeneity that is of concern here), another look at the data is warranted even though YRP does not result for the pooled table. The tables show clearly that the differences in the clinics are far more important for the survival of the infant than the amount of prenatal care received by the mother. Had we represented the data as

Table 2. An Alternate Representation of Date in Table 1

| | Less | | | More | |
|---|---|---|---|---|---|
| | Died | Survived | | Died | Survived |
| clinic A | 3 | 176 | clinic A | 4 | 293 |
| clinic B | 17 | 197 | clinic B | 2 | 23 |

we would have found the two subpopulations to be row-homogeneous. The Odds Ratios are respectively 0.2, 0.16 for the two subpopulations and 0.17 for the pooled table (7,469,19,220) indicating that the health of the infant is strongly correlated with the clinic and the relationship is similar in both the subpopulations.

We also note the effects of mis-classification in both methods of reporting the data. If one infant that died in clinic A was mis- classified according to the amount of care the mother received, then the Odds Ratio for clinic A will be either 0.67 (2.3 for the pooled table) or 2.21, (3.56 for the pooled table) according to how the mis-classification occurred. Especially in the second type of mis-classification, the table itself may give erroneous conclusions. However, if the subpopulations considered were "Less Care" and "More Care", then the same mis-classification will produce the Odds Ratios 0.26 and 0.12 or 0.13 and 0.2. We note that the pooled table will have the same Odds Ratio viz., 0.17.

The second data set, given in Table 3, is taken from Dorn (1954). It is also studied in Cox (1970), Whittemore (1978), and Cornfield (1956). The pooled table (165,5272;1401,6548) gives $P(Cancer \mid Non\text{-}smoking) = 0.105$ and $P(Cancer \mid Smoking) = 0.446$. However, we find that the subpopulations are not row-homogeneous (they are not column-homogeneous either). By inspection we split the above 14 studies into two row-homogeneous groups. Group 1 includes studies numbers 1, 3, 5, 7, 10, 12 and 14. The second group includes the rest. (This division is by no means unique.) The pooled tables for the two groups are, respectively, (89,1930;341,1881) and (76,3342;1060,4667). We find that the probabilities of interest, shown in Table 4, are strongly correlated with the groups.

Table 3. Fourteen retrospective studies on the association between smoking and lung cancer

| | Lung Cancer Patients | | Control Patients | |
|---|---|---|---|---|
| Study | Total | Non-smokers | Total | Non-smokers |
| 1 | 86 | 3 | 86 | 14 |
| 2 | 93 | 3 | 270 | 43 |
| 3 | 136 | 7 | 100 | 19 |
| 4 | 82 | 12 | 522 | 125 |
| 5 | 444 | 32 | 430 | 131 |
| 6 | 605 | 8 | 780 | 114 |
| 7 | 93 | 5 | 186 | 12 |
| 8 | 1357 | 7 | 1357 | 61 |
| 9 | 63 | 3 | 133 | 27 |
| 10 | 477 | 18 | 615 | 81 |
| 11 | 728 | 4 | 300 | 54 |
| 12 | 518 | 19 | 518 | 56 |
| 13 | 490 | 39 | 2365 | 636 |
| 14 | 265 | 5 | 287 | 28 |

Table 4. Conditional Probabilities for Two Groups

| | Group 1 | Group 2 |
|---|---|---|
| $P(Cancer \mid Non\text{-}Smoking)$ | 0.20698 | 0.0669 |
| $P(Cancer \mid Smoking)$ | 0.50643 | 0.4173 |

The Odds Ratio for Table 4 is 2.549. This indicates that the studies in group 1 may be subjected to hidden confounding factors. What these factors may be is not clear even from the details of the data given in Dorn (1954), but we do find that 47.99% of the subjects (52.13% of the smokers) were "heavy" smokers in the first group while as the corresponding percentage in the second group was 27.67 (31.56% of the smokers).

Cox (1970) gives estimates of the logistic difference for each of the 14 studies and finds strong evidence against constancy of the logistic difference. He finds large negative logistic sums for studies number 8, 6, and 11, and suggests that the largest proportional effect of smoking on lung cancer may be found in contexts in which the overall proportion of smokers is high. We do not find strong evidence of this looking at the two groups in Table 4. But at least we do not contradict this conclusion since 89.24% patients in the first group, for which the probability of cancer given smoking is somewhat larger than that for the second group, smoke compared to 87.93% in the second group. Note that the three studies 8, 6, and 11, for which the percentage of smokers is very high (95.16%), are part of the second group which overall has a smaller percentage of smokers.

Whittemore (1978) applies her criterion of collapsibility to studies numbered 1, 2, 3, 8, 9, 10, 12, and 14, and finds them collapsible. Whittemore's group of studies is not row-homogeneous. The pooled table for these studies gives $P(\mathit{Cancer} \mid \mathit{Non\text{-}smoking}) = 0.165$ and $P(\mathit{Cancer} \mid \mathit{Smoking}) = 0.491$. Cornfield (1956) eliminates the two most extreme pairs (viz. Studies 7, 11 and 4, 6) and pools the remaining 10 to get $P(\mathit{Cancer} \mid \mathit{Non\text{-}smoking}) = 0.1104$ and $P(\mathit{Cancer} \mid \mathit{Smoking}) = 0.4282$.

## ACKNOWLEDGMENT

This research was supported in part by the Office of Naval Research.

## REFERENCES

Bishop, Y. M. M., Fienberg, S. E., and Holland, P. W. (1975), *Discrete Multivariate Analysis: Theory and Practice*, MIT Press, Cambridge, MA. and London.

Cohen, M. R. and Nagel, E. (1934), *An Introduction to Logic and Scientific Method*, Harcourt, New York.

Cornfield, J. (1956), " A statistical problem arising from retrospective studies," *Proceedings of the 3rd Berkeley Symposium*, **4**, 135–148.

Cox, D. R. (1970), *Analysis of Binary Data*, Methuen, London.

Dorn, H. F. (1954), "The relationship of cancer of the lung and the use of tobacco," *The American Statistician*, **8**, 7–13.

Good, I. J. and Mittal, Y. (1987), "The amalgamation and geometry of two-by-two contingency tables," *The Annals of Statistics*, **15**, 694–711.

Lindley, D. V. and Novick, M. R. (1981), "On role of exchangeability in inference," *The Annals of Statistics*, **9**, 45–58.

Messick, D. M. and Van de Geer, J. P. (1981), "A reversal paradox," *Psychology Bulletin*, **90**, 582–593.

Shapiro, S. H. (1982), "Collapsing contingency tables — A geometric approach," *Journal of the American Statistical Association*, **36**, 43–46.

Thornton, R. J. and Innes, J. T. (1985), "On Simpson's paradox in economic statistics," *Oxford Bulletin of Economics and Statistics*, **47**, 387–394.

Yule, G. U. (1903), "Notes on the theory of association of attributes in statistics," *Biometrika*, **2**, 121–134. Reprinted in *Statistical Papers of George Udny Yule*, **71**–84, Griffin, London.

Whittemore, Alice S. (1978), "Collapsibility of multidimensional contingency tables," *Journal of the Royal Statistical Society*, Ser.B, **40** 328–340.

# 6

# A Hierarchical Bayes Approach to Small Area Estimation with Auxiliary Information


## Malay Ghosh [1]
## Parthasarathi Lahiri [2]



**Abstract**

Small area estimation is becoming increasingly popular in survey sampling. Agencies of the Federal Government have been involved in obtaining estimates of population counts, unemployment rates, per capita income, crop yields and so forth for state and local government areas. In many instances, only a few samples are available from an individual area, and an estimate of a certain area mean, or simultaneous estimates of several area means can be improved by incorporating information from similar neighboring areas.

The present paper introduces a hierarchical Bayes model incorporating auxiliary information in the context of small area estimation. One advantage of the hierarchical Bayes procedure over a classical empirical Bayes procedure is that not only can we obtain good point estimates of the small area means, but also it is possible to obtain good estimates of their variability as well.



*Key words and Phrases:* Hierarchical Bayes; Small area estimattion; Empirical Bayes; Variance components; Multivatiate normal; Auxiliary information; Covariates.


## 1. INTRODUCTION

Small area estimation is becoming increasingly popular in survey sampling. Agencies of the Federal Government have been involved in obtaining estimates of population counts, unemployment rates, per capita income, crop yields and so forth for state and local government areas. In typical instances, only a few samples are available from an individual area, and an estimate of a certain area mean or simultaneous estimates of several


[1] University of Florida
[2] University of Nebraska

area means can be improved by incorporating information from similar neighboring areas.

The need to "borrow strength" from related small areas has been recognized by many statisticians. Ericksen (1974) used a regression method for estimating population changes of local areas. Fay and Herriot (1979) used an adaptation of the James-Stein estimator to sample estimates of income for small places (that is, population less than 1000) from the 1970 census of population and housing. They fitted a regression equation to the census sample estimates, using as independent variables the county values, tax-return data for 1969, and data on housing from the 1970 census. The estimator used by them was a weighted average of the sample average and the regression estimator for each place. Battese, Harter and Fuller (1988) considered prediction of areas under corn and soybeans for 12 counties in north-central Iowa based on 1978 June Enumerative Survey and LANDSAT satellite data. They used a linear regression model defining a relationship between the survey and satellite data, and used this relationship to obtain predictors of mean crop areas per segment in the sample counties. A multivariate extension of the model considered by the latter is given in Fuller and Harter (1987).

A very common approach used by most of these authors is to obtain some form of estimated best linear unbiased predictors (BLUP's) for the unknown parameters of interest. To explain it more fully, first some mixed linear model is used for prediction purposes. Then using Henderson's (1975) result, BLUP's of the unknown parameters are obtained assuming the variance components to be known. Finally, the variance components are estimated, typically by Henderson's method of fitting of constants or the restricted maximum likelihood (REML) method. The resulting estimators, which can be called estimated BLUP's (see Harville (1977)), are used for final prediction purposes. The above fact is very clearly brought out in a series of interesting articles by Harville (1985, 1987, 1988). Prasad and Rao (1987) provide approximate estimates of the mean square errors (MSE's) of estimated BLUP's in three regression models (see also Jeske and Harville (1987)).

Another approach to this problem, first given in Fay and Herriot (1979), and later used also by Ghosh and Meeden (1986), is the so-called empirical Bayes (EB) approach. According to this procedure, first a Bayes estimate of the unknown parameter of interest is obtained by using a normal prior. The unknown parameters of the prior are then estimated by some classical methods like the method of moments, method of maximum likelihood or some combination thereof. The resulting estimator of the parameter of interest is the so-called EB estimator.

Ghosh and Meeden (1986) used a one-way ANOVA model to obtain EB estimators of means of several strata. The normality assumption was later replaced (see Ghosh and Lahiri (1987)) by the weaker assumption of posterior linearity of the strata means in the sample observations. The Monte

Carlo study undertaken in Ghosh and Lahiri (1987) clearly shows that the EB estimators of strata means, originally derived under normality, perform very satisfactorily under the binomial and Poisson models as well. It is mentioned in Prasad and Rao (1987) that the EB estimators of Ghosh and Meeden (1986) can also be viewed as estimated BLUP's. The phenomenon is more generally true. The connection between estimated BLUP's and classical EB's is brought out in a series of interesting articles by Harville (1985, 1987, 1988).

One of the major difficulties associated with the estimated BLUP or classical EB procedures is that there are no clear cut measures of standard errors associated with these estimators. Prasad and Rao (1987) have made an attempt to approximate the mean square errors (MSE's) of the estimated BLUP's by adapting an idea of Kackar and Harville (1984). These approximations work reasonably well when there exists a large number of strata. However, validity of these approximations in the presence of a small number of strata seems questionable.

We propose in this paper a hierarchical Bayes (HB) procedure as an alternative to the estimated BLUP or the classical EB procedure. Unlike the latter, the HB procedure usually provides a very natural estimate of the standard error of the estimator used. To be precise, if one estimates (as is naturally done) the parameter of interest by its posterior mean, then a very natural estimate of the risk associated with this estimator is the posterior variance. The latter, though often complicated, can be found exactly, and does not need any approximation. In a recent article, Harville (1987) has emphasized the need of a Bayesian approach in mixed linear models.

The type of HB approach which we are going to propose was introduced in the pioneering paper of Lindley and Smith (1972). However, for most of their analyses, they assumed the variance components to be known, and obtained posterior means of the unknown regression parameters involving the known variance components. For unknown variance components, they advocated use of posterior modes or type-II ML estimators of such parameters. Such estimators, when substituted in the posterior means, make the resulting estimators very close to the estimated BLUP's or the classical EB estimators.

Our approach is very close to that of Stroud (1987). However, in his most general formulation, Stroud (1987) considers only the balanced case, that is when the sample size for each small area is the same. In survey sampling, this assumption is hardly realized in practice. We have investigated the general unbalanced case which allows the possibility of the sample sizes to vary from one small area to another.

Ericksen and Kadane (1985, 1987) have used a HB model for local estimates of undercount in the 1980 census. Their approach is closely akin to that of Lindley and Smith (1972) where the variance components are either known, or are replaced by certain point estimates. Dumouchel and Harris (1983) have also used HB methods for combining the results of cancer studies in human and other species, and their HB structure is also very similar to that of Lindley and Smith (1972).

In Section 2 of this paper, we introduce a HB model where the ratio of the first and second stage variance components is known. We have considered a strict finite population set up, where the main objective is to obtain the conditional (predictive) distribution of the unobserved population units given the observed sample units. Under the assumed model in this section, this conditional distribution turns out to be a multivariate-t from which posterior means, variance and covariances are immediately obtained. Our model is also compared with a corresponding model of Stroud (1987). It is also shown in this section how our HB estimator can be obtained as a BLUP.

Section 3 introduces a more general set up where the first and second stage variance components are both unknown, and certain prior distributions are assigned to them. Our model is contrasted to that of Stroud (1987), Ericksen and Kadane (1985, 1987), Carter and Rolph (1974), Fay and Herriot (1979), Morris (1983), Hill (1980) and others. We obtain the joint posterior distribution of the unobserved units given the observed units, and obtain also certain posterior means, variances and covariances. The survey data considered in Stroud (1987) is revisited, and numerical results are provided for posterior means and variances. It is also shown that the numerical results obtained are fairly robust against a wide choice of priors.

The proof of all the technical results are deferred to the Appendix.

## 2. THE HB PROCEDURE WITH KNOWN RATIO OF VARIANCE COMPONENTS

Suppose that there are m small areas. The ith small area contains $N_i$ units (i = 1, ..., m). Let $Y_{ij}$ denote the characteristic of interest associated with the jth unit in the ith small area. Suppose that samples of sizes $n_1, \ldots, n_m$ are drawn from the m small areas. For notational convenience, we denote the sample for the ith small area by $Y_{i1}, \ldots, Y_{in_i}$(i = 1, ..., m). Consider the following hierarchical Bayes model:

a) Conditional on $\mathbf{b} = (b_1, \ldots, b_p)'$, $\boldsymbol{\theta} = (\theta_1, \ldots, \theta_m)'$ and r; $Y_{ij}$'s $(j = 1, \ldots, N_i; i = 1, \ldots, m)$ are mutually independent with $Y_{i1}, \ldots, Y_{iN_i}$'s iid $\mathrm{N}(\theta_i, r^{-1})$;

b) Conditional on $\mathbf{b}$ and r; $\Theta_1, \ldots, \Theta_m$ are mutually independent with $\Theta_i \sim \mathrm{N}\left(\mathbf{x}_i'\mathbf{b}, (\lambda r)^{-1}\right)$, where $\mathbf{x}_1, \ldots, \mathbf{x}_m$ are *known* design vectors, and $\lambda$ (> 0) is *known*;

c) Marginally $\mathbf{B}$ and R are independently distributed with $\mathbf{B} \sim$ uniform on $(R^p)$ and R $\sim$ Gamma ($\frac{1}{2}$ a, $\frac{1}{2}$ g). We say that a random variable Z has a Gamma $(\alpha, \beta)$ distribution if it has a pdf of the form

$$f(z) = \exp(-\alpha z)z^{\beta-1}\alpha^{\beta}/\Gamma(\beta); z > 0, \alpha > 0, \beta > 0.$$

We shall find it convenient to use the notation $\mathbf{Y}_i^{(1)} = (Y_{i1}, \ldots, Y_{in_i})'$ for the vector of sampled values and $\mathbf{Y}_i^{(2)} = (Y_{in_i+1}, \ldots, Y_{iN_i})'$ for the vector of unsampled values for the ith small area. Let $\mathbf{Y}_i' = (\mathbf{Y}_i^{(1)\prime}, \mathbf{Y}_i^{(2)\prime})$ and $\mathbf{Y}' = (\mathbf{Y}^{(1)\prime}, \mathbf{Y}^{(2)\prime})$. Our goal is to find the conditional (predictive) distribution of $(\mathbf{Y}_1^{(2)}, \ldots, \mathbf{Y}_m^{(2)})$ given $(\mathbf{Y}_1^{(1)}, \ldots, \mathbf{Y}_m^{(1)})$, and make inference on the basis of this predictive distribution. However, it is also possible to find the posterior distribution of $\mathbf{\Theta} = (\Theta_1, \ldots, \Theta_m)'$ given $\mathbf{Y}_1^{(1)}, \ldots, \mathbf{Y}_m^{(1)}$.

Stroud (1987) in his Section 2 considered a parallel model with *both* $\lambda$ ($> 0$) and r ($> 0$) *known*. The notations of Stroud are different from ours. In fact, Stroud considered only a simple linear regression model, but the model can easily be extended to a general linear regression model. However, Stroud obtained only the posterior distribution of $\mathbf{\Theta}$ given $\mathbf{Y}_1^{(1)}, \ldots, \mathbf{Y}_m^{(1)}$, but did not find the predictive distribution of $\mathbf{Y}_1^{(2)}, \ldots, \mathbf{Y}_m^{(2)}$.

We shall find it convenient at this stage to introduce a multivariate t-distribution. A p-dimensional random vector $\mathbf{T} = (T_1, \ldots, T_p)'$ is said to have a multivariate t-distribution with location parameter $\boldsymbol{\mu}$, scale parameter $\boldsymbol{\Sigma}$ and degrees of freedom $\nu$ if $\mathbf{T}$ has a pdf of the form

$$g(\mathbf{t}) \propto |\boldsymbol{\Sigma}|^{-\frac{1}{2}}\left[\nu + (\mathbf{t} - \boldsymbol{\mu})'\boldsymbol{\Sigma}^{-1}(\mathbf{t} - \boldsymbol{\mu})\right]^{-\frac{1}{2}(\nu+p)}; \tag{2.1}$$

(cf Zellner (1971, p 383) or Press (1972, p 136)). Also, assuming appropriate moments to exist,

$$E(\mathbf{T}) = \boldsymbol{\mu} \text{ and } V(\mathbf{T}) = \frac{\nu}{\nu - 2}\boldsymbol{\Sigma}. \tag{2.2}$$

The first theorem of this section provides the joint predictive distribution of $\mathbf{Y}^{(2)} = \left(\mathbf{Y}_1^{(2)\prime}, \ldots, \mathbf{Y}_m^{(2)\prime}\right)'$ given $\mathbf{Y}^{(1)} = \left(\mathbf{Y}_1^{(1)\prime}, \ldots, \mathbf{Y}_m^{(1)\prime}\right)'$. *Throughout we shall* use the notation $\mathbf{I}_u$ for an identity matrix of order u, $\mathbf{1}_u$ for a u-component column vector with each element equal to 1, and $\mathbf{J}_u$ for a $u \times u$ matrix with each element equal to 1. Also $\oplus_{j=1}^m A_j$ will denote the matrix

$$\begin{bmatrix} \mathbf{A}_1 & \ldots & \mathbf{0} \\ \vdots & \vdots & \vdots \\ \mathbf{0} & \ldots & \mathbf{A}_m \end{bmatrix}.$$

**Theorem 1**. Assume that $n_T +$ g - p $> 2$. The predictive distribution of $\mathbf{Y}^{(2)}$ given $\mathbf{Y}^{(1)} = \mathbf{y}^{(1)}$ is multivariate t with location parameter $\mathbf{h} = (\mathbf{h}_1', \ldots, \mathbf{h}_m')'$, scale parameter $\mathbf{V}$, and degrees of freedom $n_T +$ g - p, where

$$\mathbf{h}_i = \left[D_i\bar{y}_i + (1 - D_i)\mathbf{x}_i'\hat{\mathbf{b}}_{.}\right]\mathbf{1}_{N_i - n_i}(1 \le i \le m); \tag{2.3}$$

$$\mathbf{V} = (n_T + g - p)^{-1} \left[a + \mathbf{y}^{(1)\prime}\mathbf{C}\mathbf{y}^{(1)}\right] \mathbf{G}. \tag{2.4}$$

In the above

$$D_i = n_i(n_i + \lambda)^{-1}, \bar{y}_i = n_i^{-1}\Sigma_{j=1}^{n_i} y_{ij} (1 \le i \le m), n_T = \Sigma_{i=1}^{m} n_i; \tag{2.5}$$

$$\hat{\mathbf{b}}_{.} = \mathbf{D}^{*^{-1}} \left(\Sigma_{j=1}^{m} D_j \mathbf{x}_j \bar{y}_j\right); \tag{2.6}$$

$$\mathbf{C} = \oplus_{j=1}^{m} \left[\mathbf{I}_{n_j} - \lambda^{-1}(1 - D_j)\mathbf{J}_{n_j}\right] - \lambda^{-1}\mathbf{A}^*\mathbf{X}\mathbf{D}^{*^{-1}}\mathbf{X}'\mathbf{A}^{*'}; \tag{2.7}$$

$$\mathbf{G} = \oplus_{j=1}^{m} \left[\mathbf{I}_{N_j - n_j} + \lambda^{-1}(1 - D_j)\mathbf{J}_{N_j - n_j}\right] + \lambda^{-1}\mathbf{B}^*\mathbf{X}\mathbf{D}^{*^{-1}}\mathbf{X}'\mathbf{B}^{*'}, \tag{2.8}$$

where $\mathbf{D}^* = (\Sigma_{j=1}^{m} D_j \mathbf{x}_j \mathbf{x}_j')$, $\mathbf{A}^* = [\oplus_{j=1}^{m}(1 - D_j)\mathbf{1}_{n_j}]$ and $\mathbf{B}^* = [\oplus_{j=1}^{m}(1 - D_j)\mathbf{1}_{N_j - n_j}]$.

Suppose now one wants to estimate the vector of small area means $\gamma(\mathbf{Y}) = [\gamma_1(\mathbf{Y}), \ldots, \gamma_m(\mathbf{Y})]'$, where $\gamma_i(\mathbf{Y}) = N_i^{-1}\Sigma_{j=1}^{N_i} Y_{ij}$ $(1 \le i \le m)$. Writing $f_i = (N_i - n_i)/N_i$, and using (2.2) and Theorem 1, it follows that

$$E\left[\gamma_i(\mathbf{Y}) \mid \mathbf{Y}^{(1)} = \mathbf{y}^{(1)}\right] = (1 - f_i D_i)\bar{y}_i + f_i D_i \mathbf{x}_i' \hat{\mathbf{b}}, \tag{2.9}$$

and furthermore

$$V\left[\gamma_i(\mathbf{Y}) \mid \mathbf{Y}^{(1)}\right] = \frac{f_i(1 - D_i)}{N_i \lambda (n_T + g - p - 2)} \left(a + \mathbf{y}^{(1)\prime}\mathbf{C}\mathbf{y}^{(1)}\right) \times \left[N_i f_i + \lambda(1 - D_i)^{-1} + N_i f_i (1 - D_i)\mathbf{x}_i'\mathbf{D}^{*^{-1}}\mathbf{x}_i\right]; \tag{2.10}$$

$$\operatorname{Cov}\left[\gamma_i(\mathbf{Y}), \gamma_k(\mathbf{Y}) | \mathbf{Y}^{(1)} = \mathbf{y}^{(1)}\right] = \frac{\left(a + \mathbf{y}^{(1)\prime}\mathbf{C}\mathbf{y}^{(1)}\right) f_i f_k (1 - D_i)(1 - D_k)}{(n_T + g - p - 2)} \mathbf{x}_i'\mathbf{D}^{*^{-1}}\mathbf{x}_k. \tag{2.11}$$

It should be noted that the estimator given in (2.9) generalizes the estimator given in (2.8) of Ghosh and Lahiri (1987) provided $\mathbf{x}_i$'s and $\mathbf{b}$ are scalars with $x_i = 1$ (i = 1, ..., m). We should also note that in this article, we have juxtaposed the suffixes i and j of Ghosh and Lahiri (1987). The estimator given in (2.8) of Ghosh and Lahiri (1987) was motivated from a classical EB approach. It was noted there though the estimator could also be obtained using a HB approach as well.

The next theorem shows that the estimator given in (2.9) can also be viewed as a BLUP. For proving this result, we do not need to make any distributional assumptions. The proof of the theorem is omitted because the result follows from a more general result of Datta and Ghosh (1989).

**Theorem 2**. Consider the linear model

$$Y_{ij} = \mathbf{x}_i'\mathbf{b} + v_i + e_{ij} (j = 1, \ldots, N_i; i = 1, \ldots, m), \tag{2.12}$$

where $v_i$'s and $e_ij$'s are mutually independent with $v_i$'s iid $\mathrm{N}(0,\ (\lambda r)^{-1})$ and $e_{ij}$'s are iid $\mathrm{N}(0,\ r^{-1})$. Then the BLUP of $\gamma_i(\mathbf{Y}) = N_i^{-1}\Sigma_{j=1}^{N_i} Y_{ij}$ based on linear functions of the form $\Sigma_{i=1}^{m}\Sigma_{j=1}^{n_i}\ell_{ij}Y_{ij}$ is given by the estimator in the right hand side of (2.9).

We may also note, although the details will not be pursued here, that the estimator given in (2.9) can also be obtained using the linear empirical Bayes approach of Ghosh and Lahiri (1987). The estimator given in (2.9) generalizes the estimator given in (2.8) of Ghosh and Lahiri (1987) in that in the present case, the ith small area sample mean is shrunk towards a general regression surface. The connection between linear Bayes procedures and BLUP procedures is thoroughly investigated in Harville (1985, 1987) in the context of general mixed linear models.

As a final remark in this section, one can find the joint distribution of $\boldsymbol{\Theta}$ and $\mathbf{Y}^{(1)}$ from which the posterior distribution of $\boldsymbol{\Theta}$ given $\mathbf{Y}^{(1)} = \mathbf{y}^{(1)}$ can be obtained. The distribution is multivariate t in contrast to the multivariate normal distribution of Stroud (1987).

## 3. THE GENERAL HB MODEL

In this section, we consider the more realistic HB model where the ratio of the variance components need not be known. More specifically, we assume (a) and (b) of the model introduced in Section 2, but the assumption (c) is now replaced by

(c)′ marginally $\mathbf{B}$, R and $\Lambda R$ are independently distributed with $\mathbf{B}$ uniform on $R^p$, while R and $\Lambda R$ are distributed respectively as Gamma $\left(\frac{1}{2}\ a,\ \frac{1}{2}\ g\right)$ and Gamma $\left(\frac{1}{2}\ c,\ \frac{1}{2}\ d\right)$.

**Remark 2.** It seems more natural to us to assume that the two variance components R and $\Lambda R$ are marginally independently distributed, than to assume that R and $\Lambda$ are independently distributed. An assumption similar to ours appears in DuMouchel (1987).

Under the assumptions of this section, the predictive distribution of $\mathbf{Y}^{(2)}$ given $\mathbf{Y}^{(1)} = \mathbf{y}^{(1)}$ is given in the following theorem.

**Theorem 3.** The predictive distribution of $\mathbf{Y}^{(2)}$ given $\Lambda = \lambda$ and $\mathbf{Y}^{(1)} = \mathbf{y}^{(1)}$ is multivariate-t with location parameter $\mathbf{h} = (\mathbf{h}_1', \ldots, \mathbf{h}_m')'$, scale parameter $\mathbf{V}_0$ and d.f. $n_T$ + g + d - p, where $\mathbf{h}$ is defined in (2.3), $\mathbf{V}_0 = (n_T + g + d - p)^{-1}\left[a + c\lambda + \mathbf{y}^{(1)\prime}\mathbf{C}\mathbf{y}^{(1)}\right]\mathbf{G}$, where $\mathbf{C}$ and $\mathbf{G}$

are given respectively in (2.7) and (2.8). The conditional distribution of $\boldsymbol{\Lambda}$ given $\mathbf{Y}^{(1)} = \mathbf{y}^{(1)}$ has pdf

$$f(\lambda \mid \mathbf{y}^{(1)}) \propto \lambda^{\frac{1}{2}(m+d-p)-1} \left\{ \prod_{j=1}^{m} (\lambda + n_j)^{-\frac{1}{2}} \right\} \left| \sum_{j=1}^{m} n_j (n_j + \lambda)^{-1} \mathbf{x}_j \mathbf{x}_j' \right|^{-\frac{1}{2}}$$

$$\times \left( a + c\lambda + \mathbf{y}^{(1)\prime} \mathbf{C} \mathbf{y}^{(1)} \right)^{-\frac{1}{2}(n_T+g+d-p)}. \tag{3.1}$$

It is possible now to obtain the posterior means, variances and covariances of the vector of small area means $(\gamma_1(\mathbf{Y}), \ldots, \gamma_m(\mathbf{Y}))$ using the familiar iterative formulas for conditional expectations, variances and covariances, and (3.1). Such formulas though complicated can be easily implemented in a computer. In fact, later in this section, we shall see some numerical results in a real data example.

It is possible to use improper prior for $Z = \Lambda R$, the precision of the $\Theta_i$'s. One simple way to generate such a prior is to put $c = 0$ in the assumed gamma distribution.

Even in the case of equal sample sizes, that is when $n_1 = \ldots = n_T = n$ (say), our prior is quite different from that of Stroud (1987). Like ours, Stroud uses a uniform (on $R^p$) prior for $\mathbf{B}$, and a gamma prior for R. But, rather than putting a prior (proper or improper) on $\Lambda R$, he uses an improper prior on $\Psi = \left(\frac{1}{n} + \frac{1}{\Lambda}\right)^{-1} = n\Lambda/(n + \Lambda)$. He assumes also that marginally $\mathbf{B}$, $R$ and $\Psi$ are *independently* distributed.

Our model is not comparable to that of Fay and Herriot (1979) or Ericksen and Kadane (1985, 1987). In their framework, observations need not be available at the unit level, that is the individual $Y_{ij}$'s need not be available to the statistician. They set up a linear regression model for the means $n_i^{-1}\Sigma_{j=1}^{n_i} Y_{ij}$ or more specifically write $n_i^{-1}\Sigma_{j=1}^{n_i} Y_{ij} = \mathbf{x}_i'\mathbf{b} + v_i + e_i$, where the $v_i$'s and $e_i$'s are mutually independent with $e_i$'s independent $N(0, \sigma_i^2)$, while the $v_i$'s are iid $N(0, \sigma_v^2)$. They also, assume $\sigma_i^2$'s to be *known* so that $\sigma_v^2$ is the only unknown regression coefficient along with the unknown regression parameter $\mathbf{b}$. Also, these authors use point estimates of $\sigma_v^2$ by using some classical estimation method which makes their procedure very similar to a classical EB procedure. Classical EB procedures of this type are nicely reviewed in Morris (1983). Hill (1980) considers a simple linear regression model, but his prediction problem is different from ours.

Next in this section, we illustrate the use of the results of Section 3 by means of a small data set also considered in Stroud (1987). Stroud considered only Arts and Science students attending the Queen's University from eight municipalities in Canada. The students were asked the following question: "How many trips home do you estimate you will have taken by the end of the academic year?" In his enlarged data set, which Professor Stroud has kindly provided us, there are 15 municipalities (also mentioned in Stroud's paper) and response from students in other disciplines in addition to those in Arts and Science. We treated the 15 municipalities as the 15 small areas, and the total population of students in *all* disciplines as the finite population. As there was no reason to believe the Arts and Science students to be heterogeneous from the remaining students, we treated the Arts and Science students as our samples. In this way, samples of different sizes were generated from the 15 municipalities. The 15 municipalities with their population and sample sizes given within parentheses are (1) Belleville (3,3), (2) Brampton (3,2), (3) Brockville (3,2), (4) Calgary (5,4), (5) London (3,1), (6) Mississauga (4,2), (7) Montreal (4,3), (8) Oakville (3,3), (9) Oshawa (3,1), (10) Ottawa (25,11), (11) Pembroke (3,2), (12) Sault Ste. Marie (14,3), (13) Sudburg (3,1), (14) Toronto (31,17) and Vancouver (5,4). Stroud considered samples of size 3 each (Arts and Science students) from each of the eight selected municipalities, namely Belleville, Calgary, Montreal, Oakville, Ottawa, Sault Ste. Marie, Toronto and Vancouver.

As in Stroud (1987), we fit a simple linear regression. The auxiliary variable is taken as a function of the road distance between the municipality and Kingston (where Queen's University is located). He plotted the log mean number of trips versus log distances for all the 15 municipalities. The plot of 15 points exhibited a reasonably modest scatter about linearity, the least squares value of the slope being -.494. He decided that the reasonable x-variable should be the - $\frac{1}{2}$ power of the road distance of the municipality from Kingston. Professor Stroud kindly supplied us with all the 15 x-values, rather than the eight given in his paper. Following Stroud, we first took $\mathbf{x}_i' = (1, x_i)$, i = 1, ..., 15, that is we fitted a regression line with a slope as well as an intercept. Since infinite distance from Kingston implies zero trips, later we fitted a regression line without the intercept term and noticed the posterior means and the posterior variances to be very similar. In the first case p= 2, while for the second p= 1. For R and Z = $\Lambda R$, several gamma distributions were tried which involved using inverse J and Bell-shaped distributions or any combinations thereof. Improper priors were also used for R and Z. Our findings led essentially to the same values of the predictive means and s.d.'s (upto the second decimal place for a given p). We have decided, therefore, to incorporate only two tables, one for p= 2 and the other for p= 1. In each case we took a= 2, g= 2, c= 0, d= 2. For each of the 15 municipalities, we provide the true population means, the sample averages, the x values, the predictive means and the predictive s.d.'s. It should be noted that the sum of squares of the deviations between the true means and the sample means is 19.8966, while the sum of squares of the deviations between the true means and the HB estimators are given respectively by 12.8967 for p= 2 and 12.9144 for p= 1.

Table 1. True means, sample means, x-values, predictive means ($e_i$), predictive s.d.'s ($s_i$), $e_i \pm 2s_i$'s for the 15 municipalities when p= 2.

| Area (i) | True mean | Sample mean | $x_i$ | $e_i$ | $s_i$ | $e_i - 2s_i$ | $e_i + 2s_i$ |
|---|---|---|---|---|---|---|---|
| 1 | 12.33 | 12.33 | .1125 | 12.33 | 0.00 | 12.33 | 12.33 |
| 2 | 5.00 | 6.00 | .0594 | 5.81 | 1.63 | 2.55 | 9.07 |
| 3 | 10.67 | 13.50 | .1118 | 12.79 | 1.68 | 9.43 | 16.15 |
| 4 | 1.20 | 1.25 | .0171 | 1.12 | 1.00 | -0.88 | 3.12 |
| 5 | 4.00 | 3.00 | .0476 | 3.71 | 2.33 | -0.95 | 8.37 |
| 6 | 3.75 | 2.00 | .0606 | 4.52 | 1.74 | 1.04 | 8.00 |
| 7 | 3.00 | 4.00 | .0579 | 4.31 | 1.22 | 1.87 | 6.75 |
| 8 | 4.33 | 4.33 | .0587 | 4.33 | 0.00 | 4.33 | 4.33 |
| 9 | 5.33 | 6.00 | .0712 | 5.17 | 2.32 | 0.53 | 9.81 |
| 10 | 6.00 | 7.64 | .0774 | 7.54 | 0.84 | 5.86 | 9.22 |
| 11 | 3.00 | 3.50 | .0634 | 4.29 | 1.63 | 1.03 | 7.55 |
| 12 | 3.00 | 3.67 | .0337 | 3.37 | 1.23 | 0.91 | 5.83 |
| 13 | 2.67 | 4.00 | .0413 | 3.57 | 2.34 | -1.11 | 8.25 |
| 14 | 5.90 | 5.59 | .0624 | 5.66 | 0.67 | 3.72 | 7.00 |
| 15 | 1.60 | 1.25 | .0149 | 1.07 | 1.00 | -0.93 | 3.07 |

Table 2. True means, sample means, x-values, predictive means ($e_i$), predictive s.d.'s ($s_i$), $e_i \pm 2s_i$'s for the 15 municipalities when p= 1.

| Area (i) | True mean | Sample mean | $x_i$ | $e_i$ | $s_i$ | $e_i - 2s_i$ | $e_i + 2s_i$ |
|---|---|---|---|---|---|---|---|
| 1 | 12.33 | 12.33 | .1125 | 12.33 | 0.00 | 12.33 | 12.33 |
| 2 | 5.00 | 6.00 | .0594 | 5.88 | 1.62 | 2.64 | 9.12 |
| 3 | 10.67 | 13.50 | .1118 | 12.53 | 1.65 | 9.23 | 15.83 |
| 4 | 1.20 | 1.25 | .0171 | 1.32 | 0.97 | -0.62 | 3.26 |
| 5 | 4.00 | 3.00 | .0476 | 4.01 | 2.29 | -0.57 | 8.59 |
| 6 | 3.75 | 2.00 | .0606 | 4.62 | 1.73 | 1.16 | 8.08 |
| 7 | 3.00 | 4.00 | .0579 | 4.37 | 1.21 | 1.95 | 6.79 |
| 8 | 4.33 | 4.33 | .0587 | 4.33 | 0.00 | 4.33 | 4.33 |
| 9 | 5.33 | 6.00 | .0712 | 5.16 | 2.32 | 0.52 | 9.80 |
| 10 | 6.00 | 7.64 | .0774 | 7.47 | 0.83 | 5.81 | 9.13 |
| 11 | 3.00 | 3.50 | .0634 | 4.34 | 1.62 | 1.10 | 7.58 |
| 12 | 3.00 | 3.67 | .0337 | 3.55 | 1.21 | 1.13 | 5.97 |
| 13 | 2.67 | 4.00 | .0413 | 3.94 | 2.29 | -0.68 | 8.52 |
| 14 | 5.90 | 5.59 | .0624 | 5.73 | 0.64 | 4.45 | 7.01 |
| 15 | 1.60 | 1.25 | .0149 | 1.28 | 0.96 | -0.64 | 3.20 |

## APPENDIX. PROOFS OF THE THEOREMS

**Proof of Theorem 1**. It follows that under the assumed model, conditional on $\mathbf{b}$ and r, $\mathbf{Y}_i$'s are independent $N((\mathbf{x}_i'\mathbf{b})\mathbf{1}_{N_i}, r^{-1}(\mathbf{I}_{N_i} + \lambda^{-1}\mathbf{J}_{N_i}))$, while marginally $\mathbf{B}$ and R are independently distributed with $\mathbf{B} \sim$ uniform $(R^p)$ and R $\sim$ Gamma$(\frac{1}{2}\ a,\ \frac{1}{2}\ g)$. Hence, the joint (improper) pdf of $\mathbf{Y}_1, \ldots, \mathbf{Y}_m, \mathbf{B}$ and R is given by

$$f(\mathbf{y}_1, \ldots, \mathbf{y}_m, \mathbf{b}, r)$$

$$\propto r^{\frac{1}{2}(N_T+g)-1}\lambda^{\frac{1}{2}m}\prod_{j=1}^{m}(N_j+\lambda)^{-\frac{1}{2}}\exp\left[-\frac{1}{2}r\left\{a+\mathbf{u}_j'\mathbf{H}_j\mathbf{u}_j\right\}\right], \qquad (A.1)$$

where

$$\mathbf{u}_j = (\mathbf{y}_j - (\mathbf{x}_j'\mathbf{b})\mathbf{1}_{N_j}), \qquad (A.1.1)$$

and

$$\mathbf{H}_j = (\mathbf{I}_{N_j} - (N_j+\lambda)^{-1}\mathbf{J}_{N_j}). \qquad (A.1.2)$$

It follows after much algebraic simplifications that

$$\Sigma_{j=1}^{m}\mathbf{u}_j'\mathbf{H}_j\mathbf{u}_j = \lambda\left[\left\{\mathbf{b}-\mathbf{Q}^{-1}\mathbf{P}\right\}'\mathbf{Q}\left\{\mathbf{b}-\mathbf{Q}^{-1}\mathbf{P}\right\}\right]$$

$$+\sum_{j=1}^{m}\mathbf{y}_j'(\mathbf{I}_{N_j}-(N_j+\lambda)^{-1}\mathbf{J}_{N_j})\mathbf{y}_j - \lambda\mathbf{P}'\mathbf{Q}^{-1}\mathbf{P} \qquad (A.2)$$

where

$$\mathbf{Q} = (\Sigma_{j=1}^{m}N_j(N_j+\lambda)^{-1}\mathbf{x}_j\mathbf{x}_j'), \qquad (A.2.1)$$

and

$$\mathbf{P} = \sum_{j=1}^{m}(N_j+\lambda)^{-1}\mathbf{x}_j\mathbf{1}_{N_j}'\mathbf{y}_j. \qquad (A.2.2)$$

Integrating out with respect to $\mathbf{b}$ in (A.1) and using (A.2), one gets the joint (improper) pdf of $\mathbf{Y}_1, \ldots, \mathbf{Y}_m$ and R given by

$$f(\mathbf{y}_1, \ldots, \mathbf{y}_m, r) \propto r^{\frac{1}{2}(N_T+g-p)-1}$$

$$\times\exp\left[-\frac{1}{2}r\left\{a+W\right\}\right], \qquad (A.3)$$

where,

$$W = \mathbf{y}^{(1)\prime}\mathbf{D}_{11}\mathbf{y}^{(1)} + \mathbf{y}^{(2)\prime}\mathbf{D}_{22}\mathbf{y}^{(2)} - 2\mathbf{y}^{(2)\prime}\mathbf{D}_{21}\mathbf{y}^{(1)}; \qquad (A.3.1)$$

$$\mathbf{D}_{11} = \oplus_{j=1}^{m}\left[\mathbf{I}_{n_j} - (N_j+\lambda)^{-1}\mathbf{J}_{n_j}\right]$$
$$-\lambda\left[\oplus_{j=1}^{m}(N_j+\lambda)^{-1}\mathbf{1}_{n_j}\right]\mathbf{X}\mathbf{Q}^{-1}\mathbf{X}'\left[\oplus_{j=1}^{m}(N_j+\lambda)^{-1}\mathbf{1}'_{n_j}\right]; \qquad (A.4)$$

$$\mathbf{D}_{22} = \oplus_{j=1}^{m}\left[\mathbf{I}_{N_j-n_j} - (N_j+\lambda)^{-1}\mathbf{J}_{N_j-n_j}\right]$$
$$-\lambda\oplus_{j=1}^{m}\left[(N_j+\lambda)^{-1}\mathbf{1}_{N_j-n_j}\right]\mathbf{X}\mathbf{Q}^{-1}\mathbf{X}'\left[\oplus_{j=1}^{m}(N_j+\lambda)^{-1}\mathbf{1}'_{N_j-n_j}\right]; (A.5)$$

$$\mathbf{D}_{21} = \oplus_{j=1}^{m}\left[(N_j+\lambda)^{-1}\mathbf{J}_{N_j-n_j,n_j}\right]$$
$$+\lambda\left[\oplus_{j=1}^{m}(N_j+\lambda)^{-1}\mathbf{1}_{N_j-n_j}\right]\mathbf{X}\mathbf{Q}^{-1}\mathbf{X}'\left[\oplus_{j=1}^{m}(N_j+\lambda)^{-1}\mathbf{1}'_{N_j-n_j}\right]; (A.6)$$

where $\mathbf{J}_{u,v} = \mathbf{1}_u\mathbf{1}'_v$. Integrating out with respect to r, it follows from (A.3) and (A.3.1) that the joint (improper) pdf of $\mathbf{Y}_1, \ldots, \mathbf{Y}_m$ is

$$f(\mathbf{y}_1, \ldots, \mathbf{y}_m) \propto [a+W]^{-\frac{1}{2}(N_T+g-p)}. \qquad (A.7)$$

We write

$$W = \mathbf{y}^{(1)\prime}\mathbf{D}_{11.2}\mathbf{y}^{(1)}$$
$$+\left(\mathbf{y}^{(2)} - \mathbf{D}_{22}^{-1}\mathbf{D}_{21}\mathbf{y}^{(1)}\right)'\mathbf{D}_{22}\left(\mathbf{y}^{(2)} - \mathbf{D}_{22}^{-1}\mathbf{D}_{21}\mathbf{y}^{(1)}\right), \qquad (A.8)$$

where $\mathbf{D}_{11.2} = \mathbf{D}_{11} - \mathbf{D}_{12}\mathbf{D}_{22}^{-1}\mathbf{D}_{21}$. On an examination of (A.7) and (A.8) it follows that the conditional (predictive) distribution of $\mathbf{Y}^{(2)}$ given $\mathbf{Y}^{(1)} = \mathbf{y}^{(1)}$ is (proper) multivariate t with location parameter $\mathbf{D}_{22}^{-1}\mathbf{D}_{21}\mathbf{y}^{(1)}$, scale parameter $\left\{\left(a+\mathbf{y}^{(1)\prime}\mathbf{D}_{11.2}\mathbf{y}^{(1)}\right)/(n_T+g-p)\right\}\mathbf{D}_{22}^{-1}$ and degrees of freedom $n_T$ + g − p. Thus, we need to find $\mathbf{D}_{22}^{-1}$ and $\mathbf{D}_{11.2}$.

First we write $\mathbf{D}_{22} = \mathbf{E} - \lambda\mathbf{F}\mathbf{Q}^{-1}\mathbf{F}'$, where

$$\mathbf{E} = \oplus_{j=1}^{m}\left[\mathbf{I}_{N_j-n_j} - (N_j+\lambda)^{-1}\mathbf{J}_{N_j-n_j}\right],$$

$$\mathbf{F}' = \mathbf{X}'\left[\oplus_{j=1}^{m}(N_j+\lambda)^{-1}\mathbf{1}'_{N_j-n_j}\right].$$

Next using the familiar matrix inversion formula ( see Exercise 2.10, p 33 of Rao (1973)), one gets

$$\mathbf{D}_{22}^{-1} = \mathbf{E}^{-1} + \mathbf{E}^{-1}\mathbf{F}(-\mathbf{F}'\mathbf{E}^{-1}\mathbf{F} + \lambda^{-1}\mathbf{Q})^{-1}\mathbf{F}'\mathbf{E}^{-1}. \qquad (A.9)$$

It is easy to see ( using Exercise 2.9, p 33 of Rao (1973)) that

$$\mathbf{E}^{-1} = \oplus_{j=1}^{m}\left[\mathbf{I}_{N_j-n_j} + (n_j+\lambda)^{-1}\mathbf{J}_{N_j-n_j}\right]. \qquad (A.10)$$

It follows after much algebraic simplifications that

$$\mathbf{F}'\mathbf{E}^{-1}\mathbf{F} = \Sigma_{j=1}^{m}(N_j-n_j)(n_j+\lambda)^{-1}(N_j+\lambda)^{-1}\mathbf{x}_j\mathbf{x}_j'; \qquad (A.11)$$

$$\lambda^{-1}\mathbf{Q} - \mathbf{F}'\mathbf{E}^{-1}\mathbf{F} = \lambda^{-1}\Sigma_{j=1}^{m}n_j(n_j+\lambda)^{-1}\mathbf{x}_j\mathbf{x}_j' \qquad (A.12)$$

Now using (A.9), (A.10) and (A.12) it follows after much algebraic simplifications that

$$\mathbf{D}_{22}^{-1} = \mathbf{G} \qquad (A.13)$$

Using (A.13), it follows after some heavy algebra that

$$\mathbf{D}_{22}^{-1}\mathbf{D}_{21} = \lambda^{-1}\left\{\oplus_{j=1}^{m}(1-D_j)\mathbf{J}_{N_j-n_j,n_j}\right.$$

$$\left. + \mathbf{B}^*\mathbf{X}\mathbf{D}^{*^{-1}}\mathbf{X}'\mathbf{A}^{*'}\right\} \qquad (A.14)$$

Using (A.14), it follows after some simplifications that

$$\mathbf{D}_{22}^{-1}\mathbf{D}_{21}\mathbf{y}^{(1)} = (\mathbf{h}_1' \ldots \mathbf{h}_m')'; \qquad (A.15)$$

$$\mathbf{D}_{11.2} = \mathbf{D}_{11} - \mathbf{D}_{12}\mathbf{D}_{22}^{-1}\mathbf{D}_{21} = \mathbf{C}. \qquad (A.16)$$

The proof of Theorem 1 is complete.

**Proof of Theorem 3.** First note that since R and $\Lambda R$ are marginally independently distributed with R $\sim$ Gamma($\frac{1}{2}a$, $\frac{1}{2}g$) and $\Lambda$ R $\sim$ Gamma ($\frac{1}{2}c$, $\frac{1}{2}d$), the joint prior pdf of R and $\Lambda$ is given by

$$q(r,\lambda) \propto \exp\left[-\frac{1}{2}r(a+c\lambda)\right] r^{\frac{1}{2}(g+d)-1}\lambda^{\frac{1}{2}d-1}. \qquad (A.17)$$

Now using (A.17) and arguing as in (A.1), it follows that the joint (improper) pdf of $\mathbf{Y}_1, \ldots, \mathbf{Y}_m$, $\mathbf{B}$, R and $\Lambda$ is given by

$$f(\mathbf{y}_1, \ldots, \mathbf{y}_m, \mathbf{b}, r, \lambda) \propto r^{\frac{1}{2}(N_T+g+d)-1}\lambda^{\frac{1}{2}(m+d)-1}\prod_{j=1}^{m}(N_j+\lambda)^{-\frac{1}{2}}$$

$$\times \exp\left[-\frac{1}{2}r\left\{a + c\lambda + \Sigma_{j=1}^{m}\mathbf{u}_j'\mathbf{H}_j\mathbf{u}_j\right\}\right]. \qquad (A.18)$$

Now integrating with respect to **b** in (A.18) and using (A.2), it follows that the joint (improper) pdf of $\mathbf{Y}_1, \ldots, \mathbf{Y}_m$, R and $\Lambda$ is given by

$$f(\mathbf{y}_1, \ldots, \mathbf{y}_m, r, \lambda) \propto r^{\frac{1}{2}(N_T+g+d-p)-1}\lambda^{\frac{1}{2}(m+d-p)-1}\prod_{j=1}^{m}(N_j+\lambda)^{-\frac{1}{2}}$$

$$\times|\mathbf{Q}|^{-\frac{1}{2}}\exp\left[-\frac{1}{2}r\left\{a + c\lambda + W\right\}\right] \qquad (A.19)$$

where one may recall that $\mathbf{D}_{11}$, $\mathbf{D}_{22}$ and $\mathbf{D}_{21}$ all depend on $\lambda$. Now integrating with respect to r in (A.19), it follows that the joint (improper) pdf of $\mathbf{Y}_1, \ldots, \mathbf{Y}_m$ and $\Lambda$ is given by

$$f(\mathbf{y}_1, \ldots, \mathbf{y}_m, \lambda) \propto \lambda^{\frac{1}{2}(m+d+p)-1}\prod_{j=1}^{m}(N_j+\lambda)^{-\frac{1}{2}}|\mathbf{Q}|^{-\frac{1}{2}}$$

$$\times\left[a + c\lambda + W\right]^{-\frac{1}{2}(N_T+g+d-p)}. \qquad (A.20)$$

The first part of Theorem 3 now follows from (A.20) and the definition of a multivariate-t distribution as given in (2.1). Next, integrating out with respect to $\mathbf{y}^{(2)}$ in (A.20), it follows that

$$f\left(\lambda, \mathbf{y}^{(1)}\right) \propto \lambda^{\frac{1}{2}(m+d-p)-1}\prod_{j=1}^{m}(N_j+\lambda)^{-\frac{1}{2}}|\mathbf{Q}|^{-\frac{1}{2}}$$

$$|\mathbf{D}_{22}|^{-\frac{1}{2}}\left(a + c\lambda + \mathbf{y}^{(1)\prime}\mathbf{C}\mathbf{y}^{(1)}\right)^{-\frac{1}{2}(N_T+g+d-p)}. \qquad (A.21)$$

Using the definition of $\mathbf{D}_{22}$ as given in (A.5), and using Exercise 2.4, p. 32 of Rao (1973), it follows after some calculations that

$$|\mathbf{D}_{22}| = \left\{\frac{\left|\sum_{j=1}^{m} n_j(n_j+\lambda)^{-1}\mathbf{x}_j\mathbf{x}_j'\right|}{|\mathbf{Q}|}\right\}\prod_{j=1}^{m}\left\{\frac{(n_j+\lambda)}{(N_j+\lambda)}\right\}. \qquad (A.22)$$

From (A.21) and (A.22), one gets (3.1).

## ACKNOWLEDGMENTS

It is a pleasure to acknowledge our indebtness to Professor T.W.F. Stroud of Queen's University. He supplied us with the full set of data above and beyond what was given in his 1987 paper. Thanks are also due to Ms. Li Chu Lee for doing all the numerical computations. Prof. Ghosh's research was partially supported by the National Science Foundation under Grant #DMS-8701814

## REFERENCES

Battese, G.E., Harter, R.M., and Fuller, W.A. (1988). An error-components model for prediction of county crop areas using survey and satellite data. *J. Amer. Statist. Assoc.*, **83**, 28-36.

Carter, G., and Rolph, J. (1974). Empirical Bayes methods applied to estimating fire alarm probabilities. *J. Amer. Statist. Assoc.*, **69**, 880-885.

Datta, G.S., and Ghosh, M. (1989). Bayesian Prediction in Mixed Linear Models: Applications to Small Area Estimation. Tech. Report No. 332. Dept. of Statistics, University of Florida.

Dumouchel, W. (1987). A Bayesian model and a graphical elicitation procedure for multiple comparisons. In *Bayesian Statistics 3* (J.M. Bernardo, M.H. DeGroot, D.V. Lindley, and A.F.M. Smith eds.) 127-166, Oxford University Press, Oxford, England

Dumouchel, W., and Harris, J. (1983). Bayes methods for combining the results of cancer studies in human and other species. *J. Amer. Statist. Assoc.*, **78**, 297-305.

Ericksen, E.P. (1974). A regression method for estimating population of local areas. *J. Amer. Statist. Assoc.*, **69**, 867-875.

Ericksen, E.P., and Kadane, J.B. (1985). Estimating the population in a census year (with discussion). *J. Amer. Statist. Assoc.*, **80**, 98-131.

Ericksen, E.P., and Kadane, J.B. (1987). Sensitivity analysis of local estimates of undercount in the 1980 census. *Small Area Statistics: An International Symposium.* Eds. Platek, R., Rao, J.N.K., Särndal, C.E., and Singh, M.P.; Wiley, New York, pp 23-45.

Fay, R.E., and Herriot, R.A. (1979). Estimates of income for small places: an application of James-Stein procedures to census data. *J. Amer. Statist. Assoc.*, **74**, 269-277.

Fuller, W.A., and Harter, R.M. (1987). The multivariate components of variance model for small area estimation. *Small Area Statistics: An International Symposium*. Eds. Platek, R., Rao, J.N.K., Särndal, C.E., and Singh, M.P.; Wiley, New York, pp 103-123.

Ghosh, M., and Lahiri, P. (1987). Robust empirical Bayes estimation of means from stratified samples. *J. Amer. Statist. Assoc.*, **82**, 1153-1162.

Ghosh, M., and Meeden, G. (1986). Empirical Bayes estimation in finite population sampling. *J. Amer. Statist. Assoc.*, **81**, 1058-1062.

Harville, D.A. (1977). Maximum likelihood approaches to variance component estimation and to related problems. *J. Amer. Statist. Assoc.*, **72**, 320-340.

Harville, D.A. (1985). Decomposition of prediction error. *J. Amer. Statist. Assoc.*, **80**, 132-138.

Harville, D.A. (1987). BLUP and beyond. Preprint Number 87-4. Statistical Laboratory, ISU.

Harville, D.A. (1988). Mixed-model methodology: theoretical justifications and future directions. Invited paper presented at the 1988 ASA Annual Meeting.

Henderson, C.R. (1975). Best linear unbiased estimation and prediction under a selection model. *Biometrics*, **31**, 423-447.

Hill, B.M. (1980). Robust analysis of the random model and weighted least squares regression. *Evaluation of Econometric Models*. Academic Press, pp. 197- 217.

Jeske, D., and Harville, D.A. (1987). Mean squared error of estimation and prediction under a general linear model. Preprint No. 87-2. Statistical Laboratory, ISU.

Kackar, R.N., and Harville, D.A. (1984). Approximations for standard errors of estimators of fixed and random effects in mixed linear models. *J. Amer. Statist. Assoc.*, **79**, 853-862.

Lindley, D.V., and Smith, A.F.M. (1972). Bayes estimates for linear model (with discussion). *J. R. Statist. Soc.*, B, **34**, 1-41.

Morris, C.N. (1983). Parametric empirical Bayes inference: Theory and applications (with discussion). *J. Amer. Statist. Assoc.*, **78**, 47-65.

Prasad, N.G.N., and Rao, J.N.K. (1987). On the estimation of mean square error of small area estimators. Submitted to the *J. Amer. Statist. Assoc.*

Press, S.J. (1972). *Applied Multivariate Analysis.* Holt, Rinehart and Winston Inc., New York.

Rao, C.R. (1973). *Linear Statistical Inference and its Applications.* 2nd Edition. Wiley, New York.

Stroud, T.W.F. (1987). Bayes and empirical Bayes approaches to small area estimation. *Small Area Statistics: An International Symposium.* Eds. Platek, R., Rao, J.N.K., Särndal, C.E., and Singh, M.P., Wiley, New York, 124-137.

Zellner, A. (1971). *An Introduction to Bayesian Inference in Econometrics.* Wiley, New York.

## COMMENTS

Glen Meeden
Iowa State University

I wish to thank Malay for presenting an interesting application of the Hierarchical Bayes (HB) approach to a finite sampling problem. The authors have contrasted the HB approach to the Empirical Bayes (EB) approach and the mixed linear model approach to a problem in small area estimation. This serves to remind us that finite population sampling theory should be a part of any theory of statistical inference. I will briefly discuss how the work given here is related to other HB models.

Suppose $w_1, w_2, \ldots, w_N$ are $N$ unknown real quantities which we want to know more about. To this end, we plan to conduct a statistical experiment to gather information about these quantities. Before this experiment can be performed, there are four basic questions we need to answer.

The first is whether the $w_i$'s are observable or unobservable? By observable, I mean that we can actually learn the exact value of any of the $w_i$'s for some finite cost. Otherwise they are unobservable. In what follows, they will be denoted by $y_i$'s when they are observable and by $\theta_i$'s when they are unobservable. The second question we must answer is whether we will collect information directly about each $w_i$ or will we just "sample" a subset? Next we must decide on a model which relates the $w_i$'s; for without a model drawing inferences is usually impossible. Finally, we must decide what function or functions of the $w_i$'s we are interested in. Often we are interested in the mean of the $w_i$'s. Other possibilities is a measure of how much variability there is among the $w_i$'s or a good guess for the value of the largest $w_i$.

We now consider three examples where the resulting analysis depends on how we answer these four questions.

**Example 1**. We assume that the $w_i$'s are unobservable and that we can sample all of them. Hence, $\theta_1, \ldots, \theta_N$ are the parameters of interest. A simple model is that for $i = 1, \ldots, N$, $X_i|\theta_i$ is Normal $(\theta_i, \sigma^2)$ and the $\theta_i$'s $|(\mu, \zeta^2)$ are independent identically distributed (i. i. d.) Normal $(\mu, \zeta^2)$. We will assume $\sigma^2$ is known and that we wish to estimate $\theta_1, \ldots, \theta_N$. If we assume that $\mu$ and $\zeta^2$ are unknown parameters, then this is just the classical James-Stein problem which can be given an EB interpretation. If we take a Bayesian approach and let $\mu$ be Normal $(\beta, A)$ when $\beta$ and $A$ are known and $\zeta^2$ has a prior denoted by $\pi(\cdot)$, then this is the HB approach of Lindley and Smith (1972). Berger (1985) gives a nice discussion of this problem for a variety of more complicated models and compares the HB approach to the EB approach.

**Example 2**. We assume that the $w_i$'s are unobservable, but that we can only "sample" a subset. For notational convenience we will assume $\theta_1, \ldots, \theta_N$ are the sampled ones while $\theta_{n+1}, \ldots, \theta_N$ are the unsampled ones. For each fix $i \epsilon \{1, 2, \ldots, n\}$, we assume that given $\theta_i$ we will observe $X_{il}, \ldots, X_{ik}$ which are i.i.d. $N(\theta_i, \sigma^2)$. As in Example 1, for $i = 1, \ldots, N \theta_i|(\mu, \zeta^2)$ are i.i.d. Normal $(\mu, \zeta^2)$. We will assume $\sigma^2/\zeta^2$ is known. We let $\mu$ have the improper uniform distribution on the real line and $1/\sigma^2$ have a Gamma distribution. Suppose now we wish to estimate $\sum_{i=1}^N (\theta_i - \bar{\theta})^2$, a measure of the amount of variation among the parameters of interest. In Ghosh and Meeden (1984), it was argued that the above HB setup is appropriate for what frequentists usually call the Model II or Random Effects Model for a one-way analysis of variance problem. Given the $X_{ij}$'s $i = 1, \ldots, n$ and $j = 1, \ldots, k$, one can find the posterior of all the $\theta_i$'s, sampled and sampled, and from this calculate the posterior expectation of $\sum_{i=1}^N (\theta_i - \bar{\theta})^2$.

**Example 3**. We assume that the $w_i$'s are observable and that we only "sample" a subset, i.e., we are in the usual finite population setup. We will assume that there are $m$ strata or small areas with the $i^{th}$ stratum containing $N_i$ units. We let $y_{ij}$ denote the $j^{th}$ unit in the $i^{th}$ stratum. For a given $i$, a $\lambda_i$ and $\sigma^2$ we assume that $y_{il}, \ldots, y_{iN_i}$ are i.i.d. Normal $(\lambda_i, \sigma^2)$. Furthermore, given $\mu$ and $\zeta^2$ we assume that $\lambda_1, \ldots, \lambda_m$ are i.i.d. $N(\mu, \zeta^2)$. Suppose we wish to estimate the population total. If we assume that $\sigma^2, \mu$ and $\zeta^2$ are unknown parameters, then the above becomes an EB problem. This was done in Ghosh and Meeden (1984).

Point estimation based on an EB approach can often lead to quite sensible estimators. But if used naivély it tends to underestimate the amount of variability or uncertainty associated with such point estimates. For this

# 7

# On Empirical Bayes Selection Rules for Negative Binomial Populations

**Shanti S. Gupta**[1]
**TaChen Liang**[2]

**Abstract**

This paper deals with the problem of selecting good negative binomial populations as compared with a standard or a control. The main results are based on the use of the empirical Bayes approach. First we derive the monotone empirical Bayes estimators of the concerned parameters. Based on these estimators, we construct monotone empirical Bayes selection rules. Asymptotic optimality properties of the monotone empirical Bayes estimators and the monotone empirical Bayes selection rules are investigated. The respective convergence rates for the estimation problem and for the selection problem are studied, under some conditions.

*Key words and phrases*: Asymptotically optimal, Bayes rules, empirical Bayes rules, monotone estimators, monotone selection rules, parametric empirical Bayes, nonparametric empirical Bayes, prior distribution, rate of convergence, antitonic regression.

## 1. INTRODUCTION

The empirical Bayes approach in statistical decision theory is appropriate when one is confronted repeatedly and independently with the same decision problem. In such instances, it is reasonable to formulate the component problem in the sequence as Bayes decision problems with respect to an unknown prior distribution on the parameter space, and then use the accumulated observations to improve the decision rule at each stage. This approach is due to Robbins (1956, 1964). During the last twenty-five years, empirical Bayes methods have been studied extensively. Many such empirical Bayes rules have been shown to be asymptotically optimal in the sense that the risk for the $n$th decision problem converges to the minimum

[1] Purdue University
[2] Wayne State University

Bayes risk which would have been obtained if the prior distribution was known and the Bayes rule with respect to this prior distribution was used.

Empirical Bayes rules have been derived for subset selection goals by Deely (1965). Recently, Gupta and Hsiao (1983) and Gupta and Leu (1989) have studied empirical Bayes rules for selecting good populations with respect to a standard or a control with the underlying distributions being uniformly distributed. Gupta and Liang (1986, 1988) studied empirical Bayes rules for selecting binomial populations better than a standard or a control and for selecting the best among several binomial populations. In the above-mentioned papers, the authors have assumed that the form of the prior distribution is completely unknown. Hence, those approaches are referred to as nonparametric empirical Bayes. Gupta and Liang (1989a, 1989b) have also studied some other empirical Bayes selection rules, in which they assumed that the form of the prior distributions is known but the distributions depend on certain unknown hyperparameters. This approach is therefore referred to as parametric empirical Bayes. For a general formulation of multiple statistical decision procedures using empirical Bayes approach, reference should be made to Gupta and Liang (1987).

In this paper, we are concerned with the problem of selecting good negative binomial populations with respect to a standard or a control through the nonparametric empirical Bayes approach. The framework of the empirical Bayes selection problem is formulated in Section 2. Monotone empirical Bayes selection rules are proposed in Section 3. The monotone empirical Bayes selection rules are derived based on certain monotone empirical Bayes estimators of the posterior means of the concerned parameters which are also derived in Section 3. Asymptotic optimality properties of the monotone empirical Bayes estimators and the monotone empirical Bayes selection rules are studied in Section 4 and Section 5, respectively.

## 2. EMPIRICAL BAYES APPROACH - FORMULATION

For $k+1$ independent negative binomial populations $\pi_i$, $i = 0, 1, \ldots, k$, let $p_i$ denote the probability of success for each independent trial from $\pi_i$ and let $X_i$ denote the number of successes before attaining the $r$th failure. Then, conditional on $p_i, X_i|p_i$ has a negative binomial distribution with probability function $f_i(x|p_i)$ of the form

$$f_i(x|p_i) = \binom{x+r-1}{r-1} p_i^x (1-p_i)^r, \; x = 0, 1, 2, \ldots \tag{1}$$

Let $\pi_0$ be the control population. For each $i = 1, \ldots, k$, population $\pi_i$ is said to be good if $p_i \geq p_0$ and to be bad if $p_i < p_0$, where the control parameter $p_0$ is either known or unknown. Our goal is to derive empirical Bayes rules to select all good populations and exclude all bad populations. When the control parameter $p_0$ is known, the empirical Bayes framework can be formulated as follows:

1. Let $\Omega = \{\underset{\sim}{p} | \underset{\sim}{p} = (p_1, \ldots, p_k),\ p_i \in (0,1) \text{ for } i = 1, \ldots, k\}$ be the parameter space. For each $\underset{\sim}{p} \in \Omega$, define $A(\underset{\sim}{p}) = \{i | p_i \geq p_0\}$ and $B(\underset{\sim}{p}) = \{i | p_i < p_0\}$. That is, $A(\underset{\sim}{p})(B(\underset{\sim}{p}))$ is the set of indices of good (bad) populations. Let $G(\underset{\sim}{p}) = \prod_{i=1}^{k} G_i(p_i)$ be the prior distribution on the parameter space $\Omega$, where $G_i(\cdot)$ are unknown for all $i = 1, \ldots, k$.

2. Let $\mathcal{A} = \{a | a \subset \{1, 2, \ldots, k\}\}$ be the action space. When action $a$ is taken, it means that population $\pi_i$ is selected as good if $i \epsilon a$ and excluded as bad if $i \notin a$.

3. For each (fixed) parameter $\underset{\sim}{p}$ and action $a$, the loss function $L(\underset{\sim}{p}, a)$ is defined as:

$$L(\underset{\sim}{p}, a) = \sum_{i \in A(\underset{\sim}{p}) \backslash a} (p_i - p_0) + \sum_{i \in a \backslash A(\underset{\sim}{p})} (p_0 - p_i), \tag{2}$$

where the first summation is the loss due to not selecting good populations and the second summation is the loss due to selecting bad populations.

4. For each $i$, $i = 1, \ldots, k$, let $(X_{ij},\ P_{ij})$, $j = 1, 2, \ldots, n$, be independent random vectors associated with population $\pi_i$, where $X_{ij}$ is observable but $P_{ij}$ is not. $P_{ij}$ has prior distribution $G_i$. Conditional on $P_{ij} = p_{ij}$, $X_{ij} | p_{ij}$ has a negative binomial distribution with parameters $r$ and $p_{ij}$. Let the $j$th stage observations be denoted by $\underset{\sim}{X}_j$. That is, $\underset{\sim}{X}_j = (X_{1j}, X_{2j}, \ldots, X_{kj})$. From the assumptions, $\underset{\sim}{X}_1, \underset{\sim}{X}_2, \ldots \underset{\sim}{X}_n$, are mutually independent and identically distributed.

5. Let $\underset{\sim}{X} = (X_1, \ldots, X_k)$ denote the present observation. Conditional on $\underset{\sim}{p}$, $\underset{\sim}{X}$ has a joint probability function $f(\underset{\sim}{x} | \underset{\sim}{p}) = \prod_{i=1}^{k} f_i(x_i | p_i)$, where $\underset{\sim}{x} = (x_1, \ldots, x_k)$.
   Finally, since we are interested in Bayes rules, we can restrict our attention to the nonrandomized selection rules.

6. Let $D = \{d | d : \mathcal{X} \to \mathcal{A}, \text{ being measurable}\}$ be the set of nonrandomized selection rules, where $\mathcal{X}$ is the sample space of $\underset{\sim}{X}$. For each $d \in D$, let $r(G, d)$ denote the associated Bayes risk. Then $r(G) \equiv \inf_{d \in D} r(G, d)$ is the minimum Bayes risk and a rule, say $d_G$, is called a Bayes selection rule if $r(G, d_G) = r(G)$.

When $p_0$ is unknown, the indices in the associated notations should begin at 0 instead of at 1. In the sequel, (0) will be used to show this additional fact.

We now consider decision rules $d_n(\underset{\sim}{x}) \equiv d_n(\underset{\sim}{x}; \underset{\sim}{X}_1, \ldots, \underset{\sim}{X}_n)$. Let $r(G, d_n)$ denote the overall Bayes risk associated with the selection rule $d_n(\underset{\sim}{x})$. That is,

$$r(G, d_n) \equiv \sum_{\underset{\sim}{x} \in \mathcal{X}} E \int_{\Omega} L(\underset{\sim}{p}, d_n(\underset{\sim}{x})) f(\underset{\sim}{x}|\underset{\sim}{p}) dG(\underset{\sim}{p}) \tag{3}$$

where the expectation $E$ is taken with respect to $(\underset{\sim}{X}_1, \ldots, \underset{\sim}{X}_n)$.

**Definition 2.1**: *A sequence of selection rules $\{d_n(\underset{\sim}{x})\}_{n=1}^{\infty}$ is said to be asymptotically optimal (a.o.) relative to the prior distribution $G$ if $r(G, d_n) \to r(G)$ as $n \to \infty$.*

For constructing a sequence of a.o. empirical Bayes rules, we first need to find the minimum Bayes risk and the associated Bayes rule $d_G$. From (2), the Bayes risk associated with the selection rule $d$ is:

$$r(G, d) = \begin{cases} \sum\limits_{\underset{\sim}{x} \in \mathcal{X}} \sum\limits_{i \in d(\underset{\sim}{x})} [p_0 - \varphi_i(x_i)] f(\underset{\sim}{x}) + C & \text{if } p_0 \text{ is known,} \\ \sum\limits_{\underset{\sim}{x} \in \mathcal{X}} \sum\limits_{i \in d(\underset{\sim}{x})} [\varphi_0(x_0) - \varphi_i(x_i)] f(\underset{\sim}{x}) + C & \text{if } p_0 \text{ is unknown,} \end{cases} \tag{4}$$

where $f(\underset{\sim}{x}) = \prod_{i=m}^{k} f_i(x_i)$, $m = 0(1)$ if $p_0$ is unknown (known), $f_i(x_i) = \int_0^1 f_i(x_i|p) dG_i(p) = \beta(x_i) h_i(x_i), \beta(x_i) = \binom{x_i+r-1}{r-1}, h_i(x_i) = \int_0^1 p^{x_i}(1-p)^r dG_i(p)$, $\varphi_i(x_i) = \frac{h_i(x_i+1)}{h_i(x_i)}$ (note that $0 < \varphi_i(x_i) < 1$), and
$C = \sum\limits_{\underset{\sim}{x} \in \mathcal{X}} \sum\limits_{i=1}^{k} \int_{\Omega} (p_i - p_0) I_{(p_0,1)}(p_i) f(\underset{\sim}{x}|\underset{\sim}{p}) dG(\underset{\sim}{p})$.

Note that in (4), $C$ is a constant and does not affect the determination of the Bayes rule. Thus, the nonrandomized Bayes selection rule $d_G$ can be obtained as follows:

$$d_G(\underset{\sim}{x}) = \{i | \Delta_{iG}(\underset{\sim}{x}) \leq 0\}, \tag{5}$$

where

$$\Delta_{iG}(\underset{\sim}{x}) = \begin{cases} p_0 - \varphi_i(x_i) & \text{if } p_0 \text{ is known,} \\ \varphi_0(x_0) - \varphi_i(x_i) & \text{if } p_0 \text{ is unknown.} \end{cases} \tag{6}$$

Now, for each $i = (0), 1, \ldots, k$, based on the past data $X_{i1}, \ldots, X_{in}$, and the present observation $x_i$, let $\varphi_{in}(x_i) \equiv \varphi_{in}(x_i; X_{i1}, \ldots, X_{in})$ be an estimator of $\varphi_i(x_i)$, and let

$$\Delta_{in}(\underset{\sim}{x}) = \begin{cases} p_0 - \varphi_{in}(x_i) & \text{if } p_0 \text{ is known,} \\ \varphi_{0n}(x_0) - \varphi_{in}(x_i) & \text{if } p_0 \text{ is unknown.} \end{cases} \tag{7}$$

We then define an empirical Bayes selection rule $d_n(\underset{\sim}{x})$ as follows:

$$d_n(\underset{\sim}{x}) = \{i | \Delta_{in}(\underset{\sim}{x}) \leq 0\}. \tag{8}$$

If $\varphi_{in}(x) \xrightarrow{\text{P}} \varphi_i(x)$ for all $x = 0, 1, 2, \ldots$ and $i = (0), 1, \ldots, k$, then $\Delta_{in}(\underset{\sim}{x}) \xrightarrow{\text{P}} \Delta_{iG}(\underset{\sim}{x})$ for all $\underset{\sim}{x} \in \mathcal{X}$. Therefore, from Corollary 2 of Robbins

(1964), it follows that $r(G, d_n) \to r(G)$ as $n \to \infty$. So, the sequence of empirical Bayes selection rules $\{d_n(\underset{\sim}{x})\}$ defined in (8) is asymptotically optimal. Hence, we have only to find sequences of estimators $\{\varphi_{in}(x)\}$, $i = (0), 1, \ldots, k$, possessing the above mentioned convergence property.

## 3. PROPOSED EMPIRICAL SELECTION RULES

Before we proceed to construct empirical Bayes estimators $\{\varphi_{in}(x)\}$, we describe some properties of the Bayes selection rule $d_G$ defined in (5) and (6).

**Definition 3.1**: *Let $\underset{\sim}{x}, \underset{\sim}{y} \in \mathcal{X}$ such that $x_i \le y_i$ for $i = (0), 1, \ldots, k$.*

*a) When $p_0$ is known, a selection rule $d$ is said to be monotone if $d(\underset{\sim}{x}) \subseteq d(\underset{\sim}{y})$.*

*b) When $p_0$ is unknown, a selection rule $d$ is said to be monotone if the following two conditions are satisfied: (b1) If $x_0 = y_0$, then $d(\underset{\sim}{x}) \subseteq d(\underset{\sim}{y})$. (b2) If $x_0 < y_0$ and $x_i = y_i$ for all $i = 1, \ldots, k$, then $d(\underset{\sim}{x}) \supseteq d(\underset{\sim}{y})$.*

Note that for each $i = (0), 1, \ldots, k$, $\varphi_i(x) = \frac{h_i(x+1)}{h_i(x)}$ is increasing in $x$. Also, note that $\varphi_i(x)$ is the posterior mean of $p_i$ given $X_i = x$, and the Bayes estimator of $p_i$ given $X_i = x$ for squared error loss. By the monotone property of $\varphi_i(x)$, $i = (0), 1, \ldots, k$, the Bayes selection rule $d_G$ is a monotone selection rule for both cases, where the control parameter $p_0$ is either known or unknown.

Under the squared error loss, the problem of estimating the probabiltiy of success in a negative binomial distribution is a monotone estimation problem. From Berger (1985), for a monotone estimation problem, the class of monotone estimators form an essentially complete class. Also, under the linear loss given in (2), the concerned selection problem is a monotone decision problem. Again, from Berger (1985), the class of monotone selection rules form an essentially complete class. Now, one can see that if the estimators $\varphi_{in}(x)$, $i = (0), 1, \ldots, k$, are monotone, then the empirical Bayes selection rule given in (7) and (8) is also monotone. From these considerations, it is reasonable to require that the concerned estimators $\{\varphi_{in}(x)\}$ possess the above-mentioned monotone property.

Proposed Monotone Empirical Bayes Selection Rules

Let $F_i(x)$ denote the marginal distribution function of the random variable $X_i$. Then,

$$F_i(x) = \sum_{y=0}^{x} f_i(y) = \sum_{y=0}^{x} \beta(y) h_i(y), \ x = 0, 1, 2, \ldots . \tag{9}$$

The form of (9) will be used to construct our empirical Bayes estimators.

For each $i = (0), 1, \ldots, k$, and $x = 0, 1, 2, \ldots$ based on the past data $X_{i1}, \ldots, X_{in}$, let

$$f_{in}(x) = \frac{1}{n} \sum_{j=1}^{n} I_{\{x\}}(X_{ij}), \tag{10}$$

$$h_{in}(x) = \frac{f_{in}(x)}{\beta(x)}. \tag{11}$$

In view of the decreasing property of the function $h_i(y)$, it is desirable that the corresponding estimator possess the same property. To meet this requirement, we let $\{h_{in}^*(x)\}_{x=0}^{\infty}$ be the antitonic regression of $\{h_{in}(x)\}_{x=0}^{\infty}$ with weight $\{\beta(x)\}_{x=0}^{\infty}$. Then, let

$$F_{in}^*(x) = \sum_{y=0}^{x} f_{in}^*(y), \tag{12}$$

where

$$f_{in}^*(y) = \beta(y) h_{in}^*(y), y = 0, 1, 2 \ldots. \tag{13}$$

Note that $h_{in}^*(x)$ is nonincreasing in $x$, and if $x > \max(X_{i1}, \ldots, X_{in})$ then, $h_{in}^*(x) = 0$. By the decreasing property of the function $h_i(x)$, from Barlow, et al (1972),

$$F_{in}^*(x) \geq F_{in}(x) \text{ and } \sup_{x \geq 0} |F_{in}^*(x) - F_i(x)| \leq \sup_{x \geq 0} |F_{in}(x) - F_i(x)| \tag{14}$$

where $F_{in}(x) = \frac{1}{n} \sum_{j=1}^{n} I_{[0,x]}(X_{ij}) = \sum_{y=0}^{[x]} \beta(y) h_{in}(y)$.

Let $N_{in} = \max(X_{i1}, \ldots, X_{in}) - 1$. For each $x = 0, 1, 2, \ldots, N_{in}$, define

$$\varphi_{in}(x) = \frac{h_{in}^*(x+1)}{h_{in}^*(x)}. \tag{15}$$

Note that $0 \leq \varphi_{in}(x) \leq 1$ for $x = 0, 1, \ldots, N_{in}$. However, $\varphi_{in}(x)$ does not possess the monotone property. A smoothed version of $\varphi_{in}(x)$ is given as follows: Let

$$\varphi_{in}^*(x) = \begin{cases} \max_{y \leq x} \varphi_{in}(y) & \text{if } x \leq N_{in}, \\ \varphi_{in}^*(N_{in}) & \text{if } x > N_{in}. \end{cases} \tag{16}$$

From (16), it is easy to see that $\varphi_{in}^*(x)$ possesses the monotone property. Now, for each $\underset{\sim}{x} \in \mathcal{X}$, define

$$\Delta_{in}^*(\underset{\sim}{x}) = \begin{cases} p_0 - \varphi_{in}^*(x_i) & \text{if } p_0 \text{ is known,} \\ \varphi_{0n}^*(x_0) - \varphi_{in}^*(x_i) & \text{if } p_0 \text{ is unknown.} \end{cases} \tag{17}$$

We then propose a monotone empirical Bayes selection rule, say $d_n^*$, as follows:

$$d_n^*(\underset{\sim}{x}) = \{i|\Delta_{in}^*(\underset{\sim}{x}) \le 0\}. \tag{18}$$

<u>Asymptotic Optimality of the Selection Rules</u> $\{d_n^*\}$

As mentioned above, to prove the asymptotic optimality of the sequence of empirical Bayes selection rules $\{d_n^*\}$, it suffices to prove that $\varphi_{in}^*(x) \xrightarrow{P} \varphi_i(x)$ for all $x = 0, 1, 2, \ldots$, and $i = (0), 1, \ldots, k$. For each $x = 0, 1, 2, \ldots$, let $t$ be a number such that $0 < t < \min(\varphi_i(x), 1 - \varphi_i(x))$. We need to prove that $P\{|\varphi_{in}^*(x) - \varphi_i(x)| > t\} \to 0$ as $n \to \infty$. Now,

$$\begin{aligned}
& P\{|\varphi_{in}^*(x) - \varphi_i(x)| > t\} \\
= \;& P\{\varphi_{in}^*(x) - \varphi_i(x) < -t, N_{in} \ge x\} + P\{\varphi_{in}^*(x) - \varphi_i(x) > t, N_{in} \ge x\} \\
+ \;& \{|\varphi_{in}^*(x) - \varphi_i(x)| > t, N_{in} < x\}. \qquad (19)
\end{aligned}$$

By the definition of $N_{in}, P\{|\varphi_{in}^*(x) - \varphi_i(x)| > t, N_{in} < x\} \le [F_i(x)]^n$ which tends to 0 as $n \to \infty$.

Note by its definition, $\varphi_{in}^*(x) \ge \varphi_{in}(x)$. Also, note that $h_{in}^*(x) > 0$ as $x \le N_{in} + 1$. Let $p(x,t) = \frac{f_i(x+1)}{f_i(x)} - \frac{t\beta(x+1)}{\beta(x)}$ and $q(x,t) = \frac{t\beta(x+1)f_i(x)}{\beta(x)}$. Then $p(x,t) > 0$ since $t$ is such that $0 < t < \varphi_i(x)$. From (9), (12) – (15), letting $\Delta_i(x,n) = F_{in}^*(x) - F_i(x)$, we have

$$\begin{aligned}
& P\{\varphi_{in}^*(x) - \varphi_i(x) < -t \text{ and } N_{in} \ge x\} \\
& \le P\{h_{in}^*(x+1) - h_{in}^*(x)[\varphi_i(x) - t] < 0\} \\
& = P\{F_{in}^*(x+1) - F_{in}^*(x)[1 + p(x,t)] + F_{in}^*(x-1)p(x,t) < 0\} \qquad (20) \\
= \;& P\{\Delta_i(x+1,n) - \Delta_i(x,n) + \Delta_i(x-1,n)\frac{p(x,t)}{[1+p(x,t)]} < \frac{-q(x,t)}{[1+p(x,t)]}\} \\
& \le 3P\{\sup_{y \ge 0} |F_{in}^*(y) - F_i(y)| \ge \frac{q(x,t)}{3[1+p(x,t)]}\} \\
& \le 3P\{\sup_{y \ge 0} |F_{in}(y) - F_i(y)| \ge \frac{q(x,t)}{3[1+p(x,t)]}\}
\end{aligned}$$

which tends to zero as $n$ tends to infinity, by the Glivenko-Cantelli Theorem. Similarly,

$$\begin{aligned}
& P\{\varphi_{in}^*(x) - \varphi_i(x) > t \text{ and } N_{in} \ge x\} \\
= \;& P\{\varphi_{in}(y) - \varphi_i(x) > t \text{ for some } y \le x \text{ and } N_{in} \ge x\} \\
\le \;& P\{\varphi_{in}(y) - \varphi_i(y) > t \text{ for some } y \le x, f_{in}^*(y) > 0\} \qquad (21) \\
\le \;& P\{f_{in}^*(y+1) - f_{in}^*(y)\frac{\beta(y+1)}{\beta(y)}[\varphi_i(y) + t] > 0 \text{ for some } y \le x\} \\
= \;& P\{[F_{in}^*(y+1) - F_{in}^*(y)]
\end{aligned}$$

$$- [F^*_{in}(y) - F^*_{in}(y-1)]\frac{\beta(y+1)}{\beta(y)}[\varphi_i(y)+t] > 0 \text{ for some } y \le x\}$$

$$\le \quad 3P\{\sup_{y\ge 0} |F^*_{in}(y) - F_i(y)| \ge \min_{y\le x} \frac{|f_i(y+1) - f_i(y)\frac{\beta(y+1)}{\beta(y)}[\varphi_i(y)+t]|}{3[1+\frac{\beta(y+1)}{\beta(y)}[\varphi_i(y)+t]]}\},$$

which tends to zero as $n$ tends to infinity.

Based on the above discussions, we have shown that $\varphi^*_{in}(x) \xrightarrow{P} \varphi_i(x)$ for each $x = 0, 1, \ldots$ and each $i = (0), 1, \ldots, k$. Therefore, the sequence of empirical Bayes selection rules $\{d^*_n\}$ is asymptotically optimal.

## 4. ASYMPTOTIC OPTIMALITY OF THE MONOTONE ESTIMATORS

In this section, we study the asymptotic optimality property of the estimators $\varphi^*_{in}(x)$. It is known that under the squared error loss, $\varphi_i(x)$ is the Bayes estimator of $p_i$ given $X_i = x$. The associated Bayes risk is $R_i(G_i) = E[(P_i - \varphi_i(X_i))^2]$. Let $\psi_i(\cdot)$ be any estimator of $p_i$ with the associated Bayes risk $R_i(G_i, \psi_i)$. Then

$$R_i(G_i, \psi_i) - R_i(G_i) = E[(\psi_i(X_i) - \varphi_i(X_i))^2]. \tag{22}$$

Let $\{\psi_{in}(x; X_{i1}, \ldots, X_{in}) \equiv \psi_{in}(x)\}$ be a sequence of empirical Bayes estimators based on $(x; X_{i1}, \ldots, X_{in})$.

**Definition 4.1**: *a) A sequence of empirical Bayes estimators $\{\psi_{in}\}_{n=1}^{\infty}$ is asymptotically optimal relative to the prior distribution $G_i$ if $R_i(G_i, \psi_{in}) \to R_i(G_i)$ as $n \to \infty$.*

*b) A sequence of empirical Bayes estimators $\{\psi_{in}\}_{n=1}^{\infty}$ is asymptotically optimal of order $\alpha_n$ relative to the prior distribution $G_i$ if $R_i(G_i, \psi_{in}) - R_i(G_i) \le O(\alpha_n)$ as $n \to \infty$, where $\{\alpha_n\}$ is a sequence of positive values such that $\lim_{n\to\infty} \alpha_n = 0$.*

From Section 3, $\varphi^*_{in}(x) \xrightarrow{P} \varphi_i(x)$ for all $x = 0, 1, 2, \ldots$. Thus, $\{\varphi^*_{in}\}_{n=1}^{\infty}$ is asymptotically optimal. However, the usefulness of empirical Bayes estimators in practical applications clearly depends on the convergence rates with which the risks for the successive estimation problems approach the optimal Bayes risk. Hence in the following, we study the convergence rates of the sequence of empirical Bayes estimators $\{\varphi^*_{in}\}$.

For $0 < \varepsilon < 1$, let $A_i(\varepsilon) = \{x | f_i(x) < \varepsilon\}$, $B_i(\varepsilon) = \{x | f_i(x) \ge \varepsilon\}$. Assumption A: A1. There exist $t_i \epsilon (0, 1]$ and a positive constant $c_i$ such that $P(A_i(\varepsilon)) \le c_i \varepsilon^{t_i}$ for all $\varepsilon \in (0, 1)$. A2. There exists a positive integer $N_i$ such that $f_i(x)$ is decreasing in $x$ for $x \ge N_i$.

**Remark 4.1**: *An example where Assumption A holds is given in Lin (1972).*

**Theorem 4.1**: *Let $\{\varphi_{in}^*\}_{n=1}^{\infty}$ be the sequence of empirical Bayes estimators defined in (16). Then, under Assumption A, $R_i(G_i, \varphi_{in}^*) - R_i(G_i) \leq O(n^{-t_i/(2+t_i)})$.*

The proof of Theorem 4.1 can be obtained based on the following arguments.

For the empirical Bayes estimator $\varphi_{in}^*$, straight computation leads to

$$\begin{aligned} 0 &\leq R_i(G_i, \varphi_{in}^*) - R_i(G_i) \\ &= \sum_{x \in A_i(\delta_n)} E[(\varphi_{in}^*(x) - \varphi_i(x))^2 | X_i = x] f_i(x) \\ &\quad + \sum_{x \in B_i(\delta_n)} E[(\varphi_{in}^*(x) - \varphi_i(x))^2 | X_i = x] f_i(x), \end{aligned} \tag{23}$$

where $\delta_n = n^{-\alpha_i}$ and $\alpha_i = \frac{1}{2+t_i}$.

**Lemma 4.1**: *Under Assumption A,*

$$\sum_{x \in A_i(\delta_n)} E[(\varphi_{in}^*(x) - \varphi_i(x))^2 | X_i = x] f_i(x) \leq O(n^{-t_i/(2+t_i)}).$$

**Proof**: Note that $0 \leq \varphi_{in}^*(x),\ \varphi_i(x) \leq 1$. Thus, by the definition of $A_i(\delta_n)$ and Assumption A1,

$$\begin{aligned} &\textstyle\sum_{x \in A_i(\delta_n)} E[(\varphi_{in}^*(x) - \varphi_i(x))^2 | X_i = x] f_i(x) \\ &\leq P(A_i(\delta_n)) \leq c_i n^{-\alpha_i t_i} = O(n^{-t_i/(2+t_i)}). \end{aligned}$$

Now, for each $x \in B_i(\delta_n)$, by letting $I_{in}(x) = 1(0)$ if $f_{in}(x) > (=)0$, we obtain

$$\begin{aligned} &E[(\varphi_{in}^*(x) - \varphi_i(x))^2 | X_i = x] \\ =\ &E[(\varphi_{in}^*(x) - \varphi_i(x))^2 I_{in}(x) | X_i = x] \\ +\ &E[(\varphi_{in}^*(x) - \varphi_i(x))^2 (1 - I_{in}(x)) | X_i = x], \end{aligned} \tag{24}$$

**Lemma 4.2**: *For $x \in B_i(\delta_n)$,*

$$E[(\varphi_{in}^*(x) - \varphi_i(x))^2 (1 - I_{in}(x)) | X_i = x] \leq O(n^{-t_i/(2+t_i)}).$$

**Proof**: Note that

$$\begin{aligned} &E[(\varphi_{in}^*(x) - \varphi_i(x))^2 (1 - I_{in}(x)) | X_i = x] \\ &\leq P\{f_{in}(x) - f_i(x) \leq -f_i(x)\} \\ &\leq \exp\{-2n f_i^2(x)\} \ \text{(by Theorem 1 of Hoeffding (1963))} \\ &\leq \exp\{-2n\delta_n^2\} \ (\text{since } x \in B_i(\delta_n)) \\ &\leq O(n^{-t_i/(2+t_i)}). \end{aligned}$$

Next, a straight computation leads to:

$$E[(\varphi^*_{in}(x)-\varphi_i(x))^2 I_{in}(x)|X_i=x]$$
$$=\int_0^{\varphi_i(x)} 2sP\{\varphi^*_{in}(x)-\varphi_i(x)<-s, f_{in}(x)>0\}ds \tag{25}$$
$$+\int_0^{1-\varphi_i(x)} 2sP\{\varphi^*_{in}(x)-\varphi_i(x)>s, f_{in}(x)>0\}ds.$$

**Lemma 4.3**: *a) For $x\in B_i(\delta_n)$, and $s\in(0,\varphi_i(x))$,*

$$P\{\varphi^*_{in}(x)-\varphi_i(x)<-s \text{ and } f_{in}(x)>0\}\le 3d_i\ exp\{-2n[\frac{q(x,s)}{3(1+p(x,s))}]^2\}$$

*for some positive constant $d_i$, where $p(x,s)=\frac{f_i(x+1)}{f_i(x)}-\frac{s\beta(x+1)}{\beta(x)}>0$ and $q(x,s)=\frac{s\beta(x+1)f_i(x)}{\beta(x)}>0$.*

*b) For $x\in B_i(\delta_n)$, $\int_0^{\varphi_i(x)} 2sP\{\varphi^*_{in}(x)-\varphi_i(x)<-s \text{ and } f_{in}(x)>0\}ds\le O(n^{-t_i/(2+t_i)})$ and the upper bound is independent of $x$ for all $x\in B_i(\delta_n)$.*

**Proof**: a) First, it is trivial that $p(x,s)>0$ and $q(x,s)>0$. Next, from the definitions of $f^*_{in}(x)$ and $h^*_{in}(x)$ and that $f_{in}(x)>0$, it follows that $f^*_{in}(x)>0$ and $h^*_{in}(x)>0$. Also, $\varphi^*_{in}(x)\ge\varphi_{in}(x)$, by the definition of $\varphi^*_{in}(x)$. Then, following (20), we obtain: For $x\in B_i(\delta_n)$ and $s\in(0,\varphi_i(x))$,

$$\begin{aligned}&P\{\varphi^*_{in}(x)-\varphi_i(x)<-s \text{ and } f_{in}(x)>0\}\\ \le\ & 3P\{\sup_{y\ge 0}|F_{in}(y)-F_i(y)|\ge\frac{q(x,s)}{3[1+p(x,s)]}\}\\ \le\ & 3d_i\ \exp\{-2n[\frac{q(x,s)}{3[1+p(x,s)]}]^2\},\end{aligned}$$

where the last inequality follows from Lemma 2.1 of Schuster (1969).

b) By using the fact that $0<\frac{h_i(x+1)}{h_i(x)}<1$ and $\frac{\beta(x+1)}{\beta(x)}\le r$ for all $x\ge 0$, we have $1+p(x,s)\le 1+r$. Then, from the result of part a) of this Lemma, we obtain

$$\begin{aligned}&\int_0^{\varphi_i(x)} 2sP\{\varphi^*_{in}(x)-\varphi_i(x)<-s, f_{in}(x)>0\}ds\\ \le\ & \int_0^{\varphi_i(x)} 6d_i\ s\ \exp\{-\frac{2n}{9}[\frac{sf_i(x)\frac{\beta(x+1)}{\beta(x)}}{1+r}]^2\}ds\\ \le\ & 13.5d_i\frac{(1+r)^2\beta^2(x)}{\beta^2(x+1)}\times\frac{1}{nf_i^2(x)}\\ \le\ & 13.5d_i(1+r)^2\frac{1}{n\delta_n^2}\quad \text{(since } x\in B_i(\delta_n)\text{) and therefore, } f_i(x)\ge\delta_n)\\ =\ & O(n^{-t_i/(2+t_i)}).\end{aligned}$$

Note this upper bound is independent of $x \in B_i(\delta_n)$.

By Assumption A2, $f_i(x)$ is decreasing in $x$ for all $x \geq N_i$. In the following, we only consider the case where $n$ is large enough such that $\delta_n \equiv n^{-\frac{1}{2+t_i}} \leq f_i(y)$ for all $y \leq N_i$. Thus as $x \in B_i(\delta_n)$, then $f_i(y) \geq \delta_n$ for all $y \leq x$ (this holds true for either $x \leq N_i$ or $x > N_i$). Therefore, analogous to (21), we obtain: For $s \in (0, 1-\varphi_i(x))$,

$$\begin{aligned}
&P\{\varphi_{in}^*(x) - \varphi_i(x) > s, f_{in}(x) > 0\} \\
&\leq 3P\{\sup_{y \geq 0} |F_{in}^*(y) - F_i(y)| \geq -\min_{y \leq x} H(y)\} \\
&\leq 3d_i \exp\{-2n[-\min_{y \leq x} H(y)]^2\} \qquad (26) \\
&\quad \text{(by Lemma 2.1 of Schuster (1969)),}
\end{aligned}$$

where

$$H(y) = \frac{-sf_i(y)\frac{\beta(y+1)}{\beta(y)}}{3[1+\frac{\beta(y+1)}{\beta(y)}]} \leq \frac{-sf_i(y)}{3[1+r]} \leq \frac{-s\delta_n}{3[1+r]} < 0. \qquad (27)$$
$$\text{(since } f_i(y) \geq \delta_n\text{)}$$

**Lemma 4.4**: *For $n$ sufficiently large, and $x \in B_i(\delta_n)$,*

$$\int_0^{1-\varphi_i(x)} 2sP\{\varphi_{in}^*(x) - \varphi_i(x) > s, f_{in}(x) > 0\}ds \leq O(n^{-t_i/(2+t_i)}).$$

**Proof**: From (26) and (27), for $n$ sufficiently large, as $x \in B_i(\delta_n)$,

$$\begin{aligned}
&\int_0^{1-\varphi_i(x)} 2sP\{\varphi_{in}^*(x) - \varphi_i(x) > s, f_{in}(x) > 0\}ds \\
&\leq \int_0^{1-\varphi_i(x)} 6sd_i \exp\{-\frac{2n\delta_n^2 s^2}{9(1+r)^2}\}ds \\
&= O(n^{-t_i/(2+t_i)}).
\end{aligned}$$

From Lemmas 4.2, 4.3, 4.4 and (25), we have: For $x \in B_i(\delta_n)$,

$$E[(\varphi_{in}^*(x) - \varphi_i(x))^2 | X_i = x] \leq O(n^{-t_i/(2+t_i)}).$$

This upper bound is independent of $x \in B_i(\delta_n)$. Therefore, we conclude that

$$\sum_{x \in B_i(\delta_n)} E[(\varphi_{in}^*(x) - \varphi_i(x))^2 | X_i = x] f_i(x) \leq O(n^{-t_i/(2+t_i)}). \qquad (28)$$

Then, Lemma 4.1, (23) and (28) together complete the proof of Theorem 4.1.

## 5. ASYMPTOTIC OPTIMALITY OF THE EMPIRICAL BAYES SELECTION RULES

Let $\{d_n\}_{n=1}^{\infty}$ be a sequence of empirical Bayes selection rules relative to the prior distribution $G$. Let $r(G, d_n)$ denote the expected Bayes risk of the selection rule $d_n$. Since the Bayes rule $d_G$ achieves the minimum Bayes risk $r(G)$, $r(G, d_n) - r(G) \geq 0$ for all $n = 1, 2, \ldots$. Thus, the nonnegative difference $r(G, d_n) - r(G)$ is used as a measure of the optimality of the sequence of empirical Bayes selection rules $\{d_n\}$.

**Definition 5.1**: *The sequence of empirical Bayes selection rules* $\{d_n\}_{n=1}^{\infty}$ *is asymptotically optimal of order* $\alpha_n$ *relative to the prior distribution* $G$ *if* $r(G, d_n) - r(G) \leq O(\alpha_n)$ *as* $n \to \infty$, *where* $\{\alpha_n\}$ *is a sequence of positive numbers such that* $\lim_{n \to \infty} \alpha_n = 0$.

In the following, we evaluate the asymptotic behavior of the sequence of empirical Bayes rules $\{d_n^*\}$ according to whether the control parameter $p_0$ is known or unknown.

Convergence Rates of $\{d_n^*\}$ for $p_0$ Known Case For each $i = 1, \ldots, k$, let $S_i = \{x | \varphi_i(x) < p_0\}$ and $T_i = \{x | \varphi_i(x) > p_0\}$. Define

$$M_i = \begin{cases} \min T_i & \text{if } T_i \neq \phi, \\ \infty & \text{if } T_i = \phi, \end{cases} \tag{29}$$

$$m_i = \begin{cases} \max S_i & \text{if } S_i \neq \phi, \\ -1 & \text{if } S_i = \phi. \end{cases} \tag{30}$$

By the increasing property of $\varphi_i(x)$ with respect to the variable $x = 0, 1, \ldots, m_i \leq M_i$; also $m_i < M_i$ if $T_i \neq \phi$. Furthermore, $x \leq m_i$ iff $\varphi_i(x) < p_0$ and $y \geq M_i$ iff $\varphi_i(y) > p_0$.

Note that when $T_i = \phi$, it means that in terms of its quality, the population $\pi_i$ is bad. Also, for $S_i = \phi$, it means that in terms of its quality population $\pi_i$ is as good as the control. We exclude these extreme cases, in the following, and study the asymptotic behavior of the sequence of empirical Bayes selection rules $\{d_n^*\}$ under Assumption B.

Assumption B: $T_i \neq \phi$ and $S_i \neq \phi$ for all $i = 1, \ldots, k$.

Now, for the empirical Bayes selection rule $d_n^*$, we have

$$0 \leq r(G, d_n^*) - r(G) = \sum_{i=1}^{k} D_i(G, d_n^*), \tag{31}$$

where

$$\begin{aligned} D_i(G, d_n^*) = \quad & \sum_{x=0}^{m_i} [p_0 - \varphi_i(x)] P\{\varphi_{in}^*(x) \geq p_0\} f_i(x) \\ + \quad & \sum_{x=M_i}^{\infty} [\varphi_i(x) - p_0] P\{\varphi_{in}^*(x) < p_0\} f_i(x). \end{aligned} \tag{32}$$

By the nondecreasing property of the estimator $\varphi^*_{in}$, we have

$$\begin{cases} P\{\varphi^*_{in}(x) \geq p_0\} \leq P\{\varphi^*_{in}(m_i) \geq p_0\} & \text{for all } x \leq m_i, \text{ and} \\ P\{\varphi^*_{in}(y) < p_0\} \leq P\{\varphi^*_{in}(M_i) < p_0\} & \text{for all } y \geq M_i. \end{cases} \tag{33}$$

Thus, from (32) and (33),

$$D_i(G, d^*_n) \leq P\{\varphi^*_{in}(m_i) \geq p_0\}b_{i1} + P\{\varphi^*_{in}(M_i) < p_0\}b_{i2}, \tag{34}$$

where $b_{i1} = \sum_{x=0}^{m_i} [p_0 - \varphi_i(x)]f_i(x)$, and $b_{i2} = \sum_{x=M_i}^{\infty} [\varphi_i(x) - p_0]f_i(x)$ and $0 \leq b_{i1},\ b_{i2} \leq 1$.

Now,

$$\begin{aligned} &P\{\varphi^*_{in}(M_i) < p_0\} \\ &= P\{\varphi^*_{in}(M_i) < p_0, N_{in} < M_i\} + P\{\varphi^*_{in}(M_i) < p_0, N_{in} \geq M_i\} \\ &\leq [F_i(M_i)]^n + P\{\varphi^*_{in}(M_i) < p_0, N_{in} \geq M_i\} \end{aligned} \tag{35}$$

Analogous to Lemma 4.3. a), we can obtain

$$\begin{aligned} &P\{\varphi^*_{in}(M_i) < p_0 \text{ and } N_{in} \geq M_i\} \\ &\leq P\{h^*_{in}(M_i+1) < h^*_{in}(M_i)p_0 \text{ and } N_{in} \geq M_i\} \\ &\leq 3P\{\sup_{y\geq 0} |F^*_{in}(y) - F_i(y)| > \frac{\Delta(M_i, p_0)}{3[1 + \frac{\beta(M_i+1)}{\beta(M_i)}p_0]}\} \\ &\leq 3d_i \exp\{-\frac{2}{9}n\Delta^2(M_i, p_0)/[1 + \frac{\beta(M_i+1)}{\beta(M_i)}p_0]^2\}, \end{aligned} \tag{36}$$

where $\Delta(M_i, p_0) = f_i(M_i+1) - f_i(M_i)\frac{\beta(M_i+1)}{\beta(M_i)}p_0 > 0$.

From (35) and (36), by letting $\tau_{i1} = \min\left(\frac{2\Delta^2(M_i, p_0)}{9[1+\frac{\beta(M_i+1)}{\beta(M_i)}p_0]^2},\ \ell n \frac{1}{F_i(M_i)}\right)$, we have

$$P\{\varphi^*_{in}(M_i) < p_0\} \leq O(\exp(-\tau_{i1}n)). \tag{37}$$

Also,

$$\begin{aligned} &P\{\varphi^*_{in}(m_i) \geq p_0\} \\ &= P\{\varphi^*_{in}(m_i) \geq p_0, N_{in} < m_i\} + P\{\varphi^*_{in}(m_i) \geq p_0, N_{in} \geq m_i\} \\ &\leq [F_i(m_i)]^n + P\{\varphi^*_{in}(m_i) \geq p_0, N_{in} \geq m_i\} \end{aligned} \tag{38}$$

Analogous to (21), we obtain

$$\begin{aligned} &P\{\varphi^*_{in}(m_i) \geq p_0,\ N_{in} \geq m_i\} \\ &= P\{\varphi_{in}(y) \geq p_0 \text{ for some } y \leq m_i, N_{in} \geq m_i\} \\ &\leq 3P\{\sup_{x\geq 0} |F^*_{in}(x) - F_i(x)| \geq \min_{y\leq m_i} \frac{|\Delta(y, p_0)|}{3[1 + \frac{\beta(y+1)}{\beta(y)}p_0]}\} \end{aligned}$$

$$\leq 3d_i \exp\left\{-\frac{2n}{9}\min_{y\leq m_i}\left[\frac{|\Delta(y,p_0)|}{1+\frac{\beta(y+1)}{\beta(y)}p_0}\right]^2\right\}, \tag{39}$$

where $\Delta(y,p_0) = f_i(y+1) - f_i(y)\frac{\beta(y+1)}{\beta(y)}p_0$. Note that $\Delta(y,p_0) < 0$ for all $y \leq m_i$.

Thus, by letting $\tau_{i2} = \min\left(\frac{2}{9}\min_{y\leq m_i}\left\{\frac{|\Delta(y,p_0)|^2}{[1+\frac{\beta(y+1)}{\beta(y)}p_0]^2}\right\}, \ell n\frac{1}{F_i(m_i)}\right)$, we have

$$P\{\varphi_{in}^*(m_i) \geq p_0\} \leq O(\exp\{-\tau_{i2}n\}). \tag{40}$$

Let $\tau_i = \min(\tau_{i1},\tau_{i2})$, and $\tau = \min(\tau_1,\ldots,\tau_k)$. Note that $\tau > 0$, since $\tau_i > 0$ for each $i = 1,\ldots,k$. From the above results the following theorem follows:

**Theorem 5.1**: *Under Assumption B, we have:*

a) $D_i(G,d_n^*) \leq O(\exp(-\tau_i n))$ for each $i = 1\ldots k$, and

b) $r(G,d_n^*) - r(G) \leq O(\exp(-\tau n))$.

Convergence Rates of $\{d_n^*\}$ for $p_0$ Unknown Case

When the parameter $p_0$ is unknown, the convergence rates of the sequence of empirical Bayes selection rules $\{d_n^*\}_{n=1}^\infty$ is evaluated under Assumption A. Without loss of generality, in this section, we assume that $c_i = c > 0$, and $t_i = t \in (0,1]$ for all $i = 0,1,\ldots,k$, where the parameters $c_i,\ t_i,\ i = 0,1\ldots k$ are given in Assumption A1.

For each $i = 1,\ldots,k$, let $\psi(x_o,x_i) = \varphi_i(x_i) - \varphi_0(x_0)$ and let $S_i = \{(x_0,x_i)|\psi(x_0,x_i) < 0\}$, $T_i = \{(x_0,x_i)|\psi(x_0,x_i) \geq 0\}$. In addition, let $E_{in} = \{(x_0,x_i)||\psi(x_0,x_i)| \leq \varepsilon_n\}$ and $E_{in}^c = \{(x_0,x_i)||\psi(x_0,x_i)| > \varepsilon_n\}$. For $\varepsilon_n > 0$, let let $I_1 = T_i \cap E_{in}$, $I_2 = T_i \cap E_{in}^c \cap A_0(\delta_n) \cap A_i(\delta_n)$, $I_3 = T_i \cap E_{in}^c \cap A_0(\delta_n) \cap B_i(\delta_n)$, $I_4 = T_i \cap E_{in}^c \cap B_0(\delta_n) \cap A_i(\delta_n)$, $I_5 = T_i \cap E_{in}^c \cap B_0(\delta_n) \cap B_i(\delta_n)$, $J_1 = S_i \cap E_{in}$, $J_2 = S_i \cap E_{in}^c \cap A_0(\delta_n) \cap A_i(\delta_n)$, $J_3 = S_i \cap E_{in}^c \cap A_0(\delta_n) \cap B_i(\delta_n)$, $J_4 = S_i \cap E_{in}^c \cap B_0(\delta_n) \cap A_i(\delta_n)$ and $J_5 = S_i \cap E_{in}^c \cap B_0(\delta_n) \cap B_i(\delta_n)$. Furthermore let

$$\begin{cases} Q_i(x_0,x_i) = [\varphi_i(x_i) - \varphi_0(x_0)]P\{\varphi_{in}^*(x_i) < \varphi_{0n}^*(x_0)\}f_i(x_i)f_0(x_0) \\ R_i(x_0,x_i) = [\varphi_0(x_0) - \varphi_i(x_i)]P\{\varphi_{in}^*(x_i) \geq \varphi_{0n}^*(x_0)\}f_i(x_i)f_0(x_0), \end{cases}$$

$Q_i(x_0,x_i,I_j) = \sum_{I_j} Q_i(x_0,x_i)$ and $R_i(x_0,x_i,J_j) = \sum_{J_j} R_i(x_0,x_i)$. Then

$$0 \leq r(G,d_n^*) - r(G) = \sum_{i=1}^k D_i^*(G,d_n^*) \tag{41}$$

where $D_i^*(G,d_n^*) = \sum_{j=1}^5 Q_i(x_0,x_i,I_j) + \sum_{j=1}^5 R_i(x_0,x_i,J_j)$.

Careful examination leads to the following results: $Q_i(x_0, x_i, I_1) \leq O(\varepsilon_n)$, $R_i(x_0, x_i, J_1) \leq O(\varepsilon_n)$, $Q_i(x_0, x_i, I_2) \leq O(\delta_n^{2t})$, $R_i(x_0, x_i, J_2) \leq O(\delta_n^{2t})$, $Q_i(x_0, x_i, I_3) \leq O(\delta_n^t)$, $R_i(x_0, x_i, I_3) \leq O(\delta_n^t)$, $Q_i(x_0, x_i, I_4) \leq O(\delta_n^t)$ and $R_i(x_0, x_i, J_4) \leq O(\delta_n^t)$.

Now, for $(x_0, x_i) \in T_i \cap E_{in}^c \cap B_0(\delta_n) \cap B_i(\delta_n)$, $f_i(x_i) \geq \delta_n$, $f_0(x_0) \geq \delta_n$, and $\varepsilon_n < \varphi_i(x_i) - \varphi_0(x_0) < 1$. Thus,

$$
\begin{aligned}
&P\{\varphi_{in}^*(x_i) < \varphi_{0n}^*(x_0)\} \\
&= P\{[\varphi_{in}^*(x_i) - \varphi_i(x_i)] - [\varphi_{0n}^*(x_0) - \varphi_0(x_0)] < \varphi_0(x_0) - \varphi_i(x_i)\} \\
&\leq P\{\varphi_{in}^*(x_i) - \varphi_i(x_i) < -\frac{\varepsilon_n}{2}\} + P\{\varphi_{0n}^*(x_0) - \varphi_0(x_0) > \frac{\varepsilon_n}{2}\}. \qquad (42)
\end{aligned}
$$

Now,

$$
\begin{aligned}
&P\{\varphi_{in}^*(x_i) - \varphi_i(x_i) < -\frac{\varepsilon_n}{2}\} \\
&= P\{\varphi_{in}^*(x_i) - \varphi_i(x_i) < -\frac{\varepsilon_n}{2}, f_{in}(x_i) = 0\} \\
&+ P\{\varphi_{in}^*(x_i) - \varphi_i(x_i) < -\frac{\varepsilon_n}{2}, f_{in}(x_i) > 0\},
\end{aligned}
$$

where

$$
P\{\varphi_{in}^*(x_i) - \varphi_i(x_i) < -\tfrac{\varepsilon_n}{2}, f_{in}(x_i) = 0\} \leq P\{f_{in}(x_i) = 0\} \leq \exp\{-2n\delta_n^2\}
$$
( since $x_i \in B_i(\delta_n)$; and see Lemma 4.2),

and

$$
\begin{aligned}
&P\{\varphi_{in}^*(x_i) - \varphi_i(x_i) < -\frac{\varepsilon_n}{2}, f_{in}(x_i) > 0\} \\
&\leq 3d_i \exp\{-\frac{\beta^2(x_i+1)}{18(1+r)^2\beta^2(x_i)} n\varepsilon_n^2\delta_n^2\} \\
&\leq O(\exp\{-\frac{n\varepsilon_n^2\delta_n^2}{18(1+r)^2}\}).
\end{aligned}
$$

(The proof is analogous to that of Lemma 4.3.a). Therefore,

$$
P\{\varphi_{in}^*(x_i) - \varphi_i(x_i) < -\frac{\varepsilon_n}{2}\} \leq O(\exp\{-\frac{n\varepsilon_n^2\delta_n^2}{18(1+r)^2}\}). \qquad (43)
$$

Next, under Assumption A2, along the line of (26) and the argument given there, for $n$ sufficiently large, we have

$$
P\{\varphi_{0n}^*(x_0) - \varphi_0(x_0) > \frac{\varepsilon_n}{2}\} \leq O(\exp\{-\frac{n\varepsilon_n^2\delta_n^2}{18(1+r)^2}\}). \qquad (44)
$$

Note that the convergence rates obtained at (43) and (44) are independent of $(x_0, x_i) \in T_i \cap E_{in}^c \cap B_0(\delta_n) \cap B_i(\delta_n)$. Therefore, from (42)

$$
\sum_{T_i \cap E_{in}^c \cap B_0(\delta_n) \cap B_i(\delta_n)} Q_i(x_0, x_i) \leq O(\exp\{-\frac{n\varepsilon_n^2\delta_n^2}{18(1+r)^2}\}). \qquad (45)
$$

Similarly, we can also conclude that

$$\sum_{S_i \cap E_{in}^c \cap B_0(\delta_n) \cap B_i(\delta_n)} R_i(x_0, x_i) \leq O(\exp\{-\frac{n\varepsilon_n^2 \delta_n^2}{18(1+r)^2}\}). \tag{46}$$

By letting $\delta_n = \left[\frac{18(1+r)^2 \ell n\ n}{(2+2t)n}\right]^{\frac{1}{2+2t}}$ and $\varepsilon_n = \delta_n^t$, based on the preceding discussions we have the following theorem.

**Theorem 5.2**: *Under Assumption A, we have*

a) $D_i^*(G, d_n^*) \leq O(\delta_n^t)$ for $i = 1, \ldots, k$, and

b) $r(G, d_n^*) - r(G) \leq O(\delta_n^t)$.

## ACKNOWLEDGMENT

This research was supported in part by the Office of Naval Research Contract N00014-88-K-0170 and NSF Grants DMS-8606964, DMS-8702620, DMS-8717799 at Purdue University. Reproduction in whole or in part is permitted for any purpose of the United States Government.

## REFERENCES

[1] Barlow, R.E., Bartholomew, D.J., Bremner, J.M. and Brunk, H.D. (1972). *Statistical Inference under Order Restrictions.* Wiley, New York.

[2] Berger, J.O. (1985). *Statistical Decision Theory and Bayesian Analysis.* Springer-Verlag, New York.

[3] Deely, J.J. (1965). Multiple decision procedures from an empirical Bayes approach. Ph.D. Thesis (Mimeo. Ser. No. 45), Department of Statistics, Purdue University, West Lafayette, Indiana.

[4] Gupta, S.S. and Hsiao, P. (1983). Empirical Bayes rules for selecting good populations. *J. Statist. Plann. Infer.* **8**, 87–101.

[5] Gupta, S.S. and Leu, L.Y. (1989). On Bayes and empirical Bayes rules for selecting good populations. To appear in the *Proceedings of the 1990 International Symposium on Statistics*, held in Taipei, June 28–30, 1990.

[6] Gupta, S.S. and Liang, T. (1986). Empirical Bayes rules for selecting good binomial populations. *Adaptive Statistical Procedures and Related Topics* (Ed. J. Van Ryzin), IMS Lecture Notes-Monograph Series, **Vol. 8**, 110–128.

[7] Gupta, S.S. and Liang, T. (1987). On some Bayes and empirical Bayes selection procedures. *Probability and Bayesian Statistics* (Ed. R. Viertl), Plenum Press, New York, 233–246.

[8] Gupta, S.S. and Liang, T. (1988). Empirical Bayes rules for selecting the best binomial population. *Statistical Decision Theory and Related Topics-IV* (Eds. S. S. Gupta and J. O. Berger), Springer-Verlag, **Vol. I**, 213–224.

[9] Gupta, S.S. and Liang, T. (1989a). Selecting the best binomial population: parametric empirical Bayes approach. *J. Statist. Plann. Infer.* **23**, 21–31.

[10] Gupta, S.S. and Liang, T. (1989b). Parametric empirical Bayes rules for selecting the most probable multinomial event. (Chapter 18) in *Contributions to Probability and Statistics: Essays in Honor of Ingram Olkin*, Springer-Verlag.

[11] Hoeffding, W. (1963). Probability inequalities for sums of bounded random variables. *J. Amer. Statist. Assoc.* **58**, 13–30.

[12] Lin, P-E. (1972). Rates of convergence in empirical Bayes estimation problems: discrete case. *Ann. Inst. Statist. Math.* **23**, 319–325.

[13] Robbins, H. (1956). An empirical Bayes approach to statistics. *Proc. Third Berkeley Symp. Math. Statist. Probab.*, **1**, University of California Press, 157–163.

[14] Robbins, H. (1964). The empirical Bayes approach to statistical decision problems. *Ann. Math. Statist.* **35**, 1–20.

[15] Schuster, E. F. (1969). Estimation of a probability density function and its derivatives. *Ann. Math. Statist.* **40**, 1187–1195.

## COMMENTS

Saul Blumenthal
Ohio State University

The development of smoothed, monotone empirical Bayes selection procedures is a very nice idea, and one that is carefully and correctly executed by the authors. My initial comments will be aimed at emphasizing a few simple points which may aid the reader.

It is easily verified that the loss function (2) can be rewritten as

$$\sum_A (p_i - p_0) - \sum_a (p_i - p_0).$$

Since the sum over $A$ is not in the statistician's control, the goal of the procedure is quite obviously to choose $a$ to maximize the second sum, i.e. include exactly those populations for which $p_i > p_0$. This formulation of the loss appears implicitly in (4).

In (6); $\varphi_i(x)$ is $E(p_i|x)$, which is mentioned on p. 130 of the paper, but in conjunction with (6) and the comment above, it is apparent that the Bayes rule simply selects those populations which have $E(p_i|x) > p_0$.

In many empirical Bayes formulations, several component problems are tied together by the model that the parameters in each component are generated from the same distribution. That is not the case here. The components, or $k$ populations, are kept statistically isolated by the assumption that the $p$ for the $i$th population is generated by a distribution $G_i$, which is in no way related to any of the other $G$'s. Thus there can be no borrowing of information about the parameter of the $i$th population using data from the other populations, and we have $k$ distinct, unrelated estimation problems. What is related is the sequence of observations from any one population. In fact, the model is essentially the classical i.i.d. model for any one population, except that each observation is now a pair $(X, P)$ and the second component is not observable.

My final comments are in the spirit of suggestions for possible future investigations which have come to mind in reading this interesting paper. Although the flavor of the paper is Bayesian and the emphasis is on evaluation of Bayes risks, one might ask about the traditional frequentist properties of the proposed procedure, i.e. what are the chances of including a "good" population (one with $p_i > p_0$), or a "bad" one (one with $p_i < p_0$) and what is the expected size of the selected subset? Also, how would the procedure change with a more traditional loss function, one which merely counts how many good populations are included?

The problem considered in this paper in its broadest sense is choosing from $k$ populations, from which one can observe independent Bernoulli random variables, those which have a success probability exceeding the standard. Looked at in the broader sense, the problem has some interesting design aspects. If one is using negative binomial sampling, is there an optimal choice of $r$, the number of successes at stopping (with some appropriate sampling cost function)? If using ordinary binomial sampling, what is the best choice of sample size? Is negative binomial sampling really better than ordinary binomial sampling?

In their Introduction, the authors state that the empirical Bayes approach "is appropriate when one is confronted repeatedly and independently with the same decision problem." This presumably means that the statistician is dealing with the same experimenter on a continuing basis and could exert some influence over the type of experiment performed, so that the questions raised above should be of general relevance. If the statistician has no influence on the choice of experiment, then the present paper makes a major contribution to answering the question of how to handle the available data.

## REJOINDER

We thank Professor S. Blumenthal for a stimulating discussion of our paper. In the paper, it is assumed that the $k$ prior distributions $G_1, \ldots, G_k$ are completely unknown. It is, of course, possible to allow that (i) $G_1, \ldots, G_k$ are identically distributed, or (ii) for each $i$, $G_i$ depends on some unknown parameter $\lambda_i$, where $\lambda_i$ is a realization of a random variable $\Lambda_i$, $i = 1, \ldots, k$, and $\Lambda_1, \ldots, \Lambda_k$ are exchangeable. Under each of the above assumptions, the $k$ populations share a certain similarity. Thus, one can incorporate information from different populations for making a decision for each of the $k$ component decision problems. In a recent paper of the authors (Gupta and Liang (1989)), some research work has been done along this direction.

Now we address the points raised in the last three paragraphs of Dr. Blumenthal's discussion. In the empirical Bayes approach considerd in this paper, we first represent the Bayes procedure in terms of the unknown priors. Then, we use the data to estimate the Bayes procedure directly. Another possible approach is: One may estimate the prior distributions based on the past observations, and then proceed to a typical Bayesian analysis with respect to the estimated priors. With some appropriate sampling cost function, one can make an "optimal" choice of the number $r$ with respect to the estimated prior(s). One can also compare the relative performance of the two sampling methods, namely, the binmomial sampling and the negative binomial sampling, relative to the estimated prior(s). Also the "posterior" probabilities that a good population is not selected and/or a bad population is included in the selected subset, can similarly, be computed. Of course, whether the performance of the procedure is good or not may heavily depend on the consistency of the estimated priors. Further investigation is needed along these lines.

## ADDITIONAL REFERENCES

[1] Gupta, S. S. and Liang, T. (1989). On the asymptotic optimality of certain empirical Bayes simultaneous testing procedures. Technical Report #89-22C, Department of Statistics, Purdue University.

# 8

# Empirical Hierarchical Bayes Estimation


## C. G. Khatri & C. Radhakrishna Rao[1]



### Abstract

It is well known that the James-Stein estimates of mean values of several populations can be derived as empirical Bayes estimates assuming a common prior distribution for all the mean values. But the superiority of such estimates over the usual unbiased estimates diminishes as the variability of the true mean values between populations increases. In such cases it is suggested that the populations may be split into two or more homogeneous groups and the James-Stein procedure applied to the mean values in each group separately. In this paper, we introduce a hierarchical prior distribution by considering the mean values within a group to have a common prior with some hyperparameters which are different from group to group. The hyperparameters in different groups are themselves considered to have a common prior distribution possibly with hyper-hyperparameters. Under some conditions on variabilities between and within groups, it is shown that the empirical Bayes estimates derived from a two stage prior distribution on the mean values are better than those obtained by applying the James-Stein procedure to all the mean values in one step or to the mean values in individual groups separately.



*Keywords and Phrases:* Empirical Bayes; Hierarchical Bayes; James-Stein Estimation; Mean dispersion error; Bayes mean dispersion error; Compound Mean Squared error.


## 1. INTRODUCTION

Consider a situation where we have independent samples drawn from a number of populations which can be grouped into a smaller number of clusters such that the populations within a cluster are more homogeneous than those between. The observed data, unknown parameters and notations used are as shown in Table 1.


[1] Penn State University

Table 1. Parameters and samples for populations in the $i$-th cluster.

| Populations in $i$-th cluster | Unknown Parameters | Sample |
|---|---|---|
| 1 | $\mu_{i1}, \xi$ | $X_{i1}$ |
| 2 | $\mu_{i2}, \xi$ | $X_{i2}$ |
| $\vdots$ | $\vdots$ | $\vdots$ |
| $k_i$ | $\mu_{ik_i}, \xi$ | $X_{ik_i}$ |

The parameters and observations can be vectors. Some of the parameters like $\xi$ may be common to all the populations in all the clusters. The object is to estimate all the parameters

$$\xi, \mu_{ij}, \quad j = 1, \ldots, k_i, \quad i = 1, \ldots, p \tag{1.1}$$

related to the populations in all the $p$ clusters.

The traditional method is to consider the likelihood of all the parameters

$$\prod_{i=1}^{p} \prod_{j=1}^{k_i} p(X_{ij} \mid \mu_{ij}, \xi) \tag{1.2}$$

where $p(X_{ij} \mid \mu_{ij}, \xi)$ is the density function in the $(i, j)$-th population given the parameter values $\mu_{ij}$ and $\xi$, and maximize it with respect to all the unknown parameters (1.1). When the number of unknown parameters is large, the maximum likelihood estimates so obtained may not have some desirable optimum properties. It is found that in simultaneous estimation, there is some advantage in considering at least the varying parameters such as the $\mu_{ij}$'s as having a prior distribution and obtaining their Bayes estimates. Such Bayes estimates may involve the $\xi$ and other parameters needed to specify the prior distribution of the $\mu_{ij}$'s. If they are unknown, we have two alternatives. One is the strict Bayes approach: We choose appropriate priors for these unknowns and obtain the posterior distribution of the $\mu_{ij}$'s depending only on the $X_{ij}$'s and derive the Bayes estimates. Another is the empirical Bayes approach: We estimate the unknowns from given data and substitute these estimates for the unknowns in the original Bayes estimates.

A typical example of empirical Bayes estimates is the James-Stein estimates of the means of several normal populations assuming a common prior for all the unknown mean values as shown by Efron and Morris (1972, 1973a) and Rao (1953, 1975, 1976, 1977).

The advantage of the James-Stein estimates is lost if the parameters under estimation have a large variation. In such a case it may be profitable to consider some natural classification of the parameters into two or more groups and apply the James-Stein procedure separately for the parameters in each group. Efron and Morris (1973b) and Rao (1975) raised this problem and suggested some procedures. In this paper, we suggest a new approach by introducing a hierarchical prior distribution for all the parameters and obtaining empirical Bayes estimates. A natural choice for the hierarchical prior is as follows:

(i) $\mu_{i1}, \ldots, \mu_{ik_i}$ (i.e., the parameters of interest in the $i$-th cluster) are iid with a common probability density $p(\bullet \mid \lambda_i, \eta)$, $\quad i = 1, \ldots, p$ depending on a varying parameter $\lambda_i$ and common parameter $\eta$.

(ii) The cluster parameters $\lambda_1, \ldots, \lambda_p$ are iid with a common probability density $p(\bullet \mid \kappa)$ depending on a parameters $\kappa$.

We consider two cases, one when the parameters $\xi, \eta$ and $\kappa$ are known, and another when some or all parameters are unknown but the unknowns are estimable from given data.

We consider two types of risk functions as defined below. Let $\boldsymbol{\theta} = (\theta_1, \ldots, \theta_m)'$ be a vector of parameters estimated by $\hat{\boldsymbol{\theta}} = (\hat{\theta}_1, \ldots, \hat{\theta}_m)'$. One is the mean dispersion error (MDE) in estimation

$$\text{MDE } (\hat{\boldsymbol{\theta}}) = E[(\hat{\boldsymbol{\theta}} - \boldsymbol{\theta})(\hat{\boldsymbol{\theta}} - \boldsymbol{\theta})' \mid \boldsymbol{\theta}] \tag{1.3}$$

which is a matrix function of $\boldsymbol{\theta}$. Another is the compound mean square error (CMSE)

$$\text{CMSE } (\hat{\boldsymbol{\theta}}) = \text{ trace MDE } (\hat{\boldsymbol{\theta}}) = \sum_{i=1}^{m} E[(\hat{\theta}_i - \theta_i)^2 \mid \boldsymbol{\theta}] \tag{1.4}$$

which is a scalar function of $\boldsymbol{\theta}$.

We define the Bayes MDE to be

$$\text{BMDE } (\hat{\boldsymbol{\theta}}) = E[\text{ MDE } (\hat{\boldsymbol{\theta}})] \tag{1.5}$$

where the expectation is taken over a specified prior distribution of $\boldsymbol{\theta}$. Similarly we define

$$\text{BCMSE } (\hat{\boldsymbol{\theta}}) = E[\text{ CMSE } (\hat{\boldsymbol{\theta}})] \tag{1.6}$$

with respect to a specified prior for $\boldsymbol{\theta}$. We compare alternative estimates in terms of the risk functions, MDE, CMSE and BMDE, BCMSE.

C. G. Khatri, my co-author, unfortunately died while this paper was under preparation. He and I have collaborated in research work over the last 20 years, and he was indeed a delightful and remarkable person to work with. His death is a personal loss to me.

## 2. JAMES-STEIN ESTIMATION

First, we review the problem of James-Stein as a case of empirical Bayes estimation using a common prior for all the parameters. We derive some results useful in judging the relative efficiency of the James-Stein estimates compared to the alternative estimates proposed in Section 3 of the paper.

Let us suppose that we have independent unbiased estimates $x_1, \ldots, x_k$ of parameters $\mu_1, \ldots, \mu_k$ such that

$$x_i \sim N_1(\mu_i, \sigma_0^2), \quad i = 1, \ldots, k \tag{2.1}$$

and a statistic $S_0$ independent of $x_i$ such that

$$S_0 \sim \sigma_0^2 \chi^2(f). \tag{2.2}$$

[In the case we have independent samples of the same size $m$ from each of $k$ normal populations with mean values $\mu_1, \ldots \mu_k$ and common $\sigma^2$, and if $\bar{X}_i$ and $S_i$ are the mean and corrected sum of squares for the $i$-th sample, then $x_i = \bar{X}_i, \sigma_0^2 = \sigma^2/m$, $S_0 = m^{-1}(S_1 + \ldots + S_k)$ and $f = k(m-1)$.] Let us assume that $\mu_1, \ldots, \mu_k$ are iid with a common distribution, $N_1(\lambda, \sigma_1^2)$. Then, denoting

$$\begin{aligned} \rho_1 &= \sigma_1^2/(\sigma_0^2 + \sigma_1^2), \quad \mathbf{1} = (1, \ldots, 1)' \\ \boldsymbol{\mu} &= (\mu_1, \ldots, \mu_k)', \quad \boldsymbol{x} = (x_1, \ldots, x_k)' \\ \bar{x} &= (x_1 + \ldots + x_k)/k, \quad S_1 = (x_1 - \bar{x})^2 + \ldots + (x_k - \bar{x})^2 \end{aligned}$$

the Bayes estimate of $\boldsymbol{\mu}$ which minimizes the dispersion error is

$$\boldsymbol{\mu}^{(b)} = E(\boldsymbol{\mu} \mid \boldsymbol{x}) = \boldsymbol{x} - (1 - \rho_1)(\boldsymbol{x} - \lambda \mathbf{1}) \tag{2.3}$$

when $\rho_1$ and $\lambda$ are known. If $\lambda$ is unknown, substituting $\bar{x}$ for $\lambda$, we have a partial empirical Bayes estimate

$$\boldsymbol{\mu}^{(e_1)} = \boldsymbol{x} - (1 - \rho_1)(\boldsymbol{x} - \bar{x}\mathbf{1}). \tag{2.4}$$

If both $\lambda$ and $\rho_1$ are unknown then we have the full empirical Bayes estimate

$$\boldsymbol{\mu}^{(e)} = \boldsymbol{x} - c\frac{S_0}{S_1}(\boldsymbol{x} - \bar{x}\mathbf{1}) \tag{2.5}$$

where $c = (k-3)/(f+2)$, a suitably chosen constant.

The following computations for the estimators, $\boldsymbol{x}, \boldsymbol{\mu}^{(b)}, \boldsymbol{\mu}^{(e_1)}, \boldsymbol{\mu}^{(e)}$, of $\boldsymbol{\mu}$ are well known (see for example Rao (1976)).

Table 1. ($\boldsymbol{\mu}$ random)

| Estimator | $\sigma_0^{-2}$( BMDE ) |
|---|---|
| $\boldsymbol{x}$ | $I$ |
| $\boldsymbol{\mu}^{(b)}$ | $I - (1 - \rho_1)I$ |
| $\boldsymbol{\mu}^{(e_1)}$ | $I - (1 - \rho_1)I + (1 - \rho_1)V_1$ |
| $\boldsymbol{\mu}^{(e)}$ | $I - (1 - \rho_1)I + (1 - \rho_1)V_1 + (1 - \rho_1)(1 - \frac{k-3}{k-1}\frac{f}{f+2})W_1$ |

where $I$ is the identity matrix of order $k$ and

$$V_1 = k^{-1}\mathbf{1}\ \mathbf{1}', \quad W_1 = I - V_1.$$

It is seen from the expressions in Table 1 that in terms of the BMDE

$$\boldsymbol{\mu}^{(b)} >> \boldsymbol{\mu}^{(e_1)} >> \mu^{(e)} >> \boldsymbol{x} \tag{2.6}$$

where the symbol >> is used for greater or equal efficiency. The same inequalities hold in terms of the BCMSE.

The computations in Table 2, for the case when $\boldsymbol{\mu}$ is a fixed effect, are based on the results given in Rao and Shinozaki (1978). In this case, there are no definite inequality relationships like (2.6) for the MDE criterion. But with respect to the CMSE, $\boldsymbol{\mu}^{(e)} >> \boldsymbol{x}$, which is known as the Stein phenomenon. It is interesting to note that $\boldsymbol{\mu}^{(e)} >> \boldsymbol{\mu}^{(e_1)}$ unless $\delta$ and $\rho_1$ are very small so that the actual knowledge of $\rho_1$, when applicable, may not necessarily increase the efficiency of the estimates in terms of the CMSE.

Table 2 ($\boldsymbol{\mu}$ fixed)

| Esti- | Criterion | |
|---|---|---|
| mator | $\sigma_0^{-2}$( MDE ) | $\sigma_0^{-2}$( CMSE ) |
| $\boldsymbol{x}$ | $I$ | $k$ |
| $\boldsymbol{\mu}^{(b)}$ | $I - \rho^\star(I - \Delta_*)$ | $k - \rho^\star(k - \delta_*)$ |
| $\boldsymbol{\mu}^{(e_1)}$ | $I - \rho^\star(I - V_1) + (1-\rho_1)^2\Delta$ | $k - \rho^\star(k-1) + (1-\rho_1)^2\delta$ |
| $\boldsymbol{\mu}^{(e)}$ | $I - \frac{f(k-3)^2}{f+2}\frac{\Delta}{\delta}d(k)$ $-\frac{f(k-3)}{f+2}(I - V_1 - \frac{(k-1)\Delta}{\delta})c(k)$ | $k - \frac{f(k-3)^2}{f+2}d(k)$ |

In Table 2, we have the following notations:

$$\begin{aligned}
\Delta_* &= \sigma_0^{-2}(\boldsymbol{\mu} - \lambda\mathbf{1})(\boldsymbol{\mu} - \lambda\mathbf{1})', \quad \Delta = \sigma_0^{-2}(\boldsymbol{\mu} - \bar{\mu}\mathbf{1})(\boldsymbol{\mu} - \bar{\mu}\mathbf{1})' \\
\delta_* &= \sigma_0^{-2}(\boldsymbol{\mu} - \lambda\mathbf{1})'(\boldsymbol{\mu} - \lambda\mathbf{1}), \quad \delta = \sigma_0^{-2}(\boldsymbol{\mu} - \bar{\mu}\mathbf{1})'(\boldsymbol{\mu} - \bar{\mu}\mathbf{1}) \\
\rho^\star &= (1-\rho_1^2), K \sim P(\delta/2), \text{ i.e., a Poisson with parameter } \delta/2, \\
c(k) &= E\frac{k-3+4K}{(k-1+2K)(k-3+2K)}, \text{ and } d(k) = E(k-3+2K)^{-1}
\end{aligned}$$

It is seen from Tables 1 and 2, that the gain in efficiency per parameter of the James-Stein estimates $\boldsymbol{\mu}^{(e)}$ over the unbiased estimates $\boldsymbol{x}$, in terms of $\sigma_0^{-2}$ (BCMSE) is

$$\frac{(k-3)f}{k(f+2)}(1-\rho_1) = \frac{(k-3)f}{k(f+2)}\frac{1}{1+(\sigma_1/\sigma_0)^2}, \tag{2.7}$$

and, in terms of $\sigma_0^{-2}$ (CMSE), it is

$$\frac{(k-3)^2 f}{k(f+2)}E\frac{1}{k-3+2K} \approx \frac{(k-3)^2 f}{k(f+2)}\frac{1}{k-3+\delta}, \tag{2.8}$$

the approximation in (2.8) being on the lower side. If $(\sigma_1/\sigma_0)^2$ or $\delta$ is large, i.e., when there is a large variation in the parameters to be estimated, there is not much gain. This raises the following question.

Suppose that there are $2k$ parameters, $\mu_1, \ldots, \mu_{2k}$, to be estimated and we split them into two sets $\mu_1, \ldots, \mu_k$ and $\mu_{k+1}, \ldots, \mu_{2k}$ and estimate them by applying the James-Stein procedure on each set separately. Do we have better efficiency per parameter than by applying the James-Stein procedure on the entire set?

Let $\bar{\mu}_1 = (\mu_1 + \ldots + \mu_k)/k$, $\bar{\mu}_2 = (\mu_{k+1} + \ldots + \mu_{2k})/k$ and

$$\sigma_0^{-2} \sum_1^k (\mu_i - \bar{\mu}_1)^2 = \delta_1, \quad \sigma_0^{-2} \sum_{k+1}^{2k} (\mu_i - \bar{\mu}_2)^2 = \delta_2$$

$$2^{-1}\sigma_0^{-2} k(\bar{\mu}_1 - \bar{\mu}_2)^2 = \delta_{12}, \quad \delta = \delta_1 + \delta_2 + \delta_{12}.$$

Then $\sigma_0^{-2}$( CMSE ) per parameter by considering the two sets separately is

$$1 - \frac{f(k-3)^2}{2(f+2)k}[E(k-3+2K_1)^{-1} + E(k-3+2K_2)^{-1}] \tag{2.9}$$

where $K_i \sim P(\delta_i/2)$, $i = 1, 2$. An upper bound to (2.9) is

$$1 - \frac{f(k-3)^2}{2k(f+2)} \left( \frac{1}{k-5+\delta_1} + \frac{1}{k-5+\delta_2} \right). \tag{2.10}$$

The $\sigma_0^{-2}$ (CMSE) per parameter by estimating all the $2k$ parameters together is

$$1 - \frac{f(2k-3)^2}{2k(f+2)} E(2k-3+2K)^{-1}$$

where $K \sim P(\delta/2)$. A lower bound to this is

$$1 - \frac{f(2k-3)^2}{2k(f+2)} \frac{1}{2k-5+\delta}. \tag{2.11}$$

A sufficient condition for the estimates obtained by splitting to be more efficient is that (2.10) $\leq$ (2.11), i.e.,

$$(k-3)^2[(k-5+\delta_1)^{-1} + (k-5+\delta_2)^{-1}] \geq (2k-3)^2(2k-3+\delta)^{-1}. \tag{2.12}$$

If $\delta_1 \approx \delta_2$, the condition (2.12) is approximately the same as

$$2(k-3)^2(k-5+\delta_1)^{-1} \geq (2k-3)^2(2k-3+2\delta_1+\delta_{12})^{-1}$$

or

$$2k-3+2\delta_1+\delta_{12} \geq \frac{(2k-3)^2}{2(k-3)^2}(k-5+\delta_1) \tag{2.13}$$

which holds when $\delta_{12}$ is large compared to $\delta_1$.

In the next section, we obtain some alternative estimators which seem to have better properties than those considered above after splitting.

## 3. EMPIRICAL HIERARCHICAL BAYES METHOD

Let us suppose that the parameters under estimation can be grouped in a natural way into clusters and denote the parameters in the $i$- th cluster by

$$\boldsymbol{\mu}_i = (\mu_{i1}, \ldots, \mu_{ik_i}), \quad i = 1, \ldots, p. \tag{3.1}$$

Further let

$$\boldsymbol{x}_i = (x_{i1}, \ldots, x_{ik_i}), \quad i = 1, \ldots, p \tag{3.2}$$

be a $k_i$-variate normal statistic such that

$$\boldsymbol{x}_i \sim N_{k_i}(\boldsymbol{\mu}_i, \sigma_0^2 C_i^{-1}), \tag{3.3}$$

and $S_0$ be an independent statistic such that

$$S_0 \sim \sigma_0^2 \chi^2(f). \tag{3.4}$$

The problem we consider is that of estimating simultaneously all the parameters $\boldsymbol{\mu}_i$, $i = 1, \ldots, p$, given $\boldsymbol{x}_i$, $i = 1, \ldots, p$, and $S_0$.
Note. In the special case when we have an independent sample of size $n_{ij}$ from the population with mean value $\mu_{ij}$, we can replace the sample by its mean, $\bar{x}_{ij}$ and sum of squares of the deviations from the mean, $S_{ij}$ and choose

$$x_{ij} = \bar{x}_{ij}, \quad S_0 = \sum_1^p \sum_1^{k_i} S_{ij}, \quad f = \sum_1^p \sum_1^{k_i} (n_{ij} - 1) \tag{3.5}$$

and $C_i$ as a diagonal matrix with diagonal entries $n_{ij}, \ldots, n_{ik_i}$.

A typical example is where $\mu_{ij}$ represents a characteristic of the $j$-th town in the $i$-th county, for which an unbiased estimator $x_{ij}$ is available. In such a case, the towns within a county may be considered as forming a cluster and the different counties as different clusters.

We provide a solution to the problem of simultaneous estimation of the parameters (3.1) through an empirical Bayes procedure under a hierarchical type of prior for the parameters instead of a common prior as discussed in Section 2 in connection with the James-Stein problem. The following hierarchical prior is a natural one in situations we consider:

(i) $\mu_{i1}, \ldots, \mu_{ik_i}$ are iid random variables having a common prior distribution

$$N_1(\lambda_i, \sigma_1^2), \quad i = 1, \ldots, p. \tag{3.6}$$

(ii) $\lambda_1, \ldots, \lambda_p$ are iid random variables having a common prior distribution

$$N_1(\kappa, \sigma_2^2). \tag{3.7}$$

Using the observations (3.2) and (3.4) and the model (3.6) and (3.7) for the prior distribution of the parameters, we estimate the parameters $\mu_{ij}$ by

(i) applying the J-S procedure separately on the parameters in each cluster,

(ii) considering the parameters in all clusters together and applying the J-S procedure in a single step, and

(iii) the method described in Section 3.1 based on the hierarchical priors (3.6) and (3.7),

and compare their relative efficiencies.

### 3.1 *A general solution under hierarchical priors*

Under the assumptions (3.6) and (3.7) on the parameters, the joint distribution of $(\boldsymbol{x}_i', \boldsymbol{\mu}_i')$ is $2k_i$-variate normal with the mean vector and dispersion matrix as

$$\begin{pmatrix} \kappa\mathbf{1}_{(i)} \\ \kappa\mathbf{1}_{(i)} \end{pmatrix} \text{ and } \begin{pmatrix} \sigma_0^2 C_{(i)}^{-1} + \sigma_1^2 I_{(i)} + \sigma_2^2 U_{(i)} & \sigma_1^2 I_{(i)} + \sigma_2^2 U_{(i)} \\ \sigma_1^2 I_{(i)} + \sigma_2^2 U_{(i)} & \sigma_1^2 I_{(i)} + \sigma_2^2 U_{(i)} \end{pmatrix} \tag{3.1.1}$$

where

$$\begin{aligned} \mathbf{1}_{(i)}' &= (1, \ldots, 1), \text{ a } k_i \times 1 \text{ vector of ones,} \\ U_{(i)} &= \mathbf{1}_{(i)}\mathbf{1}_{(i)}', \text{ a } k_i \times k_i \text{ matrix of ones.} \end{aligned}$$

If $\kappa, \sigma_0^2, \sigma_1^2$ and $\sigma_2^2$ are known, then the Bayes estimator of $\boldsymbol{\mu}_i$ which minimizes the BMDE under the hierarchical prior (3.6) and (3.7) is

$$\begin{aligned} \boldsymbol{\mu}_{iH}^{(b)} &= E(\boldsymbol{\mu}_i \mid \boldsymbol{x}_i, \kappa, \sigma_0^2, \sigma_1^2, \sigma_2^2) \\ &= \kappa\mathbf{1}_{(i)} + E_i B_i^{-1}(\boldsymbol{x}_i - \kappa\mathbf{1}_{(i)}) \\ &= \boldsymbol{x}_i - \sigma_0^2 C_i^{-1} B_i^{-1}(x_i - \kappa\mathbf{1}_{(i)}) \end{aligned} \tag{3.1.2}$$

where

$$E_i = \sigma_1^2 I_{(i)} + \sigma_2^2 U_{(i)}, \quad B_1 = E_i + \sigma_0^2 C_i^{-1}.$$

The BMDE of (3.1.2) is

$$\begin{aligned} &E(\boldsymbol{\mu}_{iH}^{(b)} - \boldsymbol{\mu}_i)(\boldsymbol{\mu}_{iH}^{(b)} - \boldsymbol{\mu}_i)' \\ &= \sigma_0^2 C_i^{-1} - \sigma_0^2 C_i^{-1} B_i^{-1} C_i^{-1} \sigma_0^2 \end{aligned} \tag{3.1.3}$$

$$\begin{aligned} &= \sigma_0^2 C_i^{-1} - \sigma_0^2 C_i^{-1} E_i^{-1} C_i^{-1} \sigma_0^2 \\ &\quad + \sigma_2^2 \frac{\sigma_0^2 C_i^{-1} E_i^{-1} U_{(i)} E_i^{-1} C_i^{-1} \sigma_0^2}{1 + \sigma_2^2 \mathbf{1}_{(i)}' E_i^{-1} \mathbf{1}_{(i)}} \end{aligned} \tag{3.1.4}$$

If $\lambda_i$, the mean of the common prior distribution of $\mu_{ij}, \ j = 1, \ldots, k_i$, is known, then the Bayes estimate of $\boldsymbol{\mu}_i$ is

$$\boldsymbol{\mu}_{iG}^{(b)} = E(\boldsymbol{\mu}_i \mid \boldsymbol{x}_i, \lambda_i, \sigma_0^2, \sigma_1^2) = \boldsymbol{x}_i - \sigma_0^2 C_i^{-1} E_i^{-1}(\boldsymbol{x}_i - \lambda_i \mathbf{1}_{(i)}) \tag{3.1.5}$$

with the BMDE

$$\sigma_0^2 C_i^{-1} - \sigma_0^2 C_i^{-1} E_i^{-1} C_i^{-1} \sigma_0^2. \tag{3.1.6}$$

Comparing (3.1.6) with (3.1.4), it is seen that lack of knowledge of $\lambda_i$ increases the BMDE by the extra term in (3.1.4).

The estimators (3.1.2) and (3.1.5) involve the parameters $\lambda_i, \kappa, \sigma_0^2, \sigma_1^2, \sigma_2^2$ associated with the prior distributions. If they are not known, then we can obtain the empirical versions of (3.1.2) and (3.1.5) by substituting for the unknowns suitable estimates. The following estimates are suggested:

$$\begin{cases} \hat{\lambda}_1 &= \mathbf{1}_{(i)}' C_i \boldsymbol{x}_i / \mathbf{1}_{(i)}' C_i \mathbf{1}_{(i)} \\ \hat{\kappa} &= \sum_{i=1}^p \mathbf{1}_{(i)}' C_i \boldsymbol{x}_i / \sum_{i=1}^p \mathbf{1}_{(i)}' C_i \mathbf{1}_{(i)} \end{cases} \tag{3.1.7}$$

$$S_0^2 = f \hat{\sigma}_0^2 \tag{3.1.8}$$

$$\begin{aligned} \sum_{i=1}^p \quad & (\boldsymbol{x}_i - \hat{\lambda}_i \mathbf{1}_{(i)})' C_i (\boldsymbol{x}_i - \hat{\lambda}_i \mathbf{1}_{(i)}) \\ & = \hat{\sigma}_0^2 \sum_{i=1}^p (k_i - 1) + \hat{\sigma}_1^2 \sum_{i=1}^p \left( \text{tr } C_i - \frac{\mathbf{1}_{(i)}' C_i^2 \mathbf{1}_{(i)}}{\mathbf{1}_{(i)}' C_{(i)} \mathbf{1}_{(i)}} \right) \end{aligned} \tag{3.1.9}$$

$$\begin{aligned} \sum_{i=1}^p \quad & \left( \frac{(\mathbf{1}_{(i)}' C_i \boldsymbol{x}_i)^2}{\mathbf{1}_{(i)}' C_i \mathbf{1}_{(i)}} - \frac{(\Sigma \mathbf{1}_{(i)}' C_i \mathbf{1}_i)^2}{\Sigma \mathbf{1}_{(i)}' C_i \mathbf{1}_{(i)}} \right) \\ & = \hat{\sigma}_0^2 (p-1) + \hat{\sigma}_1^2 \left( \Sigma \frac{\mathbf{1}_{(i)}' C_i^2 \mathbf{1}_{(i)}}{\mathbf{1}_{(i)}' C_i \mathbf{1}_{(i)}} - \frac{\Sigma \mathbf{1}_{(i)}' C_i^2 \mathbf{1}_{(i)}}{\Sigma \mathbf{1}_{(i)}' C_i \mathbf{1}_{(i)}} \right) \\ & \quad + \hat{\sigma}_2^2 \left( \Sigma \mathbf{1}_{(i)}' C_i \mathbf{1}_{(i)} - \frac{\Sigma (\mathbf{1}_{(i)}' C_i \mathbf{1}_{(i)})^2}{\Sigma \mathbf{1}_{(i)}' C_i \mathbf{1}_{(i)}} \right) \end{aligned} \tag{3.1.10}$$

The estimates $\hat{\sigma}_0^2, \hat{\sigma}_1^2$ and $\hat{\sigma}_2^2$ are obtained by solving the equations (3.1.8) - (3.1.10).

In order to compare the relative efficiencies of various empirical estimators it is necessary to compute their MDE and CMSE which is somewhat complicated when $C_i$'s and $k_i$'s are not equal. In the subsequent sections we shall compute the MDE and CMSE in each case assuming that $C_i = cI$ or simply $I$ (without loss of generality) for all $i$, and $k_i = k$ for all $i$ and make the necessary comparisons. This will throw some light on the properties of the estimators in the general case.

### 3.2 *Balanced Case:* $\sigma_0^2, \sigma_1^2, \sigma_2^2$ known, $\kappa$ unknown

Let $k_i = k$ and $C_i = I$ for all $i$. If $\kappa$ is known then the Bayes estimator of $\boldsymbol{\mu}_i$ under the hierarchical prior (3.6) and (3.7) simplifies to

$$\boldsymbol{\mu}_{iH}^{(b)} = \boldsymbol{x}_i - \sigma_0^2 B^{-1} (\boldsymbol{x}_i - \kappa \mathbf{1}) \tag{3.2.1}$$

where

$$\mathbf{1} = (1, \ldots, 1)', \quad B = (\sigma_0^2 I + \sigma_1^2 I + \sigma_2^2 U), \quad U = \mathbf{1}\mathbf{1}'.$$

An empirical version of (3.2.1) when $\kappa$ is unknown is

$$\boldsymbol{\mu}_{iH}^{(e_1)} = \boldsymbol{x}_i - \sigma_0^2 B^{-1}(\boldsymbol{x}_i - \bar{x}_{\bullet\bullet}\mathbf{1}) \tag{3.2.2}$$

where $\bar{x}_{\bullet\bullet} = p^{-1}(\bar{x}_{1\bullet} + \ldots + \bar{x}_{p\bullet})$, $\bar{x}_{i\bullet} = k^{-1}(x_{i1} + \ldots + x_{ik})$. Using the identity (see Rao (1973), p. 33)

$$\sigma_0^2 B^{-1} = (1-\rho_1)I - \frac{\rho_2(1-\rho_1)}{1+k\rho_2}U \tag{3.2.3}$$

the expression (3.2.2) can be written as

$$\boldsymbol{\mu}_{iH}^{(e_1)} = \boldsymbol{x}_i - (1-\rho_1)(\boldsymbol{x}_i - \bar{x}_{i\bullet}\mathbf{1}) - \frac{1-\rho_1}{1+k\rho_2}(\bar{x}_{i\bullet} - \bar{x}_{\bullet\bullet})\mathbf{1}. \tag{3.2.4}$$

It may be noted that the first two terms on the right hand side of (3.2.4) are

$$\boldsymbol{\mu}_{iG}^{(e_1)} = \boldsymbol{x}_i - (1-\rho_1)(\boldsymbol{x}_i - \bar{x}_{i\bullet}\mathbf{1}) \tag{3.2.5}$$

the empirical version of the Bayes estimator based on the observations $\boldsymbol{x}_i$ of the $i$-th group only. The last term in (3.2.4) is the correction to (3.2.5) based on the information (3.6) and (3.7) on hierarchical priors.

Writing

$$\boldsymbol{\mu}_H^{(e_1)} = \begin{pmatrix} \boldsymbol{\mu}_{1H}^{(e_1)} \\ \vdots \\ \boldsymbol{\mu}_{pH}^{(e_1)} \end{pmatrix}, \boldsymbol{\mu}_G^{(e_1)} = \begin{pmatrix} \boldsymbol{\mu}_{1G}^{(e_1)} \\ \vdots \\ \boldsymbol{\mu}_{pG}^{(e_1)} \end{pmatrix}, \boldsymbol{\mu} = \begin{pmatrix} \boldsymbol{\mu}_1 \\ \vdots \\ \boldsymbol{\mu}_p \end{pmatrix} \tag{3.2.6}$$

which represent the estimates and all the parameters we find by routine computation, we have

$$\begin{aligned} &\sigma_0^{-2}\ \mathrm{BMDE}(\boldsymbol{\mu}_H^{(e_1)}) = \sigma_0^{-2} E(\boldsymbol{\mu}_H^{(e_1)} - \boldsymbol{\mu})(\boldsymbol{\mu}_H^{(e_1)} - \boldsymbol{\mu})' \\ &= \rho_1 I_{pk} + \tfrac{k\rho_2(1-\rho_1)}{1+k\rho_2}(I_p - V_2)\otimes V_1 + (1-\rho_1)V_2 \otimes V_1 \end{aligned} \tag{3.2.7}$$

where $V_1 = k^{-1}(\mathbf{1}_k\mathbf{1}_k')$, $V_2 = p^{-1}(\mathbf{1}_p\mathbf{1}_p')$ so that $V_2 \otimes V_1 = (pk)^{-1}(\mathbf{1}_{pk}\mathbf{1}_{pk}')$. Also,

$$\begin{aligned} &\sigma_0^{-2}\ \mathrm{BMDE}(\boldsymbol{\mu}_G^{(e_1)}) = \sigma_0^{-2} E(\boldsymbol{\mu}_G^{(e_1)} - \boldsymbol{\mu})(\boldsymbol{\mu}_G^{(e_1)} - \boldsymbol{\mu})' \\ &= \rho_1 I_{pk} + (1-\rho_1)(I_p - V_2)\otimes V_1 + (1-\rho_1)V_2 \otimes V_1 \end{aligned} \tag{3.2.8}$$

A comparison of (3.2.7) and (3.2.8) shows that $\boldsymbol{\mu}_H^{(e_1)} >> \boldsymbol{\mu}_G^{(e_1)}$ with respect to BMDE indicating that the introduction of the hierarchical prior is useful. In terms of the BMDE, the efficiency relationships are

$$\begin{gathered}\boldsymbol{\mu}_G^{(b)} >> \boldsymbol{\mu}_H^{(b)} >> \boldsymbol{x},\\ \boldsymbol{\mu}_G^{(b)} >> \boldsymbol{\mu}_G^{(e_1)} >> \boldsymbol{x}, \boldsymbol{\mu}_H^{(b)} >> \boldsymbol{\mu}_H^{(e_1)} >> \boldsymbol{x},\\ \boldsymbol{\mu}_H^{(e_1)} >> \boldsymbol{\mu}_G^{(e_1)} >> \boldsymbol{x}.\end{gathered} \tag{3.2.9}$$

### 3.3 *Balanced case: all parameters unknown*

In Section 3.1, we have considered empirical estimators (3.2.4) and (3.2.5) when the variances are known and the prior means $\kappa$ and $\lambda_i$ are unknown. In order to obtain fully empirical estimators, we have to obtain suitable estimators of $\sigma_0^2, \sigma_1^2$ and $\sigma_2^2$ and substitute them for the unknowns in (3.2.4) and (3.2.5). For this purpose we consider the statistics.

$$S_W = \sum_{i=1}^{p} (\boldsymbol{x}_i - \bar{x}_{i\bullet}\mathbf{1})'(\boldsymbol{x}_i - \bar{x}_{i\bullet}\mathbf{1}) \sim (\sigma_0^2 + \sigma_1^2)\chi^2(p(k-1) \tag{3.3.1}$$

$$S_b = \sum_{i=1}^{p} k(\bar{x}_{i\bullet} - \bar{x}_{\bullet\bullet})^2 \sim (\sigma_0^2 + \sigma_1^2 + k\sigma_2^2)\chi^2(p-1) \tag{3.3.2}$$

$$S_0^2 \sim \sigma_0^2\chi^2(f). \tag{3.3.3}$$

Substituting these estimates of the variances in the estimates $\boldsymbol{\mu}_{iH}^{(e_1)}$ and $\boldsymbol{\mu}_{iG}^{(e_1)}$ we obtain the full empirical Bayes estimates, described below in (i)-(iii), with the optimal values of the constants $c_1 - c_5$ given in Table 3.

**(i)** James-Stein separately on parameters in individual clusters:

$$\text{(ia)} \quad \boldsymbol{\mu}_{iG}^{(e')} = \boldsymbol{x}_i - c_1\frac{S_0}{S_w}(\boldsymbol{x}_i - \bar{x}_{i\bullet}\mathbf{1}_k), \quad i = 1, \ldots, p$$

$$\text{(ib)} \quad \boldsymbol{\mu}_{iG}^{(e)} = \boldsymbol{x}_i - c_2\frac{S_0}{S_{iw}}(\boldsymbol{x}_i - \bar{x}_{i\bullet}\mathbf{1}_k), \quad i = 1, \ldots, p$$

where $\bar{x}_{i\bullet} = (x_{i1} + \ldots + x_{ik})/k$,
$S_{iw} = (\boldsymbol{x}_i - \bar{x}_{i\bullet}\mathbf{1}_k)'(\boldsymbol{x}_i - \bar{x}_{i\bullet}\mathbf{1}_k)$, and $S_w = S_{1w} + \ldots + S_{pw}$.

**(ii)** James-Stein on all the parameters in one step:

$$\boldsymbol{\mu}_J^{(e)} = \boldsymbol{x} - c_3\frac{S_0}{S_b + S_w}(\boldsymbol{x} - \bar{x}_{\bullet\bullet}\mathbf{1}_{pk})$$

where $\boldsymbol{x}' = (\boldsymbol{x}_1', \ldots, \boldsymbol{x}_p'), \bar{x}_{\bullet\bullet} = (\bar{x}_{1\bullet} + \ldots + \bar{x}_{p\bullet})/p$,
and $S_b^2 = k\Sigma(\bar{x}_{i\bullet} - \bar{x}_{\bullet\bullet})^2$.

**(iii)** Empirical hierarchical Bayes:

$$\boldsymbol{\mu}_H^{(e)} = \boldsymbol{x} - c_4 \frac{S_0}{S_w}(\boldsymbol{x} - \bar{\boldsymbol{x}}_\bullet \otimes \mathbf{1}_k) - c_5 \frac{S_0}{S_b}(\bar{\boldsymbol{x}}_\bullet - \bar{x}_{\bullet\bullet}\mathbf{1}_p) \otimes \mathbf{1}_k$$

where $\bar{\boldsymbol{x}}_\bullet' = (\bar{x}_{1\bullet}, \ldots, \bar{x}_{p\bullet})$.

The BMDE of the above estimators can be computed on standard lines using the method described in Rao (1976). We give the expressions for the BMDE of the estimates with the optimum choice of constants $c_1, \ldots, c_5$, omitting all the detailed computations. We denote by $\boldsymbol{\mu}_G^{(e')}$ and $\boldsymbol{\mu}_G^{(e)}$, the combined estimates of $\boldsymbol{\mu}_{iG}^{(e')}$, $i = 1, \ldots, p$ and of $\boldsymbol{\mu}_{iG}^{(e)}$, $i = 1, \ldots, p$ respectively.

Table 3. Expression for the BMDE

| Estimate | Constants | $\sigma_0^{-2}$ (BMDE) |
|---|---|---|
| $\boldsymbol{\mu}_G^{(e')}$ | $c_1 = \frac{(pk-p-2)}{(f+2)}$ | $I_{pk} - \frac{f(1-\rho_1)}{f+2}\frac{pk-p-2}{p(k-1)} I_p \otimes (I_k - V_1)$ |
| $\boldsymbol{\mu}_G^{(e)}$ | $c_2 = \frac{(k-3)}{(f+2)}$ | $I_{pk} - \frac{f(1-\rho_1)}{f+2}\frac{k-3}{k-1} I_p \otimes (I_k - V_1)$ |
| $\boldsymbol{\mu}_H^{(e)}$ | $c_4 = c_1$ | $I_{pk} - \frac{f(1-\rho_1)}{f+2}\frac{pk-p-2}{p(k-1)} I_p \otimes (I_k - V_1)$ |
| | $c_5 = \frac{(p-3)}{(f+2)}$ | $-\frac{f(1-\rho_1)(p-3)}{(f+2)(1+k\rho_2)(p-1)}(I_p - V_2) \otimes V_1$ |

A comparison of the expressions in Table 3 gives the efficiency relationships with respect to the BMDE

$$\boldsymbol{\mu}_H^{(e)} >> \boldsymbol{\mu}_G^{(e')} >> \boldsymbol{\mu}_G^{(e)} \geq \boldsymbol{x}. \tag{3.3.4}$$

The same relationships hold with respect to the BCMSE.

We shall now compute the MDE and CMSE of the estimates given in Table 3 for fixed values of the unknown parameters and compare their efficiencies. The computations are made using the results given in the Appendix to Rao and Shinozaki (1978). The multiplier $\sigma_0^2$ is omitted in all the expressions given below to avoid using large brackets.

(ia) The estimate $\boldsymbol{\mu}_G^{(e')}$ of $\boldsymbol{\mu}$.

$$\begin{aligned} \text{MDE}(\boldsymbol{\mu}_G^{(e')}) = I_{pk} - fc_1 \Big[ &\tfrac{1}{\lambda_1} B(Q_1, \boldsymbol{\mu}) E(d_1 + 2K_1)^{-1} \\ &+ (Q_1 - \tfrac{p(k-1)}{\lambda_1} B(Q_1, \boldsymbol{\mu})) E \tfrac{d_1 + 4K_1}{(d_1+2+2K_1)(d_1+2K_1)} \Big] \end{aligned} \tag{3.3.5}$$

where

$$Q_1 = I_p \otimes (I_k - V_1), \lambda_1 = \frac{\boldsymbol{\mu}' Q_1 \boldsymbol{\mu}}{2\sigma_0^2}, B(A, \eta) = \frac{A\eta\eta' A}{2\sigma_0^2},$$

$d_1 = pk - p - 2$ and $K_1 \sim P(\lambda_1)$.

From (3.3.5), we find that for any given $i$,

$$\begin{aligned} \text{CMSE } (\mu_{iG}^{(e')}) &= k - f(f+2)c_1^2 \tfrac{\lambda_{1i}}{\lambda_1} E(d_1 + 2K_1)^{-1} \\ &- fc_1(k-1)\tfrac{\lambda_1 - p\lambda_{1i}}{\lambda_1} E\tfrac{d_1+4K_1}{(d_1+2+2K_1)(d_1+2K_1)}, \end{aligned} \tag{3.3.6}$$

where $\lambda_{1i} = [(\mu_{i1} - \bar{\mu}_{i\bullet})^2 + \ldots + (\mu_{ik} - \bar{\mu}_{i\bullet})^2]/2\sigma_0^2$, and

$$\text{CMSE } (\mu_G^{(e')}) = \text{ pk } - f(f+2)c_1^2 E(d_1 + 2K_1)^{-1}. \tag{3.3.7}$$

(ib) The estimate $\mu_G^{(e)}$ of $\mu$.

$$\text{CMSE } (\mu_{iG}^{(e)}) = k - f(f+2)c_2 E(k - 3 + 2K_{1i})^{-1} \tag{3.3.8}$$

where $K_{1i} \sim P(\lambda_{1i})$, and

$$\text{CMSE } (\mu_G^{(e)}) = \sum_1^p \text{ CMSE } (\mu_{iG}^{(e)}). \tag{3.3.9}$$

(ii) The estimate $\mu_J^{(e)}$ of $\mu$.

$$\begin{aligned} \text{MDE}(\mu_J^{(e)}) &= I_{pk} \\ &- f(f+2)c_3^2(2\sigma_0^2)B(Q,\mu)E(pk - 3 + 2K)^{-1} \\ &- fc_3(Q - \tfrac{p(k-1)}{\lambda} B(Q,\mu))E\tfrac{pk-3+4K}{(pk-1+2K)(pk-3+2K)} \end{aligned} \tag{3.3.10}$$

where $c_3 = \dfrac{(pk-3)}{f+2}, Q = I_{pk} - V_2 \otimes V_1, K \sim P(\lambda), \lambda = \mu' Q \mu / 2\sigma_0^2$, and

$$\text{CMSE } (\mu_J^{(e)}) = \text{ pk } - f(f+2)c_3^2 E(pk - 3 + 2K)^{-1} \tag{3.3.11}$$

(iii) The estimate $\mu_H^{(e)}$ of $\mu$.

$$\begin{aligned} \text{MDE}(\mu_H^{(e)}) &= \text{MDE}(\mu_G^{(e')}) \\ &- f(f+2)c_5^2 \tfrac{1}{\lambda_2} B(Q_2,\mu)E(p - 3 + 2K_2)^{-1} \\ &+ 2f(f+2)c_4c_5(B(Q_1,\mu) + B(Q_2,\mu)) \\ &\quad \times E(p-3+2K_2)^{-1}E(p-1+2K_1)^{-1} \\ &- fc_5(Q_2 - \tfrac{(p-1)}{\lambda_2} B(Q_2,\mu))E\tfrac{p-3+4K_2}{(p-3+2K_1)(p-1+2K_1)} \end{aligned} \tag{3.3.12}$$

where

$$Q_2 = (I_p - V_2) \otimes V_1, \ \lambda_2 = K \textstyle\sum_{i=1}^p (\bar{\mu}_i - \bar{\mu})^2 / 2\sigma_0^2 \text{ and } K_2 \sim P(\lambda_2).$$

Finally,

$$\text{CMSE }(\boldsymbol{\mu}_H^{(e)}) = \text{ CMSE }(\mu_G^{(e)}) - f(f+2)c_5^2 E(p-3+2K_2)^{-1}. \quad (3.3.13)$$

There are no clear cut inequalities in the efficiencies of the various estimates in terms of the MDE and CMSE except that they are all better than the usual unbiased estimators and that when $p \geq 3, \boldsymbol{\mu}_H^{(e)} >> \boldsymbol{\mu}_G^{(e')}$, i.e., the empirical estimates under the hierarchical prior for the parameters are better than the estimates derived by considering each group separately, when the within group variability is of the same order in each group.

A comparison of some interest is the CMSE $(\boldsymbol{\mu}_H^{(e)})$ given in (3.3.13) with the CMSE $(\boldsymbol{\mu}_J^{(e)})$ given in (3.3.11). The former is smaller than the latter if

$$c_4^2 E(pk-p-2+2K_1)^{-1} + c_5^2 E(p-3+2K_2)^{-1} \geq c_3^2 E(pk-3+2K)^{-1}. \quad (3.3.14)$$

Using the inequality

$$(q+2\nu)^{-1} \leq E(q+2Z)^{-1} \leq (q-2+2\nu)^{-1}$$

where $Z \sim P(\nu)$, a sufficient condition for (3.3.14) to hold is

$$\frac{c_4^2}{pk-p-2+2\lambda_1} + \frac{c_5^2}{p-3+2\lambda_2} \geq \frac{c_3^2}{pk-5+2\lambda} \quad (3.3.15)$$

which is satisfied when $\lambda_2$ representing the between group variability is large.

We have not compared the bias and mean square error in the estimates of individual parameters by the various methods. Such comparisons may indicate further advantages in using the hierarchical Bayes approach.

The results of our paper may appear similar to those of Ghosh and Lahiri (1988), but there are some differences in approach and final results.

## ACKNOWLEDGMENT

Research sponsored by the Air Force Office of Scientific Research under Grant AFOSR-89-0279. The United States Government is authorized to reproduce and distribute reprints for governmental purposes notwithstanding any copyright notation hereon.

## REFERENCES

Efron, B. and Morris, C. (1972). Empirical Bayes on vector observations. *Biometrika*, **55**, 335-347.

Efron, B. and Morris, C. (1973a). Stein's estimation rule and its competitors - an empirical Bayes approach. *Jour. Amer. Statist. Assoc.*, **68**, 117-130.

Efron, B. and Morris, C. (1973b). Combining possibly related estimation problems (with discussion). *Jour. Royal Statist. Soc.*, **B35**, 379-421.

Ghosh, M. and Lahiri, P. (1988). Bayes and empirical Bayes analysis in multistage sampling. In *Statistical Decision Theory and Related Topics IV*, Volume 1, (eds. S. S. Gupta and J. O. Berger), 195-212. Springer Verlag, New York.

James, W. and Stein, C. (1961). Estimation with quadratic loss. *Proc. 4th Berkeley Symp*, **1**, 362-379..

Rao, C. Radhakrishna (1953). Discriminant function for genetic differentiation and selection. *Sankhyā*, **12**, 229-246.

Rao, C. Radhakrishna (1973). *Linear Statistical Inference and its Applications* (Second Edition). Wiley, New York.

Rao, C. Radhakrishna (1975). Simultaneous estimation of parameters in different linear models and applications to biometric problems. *Biometrics*, **31** 545-554.

Rao, C. Radhakrishna (1976). Characterization of prior distribution and solution to a compound decision problem. *Amm. Statist*, **4**, 823-835.

Rao, C. Radhakrishna (1977). Simultaneous estimation of parameters-a compound decision problem. In *Decision Theory and Related Topics*, (eds. S. S. Gupta and D. S. Moore), 327- 350. Academic Press, New York.

Rao, C. Radhakrishna and Shinozaki, N. (1978). Precision of individual estimators in simultaneous estimations of parameters. *Biometrika*, **65** 23-30.

# 9

# Simulation Comparison of Methods for Bayesian Contextual Classification of Remotely Sensed Data

**Ruben Klein**[1]
**S. James Press**[2]

**Abstract**

This paper focusses on a simulation comparison of the Adaptive Bayesian Classification (ABC) and ICM (iterated Conditional Mode) spatial classification procedures for image reconstruction. The ABC procedure is described, since it is less well-known. The same geometrical configuration of populations is used for several examples involving normal populations with increasingly different variances. It is found that for the cases studied, ABC has considerably smaller variances than ICM in percentage of correct classifications obtained. Moreover, ABC has about the same distribution of percentage of correct classifications obtained whether we examine results for each population separately, or whether we examine the overall result. The comparison is made using for ICM the stated ICM initial classification, which is a "maximum likelihood classifier"; and using for ABC, the ABC initial classification, which is a predictive, contextual Bayesian classification procedure.

*Key Words and Phrases*: Adaptive Bayesian Classification, Iterative Conditional Mode, Remote Sensing, Contextual Classification, Spatial Statistics, Pattern Recognition, ABC, Image Reconstruction, Segmentation.

## 1. INTRODUCTION

A scene of interest on the ground is divided into a rectangular grid, sometimes called a "map". The scene is remotely viewed by a sensor aboard a satellite. The sensor produces images of the scene in each of $p$ frequencies of the electromagnetic spectrum. Each cell of the grid is rectangular, and is called a picture element, or "pixel". The $p$ signals associated with a given pixel, taken together, are called the observation vector for that pixel. Each

[1]LNCC, Rio de Janeiro & University of California, Riverside
[2]University of California, Riverside

rectangular pixel has an area equivalent to a square pixel 270 feet on a side, for LANDSAT 5 satellites (using the multispectral scanner sensor; it is 98 ft. for the thematic mapper sensor which is also carried aboard LANDSAT 5).

We assume there are exactly $K$ distinct populations that might be present in a scene; for example, in an agricultural context we might have rice, wheat, and corn. We assume also that there is only one population in any given pixel.

Our problem is to reconstruct the true ground scene by correctly classifying an entire map, that is, to correctly classify every pixel in a scene of interest (and to classify many maps along the ground track of the satellite). We assume there is training data available; i.e., observation vectors from each of the $K$ populations, that have been correctly classified on the ground. These "ground truth" data help to estimate the parameters of the underlying population distributions, and may be used to help classify observations correlated with these "known" data (to see how to accomplish this, see Klein and Press, 1989).

There is a large literature that addresses various aspects of the image reconstruction and classification problems in a spatial context. We will not attempt to review that literature here but will bring it to bear on the current work as it is required.

In this paper we adopt a multiple stage (recursive) procedure to reconstruct this scene of interest. For the first stage of classification we apply a fully Bayesian method to each pixel in the scene using only local information. In the second stage we use the map obtained from the first stage as the basis for how the underlying populations are relatively configured, and we then improve the map classification by reclassifying each pixel in the map based upon the global information, using an empirical Bayes procedure. We then iterate the reclassification process recursively until it stabilizes to a final, unchanging configuration for the entire map. This iterative process is called the ABC procedure, ( see Klein & Press, 1988).

We will describe a simulation which compares the results of classifying and reconstructing with the ABC procedure with the results of classifying with the ICM procedure of Besag (1986). The comparison will be made using several examples.

The plan of the paper is the following. In Section 2 we present the methodololgy for generating an initial reconstruction. In Section 3 we show how to use the ABC recursive procedure to improve upon the initial reconstruction. In Section 4 we compare the ABC & ICM procedures using simulation.

## 2. INITIAL RECONSTRUCTION

### 2.1. Notation

Let the observation $Z(s) : (p \times 1)$ be a realization of the covariance stationary, spatial stochastic process $\{Z(s) : s \in R^2\}$, where $s$ is a point on the ground measured in a two- dimensional, coordinate system, say latitude and longitude. (Note that in other contexts to which this same model applies with minimal modification, such as in medical imaging, or meterology, $s$ is a three-dimensional vector.) Denote the observation to be classified as $z_0 \equiv Z(s_0)$. Denote the populations to which $z_0$ might belong by $\pi_i \equiv N(\theta_i, \sum_i), \sum_i > 0, i = 1, \ldots, K$. $\sum_i > 0$ means that $\sum_i$ is assumed to be a symmetric, positive definite matrix. Elphinstone et al., 1985 have found that in the context of satellite imagery using agricultural LANDSAT data, the normality assumption is quite adequate, and that the population convariance matrices should be assumed to be unequal. We will adopt this unequal covariance matrix case throughout although we have also evaluated the model for the more unusual case of equal covariance matrices (see Klein and Press, 1989a,b), and when necessary we will use the equal covariance result. The dimension of the observation vectors, $p$, is currently no more than 7 for LANDSAT. We assume there is training data $D$ available to help to estimate the population parameters. The training data includes $n_i$ observation vectors known to come from population $\pi_i, i = 1, \ldots, K$.

We will consider two types of neighbor configurations: 4-neighbor and 8-neighbor. Let $N$ (for neighbors) denote the set of all immediately neighboring observations, and let them be located at the closest lattice points to the main pixel to be classified.

By covariance stationarity of the process we take, for $s \neq 0$,

$$cov[X_j(h), X_j(s+h)] = \Sigma_j(s), \quad j = 1, \ldots, K,$$

and

$$var[X_j(s)] = \Sigma_j, \quad \Sigma_j \text{ not depending upon } s.$$

Moreover, for simplicity in this paper assume the training data vectors are independent.

The population parameters of the $\pi_j$'s are given jointly by

$$P \equiv (\theta_1, \ldots, \theta_K, \Sigma_1, \ldots, \Sigma_K).$$

The Jeffreys invariant (vague) prior density is given by

$$p(P) \propto \prod_{j=1}^{K} |\Sigma_j|^{-(p+1)/2}.$$

The extension of our vague prior results to natural conjugate priors is immediate. These parameters may also be estimated from the current set of training data by empirical Bayes methods described in Klein and Press, (1987).

### 2.2. Conditional Configuration Probability

Denote the $r$ immediately neighboring observations of $z_0$ by $z_1, \ldots, z_r$. The conditional configuration probability is denoted by

$$P\{z_1 \in \pi_{\rho_1}, \ldots, z_r \in \pi_{\rho_r} | z_0 \in \pi_k\}.$$

In order to get the ABC process started we need an initial classification. We adopt the assumption of "local spatial continuity" for the initial classification, and we'll drop it later. (This assumption was proposed by Switzer, 1980, and was adopted by Mardia, 1984.) The assumption requires that if $z_0 \in \pi_k$, then the immediate neighbors of $z_0$ also belong to $\pi_k$.

### 2.3. Predictive Bayesian Classification

We adopt the predictive Bayesian approach to classification (see Geisser, 1964; 1966). The prediction-of-observables concept derives from de Finetti's theorem (de Finetti, 1937).

Let $\tau_k$ denote the posterior predictive classification probability for the event that $z_0$ belongs to $\pi_k$. That is,

$$\tau_k \equiv P\{z_0 \in \pi_k | z_0, N, D\},$$

where $N$ denotes the observations from the immediate neighbors of $z_0$, and $D$ denotes the training data. If $p_k(s_0) \equiv P\{z_0 \equiv Z(s_0) \in \pi_k\}$, and $g(\cdot)$ denotes a density function, then by Bayes theorem

$$\tau_k \propto p_k(s_0) g(z_0, N, D | z_0 \in \pi_k).$$

Conditioning on $D$, and on the classifications of the neighbors, and, for brevity of notation, denoting the collection $z_1 \in \pi_{\rho_1}, \ldots, z_r \in \pi_{\rho_r}$ by $Z_{(1,r)}$, gives

$$\tau_k \propto p_k(s_0) \sum_{\rho_1=1}^{K} \cdots \sum_{\rho_r=1}^{K} f(z_0, N | D, z_0 \in \pi_k, Z_{(1,r)}) P\{Z_{(1,r)} | z_0 \in \pi_k\}, \quad (2.1)$$

where $f(\cdot)$ denotes the joint probability density for $z_0$ and its neighbors, conditional upon the training data, and the classifications of $(z_0, N)$. The second term in the sum is the conditional configuration probability. Invoking the assumption of local spatial continuity for the conditional configuration probability in the last equation gives (since the summations collapse into one term when all the $r$ neighbors belong to $\pi_k$)

$$\tau_k \propto p_k(s_0) f(z_0, N | D, z_0 \in \pi_k, z_{(1,r)}). \quad (2.2)$$

Noting that $f(\cdot)$ is an average over the population parameters gives

$$\tau_k \propto p_k(s_0) \int f_P(z_0, N | D, z_0 \in \pi_k, Z_{(1,r)}, P) h(P | D, z_0 \in \pi_k, Z_{(1,r)}) dP, \quad (2.3)$$

where: $f_P(\cdot)$ denotes the density of $(z_0, N)$ conditional on $D$, the classifications of $z_0$ and its neighbors, and $P$; and $h(\cdot)$ denotes the posterior density of $P$.

Eqn. (2.3) will be evaluated differently depending upon the assumption made about how $z_0$ and $N$ are distributed. We might assume:

(a) $z_0$ and its immediate neighbors are located far enough apart from the training data, $D$ so that they may be treated as independent (see Klein and Press, 1989);

(b) $z_0$ and its immediate neighbors are located sufficiently close to the training data (but the neighbors are distinct from the training set), so that their spatial correlation must be taken into account (see Klein and Press, 1990a).

(c) Some of $z_0$'s immediate neighbors are part of the training data (see Klein and Press, 1990b).

Under the assumption of normality of the underlying spatial process the integral in eqn. (2.3) can be evaluated exactly, in closed form. It was shown in Klein and Press, 1989, 1990a, 1990b that regardless of whether the assumption in (a), (b), or (c) is made, the result is that the joint predictive density of $z_0$ and the neighboring observations is matrix $T$; the parameters of this matrix $T$ differ depending upon which of the three assumptions is made. The final result is that

$$\tau_k \propto p_k(s_0) \times \text{ (predictive matrix } T\text{- density)}.$$

## 2.4. Classification Rule

We now need a rule for classification. If we have a prespecified loss function we can classify $z_0$ into the $\pi_k$ for which the expected loss is minimized. In some other situations where we don't have a natural loss function, we might want to classify $z_0$ into that $\pi_k$ for which $\tau_k$ is a maximum (a modal estimator of the posterior distribution), or into that $\pi_k$ for which $\tau_k$ exceeds, and is closest to, some prespecified value. We complete this procedure for every pixel in the scene of interest, so that we obtain an initial reconstruction for the entire map.

## 3. ADAPTIVE BAYESIAN CLASSIFICATION (ABC)

Unfortunately, the initial map reconstruction is less than ideal because we really ignored any global patterns there may have been in the scene by classifying each pixel separately, except for using the local relationship of each pixel to its immediate neighbors using the local spatial continuity assumption. We can improve upon the initial reconstruction of the map in the following way. We drop the artificial assumption of local spatial continuity and reclassify each pixel using the initial reconstruction of the map to estimate how the neighbors are likely to be relatively configured. This will yield a new (second stage) reconstruction of the entire map. But now this reclassification procedure can be repeated recursively until the process stabilizes, in the sense that the map reconstruction changes little from one stage to the next. Then we stop. The basic classification formula we use is that in (2.1) which gives the predictive classification probability for a given pixel.

We don't actually know the conditional configuration probability in eqn. (2.1), so we approximate it using what we have learned from the initial reconstruction. The approximation we use in our numerical examples in Section 4 uses the proportions of pixels that estimate this probability in the initial classification. In the "proportions" approximation, we have taken two configurations of neighbors to be equivalent if one can be obtained from the other through a sequence of reflections along the horizontal, vertical, and diagonal axes. For instance, the two configurations in Fig. 3.1 are equivalent. We estimate the conditional configuration probability for the distinct configurations.

| | | | | | | |
|---|---|---|---|---|---|---|
| x | o | o | | o | o | x |
| x | o | o | | o | o | x |
| x | o | o | | o | o | x |

Fig. 3.1: Equivalent Neighbor Configurations

For instance, for $K = 2$, and two neighbors, there are three distinct configurations (see Fig. 3.2) for $z_0$ belonging to a given population.

| | | | | | | | | | | |
|---|---|---|---|---|---|---|---|---|---|---|
| o | o | o | | x | o | o | | x | o | x |
| $z_1$ | $z_0$ | $z_2$ | | $z_1$ | $z_0$ | $z_2$ | | $z_1$ | $z_0$ | $z_2$ |

Fig. 3.2: Distinct Configurations for $k = 2, r = 2$

Thus, if there were 100 times that $z_0$ was classified into $\pi_1$, and of these, there were 10 times that $z_1 \in \pi_2$, and $z_2 \in \pi_1$, and 5 times that $z_1 \in \pi_1$ and $z_2 \in \pi_2$, then we estimate

$$P\{z_1 \in \pi_1, z_2 \in \pi_2 | z_0 \in \pi_1\} = P\{z_1 \in \pi_2, z_2 \in \pi_1 | z_0 \in \pi_1\}$$

by $(10+5)/100 = 0.15$.
The larger the scene, the more accurate will be the approximation. Note that:

(1) we don't need the Markov random field assumption;

(2) we do need a spatial ergodicity assumption so that spatial averagaes tend to population averages.

When we fix the conditional configuration probabilities based upon the initial map classification, eqn. (2.1) is approximated by

$$\tau_k \propto p_k(s_0) f(z_0, N | D, z_0 \in \pi_k, Z_{(1,r)}) \hat{P}\{Z_{(1,r)} | z_0 \in \pi_k\}, \tag{3.1}$$

where $\hat{P}(\cdot)$ denotes the approximated conditional configuration probability. Next note that $f(\cdot)$ may be rewritten as

$$f(z_0, N | D, z_0 \in \pi_k, Z_{(1,r)}) = f_1(z_0 | N, D, z_0 \in \pi_k, Z_{(1,r)})$$
$$\times f_2(N | D, z_0 \in \pi_k, Z_{(1,r)}).$$

Note that $f_1$ is just the predictive density of $z_0$, and under normality, it is just a normal density. But the density of $N$ depends only upon the classification of the neighbors, and not upon the event $\{z_0 \in \pi_k\}$, so $f_2(\cdot)$ does not depend upon $k$. So $f_2(\cdot)$ may be absorbed into the proportionality constant, and equation (3.1) becomes

$$\tau_k \propto p_k(s_0) f_1(z_0 | N, D, z_0 \in \pi_k, Z_{(1,r)}) \hat{P}\{Z_{(1,r)} | z_0 \in \pi_k\}, \tag{3.2}$$

where

$$\tau_k \equiv P\{Z(s_0) \in \pi_k | z_0, N, D\}.$$

Note that pixels on the edge of a scene have fewer immediate neighbors than interior points have, and are analyzed accordingly.

### 3.1. Updating

What should we use for $p_k(s_0)$ at stage 2? Bayes theorem asserts that the best information we have about $z_0$ at stage 2 is what we learned at stage 1. Accordingly, at stage 2 we replace $p_k(s_0)$ by $\tau_k$ at stage 1, the last posterior probability for classifying $z_0$. We then compute $\tau_k$ for stage 2 from eqn. (3.2), replacing $p_k(s_0)$ by the $\tau_k$ from stage 1. We can now reclassify $z_0$ at stage 2. Repeating the process for every pixel in the scene generates a stage 2 reconstruction for the map. We now move to stage 3 and reclassify in the same way. We iterate the procedure until it stabilizes. The initial classification is strictly a Bayes procedure. Since the updating procedure depends upon an approximation, and upon a prior probability that depends upon the current data, however, the updating is really an empirical Bayes procedure (which approximates the Bayes procedure in large samples).

We have tried the ABC procedure for simulated data (see Section 4) with univariate observations ($p = 1$) distributed independently given the map. We stopped iterating if the number of changes in classification was smaller than 0.0025 times the number of pixels, or if the number of iterations equals 20.

The ABC procedure is stable, and it converges to a final map reconstruction typically in less than 15 iterations (for independent data). The algorithm can be shown to converge if there are no approximations. In spite of our having made approximations, however, in our experience the recursive procedure has always converged. A few more iterations may be required for very large maps that have populations that are not very well separated. On the VAX-8700, small map problems ($20 \times 20$) of adaptive classification can be run in a few seconds (including the generation of the data) while larger ($80 \times 100$), or larger) problems which use 8 neighbors for classification, and which are high dimensional, might require a minute or more, including the generation of the data (for independent data). The case of correlated data would take longer.

## 4. SIMULATION COMPARISON OF ABC WITH ICM

In this section we will present several examples (that use the same scene) in which we compare the Adaptive Bayesian Classification (ABC) with the ICM procedure of Besag (1986) by means of simulation. The simulations we carried out differ from most other simulated examples in this area of research in that instead of running our problem for one or several trials, in each case we have run it for 100 trials, and we have studied the empirical distribution of the simulated results in each case. As a consequence we can study the variability of the classification procedure, as well as its mean or median probability of correct classification, for each of the populations present in the scene. All examples are one- dimensional, the observations are assumed to be independent, given the map, and in all cases, the criterion of quality of the procedure that we use is: probability of correct classification, which we estimate by the percentage of correct classifications (PCC). We also evaluate the PCC for each population separately. For simplicity, we assume the pixels are square. We study cases of both equal and unequal variances in the populations.

Our examples involve the populations: $\pi_1 = N(0, \sigma^2), \pi_2 = N(1, 1)$, for $\sigma^2 = 1, 2.25, 4, 9$. For $\sigma^2 = 1$ we will use results applicable for populations with equal variances, whereas for $\sigma^2 > 1$ we will use results applicable to the unequal variances case. We will assume for the initial stage of reconstruction the probability $p_k(s)$, does not depend upon $s$, and it is the same for both populations, namely, $p_1(s) = p_2(s) = 0.5$. The prior will of course change at the second stage where we reclassify, and again at each iteration. We assume that the data are all uncorrelated given the map. By assuming equal costs of misclassification we can find the classical probabilities of correct classification (zero neighbors, linear or quadratic discriminant functions, known population parameters). They are as given in Table 4.1, below.

Table 4.1: Classical Probabilities of Correct Classification

| $\sigma^2$ | 1 | 2.25 | 4 | 9 |
|---|---|---|---|---|
| pcc | 0.691 | 0.673 | 0.695 | 0.753 |

Note that a linear discriminant function has been used for $\sigma^2 = 1$, and a quadratic discriminant function has been used for $\sigma^2 > 1$.

A major advantage of a simulated example is that we can know "truth", that is, what the correct ground scene should be. In our case, we take as the true ground map a square scene which is $20 \times 20$ pixels. The map is taken from Mardia, 1984. It is shown in Fig. 4.1. In this map there are 200 observations (pixels) for each population. Moreover, our training data is assumed to include observations from all pixels in columns 2, 7, 12, 17 in Fig. 4.1. Thus, there are 42 observations from $\pi_1$ and 38 from $\pi_2$. For the updating, adaptive portion of the ABC procedure we used a four-neighbor configuration. For the initial reconstruction we used an eight-neighbor configuration.

We assume that the ICM (Iterated Conditional Mode) procedure of Besag (1986) is known, so there is no need to describe it here. We use its recommended initial solution, a maximum likelihood classifier. We applied the ICM procedure 100 times, using a simulation protocol to the data sets described above, and we evaluated the distribution of the map classifications. We also applied the ABC procedure to the same data 100 times using simulation and classified the map each time. For ABC we used a predictive Bayesian contextual initial solution. Both distributions of results are shown in Figures 4.2 and 4.3, for the cases of $\pi_1 = N(0,1)$ and $N(0,4)$, for $\pi_2 = N(1,1)$. (We have omitted showing results for $\pi_1 = N(0,2.25)$ and $N(0,9)$ vs. $\pi_2 = N(1,1)$, since they are very similar to those shown.) Results are presented in the form of Box-type plots. Moreover, separate results are presented for $\pi_1$, for $\pi_2$, and overall, or total.

```
. . . . . . ; ; ; ; ; ; ; ; ; ; ; ; ; ;
. . . . . . . ; ; ; ; ; ; ; ; ; ; ; ; ;
. . . . . . . . ; ; ; ; ; ; ; ; ; ; ; ;
. . . . . . . . . ; ; ; ; ; ; ; ; ; ; ;
. . . . . . . . . . ; ; ; ; ; ; ; ; ; ;
. . . . . . . . . . . ; ; ; ; ; ; ; ; ;
. . . . . . . . . . . . ; ; ; ; ; ; ; ;
. . . . . . . . . . . . . ; ; ; ; ; ; ;
. . . . . . . . . . . . . . ; ; ; ; ; ;
. . . . . . . . . . . . . . . ; ; ; ; ;
; ; ; ; ; ; ; ; ; ; . . . . . ; ; ; ; ;
; ; ; ; ; ; ; ; ; ; . . . . . . . . . .
; ; ; ; ; ; ; ; ; ; . . . . . . . . . .
; ; ; ; ; ; ; ; ; ; . . . . . . . . . .
; ; ; ; ; ; ; ; ; ; . . . . . . . . . .
; ; ; ; ; ; ; ; ; ; . . . . . . . . . .
; ; ; ; ; ; ; ; ; ; . . . . . . . . . .
; ; ; ; ; ; ; ; ; ; . . . . . . . . . .
; ; ; ; ; ; ; ; ; ; . . . . . . . . . .
; ; ; ; ; ; ; ; ; ; . . . . . . . . . .
```

Fig. 4.1: True Map

We note from Figures 4.2 and 4.3 that the spread or variability of the ABC procedure appears to be substantially lower than that of the ICM procedure. Moreover, the sampling distribution of PCC for ABC is approximately the same regardless of whether we are interested in $\pi_1, \pi_2$, or the overall classification. This is not true for ICM which is highly dependent upon whether we are interested in the distribution of PCC for $\pi_1, \pi_2$, or for the overall result. Finally, the ABC procedure handles the case of unequal variances as well as the case of equal variances, achieving high PCC in both cases, with small variability; the ICM procedure becomes highly variable as the variance discrepancy between the populations increases. An important qualification of these results is that our comparison has be carried out only for one small and simple geometric configuration with independent, one-dimensional data. We therefore conclude that while our results are suggestive, caution must be exercised in generalizing to situations not yet studied.

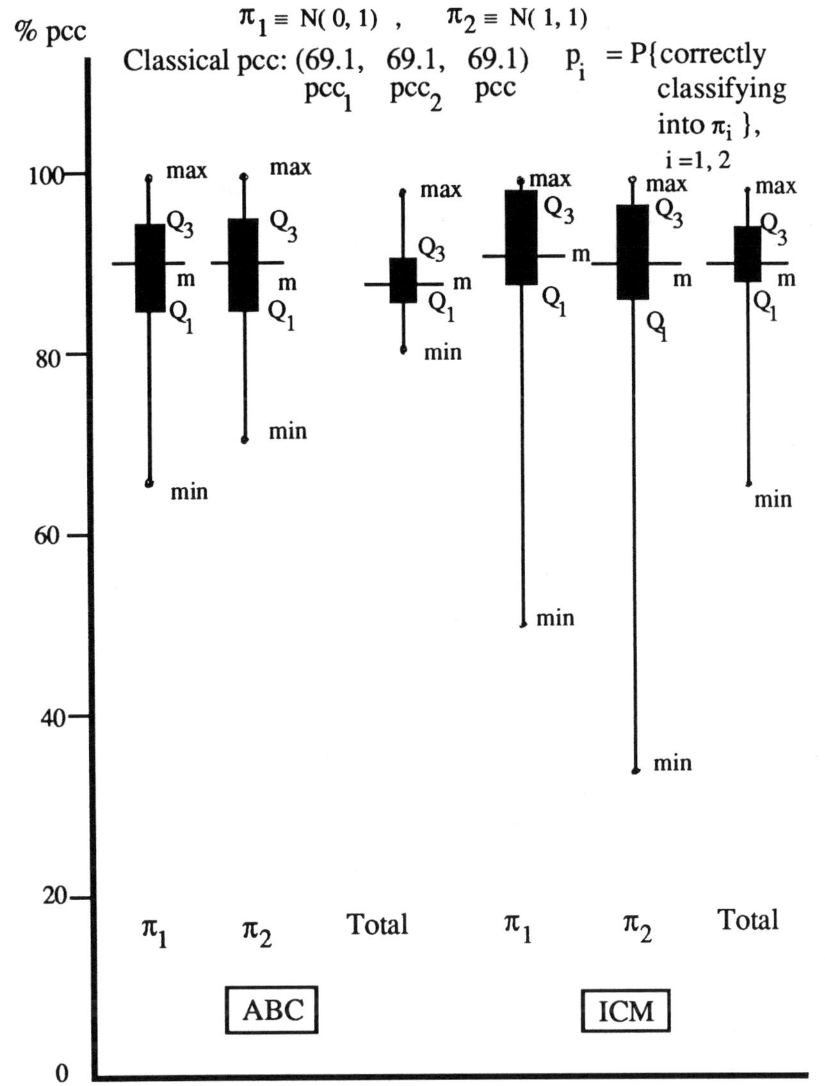


$Q_1$ = first quartile, m = mean, $Q_3$ = third quartile

Fig. 4.2: Box-type Plots for Percentage of Correct Classification (PCC) Supposing Equal Variances

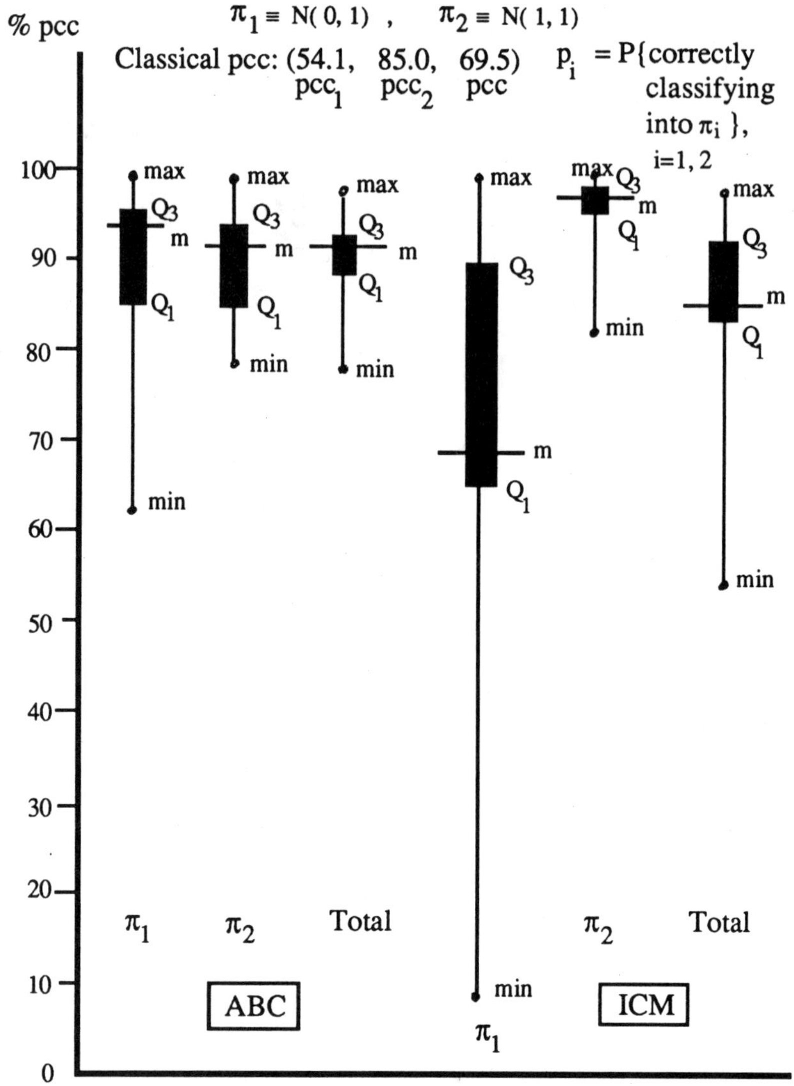


$Q_1$ = first quartile, m = mean $Q_3$ = third quartile

Fig. 4.3: Box-type Plots for Percentage of Correct Classification (PCC) Supposing Unequal Variances

## REFERENCES


Besag, J. (1986). "On the Statistical Analysis of Dirty Pictures", *Jour. Royal Statist. Soc. (B)*, **48**, 3, 259-302.

De Finetti, B. (1937). "Foresight: Its Logical Laws, Its Subjective Sources", in *Annales de l'Institut Henri Poincare*, Vol. 7, published by Gauthier-Villars; translated and reproduced in *Studies in Subjective Probability*, edited by Henry Kyburg and Howard Smokler, Krieger Publishing Co., Second Edition, 1980, 53- 118.

Elphinstone, C. D., Lanergan, A. T., Fatti, L. P. and D. M. Hawkins (1985). "An Empirical Investigation into the Application of Some Statistical Techniques to the Classification of Remotely Sensed Data." CSIRO Special Report SWISK 40, Pretoria, South Africa.

Geisser, S. (1964). "Posterior Odds for Multivariate Normal Classifications", *JRSS(B)*, **26**, 69-76.

Geisser, S. (1966). "Predictive Discrimination". *Multivariate Analysis*, ed. P. R. Krishnaiah. New York: Academic Press, 149-163.

Klein, R. and S. J. Press (1988). "The ABC (Adaptive Bayesian Classification) of Remotely Sensed Data", Technical Report No. 172, Department of Statistics, University of California, Riverside.

Klein, R. and S. J. Press (1989). "Contextual Bayesian Classification of Remotely Sensed Data", *Communications in Statistics – Theory and Methods*, **18(9)**, 3177-3202.

Klein, R. and S. J. Press (1990a). "Bayesian Contextual Classification with Neighbors Correlated with Training Data", in *Bayesian and Likelihood Methods in Statistics and Econometrics: Essays in Honor of George A. Bernard*, (eds.) S. Geisser, J. Hodges, S. J. Press, A. Zellner, New York: North Holland Pub. Co., 337-355.

Klein, R. and S. J. Press (1990b). "Bayesian Classification of Remotely Sensed Data when Training Data is Part of the Scene". *Brasileira De Probabilidade E Estatistica*, in English, **4**, No. 1.

Mardia, K. V. (1984). "Spatial Discrimination and Classification Maps", *Communications in Statistics, Theory and Methods*, **13**, No. 18, 2181-2197.

# 10

# Reference Priors in a Variance Components Problem

James O. Berger [1]
José M. Bernardo[2]

**Abstract**

The ordered group reference prior algorithm of Berger and Bernardo (1989b) is applied to the balanced variance components problem. Besides the intrinsic interest of developing good noninformative priors for the variance components problem, a number of theoretically interesting issues arise in application of the proposed procedure. The algorithm is described (for completeness) in an important special case, with a detailed heuristic motivation.

*Keywords and Phrases:* Noninformative priors; Multiparameter; Reference Priors; Bayesian Inference.

## 1. INTRODUCTION

Determination of reasonable noninformative priors in multiparameter problems is not easy; common noninformative priors, such as Jeffreys's prior, can have features that have an unexpectedly dramatic effect on the posterior. In recognition of this problem, Bernardo (1979), proposed the *reference prior* approach to development of noninformative priors, the key feature of which was a possible dependence of the reference prior on specification of parameters of interest and nuisance parameters.

This approach was further extended in Berger and Bernardo (1989a, 1989b). The first paper introduced a technically important modification: the reference prior algorithm was utilized in two stages - first for the nuisance parameter and then for the parameter of interest, as in Bernardo (1979) - but it was applied on an increasing family of compact subspaces of the parameter space, allowing a crucial first stage prior normalization to be performed. The second paper greatly extended this idea to deal with multiple nuisance parameters and parameters of interest, allowing for iterative application of the reference prior algorithm to any sequence of groupings of the parameters. This paper also contained examples and extensive discussion concerning how and when to group the coordinates.

---

[1] Purdue University
[2] Generalidad Valenciana and Universidad de Valencia, Spain

An interesting situation in which to illustrate the new algorithm is the balanced variance components model, which has for parameters a mean $\mu$, and two variances $\tau^2$ and $\sigma^2$ (see Section 4 for specifics.) The interest in considering this situation, besides the basic importance of the model in statistics, is that all the possible grouped reference priors can be determined and compared, and that interesting technical issues arise in application of the algorithm. The grouped reference priors are given in Section 4, along with a detailed derivation in the most difficult (and theoretically interesting) case. This section also gives a useful calculational expression for the ensuing posterior moments, an expression requiring only one-dimensional numerical integration.

The needed notation and results about patterned information matrices are presented in Section 2. In Section 3, a special case of the grouped reference prior algorithm, that was developed in Berger and Bernardo (1989b), ir reviewed and a detailed heuristic motivation for the algorithm is presented. In Section 5, we present concluding comments.

## 2. NOTATION AND PRELIMINARIES

The general algorithm will be presented for any parametric statistical problem in which the random observation $X$ has density $p(x|\theta)$, where $\theta \in \Theta \subset R^k$ is the unknown parameter. We assume that the Fisher information matrix

$$\boldsymbol{H}(\theta) = -E_{x|\theta}\left[\left(\frac{\partial^2}{\partial\theta_i\partial\theta_j}\log p(x|\theta)\right)\right]$$

exists and has rank $k$, so that

$$\boldsymbol{S}(\theta) = \boldsymbol{H}^{-1}(\theta)$$

also exists. Often, we will just write $\boldsymbol{H}$ and $\boldsymbol{S}$.

We assume that the $\theta_i$ are separated into $m$ groups of sizes $n_1, n_2, \ldots, n_m$, and that these groups are given by

$$\begin{aligned}\theta_{(1)} &= (\theta_1, \ldots, \theta_{n_1}), \quad \theta_{(2)} = (\theta_{n_1+1}, \ldots, \theta_{n_1+n_2}), \ldots\\ \theta_{(i)} &= (\theta_{N_{i-1}+1}, \ldots, \theta_{N_i}), \ldots \quad \theta_{(m)} = (\theta_{N_{m-1}+1}, \ldots, \theta_k),\end{aligned}$$

where $N_j = \sum_{i=1}^{j} n_i$ for $j = 1, \ldots, m$. These are the groupings to which the reference prior algorithm will be applied. (The coordinates of $\theta$ can, of course, be reordered if necessary – see Section 3.2– to achieve the desired ordered grouping.) Also we shall define, for $j = 1, \ldots, m$,

$$\theta_{[j]} = (\theta_{(1)}, \ldots, \theta_{(j)}) \quad \text{and} \quad \theta_{[\sim j]} = (\theta_{(j+1)}, \ldots, \theta_{(m)}).$$

If we write $\boldsymbol{S}$ as

$$\boldsymbol{S} = \begin{pmatrix} \boldsymbol{A}_{11} & \boldsymbol{A}_{21}^t & \cdots & \boldsymbol{A}_{m1}^t \\ \boldsymbol{A}_{21} & \boldsymbol{A}_{22} & \cdots & \boldsymbol{A}_{m2}^t \\ \vdots & \vdots & \ddots & \vdots \\ \boldsymbol{A}_{m1} & \boldsymbol{A}_{m2} & \cdots & \boldsymbol{A}_{mm} \end{pmatrix}$$

so that $\boldsymbol{A}_{ij}$ is $(n_i \times n_j)$, and define

$\boldsymbol{S}_j \equiv$ upper left $(N_j \times N_j)$ corner of $\boldsymbol{S}$, with $\boldsymbol{S}_m \equiv \boldsymbol{S}$, and $\boldsymbol{H}_j \equiv \boldsymbol{S}_j^{-1}$ then, the matrices $\boldsymbol{h}_j \equiv$ lower right $(n_j \times n_j)$ corner of $\boldsymbol{H}_j$, $j = 1, \ldots, m$ will be of central importance. Note that, if one defines $\boldsymbol{B}_j = (\boldsymbol{A}_{j1}\boldsymbol{A}_{j2}\cdots\boldsymbol{A}_{j\,j-1})$ for $j = 2, \ldots, m$, of sizes $(n_j \times N_{j-1})$, then it is straightforward to verify that, for $j = 1, \ldots, m$

$$\boldsymbol{h}_j = (\boldsymbol{A}_{jj} - \boldsymbol{B}_j \boldsymbol{H}_{j-1} \boldsymbol{B}_j^t)^{-1} \tag{2.1}$$

and

$$\boldsymbol{H}_j = \begin{pmatrix} \boldsymbol{H}_{j-1} + \boldsymbol{H}_{j-1}\boldsymbol{B}_j^t \boldsymbol{h}_j \boldsymbol{B}_j \boldsymbol{H}_{j-1} & -\boldsymbol{H}_{j-1}\boldsymbol{B}_j^t \boldsymbol{h}_j \\ -\boldsymbol{h}_j \boldsymbol{B}_j \boldsymbol{H}_{j-1} & \boldsymbol{h}_j \end{pmatrix}, \tag{2.2}$$

where any entry containing a factor of $\boldsymbol{H}_0$ is to be omitted. Thus, one may iteratively calculate $\boldsymbol{H}_1, \ldots, \boldsymbol{H}_m$, and hence $\boldsymbol{h}_1, \ldots, \boldsymbol{h}_m$.

In the important special case where each $n_j = 1$, no matrix inversions are needed above, so that calculation of the $\boldsymbol{h}_j$ is trivial if $\boldsymbol{S}$ is available. An even greater simplification occurs if, in addition,

$$\boldsymbol{B}_{i+1} = (c_i \boldsymbol{B}_i, \quad \boldsymbol{A}_{i+1\,i}) \tag{2.3}$$

for some constant $c_i$. Then, (2.1), (2.2), and (2.3) can be used to show that

$$\boldsymbol{h}_{i+1} = \left[\boldsymbol{A}_{i+1\,i+1} + c_i^2 \boldsymbol{A}_{ii} - 2c_i \boldsymbol{A}_{i+1\,i} - \boldsymbol{h}_i \left(c_i \boldsymbol{A}_{ii} - \boldsymbol{A}_{i+1\,i}\right)^2\right]^{-1}. \tag{2.4}$$

This is particularly useful when (2.3) holds for all $i$, which often occurs in patterned covariance matrices, since then (2.4) can be used to iteratively determine all the $\boldsymbol{h}_i$, starting with $\boldsymbol{h}_1 = \boldsymbol{A}_{11}^{-1}$, and defining $c_1 = 1$. Finally, if $\boldsymbol{S}$ is a block diagonal matrix, (i.e., $\boldsymbol{A}_{ij} \equiv \boldsymbol{0}$ for all $i \neq j$) then $\boldsymbol{h}_j \equiv \boldsymbol{A}_{jj}^{-1}$, $j = 1, \ldots, m$.

We will use the common symbols

$$|\boldsymbol{A}| = \text{determinant of } \boldsymbol{A}, \quad 1_\Omega(y) = \begin{cases} 1 & \text{if } y \in \Omega \\ 0 & \text{otherwise}, \end{cases}$$

and will throughout the paper adopt the conventions that $\sum_{i=l}^{l-1}(\cdot) = 0$ and $\prod_{i=l}^{l-1}(\cdot) = 1$. Also, we will often use $p(u|v)$ to generically represent the conditional density of $u$ given $v$.

## 3. THE $m$-GROUP REFERENCE PRIOR

### 3.1. The Algorithm

We suppose the $\theta_i$ have been ordered and divided into the $m$ groups $\theta_{(1)}, \ldots, \theta_{(m)}$. When the reference priors that are developed turn out to be proper, matters are straighforward. Often, however, they are improper, and care must be taken in their definition. In the improper case we proceed by specifying (see Section 4.2 for discussion) a nested sequence $\Theta^1 \subset \Theta^2 \subset \ldots$ of compact subsects of $\Theta$ such that $\cup_{l=1}^{\infty} \Theta^l = \Theta$. For simplicity, we shall assume that

$$\Theta^l = \Theta^l_{(1)} \times \Theta^l_{(2)} \times \cdots \times \Theta^l_{(m)},$$

where $\Theta^l_{(i)}$ denotes a set of possible $\theta_{(i)}$.

For a discussion on the existence of parametrizations in product form, see Kass (1989); also, the situation of arbitrary $\Theta^l$ is considered in Berger and Bernardo (1989b).

A reference prior is determined on each compact $\Theta^l$, for which the result is typically a proper prior, followed by performing a limiting operation. Specifically, one follows the following algorithm. Note that expressions for the $\boldsymbol{h}_j(\theta)$ have been given in Section 2.

*Start:* Define

$$\pi^l_m\left(\theta_{[\sim(m-1)]}|\theta_{[m-1]}\right) = \pi^l_m\left(\theta_{(m)}|\theta_{[m-1]}\right)$$

$$= \frac{|\boldsymbol{h}_m(\theta)|^{1/2} 1_{\Theta^l_{(m)}}}{\int_{\Theta^l_{(m)}} |\boldsymbol{h}_m(\theta)|^{1/2} d\theta_{(m)}}. \tag{3.1.1}$$

*Iteration:* For $j = m-1, m-2, \ldots, 1$, define

$$\pi^l_j\left(\theta_{[\sim(j-1)]}|\theta_{[j-1]}\right)$$

$$= \frac{\pi^l_{j+1}\left(\theta_{[\sim j]}|\theta_{[j]}\right) \exp\left\{\frac{1}{2} E^l_j\left[\left(\log|\boldsymbol{h}_j(\theta)|\right)|\theta_{[j]}\right]\right\} 1_{\Theta^l_{(j)}}}{\int_{\Theta^l_{(j)}} \exp\left\{\frac{1}{2} E^l_j\left[\left(\log|\boldsymbol{h}_j(\theta)|\right)|\theta_{[j]}\right]\right\} d\theta_{(j)}} \tag{3.1.2}$$

where, letting $\Theta^l_{[\sim j]} = \Theta^l_{(j+1)} \times \cdots \times \Theta^l_{(m)}$,

$$E^l_j\left[g(\theta)|\theta_{[j]}\right] = \int_{\Theta^l_{[\sim j]}} g(\theta)\ \pi^l_{j+1}(\theta_{[\sim j]}|\theta_{[j]}) d\theta_{[\sim j]}. \tag{3.1.3}$$

(Note that it is easy to check, by integrating in turn over $\theta_{(m)}, \theta_{(m-1)}, \ldots, \theta_{(j)}$, that $\pi^l_j$ defines a probability distribution.) For $j = 1$, interpret $\theta_{[\sim 0]}$ as $\theta$ and $\theta_{[0]}$ as vacuous, and write

$$\pi^l(\theta) = \pi^l_1\left(\theta_{[\sim 0]}|\theta_{[0]}\right). \tag{3.1.4}$$

*Finish:* Define the $m$-group reference prior, assuming it yields a proper posterior, by

$$\pi(\theta) = \lim_{l\to\infty} \frac{\pi^l(\theta)}{\pi^l(\theta^*)} \tag{3.1.5}$$

where $\theta^*$ is some point in $\Theta^1$.

If the integrals and expectation in (3.1.1) and (3.1.2) are finite when the "$l$" is removed (i.e., when $\Theta^l$ is replaced by $\Theta$ everywhere), then the reference prior is defined simply by $\pi_1$ (i.e., (3.1.5) is not needed). It should also be observed that the condition

$$|\boldsymbol{h}_j(\theta)| \quad \text{depends only on } \theta_{[j]}, \tag{3.1.6}$$

results (for the iteration corresponding to $j$) in (3.1.2) being replaced by

$$\pi_j^l(\theta_{[\sim(j-1)]}|\theta_{[j-1]}) = \pi_{j+1}^l(\theta_{[\sim j]}|\theta_{[j]}) \frac{|\boldsymbol{h}_j(\theta)|^{1/2}\, 1_{\Theta^l_{(j)}}}{\int_{\Theta^l_{(j)}} |\boldsymbol{h}_j(\theta)|^{1/2}\, d\theta_{(j)}}. \tag{3.1.7}$$

### 3.2. Motivation and Discussion

We suggest that in nonhierarchical models (as considered here) the ordering of the $\theta_{(i)}$ should be in terms of their inferential importance. For instance, in the variance components scenario, to be considered in Section 4, if inference concerning the "between" variance, $\tau^2$, is the primary goal, with the population mean, $\mu$, and "within" variance $\sigma^2$ being nuisance parameters, then the suggested ordering would be $\theta_1 = \tau^2, \theta_2 = \mu, \theta_3 = \sigma^2$, or maybe $\theta_1 = \tau^2, \theta_2 = \sigma^2, \theta_3 = \mu$. For inference concerning $\mu$, on the oher hand, $(\mu, \tau^2, \sigma^2)$ or $(\mu, \sigma^2, \tau^2)$ would be the suggested orderings. On the issue of grouping of coordinates, our advice is: do not group without a very good reason. Thus the $k$-group reference prior (each stage having $n_i = 1$) is generally recommended. Incidentaly, within groups the ordering of the $\theta_i$ is immaterial.

When the reference priors are proper there is no need to consider compact $\Theta^l$. And even when improper, the reference prior is often unaffected by the particular sequence $\{\Theta^l\}$ chosen. When needed, our typical choice of the $\{\Theta^l\}$ is simply a collection of nested rectangles in $\Theta$ (or other appropriate shape if $\Theta$ is not an "infinite" rectangle).

In Bernardo (1979) the motivation for the reference prior approach is discussed. The idea is basically to choose the prior which, in a certain asymptotic sense, maximizes the information in the posterior that is provided by the data. We will now summarize the argument for the simple case $k = 2$. We stress that the argument below is heuristic and that, rather than trying to make it precise, we choose to *define* the grouped reference prior as that determined by the algorithm in Section 3.1. (See Berger, Bernardo and Mendoza (1989) for an indication of some of the difficulties in making the heuristic argument precise.)

Let $x$ be the result of an experiment which provides information about $\theta_1$, according to the probability model $p(x|\theta_1, \theta_2)$, $(\theta_1, \theta_2) \in \Theta$, where $\theta_2$ is some nuisance parameter, and let $\Theta^l$ be a compact approximation to the (joint) parameter space $\Theta$. Let $z_t = \{x_1, \ldots, x_t\}$ be the result from $t$ conditionally independent replications of the original experiment; the amount of information $I_t^{\theta_1}\{p(\theta_1, \theta_2)\}$ to be expected about $\theta_1$ from $z_t$, when the prior is $p(\theta_1, \theta_2)$, is defined to be (Shannon, 1948; Lindley, 1956)

$$I_t^{\theta_1}\{p(\theta_1, \theta_2)\} = \int p(z_t) \int p(\theta_1|z_t) \log \frac{p(\theta_1|z_t)}{p(\theta_1)} d\theta_1 dz_t.$$

Using a variational argument, it may be shown under certain conditions that, for any fixed $p(\theta_2|\theta_1)$, this is maximized by a prior $\pi_t(\theta_1)$ which satisfies

$$\pi_t(\theta_1) \propto \exp\left\{\int p(z_t|\theta_1) \log p(\theta_1|z_t) dz_t\right\}, \tag{3.2.1}$$

where $p(z_t|\theta_1) = \int p(z_t|\theta_1, \theta_2) p(\theta_2|\theta_1)\, d\theta_2$.

Note that (3.2.1) only defines $\pi_t(\theta_1)$ implicitely for given $t$, since $p(\theta_1|z_t)$ depends on $\pi_t(\theta_1)$. As $t \to \infty$, perfect knowledge is approached, so that $\pi_t(\theta_1)$ approaches that prior which, given $p(\theta_2|\theta_1)$, maximizes the *missing information* about $\theta_1$; this is referred to as the *reference* prior.

Moreover, as $t \to \infty$, and under appropriate regularity conditions, $p(\theta_1|z_t)$ is approached by the asymptotic *marginal* posterior distribution of $\theta_1$, $N\{\theta_1|\hat{\theta}_1, \boldsymbol{S}_1(\hat{\theta}_1, \hat{\theta}_2)\}$, where $\boldsymbol{S}_1(\theta_1, \theta_2)$ is the upper left corner of $\boldsymbol{S}(\theta_1, \theta_2)$, the inverse of Fisher's information matrix $\boldsymbol{H}(\theta_1, \theta_2)$. For convenience, we will denote this marginal posterior by $N(\theta_1)$.

It follows that, for each conditonal prior $p(\theta_2|\theta_1)$, and for $t$ large enough, one may write

$$\begin{aligned}
\pi_t(\theta_1) &\propto \exp\left\{\int p(z_t|\theta_1) \log N(\theta_1) dz_t\right\} \\
&\simeq \exp\left\{\int\int p(\hat{\theta}_1, \hat{\theta}_2|\theta_1) \log N(\theta_1) d\hat{\theta}_1 d\hat{\theta}_2\right\} \\
&= \exp\left\{\int\int \left\{\int p(\hat{\theta}_1, \hat{\theta}_2|\theta_1, \theta_2) p(\theta_2|\theta_1) d\theta_2\right\} \log N(\theta_1)\, d\hat{\theta}_1 d\hat{\theta}_2\right\} \\
&= \exp\left\{\int p(\theta_2|\theta_1) \left[\int\int p(\hat{\theta}_1, \hat{\theta}_2|\theta_1, \theta_2) \log N(\theta_1)\, d\hat{\theta}_1 d\hat{\theta}_2\right] d\theta_2\right\} \\
&\simeq k \exp\left\{\int p(\theta_2|\theta_1) \log |\boldsymbol{S}_1(\theta_1, \theta_2)|^{-1/2} d\theta_2\right\},
\end{aligned}$$

where $k = (2\pi)^{-1} e^{-1/2}$, the last step following from

$$\int\int p(\hat{\theta}_1, \hat{\theta}_2|\theta_1, \theta_2) \log N(\theta_1)\, d\hat{\theta}_1 d\hat{\theta}_2$$

$$= \int\int p(\hat{\theta}_1, \hat{\theta}_2|\theta_1, \theta_2) \left[\log(2\pi)^{-1} - \frac{1}{2}\log|\boldsymbol{S}_1(\hat{\theta}_1, \hat{\theta}_2)| - \frac{1}{2}\frac{(\theta_1 - \hat{\theta}_1)^2}{\boldsymbol{S}_1(\hat{\theta}_1, \hat{\theta}_2)}\right] d\hat{\theta}_1 d\hat{\theta}_2$$

$$\simeq \log(2\pi)^{-1} - \frac{1}{2}\log|\boldsymbol{S}_1(\theta_1, \theta_2)| - \frac{1}{2}$$

(since, as $t \to \infty$, $p(\hat{\theta}_1, \hat{\theta}_2|\theta_1, \theta_2)$ converges to a point mass at $(\theta_1, \theta_2)$, and $\hat{\theta}_1$ is asymptotically $N\{\hat{\theta}_1|\theta_1, \boldsymbol{S}_1(\theta_1, \theta_2)\}$ ).

Furthermore, since $\boldsymbol{h}_1 \equiv \boldsymbol{H}_1 \equiv \boldsymbol{S}_1^{-1}$, $|\mathbf{h}_1|^{1/2} = |\boldsymbol{S}_1|^{-1/2}$ and hence,

$$\pi_t(\theta_1) \propto \exp\left\{\tfrac{1}{2}\int p(\theta_2|\theta_1)\log|\boldsymbol{h}_1(\theta_1, \theta_2)|d\theta_2\right\}$$

or, normalizing over $\Theta^l$,

$$\pi^l(\theta_1) = \frac{\exp\left\{\frac{1}{2}\int_{\Theta^l_{(2)}} p(\theta_2|\theta_1)\log|\boldsymbol{h}_1(\theta_1, \theta_2)|d\theta_2\right\} 1_{\Theta^l_{(1)}}}{\int_{\Theta^l_{(1)}} \exp\left\{\frac{1}{2}\int_{\Theta^l_{(2)}} p(\theta_2|\theta_1)\log|\boldsymbol{h}_1(\theta_1, \theta_2)|d\theta_2\right\} d\theta_1}, \tag{3.2.2}$$

which provides an explicit expression for the reference prior $\pi^l(\theta_1)$ corresponding to a given $p(\theta_2|\theta_1)$.

It is natural to choose $p(\theta_2|\theta_1)$ to be the *conditional* reference prior $\pi(\theta_2|\theta_1)$, i.e. that which, given $\theta_1$, maximizes the missing information about $\theta_2$. By the same argument leading to (3.2.1), this is obtained from

$$\pi_t(\theta_2|\theta_1) \propto \exp\left\{\int p(z_t|\theta_1, \theta_2)\log p(\theta_2|\theta_1, z_t)dz_t\right\}.$$

Again, as $t \to \infty$, and under appropriate regularity conditions, $p(\theta_2|\theta_1, z_t)$ is approximated by the asymptotic *conditional* posterior distribution of $\theta_2$ given $\theta_1$, $N\{\theta_2|\hat{\theta}_2, \boldsymbol{h}_2^{-1}(\hat{\theta}_1, \hat{\theta}_2)\}$ where $\boldsymbol{h}_2(\theta_1, \theta_2)$ is the lower right corner of the information matrix $\boldsymbol{H}(\theta_1, \theta_2)$. Hence, for large enough $t$, reasoning as before yields

$$\begin{aligned}
\pi_t(\theta_2|\theta_1) &\propto \exp\left\{\int p(z_t|\theta_1, \theta_2)\log N\{\theta_2|\hat{\theta}_2, \boldsymbol{h}_2^{-1}(\hat{\theta}_1, \hat{\theta}_2)\}dz_t\right\} \\
&= \exp\left\{\int\int p(\hat{\theta}_1, \hat{\theta}_2|\theta_1, \theta_2)\log N\{\theta_2|\hat{\theta}_2, \boldsymbol{h}_2^{-1}(\hat{\theta}_1, \hat{\theta}_2)\}d\hat{\theta}_1 d\hat{\theta}_2\right\} \\
&\simeq |\boldsymbol{h}_2(\theta_1, \theta_2)|^{1/2}.
\end{aligned}$$

Therefore, after normalizing,

$$\pi^l(\theta_2|\theta_1) = \frac{|\boldsymbol{h}_2(\theta_1, \theta_2|^{1/2}\, 1_{\Theta^l_{(2)}}}{\int_{\Theta^l_{(2)}} |\boldsymbol{h}_2(\theta_1, \theta_2)|^{1/2}\, d\theta_2} \,. \tag{3.2.3}$$

Finally, the joint reference prior necessary to obtain a reference posterior for $\theta_1$ will be defined to be $\pi^l(\theta_1, \theta_2) = \pi^l(\theta_2|\theta_1)\pi^l(\theta_1)$ on $\Theta^l$; hence, if $\{\Theta^l\}$ is a sequence of compacts expanding to $\Theta$, the desired reference prior is defined to be

$$\pi(\theta_1, \theta_2) = \lim_{l\to\infty} \frac{\pi^l(\theta_1, \theta_2)}{\pi^l(\theta_1^*, \theta_2^*)} \,, \tag{3.2.4}$$

where $(\theta_1^*, \theta_2^*)$ is some point in $\Theta^l$; the denominator is necessary to cancel out irrelevant multiplicative constants which may diverge as $l \to \infty$.

Equation 3.2.3 is directy analogous to "start" of the reference prior algorithm in Section 3.1, which gives $\pi^l(\theta_{(m)}|\theta_{(1)}, \ldots, \theta_{(m-1)})$; Equation 3.2.2 provides the motivation for the "iteration" steps, and Equation 3.2.4 is a version of the "finish" element in the algorithm.

Further stages (when $m > 2$) are handled in exactly the same manner yielding (3.1.2) as the stage-to-stage updating formula. The net result is $\pi^l(\theta)$, the $m$-stage reference prior on the compact $\Theta^l$.

Table 1. Reference priors for the variance components problem

| *Ordered Grouping* | *Reference Prior* |
|---|---|
| $\{(\mu, \sigma^2, \tau^2)\}$ | $\sigma^{-2}(n\tau^2 + \sigma^2)^{-3/2}$ |
| $\{(\mu, \sigma^2), \tau^2\}$ | $\sigma^{-5/2}(n\tau^2 + \sigma^2)^{-1}$ |
| $\{(\mu, \tau^2), \sigma^2\}$ | $\tau^{-3C_n/2}\sigma^{-2}\psi\left(\frac{\tau^2}{\sigma^2}\right)$ |
| $\{\sigma^2, (\mu, \tau^2)\}$ | $\sigma^{-1}(n\tau^2 + \sigma^2)^{-3/2}$ |
| $\{\tau^2, (\mu, \sigma^2)\}$ | $\tau^{-1}\sigma^{-2}(n\tau^2 + \sigma^2)^{-1/2}\psi\left(\frac{\tau^2}{\sigma^2}\right)$ |
| $\{\mu, (\sigma^2, \tau^2)\}$, $\{(\sigma^2, \tau^2), \mu\}$ | |
| $\{\mu, \sigma^2, \tau^2\}$ | $\sigma^{-2}(n\tau^2 + \sigma^2)^{-1}$ |
| $\{\sigma^2, \mu, \tau^2\}$, $\{\sigma^2, \tau^2, \mu\}$ | |
| $\{\mu, \tau^2, \sigma^2\}$, $\{\tau^2, \mu, \sigma^2\}$, $\{\tau^2, \sigma^2, \mu\}$ | $\tau^{-C_n}\sigma^{-2}\psi\left(\frac{\tau^2}{\sigma^2}\right)$ |

In Table 1, $C_n = \{1 - \sqrt{n-1}\,(\sqrt{n} + \sqrt{n-1})^{-3}\}$, and $\psi(\tau^2/\sigma^2) = [(n-1) + (1 + n\tau^2/\sigma^2)^{-2}]^{1/2}$. Grouping of parameters is indicated by parentheses.

## 4. THE VARIANCE COMPONENTS PROBLEM

The determination of $m$-group reference priors for the balanced variance components problem is of interest, not only methodologically, but also because it provides an interesting illustration of the techniques (and possible difficulties) of the general limiting derivation of $\pi(\theta)$ via (3.1.5). Section 4.1 presents the model and the $m$-group reference priors. Section 4.2 discusses some of the interesting technical issues that arose in the development. Section 4.3 briefly discusses using the reference priors in posterior calculations.

### 4.1. The Model and Reference Priors

We consider the balanced variance components model

$$X_{ij} = \mu + \alpha_i + \varepsilon_{ij} \quad , \quad i = 1, \ldots, p \quad \text{and} \quad j = 1, \ldots, n,$$

where the $\alpha_i$ are i.i.d. $N(\alpha_i|0, \tau^2)$ and, independently, the $\varepsilon_{ij}$ are i.i.d. $N(\varepsilon_{ij}|0, \sigma^2)$. The parameters $(\mu, \tau^2, \sigma^2)$ are unknown.

Since there are only $k = 3$ parameters it is easy to list *all* $m$-group reference priors. The possible ordered groupings are given in Table 1, along with the associated reference priors. Note that Jeffreys's prior is that associated with the single group $\{(\mu, \sigma^2, \tau^2)\}$; the prior suggested by Box and Tiao (1973, p. 251) is that associated with $\{\mu, (\sigma^2, \tau^2)\}$. Observe that $C_n$ is typically very near 1, and that

$$\sqrt{n-1} \leq \psi(\frac{\tau^2}{\sigma^2}) \leq \sqrt{n};$$

thus, replacing $C_n$ by 1 and $\psi$ by a constant is reasonable for all but very small $n$.

As indicated in Section 3.2, we are most favorably disposed towards the last two reference priors in Table 1, since they correspond to the various 3-group reference priors (each group having only a single element). Note that, among the 3-group reference priors, only the order of $\sigma^2$ and $\tau^2$ affects the answer; thus there are only two 3-group reference priors instead of the possible six. Thus all that need be specified, in order to determine the 3-group reference prior, is whether $\sigma^2$ or $\tau^2$ is deemed to be of more importance.

### 4.2. Determination of the Reference Priors

To implement the algorithm in Section 3.1, compact sets $\Theta^l$ must be selected. In deriving the reference priors in Table 1, nested boxes of the form

$$\Theta^l = (a_l, b_l) \times (c_l, d_l) \times (e_l, f_l), \qquad \text{for} \quad (\mu, \sigma^2, \tau^2), \tag{4.2.1}$$

were chosen, where $a_l \to -\infty$, $c_l$ and $e_l \to 0$, and the upper endpoints $\to \infty$. This would, intuitively, correspond to a presumption of prior independence among the parameters. In most cases, the precise choice of the endpoints in (4.2.1) was immaterial to the result. Disturbing exceptions were the third and last reference priors in Table 1, where the reference prior actually depends on

$$\eta = \lim_{l\to\infty}\left[\frac{\log d_l}{\log c_l^{-1}}\right].$$

If this limit does not exist, there is no reference prior for these situations. If the limit does exist, the third and last reference priors are, in general, as given in Table 1 but with $C_n$ replaced by

$$C_n(\eta) = 1 - \lambda(\eta^2 - \lambda)(\eta - \lambda)^{-2}, \tag{4.2.2}$$

where $\lambda = \sqrt{n-1}/\sqrt{n}$. Recalling that $(c_l, d_l)$ is the range for $\sigma^2$, the implication is that we must specify the relative rate at which we are "noninformative" about $\log\sigma^2$, as $\sigma^2 \to 0$ and $\sigma^2 \to \infty$, to determine the reference prior. In Table 1 we made the natural choice $\eta = 1$, but the need to make such an extra choice is clearly unfortunate.

For the third prior in Table 1, it is indeed not even possible to choose a value of $\eta$ such that $\eta \leq \lambda(\sqrt{6\lambda+3} - 1)/(3\lambda+1)$, for then it can be shown that $C_n(\eta) \geq 4/3$ and the prior will have a nonintegrable singularity at $\tau^2 = 0$, a singularity which persists in the posterior; these values of $\eta$ thus lead to unusable reference priors. Note that $\eta = 1$ does yield a proper posterior.

Alternatives to $\Theta_l$ in (4.2.1) can also be considered. One reasonable choice is

$$\Theta^l = \left\{(\mu, \sigma^2, \tau^2);\ \mu \in (a_l, b_l),\ \sigma^2 \in (c_l, d_l),\ \frac{\tau^2}{\sigma^2} \in (e_l, f_l)\right\}.$$

The point is that it is sometimes natural to be "noninformative" about the ratio $\tau^2/\sigma^2$ rather than just $\tau^2$ (cf. Hill, 1965).

If such $\Theta^l$ are used, the reference priors are as shown in Table 1, except for the second, third, and last cases, which become $\sigma^{-3}(n\tau^2+\sigma^2)^{-1}$, $\tau^{-3}\sigma^{-2}\psi(\tau^2/\sigma^2)$, and $\tau^{-2}\sigma^{-2}\psi(\tau^2/\sigma^2)$ respectively. These last two priors have nonintegrable singularities at $\tau^2 = 0$, which persist in the posterior, and hence are not usable.

We shall present here the development of the last reference prior in Table 1, for the ordered grouping $\{\mu, \tau^2, \sigma^2\}$. The analyses for all other cases in Table 1 are similar, but simpler, and are hence omitted.

We apply the algorithm in Section 3.1, with the $\Theta^l$ defined by (4.2.1). Note that the Fisher information matrix for $(\mu, \tau^2, \sigma^2)$ is

$$\boldsymbol{H}(\mu, \tau^2, \sigma^2) = \begin{pmatrix} \frac{pn}{(n\tau^2+\sigma^2)} & 0 & 0 \\ 0 & \frac{pn^2}{2(n\tau^2+\sigma^2)^2} & \frac{pn}{2(n\tau^2+\sigma^2)^2} \\ 0 & \frac{pn}{2(n\tau^2+\sigma^2)^2} & \left[\frac{p(n-1)}{2\sigma^4} + \frac{p}{2(n\tau^2+\sigma^2)^2}\right] \end{pmatrix},$$

so that

$$S\left(\mu,\tau^2,\sigma^2\right)=\begin{pmatrix}\frac{(n\tau^2+\sigma^2)}{pn} & 0 & 0\\ 0 & \left[\frac{2\sigma^4}{pn^2(n-1)}+\frac{2(n\tau^2+\sigma^2)^2}{pn^2}\right] & -\frac{2\sigma^4}{pn(n-1)}\\ 0 & -\frac{2\sigma^4}{pn(n-1)} & \frac{2\sigma^4}{p(n-1)}\end{pmatrix}$$

In the notation of Section 2, the 3-group case $\{\mu,\tau^2,\sigma^2\}$ corresponds to $\theta_{(1)}=\theta_1=\mu$, $\theta_{(2)}=\theta_2=\tau^2$, and $\theta_{(3)}=\theta_3=\sigma^2$. Also, $S$ satisfies (2.3) with $B_2=0$, $c_2=1$, and $A_{32}=-2\sigma^4/[pn(n-1)]$, so that

$$h_1=\sigma_{11}^{-1}=\frac{pn}{n\tau^2+\sigma^2}$$

and (2.4) yields

$$\begin{aligned}h_2 &= \left[A_{22}+A_{11}-H_1A_{11}^2\right]^{-1}\\ &= \left[\frac{2\sigma^4}{pn^2(n-1)}+\frac{2(n\tau^2+\sigma^2)^2}{pn^2}\right]^{-1},\end{aligned}$$

and

$$\begin{aligned}h_3 &= \left[A_{33}+A_{22}-2A_{32}-h_2(A_{22}-A_{32})^2\right]^{-1}\\ &= \left[\frac{p(n-1)}{2\sigma^4}+\frac{p}{2(n\tau^2+\sigma^2)^2}\right].\end{aligned}$$

*Start:* To begin,

$$\begin{aligned}\pi_3^l(\sigma^2|\mu,\tau^2) &= \frac{|h_3(\mu,\tau^2,\sigma^2)|^{1/2}1_{(c_l,d_l)}(\sigma^2)}{\int_{c_l}^{d_l}|h_3(\mu,\tau^2,\sigma^2)|^{1/2}\,d\sigma^2}\\ &= \frac{\left[\frac{p(n-1)}{2\sigma^4}+\frac{p}{2(n\tau^2+\sigma^2)^2}\right]^{1/2}1_{(c_l,d_l)}(\sigma^2)}{\Psi_1(\tau^2|c_l,d_l)},\end{aligned}$$

where

$$\Psi_1(\tau^2|c_l,d_l)=\int_{c_l}^{d_l}\left[\frac{p(n-1)}{2\sigma^4}+\frac{p}{2(n\tau^2+\sigma^2)^2}\right]^{1/2}d\sigma^2.$$

*Iteration for $j=2$:*

$$E^l\left[\log|h_2||(\mu,\tau^2)\right]=\frac{\Psi_2(\tau^2|c_l,d_l)}{\Psi_1(\tau^2|c_l,d_l)},$$

where

$$\Psi_2(\tau^2|c_l,d_l) = \int_{c_l}^{d_l} \left( \log \left[ \frac{2\sigma^4}{pn^2(n-1)} + \frac{2(n\tau^2+\sigma^2)^2}{pn^2} \right]^{-1} \right) \times \left[ \frac{p(n-1)}{2\sigma^4} + \frac{p}{2(n\tau^2+\sigma^2)^2} \right]^{1/2} d\sigma^2.$$

Hence

$$\pi_2^l(\tau^2,\sigma^2|\mu) = \frac{\left[\frac{p(n-1)}{2\sigma^4} + \frac{p}{2(n\tau^2+\sigma^2)^2}\right]^{1/2} 1_{(c_l,d_l)}(\sigma^2)}{\Psi_1(\tau^2|c_l,d_l)} \times \frac{\exp\left\{\frac{1}{2}\Psi_2(\tau^2|c_l,d_l)\Big/\Psi_1(\tau^2|c_l,d_l)\right\} 1_{(e_l,f_l)}(\tau^2)}{K(c_l,d_l,e_l,f_l)},$$

where

$$K(c_l,d_l,e_l,f_l) = \int_{e_l}^{f_l} \exp\left\{\tfrac{1}{2}\Psi_2/\Psi_1\right\} d\tau^2.$$

*Iteration for* $j=1$: Since neither $\boldsymbol{h}_1$, nor $\pi_2^l$ depend on $\mu$,

$$E^l\left[\log|\boldsymbol{h}_1(\mu,\tau^2,\sigma^2)| \;\big|\; \mu\right] = k^*(c_l,d_l,e_l,f_l).$$

Hence

$$\pi^l(\mu,\tau^2,\sigma^2) = \pi_1^l(\mu,\tau^2,\sigma^2) = \frac{\pi_2^l(\tau^2,\sigma^2|\mu)}{(b_l-a_l)}\, 1_{(a_l,b_l)}(\mu).$$

*Finish*: Choosing, say, the fixed point $(\mu,\tau^2,\sigma^2) = (0,1,1)$, one obtains (ignoring multiplicative constants)

$$\pi(\mu,\tau^2,\sigma^2) = \lim_{l\to\infty}\left\{ \frac{\left[\frac{p(n-1)}{2\sigma^4} + \frac{p}{2(n\tau^2+\sigma^2)^2}\right]^{1/2}}{\Psi_1(\tau^2|c_l,d_l)/\Psi_1(1|c_l,d_l)} \times \exp\left\{ \frac{\Psi_2(\tau^2|c_l,d_l)}{2\Psi_1(\tau^2|c_l,d_l)} - \frac{\Psi_2(1|c_l,d_l)}{2\Psi_1(1|c_l,d_l)} \right\}\right\}. \qquad (4.2.3)$$

To determine this limit, note first that changing variables from $\sigma^2$ to $v = \tau^2/\sigma^2$ in the integrals defining $\Psi_1$ and $\Psi_2$ yields

$$\Psi_1(\tau^2|c_l,d_l) = \sqrt{\frac{p}{2}} \int_{\tau^2/d_l}^{\tau^2/c_l} \frac{1}{v}\left[n-1+\frac{1}{(1+nv)^2}\right]^{1/2} dv,$$

$$\Psi_2(\tau^2|c_l,d_l) = -\sqrt{\frac{p}{2}} \int_{\tau^2/d_l}^{\tau^2/c_l} \left[\log\frac{2\tau^4}{pn^2} + \log\left(\frac{1}{(n-1)v^2} + (n+v^{-1})^2\right)\right]$$
$$\times\frac{1}{v}\left[n-1+\frac{1}{(1+nv)^2}\right]^{1/2} dv.$$

Divide the integrals up into integrals over the regions $(\tau^2/d_l, \varepsilon)$, $(\varepsilon, \varepsilon^{-1})$, and $(\varepsilon^{-1}, \tau^2/c_l)$. Clearly

$$\begin{aligned}\int_{\tau^2/d_l}^{\varepsilon} \frac{1}{v}\left[n-1+\frac{1}{(1+nv)^2}\right]^{1/2} dv &= \int_{\tau^2/d_l}^{\epsilon} \left[\frac{\sqrt{n}}{v} + O(1)\right] dv \\ &= \sqrt{n}\left[\left(\log\frac{\varepsilon}{\tau^2}\right) + \log d_l\right] + O(\epsilon)\end{aligned}$$

Likewise,

$$\begin{aligned}&\int_{\tau^2/d_l}^{\varepsilon} \log\left(\frac{1}{(n-1)v^2} + (n+v^{-1})^2\right)\frac{1}{v}\left[n-1+\frac{1}{(1+nv)^2}\right]^{1/2} dv \\ =& \int_{\tau^2/d_l}^{\varepsilon} \left[\log\left(\frac{n}{(n-1)v^2}\right) + O(v)\right]\left[\frac{\sqrt{n}}{v} + O(1)\right] dv \\ =& \left(\log\frac{n}{n-1}\right)\sqrt{n}\left(\log\frac{\varepsilon}{\tau^2} + \log d_l\right) + \sqrt{n}\left[(\log\frac{\tau^2}{d_l})^2 - (\log\varepsilon)^2\right] + O(\varepsilon).\end{aligned}$$

For the range $(\varepsilon^{-1}, \tau^2/c_l)$ we get

$$\begin{aligned}\int_{\varepsilon^{-1}}^{\tau^2/c_l} \frac{1}{v}\left[n-1+\frac{1}{(1+nv)^2}\right]^{1/2} dv &= \int_{\varepsilon^{-1}}^{\tau^2/d_l} \left(\frac{\sqrt{n-1}}{v} + O\left(\frac{1}{v^3}\right)\right) dv \\ &= \sqrt{n-1}\left(\log\varepsilon\tau^2 - \log c_l\right) + O(\varepsilon^2)\end{aligned}$$

and

$$\begin{aligned}&\int_{\varepsilon^{-1}}^{\tau^2/c_l} \log\left(\frac{1}{(n-1)v^2} + (n+v^{-1})^2\right)\frac{1}{v}\left[n-1+\frac{1}{(1+nv)^2}\right]^{1/2} dv \\ =& \int_{\varepsilon^{-1}}^{\tau^2/c_l} \log\left(n^2 + 0\left(\frac{1}{v}\right)\right)\left[\frac{\sqrt{n-1}}{v} + O\left(\frac{1}{v^3}\right)\right] dv \\ =& \sqrt{n-1}(\log n^2)(\log\frac{\varepsilon\tau^2}{c_l}).\end{aligned}$$

To complete the specification, define

$$K_1(\varepsilon) = \int_{\varepsilon}^{\varepsilon^{-1}} \frac{1}{v}\left[n - 1 + \frac{1}{(1+nv)^2}\right]^{1/2} dv,$$

$$K_2(\varepsilon) = \int_{\varepsilon}^{\varepsilon^{-1}} \log\left(\frac{1}{(n-1)v^2} + (n+v^{-1})^2\right)\frac{1}{v}\left[n - 1 + \frac{1}{(1+nv)^2}\right]^{1/2} dv.$$

Collecting terms yields

$$\begin{aligned}\Psi_1(\tau^2|c_l,d_l) &= \sqrt{\frac{p}{2}}\Big\{\sqrt{n}\log d_l - \sqrt{n-1}\log c_l + (\log\varepsilon)\left(\sqrt{n}+\sqrt{n-1}\right)\\ &+ (\log\tau^2)(\sqrt{n-1}-\sqrt{n}) + K_1(\varepsilon) + O(\varepsilon)\Big\}\\ \Psi_2(\tau^2|c_l,d_l) &= -\log\frac{2\tau^4}{pn^2}\Psi_1(\tau^2|c_l,d_l) - \sqrt{\frac{p}{2}}\Big\{\sqrt{n}(\log d_l)^2\\ &- (\log d_l)2\sqrt{n}(\log\tau^2) - 2\sqrt{n-1}\log n\log c_l\\ &+ \sqrt{n}\log\frac{n}{(n-1)}\log d_l + K^*(\varepsilon,\tau^2) + O(\varepsilon)\Big\},\end{aligned}$$

where $K^*$ is a function of $\varepsilon$ and $\tau^2$.

As $l \to \infty$, $\log d_l$ and $\log c_l^{-1}$ converge to infinity, while all other terms in $\Psi_1$ and $\Psi_2$ remain constant. (The $O(\varepsilon)$ terms are uniform in $l$.) Hence, it follows immediately that

$$\lim_{l\to\infty}\frac{\Psi_1(\tau^2|c_l,d_l)}{\Psi_1(1|c_l,d_l)} = 1. \tag{4.2.4}$$

More difficult is the limit of $\Psi_2/\Psi_1$. Defining

$$\begin{aligned}R(\varepsilon,\tau^2) &= (\log\varepsilon)\left(\sqrt{n}+\sqrt{n-1}\right) + (\log\tau^2)\left(\sqrt{n-1}-\sqrt{n}\right) + K_1(\varepsilon),\\ S(\varepsilon,\tau^2,c_l,d_l) &= \sqrt{n}(\log d_l)^2 - (\log d_l)2\sqrt{n}(\log\tau^2) + \sqrt{n}\log\left(\tfrac{n}{n-1}\right)\log d_l\\ &\quad -2\sqrt{n-1}(\log n)\log c_l + K^*(\varepsilon,\tau^2),\end{aligned}$$

and $cdn = \sqrt{n}(\log d_l) - \sqrt{n-1}\log c_l$, we have

$$\begin{aligned}&\frac{\Psi_2(\tau^2|c_l,d_l)}{\Psi_1(\tau^2|c_l,d_l)} + \log\frac{2\tau^4}{pn^2}\\ =\ &\frac{-\left\{S(\varepsilon,\tau^2,c_l,d_l)+O(\varepsilon)\right\}}{\left\{cdn + R(\varepsilon,\tau^2)+O(\varepsilon)\right\}}\end{aligned}$$

$$
\begin{aligned}
&= \frac{-\{S(\varepsilon,\tau^2,c_l,d_l)+O(\varepsilon)\}}{cdn}\left[1-\frac{R(\varepsilon,\tau^2)+O(\varepsilon)}{cdn}+O\left(\frac{1}{(\log d_l/c_l)^2}\right)\right]\\
&= \frac{-\sqrt{n}\left[(\log d_l)^2+\left(\log\frac{n}{n-1}\right)\log d_l\right]}{cdn}+\frac{\left[R(\varepsilon,\tau^2)+O(\varepsilon)\right]\sqrt{n}(\log d_l)^2}{(cdn)^2}\\
&+ \frac{(2\sqrt{n}(\log\tau^2)(\log d_l)+2\sqrt{n-1}(\log n)(\log c_l))}{cdn}+O\left(\frac{1}{(\log d_l/c_l)}\right).
\end{aligned}
$$

Thus

$$
\begin{aligned}
&\frac{\Psi_2(\tau^2|c_l,d_l)}{\Psi_1(\tau^2|c_l,d_l)}-\frac{\Psi_2(1|c_l,d_l)}{\Psi_1(1|c_l,d_l)}\\
&= -\log\tau^4+\frac{\left[R(\varepsilon,\tau^2)-R(\varepsilon,1)+O(\varepsilon)\right]\sqrt{n}(\log d_l)^2}{(cdn)^2}\\
&+ \frac{2\sqrt{n}(\log\tau^2)\log d_l}{cdn}+O\left(\frac{1}{(\log d_l/c_l)}\right)\\
&= -\log\tau^4-\frac{\left[(\log\tau^2)(1-\sqrt{n-1}/\sqrt{n})+O(\varepsilon)\right]}{\left[1-(\sqrt{n-1}/\sqrt{n})(\log c_l/\log d_l)\right]^2}\\
&+ \frac{2(\log\tau^2)}{\left[1-(\sqrt{n-1}/\sqrt{n})(\log c_l/\log d_l)\right]}+O\left(\frac{1}{(\log d_l/c_l)}\right).
\end{aligned}
$$

Noting that the $O(\varepsilon)$ term was uniform in $l$, and assuming that

$$\eta=\lim_{l\to\infty}\frac{\log d_l}{\log c_l^{-1}}$$

exists, we thus have, defining $\lambda=\sqrt{n-1}/\sqrt{n}$,

$$
\begin{aligned}
&\lim_{l\to\infty}\left\{\frac{\Psi_2(\tau^2|c_l,d_l)}{\Psi_1(\tau^2|c_l,d_l)}-\frac{\Psi_2(1|c_l,d_l)}{\Psi_1(1|c_l,d_l)}\right\}\\
&= -\log\tau^4+\log\tau^2\left[\frac{(\lambda-1)}{(1+\eta^{-1}\lambda)^2}+\frac{2}{(1+\eta^{-1}\lambda)}\right]+\frac{O(\epsilon)}{(1+\eta^{-1}\lambda)^2}\\
&= -(\log\tau^2)\left[1-\frac{\lambda(\eta^2-\lambda)}{(\eta+\lambda)^2}\right]+\frac{O(\epsilon)}{(1+\eta^{-1}\lambda)^2}.
\end{aligned}
$$

Observing that $O(\varepsilon)$ can be made arbitrarily small, this equation, together with (4.2.3) and (4.2.4), yields the reference prior in Table 1, with $C_n$ replaced by (4.2.2).

### 4.3. Posterior Calculations

Note that the likelihood function is proportional to

$$l(\mu,\sigma^2,\tau^2) = \sigma^{-(n-1)p}(n\tau^2+\sigma^2)^{-p/2}\exp\left\{-\tfrac{1}{2}\left[\frac{np(\mu-\overline{\overline{x}})^2}{n\tau^2+\sigma^2} + \frac{n\Sigma(\overline{x}_i-\overline{\overline{x}})^2}{n\tau^2+\sigma^2} + \frac{\Sigma\Sigma(x_{ij}-\overline{x}_i)^2}{\sigma^2}\right]\right\}.$$

Also, all of the reference priors can be written in the form

$$\pi(\mu,\sigma^2,\tau^2) = \sigma^{-\alpha}\tau^{-\beta}(n\tau^2+\sigma^2)^{-\gamma}\psi^*\left(\frac{\tau^2}{\sigma^2}\right)$$

for certain constants $\alpha, \beta, \gamma$, and $\psi^*$ either equal to $\psi$ in Table 1, or equal to 1.

Finally, suppose one is interested in evaluating the posterior expectation of a function of the form

$$\rho(\mu,\sigma^2,\tau^2) = \mu^r\sigma^s\varphi\left(\frac{\tau^2}{\sigma^2}\right),$$

a form which clearly includes all posterior moments (and cross-moments) of $\mu, \sigma$, and $\tau$. The following lemma shows that this calculation reduces to one- dimensional integration for $r = 0, 1$, or 2. (Integer $r \geq 3$ also can be handled using only one-dimensional integration, but the formula gets messier.)

**Lemma 4.1** The posterior expectation of $\rho(\mu,\sigma^2,\tau^2)$ for $r = 0, 1$, or 2, is

$$E\Big[ho(\mu,\sigma^2,\tau^2)\Big|\,\text{data}\Big] = \frac{\Gamma\left(\frac{1}{2}(c-s)\right)}{2^{s/2}\Gamma(c/2)}$$
$$\times\frac{\int_0^\infty\big[S(v)\big]^{s-c}\left\{\overline{\overline{x}}^{\,r}+\frac{(1+nv)S^2(v)}{np(c-2-s)}1_2(r)\right\}v^{\frac{-\beta}{2}}(1+nv)^{-\gamma^*}\phi(v)\psi^*(v)\,dv}{\int_0^\infty\big[S(v)\big]^{-c}v^{\frac{-\beta}{2}}(1+nv)^{-\gamma^*}\psi^*(v)\,dv},$$

where $\gamma^* = \gamma + \frac{1}{2}(p-1)$, $c = \alpha+\beta+2\gamma+np-5$, $1_2(r)$ equals 1 if $r = 2$ and equals 0 otherwise, and

$$S(v) = \left(\frac{n\sum_i(\overline{x}_i-\overline{\overline{x}})^2}{(1+nv)} + \sum_i\sum_j(x_{ij}-\overline{x}_i)^2\right)^{1/2}.$$

**Proof** First transform to the variables $(\mu,\sigma^2,v)$, where $v = \tau^2/\sigma^2$. It is then straightforward to integrate over $\mu$, followed by $\sigma^2$, yielding the result.

## 5. CONCLUDING REMARKS

The variance components example clearly illustrates the difficulties that can be encountered in applying the grouped reference prior algorithm:

(i) the grouping and ordering of the groups will frequently result in different reference priors,

(ii) the limiting process given in the algorithm can be difficult to carry out, and

(iii) the limit can depend on the compact sets chosen (and may not even yield a properposterior.)

Before addressing these points, it should be noted that it is rather rare to encounter the second and third difficulties; indeed, the variance components problem is the most pathological we have seen for the reference prior theory.

The dependence of the reference prior on the group chosen and their order is, we feel, unavoidable. Many examples exist which illustrate that no single noninformative prior will work well for all functions of a given high-dimensional parameter. As more fully discussed in Berger and Bernardo (1989b), our own preference is actually to use the reference prior corresponding to single element groups, with the groups ordered according to the inferential importance of the parameters. That different orderings of the nuisance parameters can yield different answers even has positive aspects; one can then conduct a sensitivity study over the choice of the noninformative prior.

As to the possible technical difficulty of implementing the grouped reference prior algorithm, again most examples we have seen are much easier than the variance components example. Also, the determination of the reference prior can be thought of as the theoretician's work, to be done for all common models of statistical importance.

Difficulties with existence or uniqueness of the limit in the algorithm are more troubling. It would be nice if such never occurred, and indeed occurrence of these problems is quite rare, but there are no guarantees. As in Berger and Bernardo (1989a), one can surmount the non-uniqueness problem (when present) if there is a natural sequence of compact sets that one prefers.

At the very least, the grouped reference prior algorithm can be thought of as a method for generating interesting candidate noninformative priors, either for sensitivity studies or for investigation of their performance. While our attitude is that study of the performance of noninformative priors is certainly to be encouraged, we have found the group reference priors to generally be highly satisfactory, and we would feel reasonably confident in using them in situations in which further study is impossible.

## ACKNOWLEDGMENTS

This work was supported by the U.S.-Spain. Joint Committee for Scientific and Technological Cooperation, Grant CC B8409–025, the Spanish Ministerio de Educación y Ciencia, Project PB87-0607- C02-01, and by the National Science Foundation, Grant DMS 8702620. Part of the work was also carried out while the first author was on sabbatical at Duke University. We are grateful to Ke-Ying Ye for assisting in the determination of Table 1.

## REFERENCES

Berger, J. O. and Bernardo, J. M. (1989a). Estimating a product of means: Bayesian analysis with reference priors. *J. Amer. Statist. Assoc.* **84**, 200–207.

Berger, J. O. and Bernardo, J. M. (1989b). Ordered group reference priors with applications to multinomial and variance components problems. *Tech. Rep.* **01/89**, *Dept. Estadistica.* Presidencia de la Generalidad Valenciana, Spain.

Berger, J. O., Bernardo, J. M., and Mendoza, M. (1989). On priors that maximize expected information. *Recent Developments in Statistics and their Applications* (J. Klein and J.C. Lee, eds.). Seoul: Freedom Academy. 1–20.

Bernardo, J. M. (1979). Reference posterior distributions for Bayesian inference. *J. Roy. Statist. Soc. B* **41**, 113–147 (with discussion).

Box, G. E. P. and Tiao, G. C. (1973). *Bayesian Inference in Statistical Analysis.* Reading, Mass.: Addison-Wesley.

Hill, B. M. (1965). Inference about variance components in the one-way model. *J. Amer. Statist. Assoc.* **60**, 806–825.

Jeffreys, H. (1961). *Theory of Probability.* London: Oxford University Press.

Kass, R. E. (1989). The geometry of asymptotic inference. *Statist. Sci.* **4**, 188–234.

Lindley, D. V. (1956). On a measure of the information provided by an experiment. *Ann. Math. Statist.* **27**, 986–1005.

Shannon, C. E. (1948). A mathematical theory of communication. *Bell System Tech. J.* **27**, 379–423 and 623–656.

# 11

# An Elicitation Procedure Using Piecewise Conjugate Priors


## Glen Meeden [1]



### Abstract

Although computationally convenient, conjugate priors can represent only a rather limited set of prior beliefs. By considering piecewise conjugate priors one can represent a much larger class of prior beliefs and retain much of the computational simplicity of conjugate priors. Methods of elicitation are discussed which allow one to choose a specific piecewise conjugate prior to approximate a given set of prior beliefs. The class of prior beliefs that can be approximated this way is quite large. This should allow for increased flexibility in the application of the Bayesian paradigm.



*Key Words and Phrases:* conjugate prior; exponential family; eliciting prior beliefs; piecewise conjugate.


## 1. INTRODUCTION

For many Bayesians finding an appropriate prior distribution, when faced with a specific decision problem, can be quite difficult. First, one must decide what his or her prior beliefs about the problem at hand are. Berger (1985) gives a good summary of the various elicitation methods. Then one must choose a prior to sensibly represent these beliefs. In practice, however, these two steps really cannot be separated.

It is often recommended to choose a conjugate prior whenever possible because they are so computationally convenient. This is quite a limited family however and there are many instances where they should not be used. Dalal and Hall (1983) and Diaconis and Ylvisakar (1984) consider mixtures of conjugate priors. They showed that any prior can be well approximated by a finite mixture of conjugate priors. The limitation of this work is that in a given problem with a given set of prior beliefs it is not clear how to easily find a mixture of conjugate priors which yields a good approximation.

One possibility is to approximate your prior beliefs by a discrete distribution which puts mass on a finite number of points. In many cases this


[1] Iowa State University

should yield a reasonable approximation to the given set of prior beliefs and the usual calculations involving the posterior distribution can be done with a computer. This method is fine when computing a point estimate but interpretation of a posterior credible set becomes problematical. Moreover, it is not aesthetically pleasing to have a posterior distribution concentrated on a finite set of points when your "true" beliefs allow for a whole interval of possible values for the parameter. This suggests that one usually desires a continuous prior density over the parameter.

For a given problem, what is needed is a large class of possible prior densities from which one can easily choose a prior which reasonably approximates his or her prior beliefs. Clearly such an approach will be computer intensive and not be limited to small classes of densities like the conjugate priors. One can imagine a sophisticated Bayesian statistician using a computer graphics package to draw on the computer screen a prior density over the parameter space which is a sensible approximation to his or her prior beliefs. From some family of possible densities the computer chooses one whose graph is a good approximation to the one drawn by the statistician. The computer then uses this function to carry out the Bayesian analyses desired by the statistician. I know of no such graphics package, but it points to a major difficulty in the problem of choosing a prior distribution. A knowledgable Bayesian statistician can often sketch on paper a density which reasonably approximates his or her prior beliefs. But there is just no convenient way to express this graph in functional form and get it into the computer where it is needed.

In this paper we will consider the family of piecewise conjugate priors. This is a large class of priors and we will show how one can usually choose from this class a prior which closesly approximates his or her prior beliefs and use the computer to carry out the desired analysis. This should allow for increased flexibility in the application of Bayesian methods. In Section 2, the problem of choosing a prior for the Binomial parameter $p$ will be discussed. In Section 3, the technique will be extended to the one parameter exponential family with emphasis on the problem of choosing a prior for a Normal mean.

## 2. CHOOSING A PRIOR FOR A BINOMIAL $p$

Let $X$ be Binomial $(n,p)$ where $n$ is known and $p \in (0,1)$ is unknown. Suppose we wish to make inferences about $p$. For this problem a conjugate prior for $p$ is Beta $(\alpha,\beta)$ whose density function is

$$f(p|\alpha,\beta) = \begin{cases} \frac{\Gamma(\alpha+\beta)}{\Gamma(\alpha)\Gamma(\beta)} p^{\alpha-1}(1-p)^{\beta-1} & \text{for } 0 < p < 1 \\ 0 & \text{elsewhere} \end{cases}$$

where $\alpha > 0$ and $\beta > 0$.

We say that a density $f$ is piecewise conjugate of order 2 if there exists

$\alpha_i > 0$ and $\beta_i > 0$ for $i = 1, 2$ and $\lambda \in (0, 1)$ such that

$$\begin{aligned} f(p) \quad &\propto p^{\alpha_1 - 1}(1-p)^{\beta_1 - 1} \qquad \text{for } 0 < p \leq \lambda \\ &\propto kp^{\alpha_2 - 1}(1-p)^{\beta_2 - 1} \qquad \text{for } \lambda < p < 1 \end{aligned} \tag{1}$$

where $k = \lambda^{\alpha_1 - \alpha_2}(1-\lambda)^{\beta_1 - \beta_2}$. With this choice of $k$, $f$ is continuous at $p = \lambda$ and hence continuous on $(0, 1)$. In what follows, we will always assume this particular choice of $k$, since in most instances it seems sensible to make the prior continuous. Clearly one can extend the definition to an order $r > 2$ by breaking up the interval $(0, 1)$ into $r$ disjoint intervals. For such a prior, for each $x$, the posterior is also piecewise conjugate whose normalizing constant can easily be found in closed form. This means that for any $x$ the posterior can be found explicitly and the desired computation can be carried out by the computer.

As an example consider the prior given by (1) where $\alpha_1 = 1, \beta_1 = 5, \alpha_2 = 3, \beta_2 = 2$, and $\lambda = 1/2$. This prior has its maximum value at $p = 0$ and is strictly decreasing for $p \in (0, .5)$, strictly increasing for $p \in (.5, .6)$ and strictly decreasing for $p \in (.6, 1)$ down to its minimum of zero at $p = 1$. This prior cannot be well approximated by any conjugate prior, but because it is a piecewise conjugate prior it is easy to specify and the posterior is easy to use. Table 1 gives its posterior expectations when $n = 10$.

TABLE 1: The posterior expectations of $p$

| X | 0 | 1 | 2 | 3 | 4 | 5 | 6 | 7 | 8 | 9 | 10 |
|---|---|---|---|---|---|---|---|---|---|---|---|
| E(p—x) | .062 | .125 | .189 | .255 | .331 | .425 | .534 | .638 | .725 | .798 | .866 |

In most instances a knowledgable Bayesian should be able to choose a piecewise conjugate prior that is a good approximation to his or her beliefs. For example, suppose the mode of your posterior is at one half but your beliefs are not symmetric about one half. Then in (1) you could take $\lambda = 1/2$ and $\alpha_1 = \beta_1 = \alpha$, say, and $\alpha_2 = \beta_2 = \beta$, say, where $\alpha \neq \beta$. By letting the ratio $\alpha/\beta$ vary one has a class of skewed distributions some of which are quite far from any Beta.

Note that the flexibility of the piecewise conjugate priors increases with their order. One can approximate more and more functions well as we partition the unit interval into more and more subintervals. A simple example related to the first example is the prior

$$g(\theta) \propto \begin{cases} (1-\theta)^4 & \text{for } 0 < \theta < 1/2 \\ 1/2^4 & \text{for } 1/2 \leq \theta < 3/4 \\ (1-\theta)^2/3^2 & \text{for } 3/4 \leq \theta \end{cases}$$

which again emphasizes how easy a piecewise conjugate prior is to use.

For someone who is familiar with the mathematical properties of the Beta functions, making a rough sketch of his or her prior densities and then guessing a good piecewise conjugate approximation should lead to a sensible approximation of the prior approximation. Given the order of the piecewise conjugate approximation and the $\lambda$'s dividing the subintervals and the $\alpha_i$'s and $\beta_i$'s for each subinterval it is straightforward to find the normalizing constant and have the computer graph the approximation and compute the amount of probability assigned to each subinterval. If the statistician is satisfied with the approximation he or she can proceed with the analysis. If not, one can try to adjust it until one finds a satisfactory approximation.

For a less knowledgable user a simpler approach might work better. For example, one might ask the user to break the unit interval into some finite number of subintervals and then compare the relative amount of probability assigned to successive subintervals. We will now show how such information can be used to find a piecewise conjugate prior which should yield a good approximation to their prior beliefs.

Let $g$ denote the prior which will be constructed by considering successive subintervals of $[0, 1]$. In what follows we assume that $g$ is only being defined up to a constant. When the process is completed it will be normalized to integrate to one. Suppose $g$ has already been defined on the subinterval $[0, a]$ and $\int_0^a g(p)dp = c$ is known. The next step is to extend its definition to the subinterval $(a, b]$ where $a < b \leq 1$. Let $\gamma = g(a)$. Assume that our prior beliefs forces $g$ to be strictly increasing on $(a, b]$ and that we can assert the value of the ratio

$$r = \int_a^b g(p)dp / \int_0^a g(p)dp. \tag{2}$$

Since $g$ is to be strictly increasing on $(a, b]$ then it must be the case that $g(a)(b-a) \leq rc$, i.e. that $r \geq \gamma(b-a)/c$. The case where $g$ must be strictly decreasing on $(a, b]$ can be handled in an analogous fashion. Hence to use this method the statistician must choose some $b > a$ such that his or her prior density is monotone on $(a, b]$ and the relative amount of probability to be assigned to this interval as compared to the probability assigned to $(0, a]$ can be specified.

As we will soon see, given the above conditions defined by $\gamma, c$ and $r$ one can always find $s > 0$ and $k(s) > 0$ such that, if

$$g(p) = k(s)p^s \text{ for } a < p \leq b$$

then condition (2) is satisfied and $k(s)a^s = \gamma$, i.e., $g$ is continuous at $a$. Note also that over $(a, b], g$ is proportional to a Beta density and so the density $g$ produced by this method must be a piecewise conjugate density.

To see how the method would work in practice we consider a simple example. Suppose our prior beliefs suggest a strictly increasing prior over $(0, 1)$. We will break up the interval in three subintervals $(0, .3], (.3, .8]$ and $(.8, 1)$. We will assume that our prior $g(p)$ is proportional to $p$ over $(0, .3]$. Note $\int_0^{.3} pdp = .045$. Next we assume that our prior assigns nine times as much probability to the interval (.3, 8] as to the interval (0, .3]. Finally, we assume that our prior assigns two thirds as much probability to the interval (.8, 1) as to the interval (.3, .8]. Now there are many densities which satisfy these conditions. One such density can be found using the method outlined above and is given by

$$g_1(p) \propto \begin{cases} p & \text{for } 0 < p \leq .3 \\ 436p^{.31} & \text{for } .3 < p \leq .8 \\ 827p^{3.179} & \text{for } .8 < p < 1. \end{cases} \tag{3}$$

One can find the normalizing constant for $g_1$ but in most applications of the posterior this is not necessary. If after viewing the prior the statistician feels that it is not a good approximation to his or her prior beliefs, then a new prior can be produced by considering a finer partition of subintervals. In most instances one should be able to find a reasonable approximation to his or her prior beliefs.

If is easy to see why the method outlined above must always lead to a well specified prior which satisfies the given constraints. To see this, let $a < b$ and $\gamma > 0$ be specified real numbers. Let $f$ be a continuous real valued function defined on $[a, b]$ which is strictly increasing over the interval with $f(a) > 0$. Now given a real number $M > 0$ we would like to find a real number $s$ and a constant $k(s) > 0$ such that

$$k(s)[f(a)]^s = \gamma \tag{4}$$

and

$$k(s) \int_a^b [f(x)]^s dx = M. \tag{5}$$

Using (4) we can rewrite the left-hand side of (5) as

$$w(s) = \gamma \int_a^b [f(x)/f(a)]^s dx. \tag{6}$$

For each $x \epsilon (a, b]$ we have $f(x)/f(a) > 1$ and hence the integrand at each such $x$ is a strictly increasing function in $s$ and hence $w(s)$ is a strictly increasing function in $s$ over the real numbers. Note that $w(0) = \gamma(b - a)$ while $\lim_{s \to \infty} w(s) = \infty$ and $\lim_{s \to \infty} w(s) = 0$. Hence, for any given $M > 0$ there exists a real number $s$ and $k(s) > 0$ such that (4) and (5) are satisfied. When $0 < M < \gamma(b-a)$ then the $s$ which gives inequality in (4) is negative and the function $\gamma[f(x)/f(a)]^s$ is strictly decreasing over the interval $[a, b]$.

Returning to the binomial example we see that given $\gamma = g(a)$ and $0 < a < b \leq 1$ and $r$ and $c$ and taking $f(x) = x$ there exists a unique real number $s$, such that

$$\gamma \int_a^b (x/a)^s dx = rc. \tag{7}$$

To find the solution we need to solve the equation

$$[a\gamma/(s+1)][(b/a)^{s+1}-1]=rc \tag{8}$$

which can be easily done on a computer. This was how the density given in (3) was found.

Let us return to the simple example with the prior given in (3). Suppose that our previous specification remains unchanged over the first two intervals but that now we want the probability assigned to (.8, 1) to be one ninth the probability assigned to the interval (.3, .8]. The method gives the prior $g_2(.)$ where

$$g_2(p) \propto \begin{cases} g_1(p) & \text{for } 0 < p \leq .8 \\ 0.00948p^{-16.85} & \text{for } .8 < p < 1. \end{cases} \tag{9}$$

Since this new specification forces the prior to be strictly decreasing on (.8, 1) we could use $f(x) = 1-x$ instead of $x$ and then find the appropriate power by solving the analogous equation to (8). In fact, we should have chosen $f(x) = 1 - x$ if we wanted our prior to go to zero as $p$ approaches one. Since in this case there was so little probability assigned to (.8, 1) even with the choice of $f(x) = x$ the prior found was still very small for $p$ close to one.

If more probability were given to the interval (.8, 1) then the choices of $f(x) = 1-x$ or $f(x) = x$ would give different answers near one. Similarly, if one wanted a prior that did not go to zero as $p$ approaches zero one should take $f(x) = 1 - x$.

One might argue that the density given in (9) is not a piecewise Beta density because of the factor $p^{-16.85}$ over (.8, 1). Suppose, however, one assumes a priori that, for the problem under consideration, $p \in [t, 1]$ where $0 < t < 1$. Then following the usual definition of conjugate prior (for example, see DeGroot (1970)) one finds that a conjugate prior is proportional to $p^{\alpha}(1-p)^{\beta-1}$ for $\beta > 0$ and $-\infty < \alpha < \infty$. In this case we need no restriction on the exponent of the factor $p$ since $p$ is bounded away from zero. Similarly, if $p$ were bounded away from one we would need no restriction on the exponent of the factor $(1-p)$. So under this natural extension of the notion of being conjugate the prior given in (9) is piecewise conjugate.

It is intuitively obvious that any sufficiently smooth density on the unit interval can be approximated as closely as desired by a piecewise Beta density. In the method described above we considered only a very small subset of such piecewise Beta densities in finding our approximation. On each subinterval we consider only pieces of the form $kp^r$ or $k(1-p)^r$. One could in general consider pieces of the type $kp^{r_1}(1-p)^{r_2}$. The assumptions made concerning the function $f$ in the argument leading to equation (6) are unnecessarily restrictive. For example, $w(.)$ is still strictly increasing whenever $f(x) \geq f(a)$ for $x \in [a, b]$ with strict equality for at least one $x$. However, for the more general $f$ one cannot conveniently integrate (7) and find an equation analogous to (8). Moreover there is really no need to consider the more general case because the two special cases of $f(p) = p$ and $f(p) = 1 - p$ work quite well.

We have made the piecewise conjugate priors continuous by their definition. They typically will be differentiably everywhere except at the boundary points where two different pieces meet. If a boundary point is a local minimum of a piecewise conjugate prior then its graph may have a cusp at this point. Although this is an unappealing property, in practice it will usually have little effect on the posterior. As we will soon see, in some cases one can eliminate cusps from the prior if this is deemed necessary.

In summary, to find a piecewise conjugate prior which approximates your prior beliefs, you first partition the unit interval into a set of successive subintervals. You then pick one subinterval and specify your prior, up to a constant, over this subinterval and calculate its integral over the first subinterval. Call this value $c$. If the first subinterval has zero as its left-hand endpoint or one as its right-hand endpoint, pick its neighboring subinterval. If the first subinterval is in the interior of the unit interval, pick either one of its neighboring subintervals. For this second interval specify $r$ such that it is the quotient of the probability you assigned to the first interval. Then choose $f(x) = x($ or $f(x) = 1 - x)$ and find the value of $s$ which is the solution to equation (8) (or the analogous equation for $f(x) = 1 - x$) and find $k = k(s)$ from (4). You now have your prior defined, up to a constant, on the second interval. You next consider a subinterval which is a neighbor of one of the first two subintervals and repeat this process. You continue in this way until your prior is defined, up to a constant, over the whole interval.

In the simple example given above we considered the subintervals in increasing order. There is no need to do this however. In fact, if your prior is unimodal it might be more sensible to first consider the subinterval which contains the mode and choose your prior proportion to $p^{s_1}(1-p)^{s_2}$ for some choice of $s_1$ and $s_2$ where $s_1/(s_1 + s_2)$ is in the given subinterval. This guarantees that there will not be a cusp at your mode. Moreover you begin by assessing the prior in its center where more of your probability is apt to be situated rather than tails which could contain negligible probability.

## 3. THE ONE PARAMETER EXPONENTIAL FAMILY

The methods discussed in the previous section for approximating a prior for $p$, the probability of success in the Binomial distribution, extend easily to the one parameter exponential family. In this section, we consider the Poisson distribution and Normal distribution. We will conclude with a few general remarks about the method.

Suppose after observing a Poisson random variable with mean $\theta$ we wish to make inferences about $\theta$. A conjugate prior for this problem is a Gamma

$(\alpha, \beta)$ distribution whose density is

$$f(\theta|\alpha,\beta) = \begin{cases} \frac{\beta^\alpha}{\Gamma(\alpha)}\theta^{\alpha-1}\bar{e}^{\beta\theta} & \text{for } \theta > 0, \\ 0 & \text{elsewhere} \end{cases}$$

where $\alpha > 0$ and $\beta > 0$. If $\theta$ is restricted to a closed finite interval which does not contain zero, then we can take as the family of conjugate priors all densities which are proportional to $\theta^\alpha \bar{e}^{\beta\theta}$ over the interval where $-\infty < \alpha < \infty$ and $-\infty < \beta < \infty$. That is, we need no restrictions on $\alpha$ and $\beta$ over such an interval. Just as before a knowledgeable Bayesian statistician can approximate his prior by piecing together curves of the above form for various choices of $\alpha$ and $\beta$. Or one can use the method of the previous section where the prior is specified, up to a constant, on just one subinterval and then for the remaining subintervals only successive ratios of probabilities are specified.

On the interval $[a, b]$ where $0 < a < b$ one can take $f(x) = x$ or $f(x) = e^{-x}$ and use the fact that in each case $w(s)$, defined in equation (6), is a strictly increasing function of $s$ to show there exists some power of the function $f$ which assigns the right amount of probability to the interval. For a given $r$ and $c$, defined as before, and $f(x) = e^{-x}$ the analogous equation, which we must solve for $s$, to that given in (8) for $f(x) = x$ is

$$w(s) = \frac{\gamma}{s}(d^s - 1) = rc \tag{10}$$

where $d = \exp(b - a)$. Note that equation (10) is very similar to equation (8). As before, for a given $c, r, d$, and $\gamma$, a value of $s$ which gives equality can easily be found using the computer. Just as in the Beta case we work with only one of the two factors that define the conjugate prior. If we try to use both of them, say $f(x) = x^{t_1}e^{-t_2 x}$ then solving equation (10) becomes much more difficult.

We now turn our attention to making inferences about a normal mean. For simplicity we assume the variance of the normal distribution is known. If this is not the case we can define our prior for the mean conditional on a fixed value of the variance and then incorporate a prior for the variance as well. We let $X$ be a Normal $(\theta, 1)$ random variable. Given $X$ we wish to make inferences about $\theta$. For this problem a conjugate prior is just another normal density. Even some very simple piecewise normal densities dramatically increase the class of prior beliefs which can be easily approximated.

For example, suppose your prior beliefs are unimodal about some point, say zero, but are definitely skewed to one side or the other. A normal prior cannot be used in this case. Consider the prior given by

$$g(\theta) \propto \begin{cases} \exp(-t_1\theta^2/2) & \text{for } \theta \geq 0, \\ \exp(-t_2\theta^2/2) & \text{for } \theta \leq 0 \end{cases} \tag{11}$$

where $t_1 > 0$ and $t_2 > 0$. Such a density was considered in Fechner (1897). See Runnenburg (1978) for a discussion of his work. This density was also used in Geweke (1986). Many skewed densities (even many defined on a finite interval) can be approximated fairly well by one of the densities given in (11). As the Fechner reference makes clear examples of piecewise conjugate priors have been known for a long time. Lad(1974) also discusses such priors. Bansal (1988) is a recent example which also makes use of piecewise conjugate priors.

If $X$ is Normal $(\theta, 1)$ and $\theta$ has density $g$ given in (11) and $X = x$ is observed then

$$p(\theta|x) \propto \begin{cases} \exp\{-\frac{1+t_1}{2}(\theta - \frac{x}{1+t_1})^2 - \frac{t_1}{1+t_1}\frac{x^2}{2}\} & \text{for } \theta \geq 0 \\ \exp\{-\frac{1+t_2}{2}(\theta - \frac{x}{1+t_2})^2 - \frac{t_2}{1+t_2}\frac{x^2}{2}\} & \text{for } \theta \leq 0 \end{cases} \tag{12}$$

Note that we cannot ignore the terms $t_i x^2/2(1+t_i)$ even though they do not directly involve $\theta$ since each of them only appears in part of the range of $\theta$ . If we let $\sqrt{2\pi}K(x)$ be the normalizing constant for (12) and

$$K_i(x) = K(x)(1+t_i)^{-1/2}\exp\{-\frac{t_i}{1+t_i}\frac{x^2}{2}\}, i = 1, 2,$$

then

$$p(\theta|x) = \begin{cases} (K_1(x))^{-1}(\frac{1+t_1}{2\pi})^{1/2}\exp\{-\frac{1+t_1}{2}(\theta - \frac{x}{1+t_1})^2\} & \text{for } \theta \geq 0 \\ (K_2(x))^{-1}(\frac{1+t_2}{2\pi})^{1/2}\exp\{-\frac{1+t_2}{2}(\theta - \frac{x}{1+t_2})^2\} & \text{for } \theta \leq 0. \end{cases}$$

One can also use the method of assigning the ratio of probabilities of successive subintervals. The natural choice of the function $f$ in equations (4), (5) and (6) is $f(x) = \exp(-(x-\mu)^2/2)$. For a given subinterval $[a, b]$, where $a < b$ ,we assume that $\mu \leq a$ or $\mu \geq (a+b)/2$. Since if $\mu \leq a$ then $f(x) \leq f(a)$ for $x \in [a, b]$ while if $\mu \geq (a+b)/2$ then $f(x) \geq f(a)$ for $x \in [a, b]$. For such an $f$ and a given $\gamma > 0$ equation (6) becomes

$$w(s) = \gamma(2\pi/s)^{1/2}(f(a))^{-s}\{\Phi(\sqrt{s}(b-\mu)) - \Phi(\sqrt{s}(a-\mu))\}, \tag{13}$$

where $\Phi$ is the cumulative distribution function of a standard normal distribution. Because of the appearance of $\Phi$, setting this equation equal to a constant and solving for $s$ yields a more complicated equation than equations (8) or (10). It can be simplified a bit by taking either $\mu = a$ or $\mu = b$. However, we can choose $f(x) = e^x$ to get the simplier form of the equation for $w(s)$ and still have the posterior be proportional to some normal density. This follows from an easy to check fact that, if $X$ is Normal $(\theta, 1)$ and the prior for $\theta$ is proportional to $\exp(\theta s)$ over $[a, b]$, then

$$p(\theta|x) \propto e^{-\frac{1}{2}\{(\theta-(x+s))^2 - (x^2+(x+s)^2)\}} \text{ for } \theta \in [a, b].$$

This suggests that in the normal case, for ease of computation, one should approximate his or her prior over successive subintervals by powers of the exponential density rather than powers of a normal density.

In summary, we have suggested that one should choose a prior from the family of piecewise conjugate priors. This is a large family from which one can find a prior to approximate a wide variety of prior beliefs. In addition to this flexibility, it retains the essential computation simplicity of the family of conjugate priors. We have outlined two approaches for choosing such a prior in practice. The first, for a more sophisticated user, has the statistician choosing directly pieces of conjugate priors which reflect his or her prior beliefs. The second, for a less mathematical knowledge user, has the statistician specifying the ratios of probability of successive subintervals. In either case, after specifying the prior, the statistician should have the computer graph his or her prior to see if it is satistfactory. The second approach just chooses one specific prior from the class of priors which satisfy the given conditions. A natural concern is how much variablility there is in the class of possible posterior distributions which arise from all the priors which satisfy the given restraints. Such questions have been discussed in the literature. See, for example, O'Hagen and Berger (1988). Their calculations would be helpful in deciding the appropriate number of subintervals when using the second approach. It is straightforward to write a computer program to implement the first approach. The software necessary for the second approach can be more complicated depending on the intended audience.

## REFERENCES

Bansal, A.K. (1988). Robust Bayes analysis with stochastic mixture of prior distributions, presented at Indo-U.S. Workshop on Bayesian Analysis in Statistics and Econometrics, Bangalore, India.

Berger, J. O. (1985). *Statistical Decision Theory and Bayesian Analysis,* 2nd ed. Springer-Verlag, New York.

Dalal, S. R. and Hall, W. J. (1983). Approximating priors by mixtures of natural conjugate priors. *J. Roy. Statist. Soc.* **(Ser. B), 45**, 278-286.

DeGroot, M. H. (1970). *Optimal Statistical Decisions.* McGraw-Hill, New York.

Diaconis, P. and Ylvisaker, D. (1984). Quantifying prior opinions. In *Bayesian Statistics* II, J. M. Bernardo, M. H. DeGroot, D. V. Lindley and A.F.M. Smith (Eds.). North-Holland, Amsterdam.

Fechner, G. Th. (1897). *Kollektivmasslehre.* Leipzeg, Engelman.

Geweke, J. (1986). Bayesian inference in econometric models using Monte Carlo integration.

Lad, Frank (1974), *On the Adequacy of Natural Conjugate Distributions for Representing Prior Information*, University of Michigan Doctoral Dissertation.

O'Hagen, A. and Berger, J. O. (1988). Ranges of posterior probabilities for quasiunimodal priors with specified quantiles. *J. Amer. Stat. Assoc.*, **83**, 503-507.

Runnenburg, J. Th. (1978). Mean, median, mode. *Statistica Neerlandica,* **32**, 73-80

## COMMENTS

Bruce M. Hill
University of Michigan

Professor Meeden offers an interesting class of distributions with which Bayesians can make computations. Certainly, there is need for additional such procedures. As he notes, Lad (1974) had used similar distributions in his University of Michigan doctoral dissertation. Also, Sid Dalal has used mixtures to deal with the same type of problems. Although my mind is open, I tend to prefer the mixture approach. It represents much the same sort of thing, but in addition has a more natuiral subjectivistic interpretation. Real distributions are often representable as complex mixtures of standard distributions. See also my discussion of the article by Berger (1984, p. 135), and Hill (1987), which presents a theory of mixtures in connection with $A_n$.

It was good to see the careful historical research by Meeden. As Bayesian statistics becomes dominant, it is important that it reaches a state of maturity with respect to attention to past research. In part because of the controversy regarding Bayesian statistics, much valuable research has been neglected, even when it appears in major journals. Many of us have horror stories concerning appropriation of our results without reference, or perhaps even with reference, and it is hoped that the good example of Meeden will help to prevent such things in the future. Otherwise some will continue to reinvent the wheel, with corners, a saying I first learned from Phil Dawid.

## REJOINDER

I wish to thank Professor Hill for his thoughtful comments. He tends to prefer the mixture approach for eliciting and specifying prior distributions, since he believes they often have a more natural subjective interpretation. My own feelings about the mixture approach are somewhat different. I find it to be an elegant and appealing mathematical theory which is often quite awkward to implement in practice. Perhaps I just don't know the proper way to think about mixtures. On the other hand the elicitation procedure described here is simple and direct. This simplicity, along with its flexibility, should make it easy to find a realistic prior for a variety of situations. In any case, one need not always choose either the mixture approach or the piecewise conjugate approach. Rather, in each particular instance we should choose the one which seems most sensible and not restrict ourselves to the family of conjugate priors.

## ADDITIONAL REFERENCES

Berger, James (1984), "The robust Bayesian viewpoint," in *Robustness of Bayesian Analysis*, J. Kadane, ed., North-Holland: Amsterdam, 63-124 (with discussion).

Hill, Bruce M. (1987), "Parametric models for $A_n$: splitting processes and mixtures," unpublished University of Michigan technical report.

# 12

# Small Worlds and State Dependent Utilities


## Mark J. Schervish, Teddy Seidenfeld & Joseph B. Kadane [1]



**Abstract**

Several axiom systems for preference amongst gambles lead to the existence of a unique probability and a state independent utility such that gambles are ranked according to their expected utilities. These axioms have been used as a foundation for Bayesian decision theory and the subjective probability calculus. We note that the uniqueness of the probability is relative to the choice of what counts as a constant outcome. Although it is sometimes clear what should be considered constant, there are many cases in which there are several possible choices. Each choice can lead to a different "unique" probability and utility. By focusing attention on state dependent utilities, we hope to determine conditions under which a truly unique probability and utility can be determined from an agent's expressed preferences amongst gambles.



*Key words and Phrases:* Expected utility; State dependent utility; Preference orderings; Foundation; Decision theory; Rational behavior; Subjective probability.


## 1. INTRODUCTION

Several axiom systems for preference amongst acts lead to the existence of a unique probability and a unique (up to positive affine transformations) state independent utility such that acts are ranked according to their expected utilities. These axioms have been used as a foundation for Bayesian decision theory and the subjective probability calculus. For example, suppose that there are $n$ states of nature which form the set $S = \{s_1, \ldots, s_n\}$ and $m$ prizes in the set $Z = \{z_1, \ldots, z_m\}$. A simple type of act might be a function $f$ mapping $S$ to $Z$. That is, if $f(s_i) = z_j$, then we receive prize $z_j$ if state $s_i$ occurs. (We will consider more complicated acts than this later.) Now, suppose that there is a probability over the states such that $p_i = \Pr(s_i)$ and that there is a utility $U$ over prizes. To say that acts are


[1] Carnegie Mellon University

ranked by expected utility means that we strictly prefer act $g$ to act $f$ just in case

$$\sum_{i=1}^{n} p_i U(f(s_i)) < \sum_{i=1}^{n} p_i U(g(s_i)). \tag{1}$$

If we allow the utilities of prizes to vary conditional on which state of nature occurs, we can rewrite (1) as

$$i = 1^n p_i U_i(f(s_i)) < \sum_{i=1}^{n} p_i U_i(g(s_i)), \tag{2}$$

where $U_i(z_j)$ is the utility of prize $z_j$ given that state $s_i$ occurs. However, without restrictions on the degree to which $U_i$ can differ from $U_{i'}$ for $i \neq i'$, the uniqueness of the personal probability no longer holds. For example, let $q_1, \ldots, q_n$ be another probability over the states such that $p_i > 0$ if and only if $q_i > 0$. Then, for an arbitrary act $f$,

$$\sum_{i=1}^{n} q_i V_i(f(s_i)) = \sum_{i=1}^{n} p_i U_i(f(s_i)),$$

where $V_i(\cdot) = p_i U_i(\cdot)/q_i$ when $q_i > 0$ ($V_i$ can be arbitrary when $q_i = 0$.) In this case, it is impossible to determine what is an agent's personal probability by studying the agent's preferences amongst acts. Rubin (1987) notes this fact and develops an axiom system which does not lead to a separation of probability and utility. In this paper, we present an example in which it is impossible to tell which of several possible probabilities and state-independent utilities an agent is using to rank acts. The problem stems from ambiguity over what should count as "constant" acts. Because the theory of Savage (1954) makes heavy use of the concept of constant act, we examine the implications of the choice of constants for Savage's theory. In particular, in Section 4 we see how the problem of "Small Worlds," introduced by Savage, stems from alternative choices of what count as constant acts.

## 2. SAVAGE'S POSTULATES

Savage (1954) gives a set of postulates which imply the existence of a unique probability and state-independent utility to represent preferences amongst acts. These postulates rely on the use of prizes which can be considered as "constant" across states. Savage's most general acts are functions from states to prizes. The important features of Savage's theory, for this discussion, are the third postulate and the definition of probability. These are stated in terms of *events*, which are sets of states. Savage assumes that there is a *strict preference* relation $\prec$ amongst acts and to say that act $f_1$ is strictly prefered to act $f_2$ is denoted $f_1 \prec f_2$. *Weak preference* is denoted

$\preceq$ and $f_1 \preceq f_2$ if it is not the case that $f_2 \prec f_1$. We say that two acts are *indifferent* or *equivalent* if $f_1 \preceq f_2$ and $f_2 \preceq f_1$. The third postulate of Savage can be stated loosely as follows.

**Postulate 3** *(Non-technical statement) Suppose that the only difference between two acts $f_1$ and $f_2$ is that, on some event $B$, $f_1$ awards prize $z_1$ and $f_2$ awards prize $z_2$. Then $f_1 \preceq f_2$ if and only if $z_1 \preceq z_2$, where $z_i$ also stands for the constant act which awards prize $z_i$ in every state.*

This postulate says that the *relative* values of prizes cannot change between states. Savage's definition of probability assumes that the *absolute* values of prizes do not change from state to state.

**Definition 1** Suppose that $A$ and $B$ are events. We say that *$A$ is at least as likely as $B$* if, for each pair of prizes $z$ and $w$ with $z \prec w$ we have $f_B \preceq f_A$, where

$$f_A(s) = \begin{cases} w & \text{if } s \in A \\ z & \text{if } s \notin A, \end{cases}$$

$$f_B(s) = \begin{cases} w & \text{if } s \in B \\ z & \text{if } s \notin B. \end{cases}$$

The example in Section 3 suggests that it might be very difficult to find prizes with the property that their "absolute" values do not change from state to state.

## 3. AN EXAMPLE

Let the set of available prizes be various amounts of Dollars. We suppose that there are three states of nature, which we will describe in more detail later, and we suppose that the agent expresses preferences which satisfy the the postulates of Savage (1954). Furthermore, suppose that the agent's utility for money is linear. That is, for each state $i$, $U_i(\$cx) = cU_i(\$x)$. In particular, $U_i(\$0) = 0$. Now, we offer the agent three acts $f_1$, $f_2$, and $f_3$ whose outcomes are

| | State of Nature | | |
|---|---|---|---|
| | $s_1$ | $s_2$ | $s_3$ |
| $f_1$ | \$1 | \$0 | \$0 |
| $f_2$ | \$0 | \$1 | \$0 |
| $f_3$ | \$0 | \$0 | \$1 |

Suppose that the agent expresses indifference between these three acts. If we assume that the agent has a state independent utility, the expected utility of $f_i$ is $U(\$1)P(s_i)$. It follows from the fact that the three acts are indifferent, that $P(s_i) = 1/3$ for each $i$. Next, we alter the set of prizes to be various Yen amounts (the Japanese currency). Suppose that we offer the agent three Yen acts $f_4$, $f_5$, and $f_6$ whose outcomes are

State of Nature

| | $s_1$ | $s_2$ | $s_3$ |
|---|---|---|---|
| $f_4$ | 100Y | 0Y | 0Y |
| $f_5$ | 0Y | 125Y | 0Y |
| $f_6$ | 0Y | 0Y | 150Y |

If the agent were to express indifference between these three horse lotteries, and if we assumed that the agent used a state independent utility for Yen prizes, then

$$P(s_1)U(100\text{Y}) = P(s_2)U(125\text{Y}) = P(s_3)U(150\text{Y}).$$

Supposing that the agent's utility is linear in Yen, as it was in dollars, we conclude that

$$P(s_1) = 1.25P(s_2) = 1.5P(s_3). \tag{3}$$

Hence the probabilities are not all the same. It would seem incoherent for the agent to express both sets of indifferences since it appears that the agent is now committed to two different probability distributions over the three states. This is not correct, however, as we now see. Suppose that the three states of nature represent three different exchange rates between Dollars and Yen.

$$\begin{aligned} s_1 &= \{\$1 \text{ is worth } 100\text{Y}\} \\ s_2 &= \{\$1 \text{ is worth } 125\text{Y}\} \\ s_3 &= \{\$1 \text{ is worth } 150\text{Y}\} \end{aligned}$$

Suppose further that the agent can change monetary units at the prevailing rate of exchange without any penalty. As far as this agent is concerned, $f_i$ and $f_{3+i}$ are worth exactly the same for $i = 1, 2, 3$ since, in each state the prizes they award are worth the same amount. The problem that arises in this example is that the two probability distributions were constructed under incompatible assumptions. The discrete uniform probability was constructed under the assumption that $U(\$1)$ is the same in all three states, while the other probability was constructed under the assumption that $U(100\text{Y})$ was the same in all three states. Clearly these cannot both be true given the nature of the states. What saves Savage's theory is that it works *no matter which* of the two assumptions one makes. Unfortunately, this same fact makes the uniqueness of the probability relative to the choice of which prizes count as constants in terms of utility. There are two different representations of the agent's preferences by probability and state independent utility. But what is state independent in one representation is state dependent in the other. This example has serious implications for elicitation. Suppose we attempt to elicit the agent's probabilities over the three states by examining preferences amongst acts in Yen amounts, assuming that Yen prizes are constant. And suppose that the agent actually uses the state independent utility for Dollars and the discrete uniform distribution to rank acts, but we don't know this. Then, it is easy to prove (see Schervish, Seidenfeld, and Kadane 1989) that the agent will appear coherent, but that we will elicit a probability satisfying (3) instead of the actual probability being used by the agent! The expressed preferences, in this example, satisfy all of Savage's postulates, but we are still unable to determine the probabilities of the states based only on preferences. The problem becomes clearer if we allow both Dollar and Yen prizes at the same time. In that case, it is impossible for a single utility to be state independent for all prizes. In fact, Postulate 3 would no longer hold. A final note on elicitation is in order. Tversky and Kahneman (1981) describe instances of what they call *framing effects* for decisions. These obtain when an agent makes inconsistent choices among (assumed) equivalent options as a result of a mere redescription of the problem–a redescription that shifts the agent's reference point for judging relevant outcomes. For instance, adapting a theme of Markowitz (1952), whether states of wealth are described as gains or losses relative to the status quo as opposed to redescribing the same states of wealth as gains or losses from some alternative reference point leads agents to choose differently among acts leading to identical final states of wealth. Does the selection of a currency for describing monetary outcomes in the example of this section create a framing effect? Assume agents care only about monetary wealth. If agents are given choices among horse lotteries with dollar payoffs, will they make the same choices when these payoffs are converted to their state dependent yen equivalents? Or, will the agents act as if the states carry an independent, determinate probability and use the currency offered as a reference point for their decisions–framing their decisions so that first dollars and then yen have constant, state independent ulitities? We think that this question is open to empirical investigation.

## 4. SAVAGE'S "SMALL WORLDS" EXAMPLE

Savage (1954) introduces the topic of *small worlds* in his Section 5.5. An anomaly occurs in this discussion, and Savage seems to imply that it is an effect of the construction of the small world. In this section, we briefly introduce small worlds and then explain why we believe that the anomaly discovered by Savage is actually another example of the non-uniqueness illustrated in Section 3. The fact that it arose in the discussion of small worlds is a mere coincidence. We explain how precisely the same effect arises without any mention of small worlds. A small world can be thought of as a description of the states of nature in which each state can actually be partitioned into several smaller states, but we don't actually do the partitioning when making comparisons between acts. For a mathematical example, Savage mentions the following case. Consider the unit square $S = \{(x,y) : 0 \leq x,y \leq 1\}$ as the finest possible partition of the states of nature. Suppose, however, that we consider as states the subsets $\overline{x} = \{(x,y) : 0 \leq y \leq 1\}$ for each $x$. The problem which Savage discovers in this example is the following. It is possible to define small world prizes in a natural way and for preferences amongst small world acts to satisfy all of his axioms and, at the same time, consistently define prizes in the "grand world" consisting of the whole square $S$. The preferences amongst small world acts would be completely consistent with the preferences amongst grand world acts, while the probability measure determined from the small world preferences is not the marginal probability measure over the sets $\overline{x}$ induced from the grand world probability. As we will see, the problem which Savage discovers is due to using different prizes as constants in the two problems. It is not due to the small world but actually will appear in the grand world as well. Any grand world act can be considered a small world prize. In fact, the very reason for introducing small worlds is to deal with the case in which what we count as a prize turns out to actually be worth different amounts depending on which of the subdivisions of the small world state of nature occurs. So, suppose we let the grand world prizes be non-negative numbers and the grand world acts be all bounded measureable functions on $S$. The grand world probability is uniform over the square and the grand world utility is the numerical value of the prize. In order to guarantee that Savage's axioms hold in the small world, choose the small world prizes to be 0 and positive multiples of a single function $h$. Assuming that $U(h) = 1$, the small world probability of each event and the small world expected utility of each small world act can be calculated. As Savage shows, the small world expected utility of each act is just the ratio of the large world expected utility of the corresponding large world act, divided by the large world expected utility of $h$. On the other hand, the small world probability of an event is calculated as if the probability had a density $q$ which is not constant. Hence, small world acts are ranked in precisely the same order as their grand world counterparts, even though the small world probability is not consistent with the grand world probability. We claimed that the inconsistency of the two probabilities is due to the choice of "constants" and not to the small worlds. To see this, suppose we change the definition of large world constants to 0 and the positive multiples of $h$. Then an act $f$ in the original problem becomes an act $f^*$ with $f^*(x,y) = f(x,y)/h(x,y)$. It is easy to see that the original preferences between large world acts will be preserved if the uniform probability over the states is replaced by the probability with density proportional to $h$. Both the original representation and that using multiples of $h$ as constants are state independent, however each one is relative to a different choice of what is constant. The constants in one representation have different utilities in different states in the other representation. Both representations satisfy Savage's axioms, however. (Note that the small world probability constructed earlier is the marginal probability associated with the density proportional to $h$, so that Savage's small world problem evaporates when the definition of constant is allowed to change.) The point to remember is that the uniqueness of the probability-utility representation for a collection of preferences is relative to what counts as a constant.

## 5. DISCUSSION

The postulates of Savage (1954) concern only preference amongst acts, and are not sufficient to guarantee a representation of preference by a unique state dependent utility and probability, unless a particular set of prizes are considered constant in value. If there exist several sets of prizes whose values could each be considered constant, different "unique" probabilities could be elicited. One way to elicit unique probabilities and utilities is to take the approach of DeGroot (1970). There, the concept of "at least as likely as" is taken as primitive, and probabilities are elicited using comparisons of the likelihood of various events. Some researchers believe that preference is a more universally understood primitive and would prefer to base a theory on preference. Another method for eliciting unique probabilities and utilities was introduced by Karni, Schmeidler, and Vind (1983) and Karni (1985). This method requires that direct comparisons need to be made between acts which not only specify which prizes are awarded, but also which states of nature will occur (or will be precluded). A discussion of this method, and how it relates to the other topics described here can be found in Schervish, Seidenfeld, and Kadane (1990).

## ACKNOWLEDGMENT

This research was supported, in part, by National Science Foundation grants numbered DMS-8805676 and DMS-8705646 and Office of Naval Research contract number N00014-88-K0013.

## REFERENCES

DeGroot, M.H., (1970), *Optimal Statistical Decisions*, John Wiley, New York.

Karni, E., (1985), *Decision Making under Uncertainty*, Harvard University Press, Cambridge.

Karni, E., Schmeidler, D., and Vind, K., (1983), On state dependent preferences and subjective probabilities, *Econometrica*, **51**, 1021-31.

Markowitz, H., (1952), The utility of wealth, *Journal of Political Economy*, **60**, 453-8.

Rubin, H., (1987), A weak system of axioms for "rational" behavior and the non-separability of utility from prior, *Statistics and Decisions*, **5**, 47-58.

Savage, L.J., (1954), *Foundations of Statistics*, John Wiley, New York.

Schervish, M.J., Seidenfeld, T., and Kadane, J.B. (1990), State-dependent utilities, *Journal of American Statistical Association*, **85**, 840-847.

Tversky, A., and Kahneman, D., (1981), The Framing of decisions and the psychology of choice, *Science*, **211**, 453-8.

## COMMENTS

Bruce M. Hill
University of Michigan

I think the article by Schervish, Seidenfeld, and Kadane (SSK) is potentially extremely important. The example regarding exchange rates is a beautiful one to illustrate state dependent utility. The problem is intimately related to that of small worlds so brilliantly initiated by Jimmie Savage over thirty years ago. This is one of the most important questions for the Bayesian theory to attempt to resolve. We are forced to make our explicit considerations in very small worlds, but it is wise always to keep in mind the larger world in which our actions take place. It is very difficult to delimit a decision problem to a small world. For all we know, every action we make may reverberate down through eternity. As my article on the validity of the likelihood principle Hill (1987) suggests, many things may be connected in ways unknown to us. It is therefore best to keep in mind at least a somewhat larger world than that in which we ordinarily work. For example, the use of parametric models is a small world device to enable one to make computations. Use of such models guarantees coherency within that small world, but may be very foolish in a somewhat larger world. The problem is to balance the computational advantages and simplicity of the small world as against the greater realism of the larger world.

On the technical question as to the validity of Savage's state independent utilities, the answer is not clear to me. Savage (1972, p. 25) discussed this question and argued that it was a question of interpretation of terms. The discussion arises when he introduces acts with constant consequences, and argues that conditional on any non null event, the preference between two such acts should remain the same, irrespective of which such event is conditioned on, as recorded in Postulate P3. At first glance this does not seem possible, but as he notes there are in fact different ways to interpret the postulate, and I am hopeful that the exchange rate problem of SSK can be dealt with in this way, thus retaining the simplicity and elegance of Savage's approach. In other words, there may be a different representation of the problem, with state independent utilities, in accord with Savage's postulates and therefore simpler to deal with. The question is whether this can be done and still yield an appropriate analysis of the exchange rates problem of SSK. There is much future work to be done on this extremely important topic, and SSK are to be congratulated for their interesting example and analysis, which may serve to return statistical theory to more meaningful questions for the long run development of the subject.

## REJOINDER

We wish to thank Hill for his insightful commentary. We agree that it is difficult to delimit a decision problem to a small world and that it is wise always to keep in mind the larger world in which our actions take place. We are not quite so optimisitic as Hill that the exchange rate problem can be resolved within the existing framework of Savage's postulates. Since constant dollar acts already have the property that "conditional on any non null event, the preference between two such acts should remain the same, irrespective of which such event is conditioned on," we believe that something more is needed to resolve the problem. As mentioned in our paper, there are methods of eliciting unique state dependent utilities and probabilities by introducing an enlarged space of acts. The existing such methods use extraneaous probabilities unlike Savage's postulates. We share Hill's hope that there is a way to extend Savage's postulates to be able to handle state dependent utilities in a manner nearly so elegant as Savage's existing theory.

## ADDITIONAL REFERENCES

Hill, Bruce M. (1987), "The validity of the likelihood principle," *The American Statistician*, 41, 95-100.

13

# Learning Statistics from Counter Examples: Ancillary Statistics

D. Basu [1]

**Abstract**

Bayesian objection to the analysis of data in frequency theory terms is amplified through several counter examples in which an ancillary statistic exists and there is a temptation to choose a reference set after looking at the data. It is argued that Fisher insisted on conditioning by an ancillary statistic, because conditioning the data $\mathbf{x}$ by an ancillary $Y$ does not change the likelihood. In this sense Fisher discovered the supremacy of the likelihood function.

*Key words and Phrases:* Ancillary statistics; Conditional frequentist inference; Information; Liklihood principle; Reference set; Sufficiency principle.

## 1. INTRODUCTION

This paper is especially addressed to the statisticians who have not yet fully grasped the Bayesian objection to the analysis of data in repeated sampling terms. Let $\mathbf{x}$ be the sample, $f(\mathbf{x}|\theta)$ the model and $\theta$ the parameter. A statistic $Y = Y(\mathbf{x})$ is *ancillary* if the sampling distribution of $Y$, given $\theta$, is $\theta$-free (is the same for all values of $\theta$). A statistic $T = T(\mathbf{x})$ is *sufficient* if the distribution of the sample $\mathbf{x}$, given $T$ and $\theta$, is $\theta$-free. An ancillary statistic $Y$ by itself contains no information about the parameter, whereas a sufficient statistic $T$ is fully informative in a sense. R.A. Fisher's attempt to make sense of the notion of *information in the data* led him to these two important concepts in Statistics.

Let $L(\theta) = f(\mathbf{x}|\theta)$ be the *likelihood function* determined by the sample $\mathbf{x}$ and let $\hat{\theta}$ be the *maximum likelihood* (ML) estimate of $\theta$. If $\hat{\theta}$ is a sufficient statistic then, according to Fisher, there would be no loss of information if the performance characteristics of $\hat{\theta}$ as an estimate of $\theta$ is sought to be evaluated in terms of the sampling distribution of $\hat{\theta}$. We shall repudiate this in the end with an example.

---

[1]Indian Statistical Institute and Florida State University

If the ML estimate $\hat{\theta}$ is not a sufficient statistic then Fisher sought to recover the information lost in the sampling distribution of $\hat{\theta}$ with the help of an *ancillary complement* $Y$ to the estimator $\hat{\theta}$. The ancillary statistic $Y$ has to complement $\hat{\theta}$ in the sense that the pair $(\hat{\theta}, Y)$ is jointly sufficient. The Fisher Information $I_{\hat{\theta},Y}(\theta)$ in the sufficient statistic $(\hat{\theta}, Y)$ is then the same as the full information

$$I(\theta) = -E[\frac{\partial^2}{\partial\theta^2} \log L(\theta)]$$

in the sample $\mathbf{x}$. (Note that $I(\theta)$ does not relate to the particular sample $\mathbf{x}$ but is obtained by averaging the quantity $-\frac{\partial^2}{\partial\theta^2} \log L(\theta)$ over the sample space.) The Fisher Information in the statistic $\hat{\theta}$ is less than the full information $I(\theta)$. The cornerstone of the Fisher argument lies in the identity

$$I(\theta) = I_{\hat{\theta},Y}(\theta) = E[I_{\hat{\theta}}(\theta|Y)],$$

where $I_{\hat{\theta}}(\theta|Y)$ is the conditional information in the statistic $\hat{\theta}$, given $Y$, and the expectation on the right hand side is with respect to the ancillary statistic $Y$. Thus, the conditional information in $\hat{\theta}$, given $Y$, depends on $Y$ and can be, for a particular value of the statistic $Y$, much less or much greater than the full information $I(\theta)$. The *conditionality argument* of R.A. Fisher rests on the proposition that the performance characteristics of the estimator $\hat{\theta}$ ought to be evaluated conditionally, holding the ancillary statistic $Y$ fixed at its observed value $y$. As Fisher argued, the event $Y = y$, even though uninformative by itself, has a lot of latent information about $\theta$ in the sense that it helps us discern how good or bad the estimate $\hat{\theta}$ is in the present instance. The set $S(y) = \{\mathbf{x} : Y(\mathbf{x}) = y\}$ defines what Fisher called the *reference set.* Sir Ronald was trying to cut down the *sample space* $S$ to size. We illustrate the conditionality argument with several examples.

## 2. EXAMPLES

**Example 1:** Let $\mathbf{x} = (x_1, x_2, \ldots, x_n)$ be iid observations on a random variable that is uniformly distributed over the interval $[\theta, 2\theta]$, where $\theta > 0$ is the unknown scale parameter. With

$$m = \min x_i \text{ and } M = \max x_i,$$

the likelihood function $L(\theta)$ equals $1/\theta^n$ over the interval $[M/2, m]$ and zero outside the interval. The ML estimator $\hat{\theta} = M/2$ is not sufficient, the minimal sufficient statistic being the pair $(m, M)$. Since the two statistics $m$ and $M$ are stochastically independent in an asymptomatic sense, it is clear that there will be a substantial loss of information if we marginalize the data to the ML estimator $M/2$. Comparing the mean squared error (MSE) of $M/2$ with that of $m$ as estimators of $\theta$, we find that the former is exactly four times better than the latter. Consider, therefore, the estimator

$T = (2M + m)/5$ which is the weighted average of $M/2$ and $m$ with weights 4 and 1 respectively. Both $M/2$ and $T$ are equivariant estimators of the scale parameter $\theta$, and so their MSE's are constant multiples of $\theta^2$. It works out that the ratio of the two MSE's tends to 25/12 as the sample size $n$ tends to infinity. The ML estimator $\hat{\theta}$ can hardly be called an efficient estimate of $\theta$ in the usual sense of the term. Over thirty-six years ago, when I came upon this counterexample, it was pointed out to me by C.R. Rao that the ML estimator $\hat{\theta}$ ought to be judged conditionally after holding fixed its ancillary complement $Y = M/m$ at its observed value. That $Y$ is an ancillary statistic follows from the facts that $Y$ is scale invariant and that $\theta$ is a scale parameter. As we noted before, the likelihood mass is spread over the interval $[M/2, m]$ pinpointing the parameter $\theta$ within that interval. The statistic $Y = M/m$ varies over the range $[1, 2]$ and is indeed a measure of how good the sample is – the nearer $Y$ is to 2 the better the sample is. While evaluating the ML estimate $\hat{\theta}$ we ought to take note of the observed value $y$ of the statistic $Y$. That is, instead of referring $\hat{\theta}$ to the full sample space $S$, we ought to refer it to the *reference set* $S(y)$.

In terms of the full sample space $S$ the ML estimator $M/2$ is not sufficient. But when it is conditioned by $Y$ it suddenly becomes fully informative (sufficient, that is). Note that the other two estimators $m$ and $T$ also become fully informative when they are referred to the set $S(y)$. Indeed, the three statistics $M/2$, $m$ and $T$ become functionally related when conditioned by $Y$.

This example beautifully illustrates what Fisher meant by *recovery of ancillary information*. The next example illustrates how a weak pivotal quantity can be strengthened by proper conditioning with an ancillary statistic.

**Example 2:** Let $\mathbf{x} = (x_1, x_2, \ldots, x_n)$ be $n$ iid observations on a random variable with pdf $f(x - \theta)$, where $f$ is known but $\theta$ (the location parameter) is unknown. Consider the statistic $x_1$ and its ancillary complement $D = (x_2 - x_1, x_3, -x_1, \ldots, x_n - x_1)$. The statistic $x_1$ by itself carries very little information about $\theta$, but it becomes fully informative (sufficient) when conditioned by $D$. The conditional pdf of $x_1$, given $D$, has $\theta$ embedded in it as a location parameter. Fisher derived the *fiducial distribution* of the parameter $\theta$ by inverting the pivotal quantity $x_1 - \theta$ after conditioning it by the ancillary statistic $D$.

The previous example raises many questions. Some sample questions and answers are listed below.

Question: What is the status of the ancillary statistic $D$? Is it the *maximum ancillary* in the sense that every other ancillary statistic is a function of $D$?

Answer: No. $D$ is never the maximum ancillary. However, in some situations $D$ will be a *maximal ancillary* in the sense that no larger (with respect to the partial order of functional relationship) ancillary statistic exists. A multiplicity of maximal ancillaries is a fact of life in this situation.

Question: Is the fiducial distribution of $\theta$ in Example 2 critically dependent on the choice of the pivotal quantity $x_1 - \theta$?

Answer: No. Another pivotal quantity like, say, $\mathbf{x} - \theta$, when conditioned by $D$, will result in the same fiducial distribution of $\theta$. This is because $\bar{x} = x_1 + (\bar{x} - x_1)$ and $\bar{x} - x_1$ is a function of $D$.

Question: Can we interpret the fiducial distribution of $\theta$ probabilistically?

Answer: It was pointed out by Harold Jeffreys that the fiducial distribution of the location parameter (as derived by Fisher) coincides with the posterior distribution of $\theta$ corresponding to the uniform prior (over the entire real line) for the parameter.

In the presence of multiple ancillaries, the choice of the proper reference set is a problem. The dilemma is best exemplified by the following example.

**Example 3:** Let $(x_1, y_1), i = 1, 2, \ldots, n$, be $n$ iid observations on $(X, Y)$ whose joint distribution is Bivariate Normal with zero means, unit variances and covariance $\theta$, which is the parameter of interest. In this case both $\mathbf{x} = (x_1, \ldots, x_n)$ and $\mathbf{y} = (y_1, \ldots, y_n)$ are ancillary statistics. Note that the pair $(\mathbf{x}, \mathbf{y})$ is the entire data and therefore is sufficient. Holding the ancillary $\mathbf{x}$ as fixed and regarding $\mathbf{y}$ as the variable, we may want to estimate $\theta$ by $\sum x_i y_i / \sum x_i^2$ and then regard the estimate as unbiased with variance $(1 - \theta^2)/\sum x_1^2$. But how about holding $\mathbf{y}$ fixed and reporting that $\sum x_i y_i / \sum y_i^2$ is an unbiased estimate with variance $(1-\theta^2)/\sum y_1^2$? It is tempting to opt for the ancillary with the larger sum of squares. But would it not be a statistical heresy to choose the reference set after looking at the data?!

## 3. COX ON ANCILLARIES

D.R. Cox (1971) suggested a way to deal with the problem of multiple ancillaries. Looking back at the Fisher identity $I(\theta) = EI(\theta|Y)$, Cox argued that the basic role of the conditioning ancillary $Y$ is to discriminate between samples with varying degrees of information. So in the presence of multiple ancillaries we should choose that $Y$ for which $I(\theta|Y)$ is most variable in $Y$. So opt for the $Y$ for which Var $I(\theta|Y)$ is maximum. One snag in the Cox argument is that Var $I(\theta|Y)$ is a function of $\theta$ and so there may not exist a $Y$ that maximizes the function uniformly in $\theta$. Also note that in our Example 3 the Cox method fails because, in view of the perfect symmetry between $\mathbf{x}$ and $\mathbf{y}$, Var $I(\theta|\mathbf{x}) =$ Var $I(\theta|\mathbf{y})$.

But the real snag in the Cox argument is the meaninglessness of the notion of Fisher Information as a measure of the evidential meaning of the particular data at hand. Fisher's preoccupation with the elusive notion of information in the data led him to the likelihood function which he recognized as the carrier of all the information in the data. The likelihood was then partially summarized in the two statistics $\hat{\theta}$, the ML estimate, and $Z(\hat{\theta})$, the second derivative of $-\log L(\theta)$ at $\theta = \hat{\theta}$. Note that $Z(\hat{\theta})$ is the reciprocal of the radius of curvature of the log likelihood at its mode, the larger the value of $Z(\hat{\theta})$ the sharper is the fall of the likelihood function as $\theta$ moves away from $\hat{\theta}$. We have to stretch our minds a little to regard $Z(\hat{\theta})$ as a rough measure of the concentration of the likelihood mass around $\hat{\theta}$. The greater the concentration the more informative is the likelihood. The Fisher Information $I(\theta)$ is obtained from $Z(\hat{\theta})$ by first replacing $\hat{\theta}$ by $\theta$ and then taking the average value of $Z(\theta)$ over the whole sample space $S$. But how can we regard $I(\theta)$ as information in the data?

Why did Fisher insist that the conditioning statistic $Y$ has to be ancillary? Because, conditoning the data $\mathbf{x}$ by an ancillary $Y$ does not change the likelihood. Fisher discovered the supremacy of the likelihood but got carried away by his amazing craftmanship with sample space mathematics.

## 4. E.L. LEHMANN ON ANCILLARIES

Eric Lehmann (1981) finally recognized the conditionality argument. And now he has to cope with the disturbing presence of ancillary statistics. Invoking the *Sufficiency Principle*, Eric would reduce the data $\mathbf{x}$ to the minimal sufficient statistic $T = T(\mathbf{x})$. Since $T$ is sufficient, all reasonable inference procedures ought to depend on $\mathbf{x}$ only through $T(\mathbf{x})$. This data reduction sweeps away much of the ancillary dust under the rug. But, as in Example 1, some functions of the minimal sufficient statistic $T$ may still be recognized as ancillary statistics. Eric has yet to come out openly on the question of how to deal with such persistent ancillaries.

From what Eric writes in his 1981 article, it seems that he feels quite comfortable with statistical models for which the minimal sufficient statistic $T$ is complete. In such cases no nontrivial function of $T$ can be ancillary. Furthermore, thanks to the so called Basu Theorem, every ancillary statistic $Y$ is stochastically independent (conditionally on $\theta$) of $T$. Therefore, no $T$- based decision procedure can be altered by conditioning with an ancillary $Y$. So who needs to think of the conditionality argument when we have a complete sufficient statistic? Remember, Fisher looked for an ancillary complement to the ML estimate $\hat{\theta}$ only when the statistic $\hat{\theta}$ was not sufficient. So in the most favorable set up where $\hat{\theta}$ is a complete sufficient statistic, can anyone object if we evaluate the estimate $\hat{\theta}$ in terms of the sampling distribution of the estimator? We give an example to prove both Fisher and Lehmann wrong on this question.

**Example 4:** Consider a sequence of Bernoulli trials with parameter $p$ that results in a finite sequence $w = SFFS\dots FS$ of successes $S$ and failures $F$. Let $X(w)$ and $Y(w)$ denote, respectively, the number of $S$'s and the number of $F$'s in the sample sequence $w$. We picture $w$ as a sample path, the locus of a point that begins its journey at the original and travels through the lattice points of the positive quadrant, moving one step to the right for each $S$ and one step up for each $F$. The lattice point with coordinates $X(w)$ and $Y(w)$ is where the sample path $w$ ends. Our example relates to a particular sampling (stopping) rule **R**. Writing $(X, Y)$ for the location of the moving point, the rule is described as:

Rule **R**: Continue sampling as long as (I) $Y < 2X + 1$, (II) $Y > X - 2$, and (III) $X + Y < 100$. Alternatively, the rule may be defined as: Stop sampling as soon as the sample path hits one of the three boundary lines (i) $y = 2x + 1$, (ii) $y = x - 2$, and (iii) $x + y = 100$.

As always, the likelihood does not recognize the stopping rule and comes out as

$$L(p) = f(w|p) = p^{X(w)}q^{Y(w)}$$

where $q = 1 - p$. The pair $X(w), Y(w)$ constitute the minimal sufficient statistic. The ML estimate is $\hat{p} = X/(X + Y)$. The range of the sufficient statistic $(X, Y)$ consists of the boundary points

$$\begin{array}{lllllll} (0,1), & (1,3), & \ldots, & (33,67) & \text{on line} & (i), \\ (34,66), & (35,65), & \ldots, & (50,50) & \text{on line} & (iii), and \\ (51,49), & (50,48), & \ldots, & (2,0) & \text{on line} & (ii). \end{array}$$

The ML estimator $\hat{p} = X/(X + Y)$ monotonically increases from zero to unity as $(X, Y)$ moves through the above set of boundary points. Hence $\hat{p}$ itself is minimal sufficient. Let us assert here without proof that $\hat{p}$ is a complete sufficient statistic in this case and that no nontrivial ancillary statistic exists.

Sir Ronald is no longer with us. So let me address the following questions to my good friend Eric Lehmann who is a living legend among us for his unparalleled erudition in Statistical Mathematics. The questions relate to Example 4.

Question: What should be our criterion for the choice of an estimate of $p$?

(The unbiasedness criterion is sort of vacuous in this case. There is only one unbiased estimator, which is zero or unity depending on whether the first trial results in an $F$ or an $S$.)

Question: If ML is the chosen criterion, then how should we evaluate the estimate $\hat{p} = X/(X + Y)$? Does it make sense to evaluate $\hat{p}$ in terms of some average performance characteristics?

Question: Are all sample paths $w$ equally informative?

(Even though there are no ancillary statistics in this case, we can still detect major qualitative differences between different sample paths. For instance, short sample paths like $F$ or $SS$ have very little to say about the parameter, whereas long paths that end on line (iii) are clearly much more informative.)

Question: Why do we need to decipher what the sample $w$ has to say about the parameter $p$ in terms of a sample space? Does the sample $F$

obtained following the rule **R** say anything different from the statement: A single Bernoulli trial has resulted in a failure?

Question: Do sample space ideas like bias, variance, risk function, etc., make any sense in this case?

Question: Why not act like a Bayesian and analyze the particular likelihood function generated by the data? Isn't is quite clear in this case that all that the data has to say about the parameter is summarized in the likelihood?

## REFERENCES

Basu, D. (1988), *Statistical Information and Likelihood: A Collection of Critical Essays* by D. Basu, ed. J.K. Ghosh, Springer Verlag, New York.

Cox, D. R. (1971), The Choice Between Alternative Ancillary Statistics. *Jour. Royal Statist. Soc.* **(B)33**, 251-255.

Lehmann, E.L. (1981), An Interpretation of Completeness and Basu's Theorem. *Jour. Amer. Stat. Assoc.,* **76**, 335-340.

# 14

# The Horvitz-Thompson Estimate and Basu's Circus Revisited


## J. K. Ghosh [1]



### Abstract

The asymptotic properties of the Horvitz-Thompson estimator in an unequal probability sampling situation are examined. A design free estimate is also studied. It is shown that the later estimator makes more sense than the Horvitz-Thompson estimator in the conditional frequentist framework.



*Key words and Phrases:* Horvitz-Thompson estimator; Asymptotic properties; Conditional frequentist inference; Design free estimator.


## 1. INTRODUCTION

In a hilarious example, Basu (1988, p. 211, 212) shows how a combination of unequal probability sampling and the Horvitz-Thompson estimate may lead to absurdities. Later, Basu (1988, p. 325) again questions the face validity of the H-T estimate and asks if in some sense the estimate is consistent. The object of this note is to throw some light on these questions. Somewhat surprisingly, the H-T estimate turns out to be consistent, and, if the sample size $n$ is large, reasonable for most samples even if the probabilities of selection are inappropriately chosen. However this is not the whole story. In course of our analysis we introduce a conditional framework that seems natural even for frequentist studies and a design free estimate $T$ which seems quite reasonable from a conditional frequentist point of view. The conditional framework makes clear that $T$ makes more sense than the H-T estimate and that there will be special samples (with small probability) for which the H-T estimate could be very bad.

## 2. A CONDITIONAL FRAMEWORK AND A NEW ESTIMATE

The set up we consider is as follows. There are $N$ sampling units, the variable of interest is $y$, the label is ignored but this is partly compensated by a size variable $x$, which is believed to contain information about $y$.


[1] Indian Statistical Institute and Purdue University

Further we assume there are $k$ distinct values of $x$, namely, $x_1, x_2, \ldots, x_k$, $x_i$ repeated $N_i$ times, $\Sigma N_i = N$. It is assumed $x_i$'s and $N_i$'s are known. Then it is natural to write the $y$'s as $y_{ij}$, $i = 1, \ldots, k$, $i = 1, \ldots, N_i$, it being understood that for a fixed $i$ all units have the same value of $x$, namely, $x_i$. The sample size will be $n$, assumed much smaller than $N$ to avoid the problems of p.p.s. sampling without replacement. In view of this we will assume the $n$ units are selected with replacement. There is a fixed probability $\pi_i$ of a unit with size $x_i$ being selected in any particular draw; $\pi_i$ may be proportional to $x_i$. Clearly

$$\Sigma N_i \pi_i = 1 \tag{1}$$

The sample may be represented as $x_{i_1}, y^s_{i_1,j}, j = 1, \ldots, n_{i_1}), \ldots, (x_{i_m}, y^s_{i_m,j}, j = 1, \ldots, n_m)$, the superscript "$s$" identifies the $y$'s as belonging to the sample. Of course each $y^s_{i,j} = y_{ij'}$ for some $j'$.

Let $n_i$ = frequency of $x_i$ in the sample, $n_i > 0$ iff $i = i_1, \ldots, i_m$. Let $\mathbf{n} = (n_1, \ldots, n_k)$. A similar convention for denoting vectors will be followed throughout.

Clearly $\mathbf{n}$ is an ancillary statistic whose distribution depends only in $N_i$'s and $\pi_i$'s. We feel in the present problem even frequentist inference should be conditional on $\mathbf{n}$.

If $\mathbf{n} > 0$, i.e., each $n_i > 0$, define, as an estimate of the population mean,

$$T = \{\sum_i (\sum_j y^s_{ij}/n_i) N_i\}/N$$

which is simply a weighted mean of the sample array means

$$\overline{y}^s_i = (\sum_i y^s_{ij})/n_i.$$

If $\mathbf{n} \not> \mathbf{0}$, then the estimate $T$ would be a suitable weighted mean of the $\overline{y}^s_i$'s and an estimate for the unobserved arrays, obtained from prior knowledge or a mixture of extrapolation and interpolation. For the purposes of this note it will not be necessary to be more specific.

We first consider the Hanson-Horvitz estimate $T_H$,

$$T_H = (\sum_i \sum_j y^s_{ij}/\pi_i)/(nN)$$

where $i$ ranges over $i_1, \ldots, i_m$.

## 3. THE ASYMPTOTICS AND CONCLUSIONS

Let $\overline{\mathbf{y}} = (\overline{y}_1, \ldots, \overline{y}_k)$, $\sigma^2 = (\sigma^2_1, \ldots, \sigma^2_k)$ where

$$\overline{y}_i = \sum_i y_{ij}/N_i, \quad \sigma_i^2 = \sum_j (y_{ij} - \overline{y}_i)^2/N_i.$$

The following facts are immediate.

$$P\{\mathbf{n} \not> 0\} \leq \sum_i (1 - N_i\pi_i)^n \tag{2}$$

$$P\{|\overline{y}_i^s - \overline{y}_i| > \epsilon \text{ for some } i|\mathbf{n}\} \leq \sum_i \sigma_i^2/n_i\epsilon^2 \tag{3}$$

$$P\{|\frac{n_i}{n} - N_i\pi_i| > \epsilon \text{ for some } i|\mathbf{n}\} \leq \sum_i N_i\pi_i(1 - N_i\pi_i)/n\epsilon^2 \tag{4}$$

To do asymptotics, we now imagine that all the vectors $\mathbf{n}$, $\pi$, $\mathbf{N}$, $\overline{\mathbf{ys}}$, $\overline{\mathbf{y}}$, $\sigma^2$ have a suffix $\nu$, which will remain suppressed. Assume $\nu$ goes to infinity such that

$$\lim\ N = \infty,\quad \lim\ n = \infty,\quad \lim\ n/N = 0, \tag{5}$$

$$0 < \underline{\lim}\ N_i/N_j \leq \overline{\lim}\ N_i/N_j < \infty,\quad i \neq j \tag{6}$$

$$0 < \underline{\lim}\ \pi_i/\pi_j \leq \overline{\lim}\ \pi_i/\pi_j < \infty,\quad i \neq j \tag{7}$$

$$\overline{\lim}\ \sigma_i^2 < \infty \tag{8}$$

For fixed $k$, as assumed here, the above assumptions as well as (12) below are quite reasonable.

The above implies, among other things,

$$0 < \underline{\lim}\ N_i\pi_i \leq \overline{\lim}\ N_i\pi_i < \infty. \tag{9}$$

Then by (1), (2) and (3) and the above assumptions, (unconditionally)

$$T - (\Sigma\overline{y}_i N_i)/N \xrightarrow{p} 0 \tag{10}$$

and

$$|n_i/n - N_i\pi_i| \xrightarrow{p} 0. \tag{11}$$

If in addition we assume

$$\overline{\lim}\ |\overline{y}_i| < \infty \tag{12}$$

then

$$\begin{aligned} |T - T_H| &\leq \Sigma|\overline{y}_i^s||(N_i/N - n_i/n\pi_i N)| \\ &= \Sigma|\overline{y}_i^s||(N_i\pi_i - n_i/n)|/N\pi_i \xrightarrow{p} 0, \end{aligned} \tag{13}$$

consequently

$$|T_H - (\Sigma\overline{y}_i N_i)/N| \xrightarrow{p} 0. \tag{14}$$

We conclude that $T_H$ is consistent and that, with probability tending to 1, it approximates a reasonable design free estimate $T$, and to that extent has some face validity. However the conditional framework makes clear that $T_H$ is no longer unbiased, $T$ is the more reasonable estimate to use and when $\mathbf{n} \not> 0$ or the $n_i$'s are not close to $n\ N_i\ \pi_i$, use of $T_H$ can be disastrous.

We now note exactly the same conclusions hold for the Horvitz- Thompson estimate $T_{HT}$ [see Cochran (1977), pp. 259-261], for the population mean where

$$T_{HT} = \frac{1}{N} \sum_i \sum_j y_{ij}^s / p_i .$$

For this note that the probability of ith unit being selected is

$$\begin{aligned} p_i &= 1-(1-\pi_i)^n \\ &= n\pi_i[1+o(1)] \end{aligned}$$

where by our previous assumptions $n\pi_i$ is $O(n/N)$ and $\pi_i$ is $O(1/n)$. Consequently,

$$| T_H - T_{HT} | \le \sum_i | \bar{y}_i^s | \quad | \frac{n_i}{n} - \frac{n_i \pi_i}{p_i} | \frac{1}{N\pi_i} .$$

We now check that

$$| \frac{n_i}{n} - \frac{n_i \pi_i}{p_i} | \xrightarrow{P} 0,$$

since $p_i = n\pi_i(1+0(1))$.
It follows

$$| T_H - T_{HT} | \xrightarrow{P} 0,$$

and hence by (14),

$$| T - T_{HT} | \xrightarrow{P} 0 .$$

It has been assumed above that no special relation is known between $y$ and $x$. If more is known e.g. if $y \sim cx$, naturally the knowledge should be used to improve $T$ in the conditional framework but not to bring back $T_H$ or $T_{HT}$. It would be worth exploring if $T$ should be smoothed, as is normally done in other applications of non-parametric regression. The most difficult remaining problem is to examine what happens when $k$ also tends to infinity. The elementary methods of the present note will not work.

## REFERENCES

Basu, D. (1988). *Statistical Information and Likelihood, A Collection of Critical Essays by Dr. D. Basu,* edited by J. K.Ghosh, Lecture Notes in Statistics, **45**, Springer-Verlag, New York.

Cochran, William G. (1977) Sampling Techniques, 3rd. ed., Wiley, New York.

# 15

# Comparison of Experiments for Selection and Censored Data Models

**Prem K. Goel**[1]
**Morris H. DeGroot**[2]

**Abstract**

Statistical comparison of two experiments involving *randomly right-censored data*, such that one experiment has a stochastically larger censoring variable than the other, leads to a strong result. It is shown in Goel (1988) that the experiment corresponding to the stochastically larger censoring variable is sufficient for the other experiment. Comparison of two experiments involving *selection models data* does not lead to such a phenomena. For the family of distributions with the *mlr* property, a necessary and sufficient condition for the *pairwise-sufficiency* of one selection experiment for the other is established. Some illustrative examples are also given.

*Key Words and Phrases:* Sufficiency; More informative experiment; Stochastic randomization; Pair-wise sufficiency; Censored data models; Selection models; Truncated data models; Monotone likelihood ratio.

## 1. INTRODUCTION

A statistical experiment consists of an observable random variable $X$, defined on a measurable space $[\mathcal{X}, \mathcal{A}]$. In the parametric model framework, it is assumed that the probability distribution, $P_\theta$, of the random variable $X$, is indexed by an unknown parameter $\theta$, a member of a fixed collection of states of nature $\Theta$. Inference about the parameter $\theta$ is usually based on a random sample from the population. In many situations, however, the experimenter may not be able to collect data from this experiment and instead a restricted experiment, consisting of an observable random variable $Y$ defined on the measurable space $[\mathcal{Y}, \mathcal{B}]$, is conducted in which the observations are either censored or restricted to a *selection set* $S$, but the family of their probability distributions is indexed by the same set $\Theta$. The following question arises naturally in these situations: How does the

[1]The Ohio State University
[2]Carnegie Mellon University

information in the restricted experiment change as the amount of censoring increases or as the size of the selection set $S$ decreases?

Before discussing the specific problems, we will present a brief discussion of the general ideas involved in comparing information in various experiments. The most general methodology in a decision theoretic framework for answering these questions is the so called *comparison of experiments*, in which the notion 'the experiment $\mathcal{E}$ is *more informative than* the experiment $\mathcal{F}$' is defined as follows:

**Definition 1.1.** The experiment $\mathcal{E}$ is said to be more informative than the experiment $\mathcal{F}$, if for every decision problem involving $\theta$, given any decision procedure $\delta(Y)$ based on the experiment $\mathcal{F}$, there exists a decision procedure $\delta^*(X)$, based on $\mathcal{E}$, such that the risk functions $R(\theta, \delta(Y))$ and $R(\theta, \delta^*(X))$ satisfy, $R(\theta, \delta^*(X)) \leq R(\theta, \delta(Y))$ for all $\theta \in \Theta$.

In the Bayesian version of the above definition given below, the minimum Bayes risk $r(\pi, \delta^\pi(Y)) = E^\pi[R(\theta, \delta^\pi(Y))]$, where $\delta^\pi(Y)$ is the optimal Bayes procedure with respect to the prior distribution $\pi$ is compared with the minimun Bayes risk based on $X$.

**Definition 1.1a.** The experiment $\mathcal{E}$ is said to be more informative than the experiment $\mathcal{F}$ if for every decision problem involving $\theta$, and for every prior distribution $\pi$ on $\Theta$, the minimum Bayes risk obtainable from the experiment $\mathcal{E}$ is less than the minimum Bayes risk obtainable from the experiment $\mathcal{F}$.

An equivalent concept of sufficiency of one experiment for the other, suggested by Blackwell (1953), is defined below. This concept is the operational approach for checking the more-informativeness of one experiment than the other, whenever two experiments can be compared in the above sense.

**Definition 1.2.** The experiment $\mathcal{E}$ is sufficient for the experiment $\mathcal{F}$, denoted by $\mathcal{E} \succeq \mathcal{F}$ if there exists a *stochastic transformation* of $X$ to a random variable $Z(X)$ such that for each $\theta \in \Theta$, the random variables $Z(X)$ and $Y$ have identical distributions.

These concepts have been investigated thoroughly and been applied to many experiments and decision problems [see e.g., Hansen and Torgersen (1974), Heyer (1982), Lehmann (1988), and Torgersen (1972, 1976, 1981, 1988)]. It is also well known that if $\mathcal{E} \succeq \mathcal{F}$, then $\mathcal{E}$ has more information than $\mathcal{F}$ in terms of a large class of measures of information [see e.g., Goel and DeGroot (1979), Ferentino and Papaioannou (1982), Goel (1988), and Taneja (1987)].

In order to establish that $\mathcal{E} \succeq \mathcal{F}$, usually one must find a stochastic transformation from $X$ to $Y$. Furthermore, to show that $\mathcal{E}$ is not sufficient for $\mathcal{F}$, one must either find a decision problem for which $\mathcal{F}$ is better than $\mathcal{E}$ in the sense of attainable risk or show that for some information measure,

$\mathcal{F}$ has more information than $\mathcal{E}$. Another useful method for showing that $\mathcal{E}$ is not sufficient for $\mathcal{F}$ is a consequence of the following result proved in Goel(1988). It requires showing the existence of an unbiased estimator of a parametric function based on $\mathcal{F}$, for which there is no unbiased estimator based on $\mathcal{E}$.

**Proposition 1.1**. Suppose that $\mathcal{E} \succeq \mathcal{F}$. Let $g(\theta)$ be a function of $\theta$ for which $\exists$ an unbiased estimator $T(Y)$ based on $\mathcal{F}$. Then there must exist an unbiased estimator of $g(\theta)$ based on $\mathcal{E}$.

The comparison of information in various *censored-data* experiments was considered by Hollander, Proschan and Sconing (1985, 1987). In these problems, the standard experiment $\mathcal{E}$ consists of a random variable $X$ corresponding to the survival time of an item on life-test. However, instead of observing $X$, the restricted experiment $\mathcal{F}$ is characterized by a censoring variable $V$, distributed independently of $X$. The censoring variable $V$ may inhibit the experimenter's ability to learn about the parameter $\theta$. The observation $(Y, \delta)$ in the censored data experiment is defined by

$$Y = min(X, V), \quad and \quad \delta = I_{[X \leq V]}, \tag{1.1}$$

where $I_{[A]}$ denotes the indicator function of the set $A$. Here, $Y$ denotes either the censored or the uncensored life time of the item under test and $\delta = 1$ indicates that the observation was uncensored, while $\delta = 0$ indicates that the observation is censored. This is the well known *randomly right-censored data* model. Hollander, Proschan and Sconing (1987) show that for several well known measures of information, the censored data experiments contain decreasing amount of information as the censoring variable gets stochastically larger. However, their results do not assure us that this property holds for all measures of information. For the censored data experiments, the fact that an original experiment is always sufficient for a censored experiment follows as a corollary to the following result proved in Goel (1988).

**Theorem 1.1.** Let $\mathcal{E}$ and $\mathcal{F}$ be two randomly right- censored experiments based on the random variables $(X, V_1)$ and $(X, V_2)$ respectively where $V_1$ is stochastically larger than $V_2$. Then $\mathcal{E} \succeq \mathcal{F}$.

This result implies that for every information measure, the amount of information is a decreasing function of the stochastically increasing level of censoring. Thus the question raised initially has been answered completely for censored data models.

In the experiments involving *selection models*, it is assumed that the observation $Y$ is restricted to a specified subset $S$ of the sample space and the inference about $\theta$ is to be based on a random sample from the density

$$g(y, \theta) = \frac{f(y, \theta)}{P_\theta[X \in S]}. \tag{1.2}$$

The set $S$ is called the *selection set.* The selection models are a special case of well studied weighted distributions [see, e.g., Rao (1964, 1985), Patil (1984), and Bayarri & DeGroot (1987a)]. Fisher information in selection models has been compared with the Fisher information in the standard experiments for various distributions and a variety of selection sets in Bayarri and DeGroot(1987b). These comparisons also include the distributions belonging to the natural parameter exponential family and the selection set $S(\tau) = [\tau, \infty)$. This selection set is equivalent to a left truncated distribution. Bayarri and DeGroot show that the standard experiment has more Fisher information than the left truncated data experiment. One exception to this result is the gamma distribution with shape parameter $\alpha < 1$ for which the converse is true.

Thus, in contrast to the censored data experiments, which have complete information ordering as a function of the stochastically increasing censoring, the standard experiment $\mathcal{E}$ is not sufficient for the selection experiment $\mathcal{F}$, [see, e.g., Bayarri and DeGroot (1989)]. Thus it is possible that for some decision problems, a selection experiment is better than the standard experiment while the reverse is true for some other decision problems.

To obtain more meaningful ordering in the absence of complete ordering, Lehmann(1988) defined the concept '$\mathcal{E}$ is more effective than $\mathcal{F}$,' if for some interesting subclass of decision problems, the experiment $\mathcal{E}$ provides decision procedures with smaller risk than those based on the experiment $\mathcal{F}$. An useful subclass of decision problems is that of test of hypotheses, for which the concept of pairwise sufficiency is defined as follows:

**Definition 1.3**. An experiment $\mathcal{E}$ is *pairwise sufficient* for the experiment F, denoted by $\mathcal{E} \succeq_2 \mathcal{F}$, if for every pair $\theta_0, \theta_1$ in $\Theta$, the experiment $\mathcal{E}$ is sufficient for the experiment $\mathcal{F}$, when the parameter space consists of only two values $\theta_0, \theta_1$ .

Clearly, if $\mathcal{E} \succeq \mathcal{F}$, then $\mathcal{E} \succeq_2 \mathcal{F}$, but the converse is not true. Furthermore, if the Fisher information exists, then $\mathcal{E} \succeq \mathcal{F}$ implies that the experiment $\mathcal{E}$ has more Kullback-Leibler information than the experiment $\mathcal{F}$, which in turn implies that $\mathcal{E}$ has more Fisher information than $\mathcal{F}$, denoted by $\mathcal{E} \succeq_F \mathcal{F}$. In addition, for the experiments having the monotone likelihood ratio property, Lehmann(1988) has shown that pairwise sufficiency of the experiment $\mathcal{E}$ for the experiment $\mathcal{F}$ also implies that $\mathcal{E}$ is more effective than $\mathcal{F}$ for the class of all monotone decision problems.

In Section 2, we present some useful results concerning the monotonicity and the $TP_2$ property of the hazard function $h$. Since the standard experiment $\mathcal{E}$ and the corresponding selection experiment $\mathcal{F}$ are not necessarily comparable, it is worthwhile to learn if $\mathcal{E}$ is more effective than $\mathcal{F}$ or vice versa. In Section 3, we concentrate mainly on the pairwise sufficiency of the experiment $\mathcal{E}$ for the experiment $\mathcal{F}_\tau$, corresponding to the selection set $S(\tau)$, when the class of distributions has monotone likelihood ratio *(mlr)* property. In Section 4, we illustrate the pairwise sufficiency results for the Normal, Gamma, Binomial and Poisson distributions. In Section 5, we present a pairwise comparison of two known distributions with unknown left-truncation parameter $\theta$.

## 2. NOTATIONS AND PRELIMINARIES

Assume that the distribution $P_\theta$ has a density $f(.,\theta)$ with respect to a $\sigma$-finite measure $\mu$, and let $F(x,\theta)$ denote the cdf of $X$. The distribution of the observation $Y$, from the restricted experiment $\mathcal{F}_\tau$ based on the selection set $S(\tau)$, is given by

$$g_\theta(y,\tau) = \frac{f(y,\theta)}{\bar{F}(\tau,\theta)}, y \geq \tau, \tag{2.1}$$

where $\bar{F}(x,\theta) = P_\theta(X \geq x)$ denote the survival distribution of $X$. If $f(x,\theta)$ has the *mlr* property, then the distribution of random variables in the selection experiment, $g_\theta(y,\tau)$, defined in (2.1), also has the mlr property. Now let $f_i(.) \equiv f(.,\theta_i)$, i = 0,1. For $\theta_1 > \theta_0$ , if the likelihood ratio

$$L(x) = \frac{f_1(x)}{f_0(x)}, x \in \mathcal{X}, \tag{2.2}$$

is monotone increasing(decreasing), then $f(.,.)$ is said to be a $TP_2$ $(SR_2)$ function on $\mathcal{X} \times \Theta$.

Before developing conditions under which $\mathcal{E} \succeq_2 \mathcal{F}_\tau$, or $\mathcal{F}_\tau \succeq_2 \mathcal{E}$, we present some useful properties of families of distributions having *mlr* property. Some of the results given below are well known [see, e.g., Karlin (1968) and Marshall and Olkin (1979)]. However, the proofs given here are elementary and novel.

**Lemma 2.1**. Let $f(x,\theta)$ be a $TP_2$ $(SR_2)$ function on $\mathcal{X} \times \Theta$, and let $h(t,\theta)$ denote the hazard rate corresponding to $f(x,\theta)$ , i.e.,

$$h(t,\theta) = \frac{f(t,\theta)}{\bar{F}(t,\theta)}. \tag{2.3}$$

Then $h(t,\theta)$ is a monotone decreasing(increasing) function of $\theta$.
Proof. For $\theta_1 > \theta_0$, let $R(t)$ denote $h(t,\theta_1)/h(t,\theta_0)$, $\bar{F}_i(.) \equiv \bar{F}(x,\theta_i)$, i = 0,1, and $g_0(.,.) = g_{\theta_0}(.,.)$. Then, with $L(.)$ defined in (2.2), we can write

$$\begin{aligned} R(t) &= L(t)/[\bar{F}_1(t)/\bar{F}_0(t)] \\ &= L(t)/\int_{S(t)} L(u) g_0(u,t) d\mu(u). \end{aligned} \tag{2.4}$$

Now, if $f(x,\theta)$ is a $TP_2$ $(SR_2)$ function, then $L(u)$ is monotone increasing (decreasing) function of $u$. The result now follows from the fact that for $u \in S(t), L(u)/L(t) \geq 1$, if $f(x,\theta)$ is $TP_2$ and it is $\leq 1$ if $f(x,\theta)$ is $SR_2$.

**Remark 2.1** If a discrete distribution $f(x,\theta)$ is a $TP_2$ $(SR_2)$ function, then Lemma 2.1 also implies that $f(t,\theta)/\bar{F}(t+1,\theta)$ is a monotone decreasing (increasing) function of $\theta$.

Since L(u) is a monotone function for a family of distributions having *mlr* property, by using (2.4), we can write

$$\frac{\bar{F}_1(t)}{\bar{F}_0(t)} = E_0[L(X)|X \geq t] \tag{2.5}$$

and

$$\frac{F_1(t)}{F_0(t)} = E_0[L(X)|X \leq t] \tag{2.6}$$

The following well known result can also be proved in a manner similar to Lemma 2.1.

**Lemma 2.2**. If $f(x,\theta)$ has the $TP_2$ $(SR_2)$ property, then $F(x,\theta)$ and $\bar{F}(x,\theta)$ also have the same property. Furthermore, if $f(x,\theta)$ is a $PF_2(RR_2)$ function, i.e., for each $\theta, f(x-y,\theta)$ is a $TP_2$ $(SR_2)$ function on $\mathcal{X} \times \mathcal{X}$ , then $\bar{F}(x,\theta)$ is a $PF_2(RR_2)$ function.

**Remark 2.2**. A consequence of Lemma 2.2 is that, if $f(x,\theta)$ is a $PF_2$ $(RR_2)$ function for each fixed $\theta$, then the hazard rate $h(t,\theta)$ is monotone increasing (decreasing) in $t$.

**Remark 2.3.** Note that for the location parameter $\theta, f(x,\theta) \equiv k(x-\theta)$ is a $TP_2$ $(SR_2)$ function iff the kernel $k(.)$ satisfies $\partial^2 \log k(x)/\partial x^2 \leq (\geq) 0$, and for the scale parameter $\theta, f(x,\theta) \equiv k(x\theta)$ is a $TP_2$ $(SR_2)$ function iff the kernel $k(.)$ satisfies $\partial[x\partial \log k(x)/\partial x]/\partial x \geq (\leq) 0$. For $f(x,\theta) \equiv k(x/\theta)$, the directions are reversed in this in this inequality.

The next result is a direct consequence of Remarks 2.2 and 2.3. We also prove a general result on $TP_2$ property of the hazard function $h$ for the exponential family.

**Theorem 2.1.** If a location parameter family is $TP_2$ $(SR_2)$, then the hazard rate $h(t)$ is monotone increasing (decreasing) in $t$ and if a scale parameter family is $TP_2$ $(SR_2)$, then $\{t\ h(t)\}$ is monotone decreasing (increasing) in $t$.

**Theorem 2.2.** Let $f(x,\theta)$ belong to the exponential family of the form

$$f(x,\theta) = a(\theta)b(x)e^{\{c(\theta)x\}}, x \in X \text{ and } \theta \in \Theta. \tag{2.7}$$

If $f(x,\theta)$ is $TP_2$ (i.e. $c(\theta)$ is monotone increasing) and, in addition, it is also $PF_2(RR_2)$ in $x$ for each $\theta$, then the hazard rate $h(x,\theta)$ is $TP_2$ $(SR_2)$. Similarly, if $f(x,\theta)$ is $SR_2$, (i.e. $c(\theta)$ is monotone decreasing) and, in addition, it is also $PF_2(RR_2)$ in $x$ for each $\theta$, then the hazard rate $h(x,\theta)$ is $SR_2$ $(TP_2)$.

Proof. First let us assume that $c(\theta)$ is a monotone increasing function. For $\theta_1 > \theta_0$, note that $R(t) = h(t,\theta_1)/h(t,\theta_0)$ can be expressed in the form (2.4). However, for $u > t, L(u)/L(t) = \exp[c(\theta)\ v]$, where $v = u - t$ and $g_0(u,t)$ is a probability distribution, proportional to $b(v{+}t)\exp[c(\theta_0)(v{+}t)]$, on the positive part of the real line. Note that, if $f(.,\theta_0)$ is $RR_2(PF_2)$, $g_0(u,t)$ is $TP_2$ ($SR_2$) in $v$ and $t$. It follows from Karlin (1968) [cf. Proposition 3.1, pp. 22] that if $f(.,\theta_0)$ is $RR_2(PF_2)$, then $R(t)$ is monotone increasing (decreasing) in $t$. Hence the hazard function is $TP_2$ ($SR_2$). The proof is similar for a monotone decreasing $c(\theta)$.

## 3. SELECTION MODELS: LEFT-TRUNCATED DATA

As noted in Section 1, the selection experiments are not comparable according to the notion of sufficiency. In this section, we will concentrate on establishing the pairwise sufficiency ordering of truncated data experiments. Bayarri and DeGroot (1989) show that in case of observations from a normal distribution with mean $\theta$, the standard experiment is pairwise sufficient for the restricted experiment for selection models. Furthermore, they also show that for Binomial distribution with $n = 2$, the unrestricted experiment is pairwise sufficient for the corresponding experiment with 0-class missing. We will now generalize these results for family of distributions having *mlr* property. For these families, the admissible critical regions for testing hypotheses $\theta = \theta_0$ against $\theta = \theta_1$, based on either the standard experiment or the selection experiment, are of the form

$$C(x) = \{X \geq x\}. \tag{3.1}$$

For testing hypotheses based on an observation $X$ from the unrestricted experiment with cdf $F_\theta$, $P$ [Type I error] and the power for the critical region $C(x)$, in (3.1), are given by

$$\alpha(x) = \bar{F}_0(x) \quad \text{and} \quad P(x) = \bar{F}_1(x). \tag{3.2}$$

These functions for testing hypotheses based on the observation $Y$ from the restricted experiment $\mathcal{F}_\tau$, with distribution $g_\theta(y,\tau)$ in (2.1), can be written as

$$\alpha_\tau(x) = \frac{\bar{F}_0(x)}{\bar{F}_0(\tau)} \text{ and } P_\tau(x) = \frac{\bar{F}_1(x)}{\bar{F}_1(\tau)}, \text{ for } x \geq \tau. \tag{3.3}$$

Now, for an absolutely continuous distribution, one can write the power function $P(x)$ as an implicit function of $\alpha$ through (3.2), i.e.,

$$P(\alpha) = \bar{F}_1(\bar{F}_0^{-1}(\alpha)). \tag{3.3a}$$

Furthermore, for discrete distributions, the power function is a piece-wise linear function of $\alpha$ with the change points defined by (3.3). It is well known that the power function $P(.)$ for the most powerful test of a simple against simple hypothesis is a continuous, concave, non-decreasing function of $\alpha$.

It follows from (3.3) and (3.3a) that for the most powerful tests based on observations from the truncated experiment restricted to $S(\tau)$, we can write

$$\alpha_\tau(x) = \frac{\alpha(x)}{\alpha(\tau)} \text{ and } P_\tau(\alpha_\tau(x)) = \frac{P(\alpha(x))}{P(\alpha(\tau))}. \tag{3.4}$$

Given $\tau$, let $\alpha^*$ denote $\alpha(\tau) < 1$. It follows from (3.4) that the power curve $(\alpha, P^*(\alpha))$ for tests based on the truncated observations can be obtained from the power curve $(\alpha, P(\alpha))$ for tests based on the standard random variable, through the following change of scale on both the $\alpha$- and the $P$-axes:

$$P^*(\alpha) = \frac{P(\alpha\alpha^*)}{P(\alpha^*)}, 0 \leq \alpha \leq 1. \tag{3.5}$$

**Remark 3.1.** Note that the choice of the selection set $S(\tau)$ for *mlr* distributions amounts to deleting the subset of the sample space which corresponds to values of $\alpha$ greater than $\alpha^*$. Thus we do not necessarily need to assume the *mlr* property for the following analysis. Instead it is sufficient to assume that the selection set $S$ corresponds to tail-values of the likelihood ratio $L(x)$.

Now the difference, $D(\alpha)$, between the two power curves,

$$D(\alpha) = P^*(\alpha) - P(\alpha), 0 \leq \alpha \leq 1, \tag{3.6}$$

can be used to examine the pairwise sufficiency of one experiment for the other. It is well known that $\mathcal{E} \geq_2 \mathcal{F}_\tau$, if and only if $D(\alpha) \leq 0$ for $0 \leq \alpha \leq 1$ and $\mathcal{F}_\tau \succeq \mathcal{E}$, if and only if $D(\alpha) \geq 0$ for $0 \leq \alpha \leq 1$ [See Torgersen (1988)].

Note that if $P(\alpha^*) = 1$ then $D(\alpha) \geq 0$ for all $\alpha$ in which case $\mathcal{E} \succeq_2 \mathcal{F}_\tau$. Let us, therefore, assume that $P(\alpha^*) < 1$. Now if $P(0) > 0$, then it follows from (3.6) that $D(0) > 0$. Furthermore, it is also easy to see that for every experiment $D(1) = 0$, and if $P(0) = 0$, then $D(0) = 0$.

We shall first examine the boundary behavior of $D(\alpha)$ near 0 and 1. Note that $P'(\alpha) = \partial P(\alpha)/\partial\alpha$ is well defined for all $\alpha$, except for a set with Lebesgue measure zero (i.e., at the jump points of $F(.,\theta)$ when $X$ is a discrete random variable). In fact, for absolutely continuous distributions, $P'(\alpha)$ is equal to the likelihood ratio $L(x)$ at the point $x = \bar{F}_0^{-1}(\alpha)$. We denote this value by $L(\alpha)$. For discrete distributions, let $\alpha_j = \bar{F}_0(x_j)$, where $x_j$ are the points with positive mass. Then $P'(\alpha)$ does not exist at the mass points and $P'(\alpha) = L(\alpha_j)$ for $\alpha_j < \alpha < \alpha_{j+1}$. Now, the derivative of $D(\alpha)$, with respect to $\alpha$, can be defined almost everywhere by

$$D'(\alpha) = \alpha^* \frac{P'(\alpha\alpha^*)}{P(\alpha^*)} - P'(\alpha), \quad 0 < \alpha < 1. \tag{3.7}$$

Note that if the derivative of $P$ exists, then $\alpha P'(\alpha)/P(\alpha) = R(\alpha(x))$, where $R(x)$ is defined in (2.4).

Now, it follows from (3.7) that if $P'(0+) = M < \infty$, then $D'(0+) = M[(\alpha^*/P(\alpha^*))-1] < 0$, and therefore $D(0+) < 0$. Hence, $D(0+)$ is $\leq (\geq) 0$, if and only if, for large $x$, $R(x)$ is monotone increasing (decreasing) in $x$. Furthermore, $D'(1-) = R(\tau) - R(\tau^*)$, where $\tau^*$ corresponds to $\alpha = 1$. Again, $D'(1-)$ is well defined with $P'(\alpha^*-) = L(\tau)$. If $P'(1-) = 0$, then $D'(1-) > 0$, i.e., $D(1-) < 0$. Hence, for $P'(1-) > 0$, if $R(\tau) > R(\tau_*)$, then $D'(1-) > 0$, i.e., $D(1-) < 0$, and if $R(\tau) < R(\tau_*)$, then $D'(1-) < 0$, i.e., $D(1-) > 0$. Note that if $R(\tau)$ is monotone (i.e.,the hazard rate $h(\tau, \theta)$ is $TP_2$ ($SR_2$)) then as $\tau$ increases, or, equivalently $\alpha^*$ decreases, $D(1-)$ becomes more and more negative(positive). The above discussion can be summarized as follows :

**Lemma 3.1**. Let $f(x, \theta)$ have *mlr* property and the selection experiment be restricted to $S(\tau)$. If $P(\alpha^*) = 1$ then $\mathcal{E} \succeq_2 \mathcal{F}_\tau$. Furthermore, the difference $D(\alpha)$ in (3.6), at $\alpha = 0+$ and $1-$ satisfies the conditions stated in Table I.

Table I. Behavior of $D(\alpha)$ at $\alpha = 0+$ and $1-$.

| | $P'(1-) = 0$ | $P'(1-) > 0$ & $f(x,\theta)$ and $h(x,\theta)$ satisfy | |
|---|---|---|---|
| | | Both $TP2(SR2)$ | One $TP_2$ one $SR_2$ |
| $P'(0) = M$ | both $-ve$ | both $-ve$ | changes sign |
| $P'(0) \to \infty$ | $D(1-) < 0$ | both $-ve$ | both $+ve$ |
| $P(0) > 0$ | changes sign | changes sign | both $+ve$ |

We will first assume that the family of distributions is absolutely continuous. Then, the derivatives of $P(\alpha)$ exist for all $\alpha$. For $D(\alpha)$ to have same sign, it is sufficient to show that the function has same sign at all the stationary points of $D(\alpha)$. It follows from (3.7) that $D'(\alpha) = 0$ implies that

$$P(\alpha^*) = \alpha^* \frac{P'(\alpha\alpha^*)}{P'(\alpha)}, \tag{3.8}$$

and $D(\alpha) \leq (\geq) 0$ is equivalent to

$$\alpha^* \frac{P'(\alpha\alpha^*)}{P(\alpha\alpha^*)} \geq (\leq) \frac{P'(\alpha)}{P(\alpha)}. \tag{3.9}$$

On Multiplying both side of the inequality in (3.9) by $\alpha$, it is easy to see that $D(\alpha) \leq (\geq) 0$ equivalent to

$$\alpha \frac{P'(\alpha)}{P(\alpha)} \text{ is monotone decreasing(increasing) in } \alpha. \tag{3.10}$$

However, $\alpha P'(\alpha)/P(\alpha) = R(\alpha(x))$, where $R(x)$ is defined in (2.4). Hence (3.10) is equivalent to the fact that $R(x)$ is monotone increasing (decreasing) in $x$. Finally, note that monotonicity of $R(x)$ is equivalent to $TP_2(SR_2)$

property of $h(x,\theta)$. The following result now follows from Lemma 3.1 and the above discussion.

**Theorem 3.1.** Suppose that the random variable $X$ has an absolutely continuous distribution and that the family $f(x,\theta)$ has *mlr* property with $L(x) \to \infty$. Let the selection experiment be restricted to $S(\tau)$ and let $\alpha^* = \alpha(\tau)$. If $P(\alpha^*) = 1$ then $\mathcal{E} \succeq_2 \mathcal{F}_\tau$. Furthermore, $\mathcal{E} \succeq_2 \mathcal{F}_\tau$, iff both $f(x,\theta)$ and $h(x,\theta)$ are either $TP_2$ or $SR_2$. In addition, if one of $f(x,\theta)$ and $h(x,\theta)$ is $TP_2$ and the other is $SR_2$, then $\mathcal{F}_\tau \succeq_2 \mathcal{E}$.

Another proof for the above result can be given by treating $P^*(\alpha)$, in (3.2), as a function of $\alpha^*$ for each fixed $\alpha$, and taking its partial derivative with respect to $\alpha^*$. This examines the behavior of the power curve as the selection set changes. If the derivative of $lnP^*$ with respect to $\alpha^*$ is of same sign, then the $D(\alpha)$ has same sign and we will have one experiment sufficient for the other. In fact,

$$\frac{\partial lnP^*}{\partial\alpha^*} = \alpha\frac{P'(\alpha\alpha^*)}{P(\alpha\alpha^*)} - \frac{P'(\alpha^*)}{P(\alpha^*)} \geq (\leq)\ 0 \tag{3.11}$$

is equivalent to the monotoneness of $R(\alpha(x))$ for all $\alpha$. This proves the following generalization of Theorem 3.1.

**Theorem 3.2.** Suppose that the random variable $X$ has an absolutely continuous distribution. Let the family $f(x,\theta)$ have *mlr* property with $L(x) \to \infty$. Given $\tau > \omega$, consider two selection experiments, $\mathcal{F}_\omega$ and $\mathcal{F}_\tau$, corresponding to the selection set $S(\omega)$ and $S(\tau)$ respectively. Let $\alpha^* = \alpha(\tau)$. If $P(\alpha^*) = 1$ then $\mathcal{F}_\omega \geq_2 \mathcal{F}_\tau$ . Furthermore, $\mathcal{F}_\omega \geq_2 \mathcal{F}_\tau$, if and only if both $f(x,\theta)$ and $h(x,\theta)$ are either $TP_2$ or $SR_2$. In addition, if one of $f(x,\theta)$ and $h(x,\theta)$ is $TP_2$ and the other is $SR_2$, then $\mathcal{F}_\tau \geq_2 \mathcal{F}_\omega$.

Since for discrete distributions, $\alpha^*$ only takes discrete values, the derivative of $P^*(\alpha)$ with respect to $\alpha^*$ does not have any meaning. However, we can take derivatives of $D(\alpha)$ with respect to $\alpha$ almost everywhere. The proof for continuous distributions and the behavior of $D(\alpha)$ near 0 and 1 suggests that one can obtain the sign of $D(\alpha)$ almost everywhere. Now since $D(\alpha)$ is continuous, the sign of $D(\alpha)$ can be obtained everywhere via a continuity argument. However, we now provide a direct proof of the following result, since the indirect approach above does not highlight the effect of discreteness in the distributions.

**Theorem 3.3.** Suppose that $X$ has probability distribution $f(x,\theta), \theta \in \Theta$, and $x \in X = \{0,1,2,\ldots\}$. Let the family $f(x,\theta)$ have *mlr* in $x$ and $\theta$ and be $PF_2$ in $x$ for each fixed $\theta$, and let $\bar{F}_i(x) = \Sigma_{t\geq x} f_i(t), i = 0,1$. Define

$$h^*(x,\theta) = \frac{f(x,\theta)}{\bar{F}(x+1,\theta)}. \tag{3.12}$$

Let $h_i^*(.) = h^*(.,\theta_i), i = 0,1$. Using the notation from the analysis for continuous distributions above, the following statements hold:

(i) If $L(x) \leq M < \infty$ for all $x$, and one of $f$ and $h$ is $TP_2$, while the other is $SR_2$ or if $P(0) > 0$ and both $f$ and $h^*$ are either $TP_2$ or $SR_2$, then the power curves of $\mathcal{F}_\tau$ and $\mathcal{F}_{\tau+1}$ cross each other.

(ii) If $L(x) < \infty$, and both $f$ and $h^*$ are either $TP_2$ or $SR_2$ then

$\mathcal{F}_\tau \geq_2 \mathcal{F}_{\tau+1}$.

(iii) If one of $f$ and $h$ is $TP_2$ while the other is $SR_2$ and either $L(x) \to \infty$, or $P(0) = 0$, then $\mathcal{F}_{\tau+1} \geq_2 \mathcal{F}_\tau$.

Proof: For the experiment $\mathcal{F}_\tau$, with selection set $S(\tau)$ and the most powerful critical regions (3.1), the power function $P^*$ satisfies (3.3). For the standard experiment, the power function $P(\alpha)$ is piece-wise linear with change points at $\alpha(j) = \bar{F}_0(j), j = 0, 1, 2, \ldots$. A typical power function $P(\alpha)$ is shown in Figure 1, below.

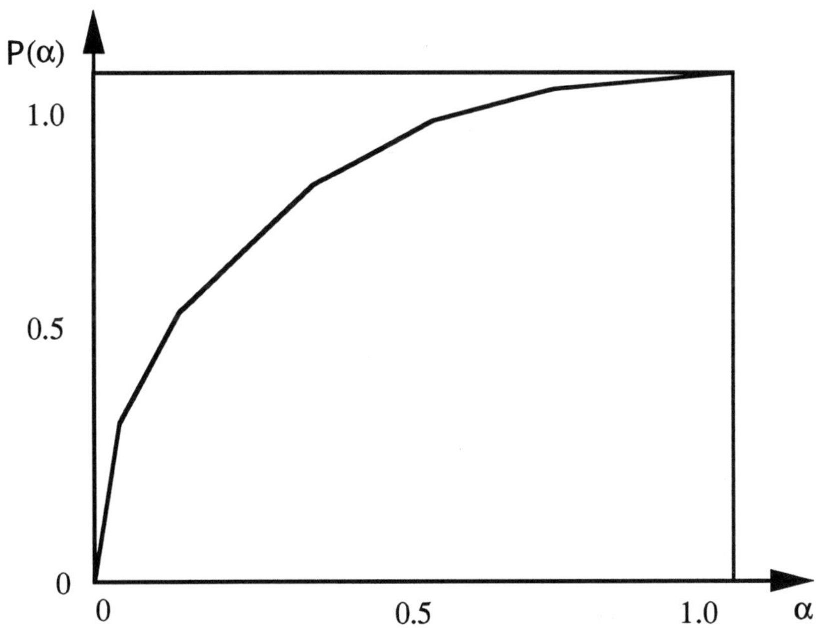


Figure 1. A Typical power curve for discrete distributions

The slope of the line segment joining the point $[\alpha_\tau(j+1),\ P(\alpha_\tau(j+1))]$ to the point $[\alpha_\tau(j), P(\alpha_\tau(j))]$ is given by

$$\gamma_\tau(j) = L(j)\frac{\bar{F}_0(\tau)}{\bar{F}_1(\tau)}. \tag{3.13}$$

Since $L(j)$ is monotone increasing in $j$, it follows that $\gamma_\tau(j)$ increases as $j$ increases. If the support set of $f$ is finite, say $X = (0, 1, \ldots, T)$, then $\gamma_\tau(T) < \infty$ unless $f_0(M) = 0$ and $f_1(M) > 0$ for some $M$. We need to examine the behavior of $(\alpha, P^*(\alpha))$ as $\tau$ changes to $\tau+1$. In this situation, there is one less segment in the $(\tau+1)$-power function than that in the $\tau$-power function. In fact, the change points of $(\tau+1)$-curve are at

$$\alpha_{\tau+1}(j) = \frac{\bar{F}_0(j)}{\bar{F}_0(\tau+1)}\ ,\ j = \tau+1,\ \tau+2, \ldots.. \tag{3.14}$$

Thus the power functions $\tilde{P}$ for the experiment $\mathcal{F}_{\tau+1}$ is obtained from the power function $P^*$ for the experiment $\mathcal{F}_\tau$ by re-scaling on $(a, P)$ axes as follows:

$$\begin{aligned} \alpha &\to \alpha[1+h_0^*(\tau)] \quad 0 \le \alpha \le \alpha^*(\tau+1) \\ P &\to P[1+h_1^*(\tau)] \quad 0 \le P \le P^*(\tau+1) \end{aligned} \tag{3.15}$$

and the line segment from $(\alpha^*(\tau+1), P^*(\tau+1))$ to $(1,1)$ is deleted. Thus, $\tilde{P}$ and $P^*$ curves are related via a constant ratio of the slopes at $\alpha_\tau(j)$ and $\alpha_{\tau+1}(j) = \tilde{\alpha}(j)$, i.e.,

$$\frac{\gamma_\tau(j)}{\gamma_{\tau+1}(j)} = \frac{1+h_0^*(\tau)}{1+h_1^*(\tau)} = \delta(\text{Say}). \tag{3.16}$$

However, it follows from Remark 2.1, that if $f(x,\theta)$ is $TP_2$, then $h_0^*(\tau) \le h_1^*(\tau)$. Hence $\delta \ge 1$.

For a random variable with a finite support set and $P^*(0) = 0$, it is clear that slope of the $P^*$-curve near $\alpha = 0$ will be larger than that for $\tilde{P}$-curve, and if $P^*(0) > 0$, then $\tilde{P} > P^*(0)$. On the other hand, if $L(x) \to \infty$, then there is no initial segment on the $P^*$-curve and the behavior of $P^*(\alpha(j))$ near zero is determined in the limit as $j \to \infty$.

Now we shall compare the slopes of the first line segments on the two power curves near $\alpha = 1$. The ratio of the two slopes is given by

$$\eta(\tau) = \frac{\gamma_\tau(\tau)}{\gamma_{\tau+1}(\tau+1)}. \tag{3.17}$$

From (3.13), it is easy to write $\eta(\tau)$ as

$$\eta(\tau) = \frac{R(\tau)}{R(\tau+1)}, \tag{3.18}$$

where

$$R(\tau) = \frac{h_1(\tau)}{h_0(\tau)}. \tag{3.19}$$

Now, if $\eta(\tau) \leq 1$ then $\tilde{P}(1-) \leq P^*(1-)$, and if $\eta(\tau) \geq 1$, then $P^*(1-) \geq \tilde{P}(1-)$. However, sign of $D(\alpha)$ should not change if we have a pairwise comparison of $\mathcal{F}_\tau$ and $\mathcal{F}_{\tau+1}$. This requires $\eta(\tau) \geq 1$ for all $\tau$, or $\leq 1$ for all $\tau$, i.e., $R(\tau)$ to be monotone in $\tau$, which is equivalent to the requirement that $h(\tau, \theta)$ be $TP_2$ $(SR_2)$. This result is similar to that for the continuous case.

Here the *mlr* property of $h(\tau, \theta)$ guarantees that for all changes in $\alpha$, $P^*(\alpha)$ changes monotonically near 1. Now, to prove that, for each $\alpha$, the power curves moves consistently in one direction, we consider an arbitrary line segment with slope $\gamma_\tau(j)$ on $P^*(\alpha)$ curve and examine its location under the transformation (3.15). We explore conditions under which the intervals on the $\alpha$-axis corresponding to line segments in $P^*$ and $\tilde{P}$ are interlaced, i.e.

$$\tilde{\alpha}(j+1) \leq \alpha^*(j). \tag{3.20}$$

It is easy to see that if $h^*(\tau, \theta)$ is monotone increasing in $\tau$, then

$$\alpha^*(j+1) \leq \tilde{\alpha}(j+1) \leq \alpha^*(j) \text{ and } P^*(j+1) \leq \tilde{P}(j+1) \leq P^*(j). \tag{3.21}$$

The conditions in (3.21) are equivalent to the fact that $\bar{F}(\tau, \theta)$ be $PF_2$. It follows from Remark 2.2 that a sufficient condition for this to hold is that $f(x, \theta)$ be $PF_2$ for each $\theta$. We then have the following mesh of constant slope segments on $P^*$ and $\tilde{P}$ curves:

$$\alpha^*(j+1) \qquad \tilde{\alpha}(j+1) \qquad \alpha^*(j) \qquad \tilde{\alpha}(j)$$

Now, in the interval $(\tilde{\alpha}(j+1), \alpha^*(j))$, the ratio of the slopes of line segments in $P^*$ and $\tilde{P}$ is equal to $\delta$ (see (3.16)). Since $\delta \geq 1$, the difference $(P^* - \tilde{P})$ will be smallest at $\alpha = \tilde{\alpha}(j+1)$. On the other hand, in the interval $(\alpha^*(j), \tilde{\alpha}(j))$, the difference $(P^* - \tilde{P})$ may be largest at either end. If $h(\tau, \theta)$ is $TP_2$, then $P^*(\alpha) \geq \tilde{P}(\alpha)$ for all $\alpha \in (\alpha^*(\tau+1), \alpha^*(\tau))$, with the largest difference at $\alpha = \alpha^*(\tau+1)$. Now, if $\tilde{P}(\tilde{\alpha}(j))$ are below the line segment joining $(\alpha^*(j), P^*(\alpha^*(j)))$ and $(\alpha^*(j-1), P^*(\alpha^*(j-1)))$ for $j = \tau+2, \tau+3, \ldots$, then the $P^*$ curve will be above the $\tilde{P}$ curve, i.e.

$$\tilde{P}(\tilde{\alpha}(j)) \leq P^*(\tilde{\alpha}(j)), \quad j = \tau+2, \ldots, \tag{3.22}$$

holds if

$$\frac{\tilde{P}(j) - P^*(j)}{\tilde{\alpha}(j) - \alpha^*(j)} \leq \gamma^*(j-1), \quad j = \tau+2, \ldots. \tag{3.23}$$

It follows from (3.13) and (3.15) that (3.23) is equivalent to

$$\frac{P^*(j)}{\alpha^*(j)} \frac{h_1^*(\tau)}{h_0^*(\tau)} \leq L(j-1) \frac{\bar{F}_0(\tau)}{\bar{F}_1(\tau)}, j = \tau+2, \ldots. \tag{3.24}$$

However, substituting the values of $P^*(j)$ and $\alpha^*(j)$ from (3.3), one gets

$$\frac{\bar{F}_1(j)}{\bar{F}_0(j)} \frac{h_1^*(\tau)}{h_0^*(\tau)} \le L(j-1),\ j = \tau+2, \tau+3, \ldots. \tag{3.25}$$

It follows from (3.12) that (3.25) is equivalent to

$$\frac{h_1^*(\tau)}{h_0^*(\tau)} \le \frac{h_1^*(j)}{h_0^*(j)}, \quad j = \tau+1, \tau+2, \ldots. \tag{3.26}$$

Therefore, if $[h_1^*(j)/h_0^*(j)]$ is monotone in $j$, then for all $j \ge \tau+1$, the change points of $\tilde{P}$ will be below the $P^*$ curve. Hence the $\tilde{P}$-curve will be below the $P^*$-curve. In other words, $\mathcal{F}_\tau \succeq_2 \mathcal{F}_{\tau+1}$. Note that, since

$$\frac{h_1^*(j)}{h_0^*(j)} = \frac{h_1(j)}{h_0(j)} \frac{(1-h_1(j))}{(1-h_0(j))}, \tag{3.27}$$

it is easy to check that

$$\text{both } f \text{ and } h\ TP_2(SR_2) \Rightarrow h^* \text{ is } TP_2(SR_2) \tag{3.28}$$

and

$$f\ SR_2(TP_2) \text{ and } h\ TP_2(SR_2) \Rightarrow h^* \text{ is } TP_2(SR_2). \tag{3.29}$$

The results now follow from the above discussion and the fact that the $TP_2$ properties of $h$ and $h^*$ are related through (3.28) and (3.29).

## 4. SOME SPECIFIC DISTRIBUTIONS

### 4.1. Normal Distribution

Let $X \sim N(\theta, 1)$ where $\theta$ is the unknown parameter. It is well known that $\phi(x-\theta)$ and $\Phi(x-\theta)$ have $TP_2$ property and that $\phi$ is $PF_2$. Furthermore, $h(t)$ is monotone increasing in $t$ (see Theorem 2.1). The following result is a consequence of Theorem 2.2.

**Lemma 4.1.** If $h(t) = \phi(t)/[1-\Phi(t)]$, then $h(t,\theta)$ is a $TP_2$ function.

Next result is a direct consequence of Theorem 3.2 and Lemma 4.1.

**Theorem 4.1.** For the normal distribution $N(\theta, \sigma^2)$, with $\sigma^2$ known, if $\tau > \omega$, then the selection experiment $\mathcal{F}_\omega$ is pairwise sufficient for the selection experiment $\mathcal{F}_\tau$.

This result implies that a standard Normal$(\theta, 1)$ experiment is pairwise sufficient for left-truncated Normal experiment, a result proved in Bayarri and DeGroot (1989).

### 4.2. Gamma Distribution

Let $X$ be a random variable with Gamma Distribution with probability density

$$f(x;\theta) = \frac{\theta^\alpha}{\Gamma(\alpha)} x^{\alpha-1} e^{-\theta x} \quad x \geq 0, \alpha \text{ fixed}, \tag{4.2}$$

It is well known that the Gamma density is $SR_2$. Now, for $\alpha = 1$, i.e., for the exponential distribution. $h(x) \equiv \theta$ and therefore, it follows $h(.,.)$ is both $TP_2$ and $SR_2$. Therefore, for $\tau > \omega, \mathcal{F}_\tau$ and $\mathcal{F}_\omega$ are pairwise equivalent. However, it is well known that pairwise equivalent experiments are equivalent. Now for $\alpha > 1, f$ is known to be $PF_2$ with a monotone increasing hazard rate and for $\alpha < 1, f$ is known to be $RR_2$. The following result is a direct consequence of Theorem 2.2.

**Lemma 4.2.** Let $X$ be a Gamma distribution with density given in (4.2). If $\alpha > 1 (\alpha < 1)$, then the corresponding hazard rate $h(x,\theta)$ is a $SR_2$ ($TP_2$) function of $x$ and $\theta$.

The following result is a direct consequence of Theorem 3.2 and Lemma 4.2.

**Theorem 4.2.** For the Gamma distribution with density given in (4.2), if $\tau > \omega$, then for $\alpha < 1, \mathcal{F}_\tau \succeq_2 \mathcal{F}_\omega$ and for $\alpha > 1, \mathcal{F}_\omega \succeq_2 \mathcal{F}_\tau$.

**Remark 4.1.** When $\theta$ is replaced by $1/\lambda$ in (4.2), then $f(x,\lambda)$ is $TP_2$ and $h(x,\theta)$ is $TP_2$ for $\alpha > 1$ and $SR_2$ for $\alpha < 1$. Thus the result in Theorem 4.2 still holds true.

The result in Theorem 4.2 implies that when $\alpha > 1$, a standard Gamma experiment is pairwise sufficient for the left-truncated Gamma experiment and when $\alpha < 1$, the converse holds. Thus we have a generalization of the Fisher information ordering for the Gamma distribution proved in Bayarri and DeGroot (1987b).

### 4.3. Normal distribution with known mean and unknown variance.

The normal density $N(0,\sigma^2)$ is given by

$$f(x,\sigma) = \frac{1}{\sigma\sqrt{2\pi}} e^{-\frac{1}{2\sigma^2}x^2}, -\infty < x < \infty.$$

Since $\frac{\partial^2}{\partial x \partial \sigma} \log f(x,\sigma) > 0$ for $x > 0$ and $< 0$ for $x < 0$, the distribution is not $TP_2$. However if we consider two truncated distributions $\mathcal{F}_\tau$ and $\mathcal{F}_\omega$ with $\tau, \omega > 0$, then the problem is essentially equivalent to a gamma variable case with $\alpha = 1/2$. Furthermore, if we have symmetric truncation of this normal distribution around zero, the problem is equivalent to the left truncation of gamma distribution. Bayarri and DeGroot (1989) also discuss the two tail truncation for this distribution.

### 4.4. Binomial and Poisson Distributions

It is well known that both the Binomial and the Poisson distributions are $TP_2$ and in addition both are $PF_2$. Thus, it follows from Theorem 2.2 that the hazard rates for both of these distributions have $TP_2$ property. Therefore we get the following pairwise comparison result as a direct consequence of Theorem 3.3.

**Theorem 4.3.** For the binomial and Poisson distributions, if $\tau > \omega$, then the selection experiment $\mathcal{F}_\omega$ is pairwise sufficient for the selection experiment $\mathcal{F}_\tau$.

The result in Theorem 4.3 again generalizes the Fisher information ordering, established in Bayarri and DeGroot(1987), for these two distributions.

## 5. ANOTHER SELECTION MODEL

Let us assume that $F$ and $G$ are two completely known distributions. However, suppose that the selection set $S = (\theta, \infty)$, where $\theta$ is an unknown parameter. We want to learn about $\theta$ from a sample of observations from either of the two restricted experiments. An interesting question is: which of the two experiments is preferable?

By transforming $X \rightarrow \bar{F}(X)$ and $Y \rightarrow \bar{G}(Y)$, the distributions of restricted observations are given by $X_\theta \sim U[0, \bar{F}(\theta)]$ and $Y_\theta \sim U[0, \bar{G}(\theta)]$. Assume that we wish to test $H_0 : \theta = \theta_1$ vs $H_1 : \theta = \theta_2, \theta_1 > \theta_2$, based on $X_\theta$ or $Y_\theta$. The power curves for tests based on an observation from $F$ or $G$ are given in Figure 2, where $P_g(0) = 1 - \frac{\bar{G}_1}{\bar{G}_2}$ and $P_f(0) = 1 - \frac{\bar{F}_1}{\bar{F}_2}$ represent the power at $\alpha = 0$.

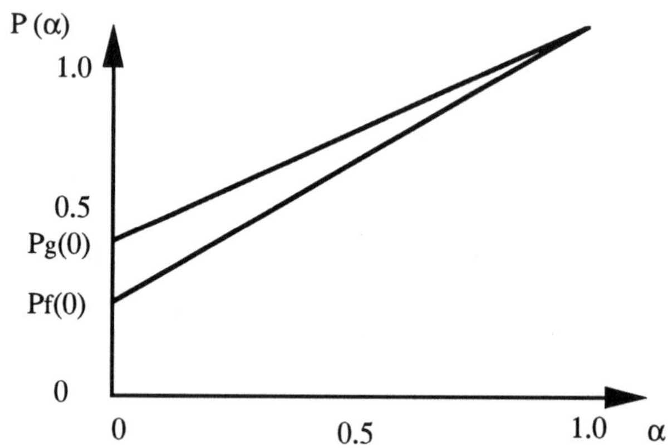


Figure 2: Power curves for F and G distributions

Now, $\mathcal{E}_G \succeq_2 \mathcal{E}_F$ if for all $(\theta_1, \theta_2)$,

$$\left[\bar{G}_2 - \frac{\bar{G}_1}{\bar{G}_2}\right] \geq \left[\bar{F}_2 - \frac{\bar{F}_1}{\bar{F}_2}\right] \Leftrightarrow \frac{\bar{F}_1}{\bar{F}_2} \geq \frac{\bar{G}_1}{\bar{G}_2} \Leftrightarrow \frac{\bar{F}_1}{\bar{G}_1} \geq \frac{\bar{F}_2}{\bar{G}_2}.$$

Thus if $\frac{\bar{F}(\theta)}{\bar{G}(\theta)}$ is monotone increasing in $\theta$, then $\mathcal{E}_G \succeq_2 \mathcal{E}_F$. However, this ratio can be written as

$$\frac{\bar{F}(\theta)}{\bar{G}(\theta)} = \frac{\int_\theta^\infty \frac{f(x)}{g(x)} g(x)\, dv(x)}{\int_\theta^\infty g(x)\, dv(x)}$$

$$= E_g\left[\frac{f(X)}{g(X)} | X \geq \theta\right].$$

Therefore, if $\frac{f(x)}{g(x)}$ is monotone in $x$, then $\frac{\bar{F}(\theta)}{\bar{G}(\theta)}$ is monotone increasing in $\theta$. Hence, $\mathcal{E}_G \succeq_2 \mathcal{E}_F$ for the selection set $S = (\theta, \infty)$. Thus, a sufficient condition for $\mathcal{E}_G \succeq_2 \mathcal{E}_F$ for the selection set $S = (\theta, \infty)$ is that $f/g$ be monotone increasing in $x$. Note that in such a situation, the smaller values of the likelihood ratio $f/g$ have been truncated.

## ACKNOWLEDGMENTS

This research is supported in part by the National Science Foundation, under grants DMS-8906787, DMS-9008067 and INT-8913294. The U.S. Government is authorized to reproduce and distribute reprints for Governmental purposes notwithstanding any copyright notation thereon. A draft version of this paper was read by Prof. DeGroot and extensively commented upon. Unfortunately, due to his failing health in 1989, he did not get a chance to look at the final version. If there are any errors or omissions, they are deemed to be my responsibility. This is one of the last few projects, I and Prof. DeGroot collaborated on. Memories of his visit to the Ohio State University are tied very heavily into this work.

## REFERENCES


Bayarri, M.J. and DeGroot, M. H. (1987a) Bayesian analysis of selection models. *The Statistician* **36** 137-146.

Bayarri, M.J. and DeGroot, M. H. (1987b) Information in selection models. In *Probability and Bayesian Statistics*, (R. Viertl, ed.) 39-51. New York: Plenum Press.

Bayarri, M.J. and DeGroot, M. H. (1989) Comparison of experiments with weighted distributions. In *Statistical Data Analysis and Inference* (Y. Dodge ed.), 185-197. Amsterdam: Elsevier Science Publishers B.V. (North Holland).

Blackwell, D. (1953) Equivalent comparison of experiments. *Annals Math. Statist.* **24** 265-272

Ferentino, K. and Papaioannou,T. (1982) Information in experiments and sufficiency. *Jour. Statist. Planning Inference*, **6** 309-317.

Goel, P. K. and DeGroot, M. H. (1979) Comparison of experiments and information measures. *Annals Statist.*, **7** 1066-1077.

Goel, P. K. (1988) Comparison of experiments and information in censored data. In *Statistical Decision Theory and Related Topics IV* (S. S. Gupta and James O. Berger, eds.), , **Vol. 2** 335- 349 New York: Springer-Verlag.

Hansen, O. H. and Torgersen, E. N. (1974) Comparison of linear normal experiments. *Annals Statist.*, *2* 367-373.

Heyer, H. (1982) *Theory of Statistical Experiments.* New York: Springer-Verlag

Hollander, M., Proschan, F. and Sconing, J. (1985) Measures of dependence for evaluating information in censored models. *Tech. Report, #M706*, Department of Statistics, Tallahassee: Florida State University.

Hollander, M., Proschan, F. and Sconing, J. (1987) Measuring information in right-censored models, *Naval Research Logistics Quarterly*, **34** 669-681.

Karlin, S. (1968) *Total Positivity, Vol. I*, Stanford: Stanford University Press

Lehmann, E. L. (1988) Comparing location experiments. *Annals Statist.*, **16** 521-533

Marshall, A. W. and Olkin, I. (1979) *Inequalities: Theory of Majorization and Its Applications*, New York: Academic Press

Patil, G. P. (1984) Studies in statistical ecology involving weighted distributions. In *Statistics: Applications and New Directions*, 475-503. Calcutta: Indian Statistical Institute.

Rao, C. R. (1965) On discrete distributions arising out of methods of ascertainment. In *Classical and Contagious Discrete Distributions*, (G.P. Patil, ed.) 320-333. Calcutta: Statistical Publishing Society.

Rao, C. R. (1985) On discrete distributions arising out of methods of ascertainment: What population does a sample represent. In *A Celebration of Statistics: The ISI Centenary Volume*, (A.G. Atkinson and S.E. Feinberg. eds.) 543-569 New York: Springer-Verlag.

Taneja, I. J. (1987) Statistical aspects of divergence measures. *Journal Statistical Planning and Inference*, **16** 137-145.

Torgersen, E. N. (1972) Comparison of translation experiments. *Annals Math. Statist.*, **43** 1383-1399.

Torgersen, E. N. (1976) Comparison of statistical experiments. *Scand. Jour. Statist .*, **3** 186-208.

Torgersen, E. N. (1981) Measure of information based on comparison with total information and with total ignorance. *Annals Statist.*, **9** 638-657.

Torgersen, E. N. (1988) *Comparison of Statistical Experiments*, Cambridge, England: Cambridge University Press.

# 16

# Jeffreys-Lindley Paradox and a Related Problem

## Sudhakar Kunte [1]

**Abstract**

The Jeffreys-Lindley paradox is resolved by imposing a reasonable condition on the classical test procedures. It is also pointed out that for the problem of testing goodness of fit, the usual fixed level $\chi^2$ test may be inadequate and some adjustment for the choice of level depending upon the size of the sample may be needed.

*Key words and Phrases:* Jeffreys-Lindley paradox; Goodness of fit; Level of significance; Sample size.

## 1. INTRODUCTION

Jeffreys-Lindley paradox points out a discrepancy between the classical Neyman Pearson type test procedures and Bayesian test procedures. Suppose that the test procedure is based on a standardized statistic which is $\sqrt{n}$-consistent. In such a case for classical test procedures the cut-off point for the critical region depends only upon the selected level of significance and not on the actual sample size $n$. In a similar situation for the Bayes test, which also depends upon the same standardized statistic, the cut-off point of the critical region depends upon $n$ and in fact goes to infinity as $n \to \infty$. This discrepancy was pointed out by Jeffreys (1961). Lindley (1957), Zellner (1971), and Berger (1985), call this discrepancy a paradox. Shafer (1982) discusses this paradox from the point of view of belief function theory. In Berger and Sellke (1987), Berger and Delempady (1988) and some other recent papers of Berger, this paradox is discussed from the point of view of evidence supplied by the data in the form of observed level of significance and a lower bound for the posterior probability for the null hypothesis to be true.

In this paper we try to resolve this paradox by putting a reasonable extra condition on the classical test procedures. We also point out that for the problem of testing goodness of fit, the usual fixed level $\chi^2$ test may be inadequate and some adjustment for the choice of level depending upon the size of the sample may be needed.

---

[1] University of Poona, Pune, India

## 2. THE PARADOX

Berger (1985, p. 149) introduces the paradox in the form of the following example.

**Example 2.1:** Let $X$ be a normally distributed random variable with mean $\theta$ and variance $\sigma^2$ known. The object is to test the simple hypothesis $H_o : \theta = \theta_o$ against the alternative $H_1 : \theta \neq \theta_o$. Berger considers the Bayesian solution by putting a prior point mass $\pi_o$ at $\theta_o$ and the remaining mass $(1-\pi_o)$ being distributed with prior density $N(\mu, \tau^2)$. The acceptance region for the Bayes test in this case is of the form

$$\left|\frac{(x-\theta_o)}{\sigma} + \frac{(\mu-\theta_o)}{\tau}\right| < (1+\frac{\sigma^2}{\tau^2})^{\frac{1}{2}}\{\ell n(\frac{\pi_o}{1-\pi_o})^2(1+\frac{\tau^2}{\sigma^2}) + \frac{(\theta_o-\mu)^2}{\tau^2}\}^{\frac{1}{2}} \tag{2.1}$$

Note that for $\mu = \theta_o$, as $\sigma \to 0$, the acceptance region is of the type

$$\left|\frac{x-\theta_o}{\sigma}\right| < Z_\sigma. \tag{2.2}$$

If $X$ represents the sample mean for a sample of size $n$, then $\sigma$ should be replaced by $\sigma/\sqrt{n}$ and thus the case $\sigma \to 0$ is equivalent to the case where $n \to \infty$. The acceptance region in this case is

$$|\sqrt{n}(\overline{x}-\theta_o)| < Z_n, \tag{2.3}$$

where $Z_n = O(\sqrt{\ell n(n)})$.

In particular, if $\sigma = 1$ and $n = \exp(25), Z_n = 5$ and then the acceptance region for the Bayes test is approximately of the type

$$\sqrt{n}|\overline{x}-\theta_o| < 5.$$

The cut-off point like 5 for the acceptance region in terms of the standardized sample mean, seems to be rather too large from the classical point of view. This difference between the classical UMPU test and the Bayes test is what Berger calls Lindley's Paradox. Berger has further given some comments on this difference between the two methods of testing. In the next example we show that this discrepancy between the classical and Bayes test is further aggravated.

**Example 2.2:** Suppose $X_1, X_2, \ldots, X_n$ are i.i.d. $N(\theta, \sigma^2)$, $\sigma^2$ known and the problem is to test $H_o : \theta = 0$ against $H_1 : \theta = \delta (\delta > 0)$. The classical most powerful test in this case has acceptance region of the type

$$\sqrt{n}\,\overline{X}_n < Z_\alpha,$$

where $Z_\alpha$ is such that $1 - \Phi(Z_\alpha) = \alpha$. This value of the cut-off point is fixed and remains the same for all $n$.

For the Bayes test if we choose $\pi_o$ to be the prior probability for $\theta = 0$ and $1 - \pi_o$ the prior probability for $\theta = \delta$, then the acceptance region for the Bayes test comes out to be

$$\sqrt{n}\,\overline{X}_n < Z_n, \tag{2.4}$$

where

$$Z_n = \frac{\sqrt{n}\,\delta}{2} - \frac{1}{\sigma\sqrt{n}}\ell n\,\frac{\pi_o}{1-\pi_o} = O(\sqrt{n}).$$

Thus, in particular, if we take $\pi_o = \frac{1}{2}$ and $\delta = 1$, we get

$$Z_{100} = 5 \text{ and } Z_{10,000} = 50,$$

and an acceptance region of the type

$$\sqrt{n}\,\overline{X}_n < 50.$$

An acceptance region of this type is really unimaginable from the usual wisdom of classical most powerful tests, whereas this is what the Bayesian theory suggests for $n = 10,000$, not an astronomically large value of $n$ like $e^{25}$.

## 3. RESOLUTION OF THE PARADOX

In the Examples mentioned in Section 2 or for that matter in any testing hypothesis problem, where the test is based upon a consistent test statistic, it is well known that for such tests with fixed level of significance, the power function goes to one as $n \to \infty$. Thus using classical tests if $n$ is large we would end up using a test for which probability of type II error (which goes to zero as $n \to \infty$) is smaller than the probability of type I error (which is fixed independent of $n$). Even from the classical point of view this cannot be a desirable property of a test procedure. Thus $\alpha$ should not be fixed without regard to the sample size. In fact $\alpha$ should depend upon $n$ in such a way that $\alpha_n \to 0$ as $n \to \infty$. This observation on the choice of $\alpha_n$ is not new, in fact, Zellner (1971) in a footnote on page 304 also makes similar suggestion. The important question is the following:

"What is an appropriate rate for $\alpha_n$ to go to zero?"

**Proposal I:** We propose that $\alpha_n \to 0$ in such a way that for all $n$, the probability of type II error should be at least as much as the probability of type I error.

With this proposed extra condition on the choice of $\alpha_n$, we further discuss Example 2.2.

**Example 2.2 (continued):** The probability of type II error for the classical most powerful test with acceptance region

$$\sqrt{n}\,\overline{X}_n < Z_{\alpha_n},$$

is given by

$$1 - \beta(\delta; n, \alpha_n) = \Phi(Z_{\alpha_n} - \sqrt{n}\delta). \tag{3.1}$$

Thus, the proposed condition for the choice of $\alpha_n$ gives us

$$\Phi(Z_{\alpha_n} - \sqrt{n}\delta) \geq 1 - \Phi(Z_{\alpha_n}) = \Phi(-Z_{\alpha_n}). \tag{3.2}$$

Since $\Phi$ is a monotone function we get the condition that

$$Z_{\alpha_n} \geq \frac{\sqrt{n}\delta}{2} \tag{3.3}$$

This is precisely the choice of $Z_n$ in (2.4) with $\pi_o = 1/2$. Thus with this extra condition on the choice of $\alpha_n$, there is no disagreement between the Bayesian and the classical most powerful test.

Our proposal I cannot be used for Example 2.1 with $\theta_o = 0$, because here even for a fixed $n$, we get

$$\lim_{|\delta| \to \infty} [1 - \beta(\delta, n, \alpha_n)] = 0.$$

Thus the condition

$$1 - \beta(\delta, n, \alpha_n) \geq \alpha_n$$

cannot be satisfied uniformly in $\delta$, unless $\alpha_n \equiv 0$. One possible modification is to have some weight function $\pi(\theta)$ for $\theta$ under $H_1$ and to demand that the weighted probability of type II error be at least as much as the probability of type I error.

**Proposal II:** In case of composite alternative hypothesis, with an appropriate weight function on the alternative parameter values, $\alpha_n \to 0$ in such a way that for all $n$, the weighted probability of type two error should be at least as much as the probability of type I error.

With this proposed extra condition on the choice of $\alpha_n$, we further discuss Example 2.1.

**Example 2.1** (continued): Let $\theta_o = 0$. Suppose the weight function on the alternative is the density function of standard normal distribution.

The classical UMPU size $\alpha_n$ test in this case has the acceptance region

$$\sqrt{n}\ |\overline{X}_n| \leq Z_{\alpha_n} \tag{3.4}$$

where $Z_{\alpha_n}$ satisfies

$$\Phi(Z_{\alpha_n}) - \Phi(-Z_{\alpha_n}) = 1 - \alpha_n.$$

The type II error probability function in this case is

$$1 - \beta(\theta, n, \alpha_n) = \Phi(Z_{\alpha_n} - \sqrt{n}\theta) - \Phi(-Z_{\alpha_n} - \sqrt{n}\theta).$$

Considering the $N(0,1)$ weight function for $\theta$, we get the weighted probability of type II error as

$$E(1-\beta(\theta,n,\alpha_n)) = \Phi(\frac{Z_{\alpha_n}}{\sqrt{n+1}}) - \Phi(-\frac{Z_{\alpha_n}}{\sqrt{n+1}}) \tag{3.5}$$

The proposed condition on the choice of $\alpha_n$ now reduces to

$$\Phi(\frac{Z_{\alpha_n}}{\sqrt{n+1}}) - \Phi(-\frac{Z_{\alpha_n}}{\sqrt{n+1}}) \geq 1 - [\Phi(Z_{\alpha_n}) - \Phi(-Z_{\alpha_n})],$$

or that

$$\Phi(\frac{Z_{\alpha_n}}{\sqrt{n+1}}) - \frac{1}{2} \geq 1 - \Phi(Z_{\alpha_n}). \tag{3.6}$$

Thus $Z_{\alpha_n} \to \infty$ as $n \to \infty$. It is easy to check that the Bayesian choice of $Z_n = 0(\sqrt{\ell n\,(n)})$, does satisfy this equation of $Z_{\alpha_n}$. Thus with an appropriate choice of $\alpha_n \to 0$, the Lindley type paradox can be resolved. The proposals of the type I and II made here seem to give this appropriate rate at which $\alpha_n$ should approach zero.

## 4. A PROBLEM

An often discussed problem in classical statistical methods is the problem of goodness of fit. For this problem the traditionally used test is the $\chi^2$ test of goodness of fit for fixed level of significance $\alpha$. This problem does not seem to be much discussed in the Bayesian theory set up. Here again the choice of $\alpha$ should depend upon the sample size in such a way that $\alpha_n \to 0$ as $n \to \infty$. What is an appropriate rate? We give here a problem in which this question would be important.

Fisher (1946) has reported the data, given in Table 1, on 53,680 families, showing the distribution of number of boys in families with eight children. He fitted the Binomial distribution leading to the $\chi^2$ value of 91.869 for seven d. f . This value is too large at the usually accepted levels of significance. There is some genetic reason to believe that $p$, the probability of a male, child may vary from couple to couple, thus having a probability distribution of its own. We fit a beta-binomial model to this data, $\chi^2$-value 53.307. This model has also been fitted to the same data by Ishii and Hayakawa (1960). For 6 d . f , this observed $\chi^2$-value is still high at the usually adopted levels of significance.

However, we should note that the sample size $n$ in this case is 53, 680. For this large $n$, what should be the appropriate level of significance for the corresponding $\chi^2$-test of goodness of fit? What would be an appropriate Bayesian analysis for this problem? These are some interesting questions which need to be resolved.

TABLE 1: Distribution of Number of Boys in Families with 8 children.

| No. of Boys | Observed Frequency | Expected Frequency | |
|---|---|---|---|
| | | Binomial | Beta-Binomial |
| 0 | 215 | 165 | 190 |
| 1 | 1,485 | 1,402 | 1,502 |
| 2 | 533 | 5,203 | 5,315 |
| 3 | 10,649 | 11,035 | 10,940 |
| 4 | 14,959 | 14,627 | 14,346 |
| 5 | 11,929 | 12,410 | 12,272 |
| 6 | 6,678 | 6,580 | 6,689 |
| 7 | 2,092 | 1,994 | 2,123 |
| 8 | 342 | 264 | 301 |
| Total | 53,680 | 53,680 | 53,680 |
| $\chi^2$ | | 91.869 | 53.307 |

## REFERENCES

Berger, J.O. (1985). *Statistical Decision Theory and Bayesian Inference, 2nd Edition* Springer-Verlag, New York.

Berger, J.O. and Sellke, T. (1987). Testing a point null hypothesis: The irreconcilability of P-value and evidence (with discussion). *J. Amer. Statist. Assoc.* **82**, 112-139.

Berger, J.O. and Delampady, M. (1988). Testing precise hypotheses (with discussion),. *Statistical Science* **2**, 317-352.

Fisher, R.A. (1946). *Statistical Methods for Research Workers.* Oliver and Byod,.London

Ishii, G. and Hayakawa (1960). On the Compound Binomial Distribution. *Annl. Inst. Stat. Math.* **12**, 69-80.

Lindley, D.V. (1957). A Statistical Paradox, *Biometrika* **44**, 187-192.

Jeffreys, H. (1961). *Theory of Probability* (3rd Ed.) Oxford University Press, London.

Shafer, G. (1982). *Belief Functions and Parametric Models* (with discussion). Jour. Royal Statist. Soc. B **44**, 322- 352.

Zellner, A. (1977). *An Introduction to Bayesian Inference in Econometrics,* John Wiley and Sons, New York.

## COMMENTS

Mohan Delampady
University of British Columbia

Professor Kunte discusses "Lindley-Jeffreys Paradox" in detail. At the end of the paper, he also suggests that there is no Bayesian approach to goodness of fit procedures. The following are some of the comments that I made during the discussion of the paper.

(i) "Lindley-Jeffreys Paradox" is actually no paradox at all. It just illustrates some of the serious drawbacks of the frequentist approach to hypothesis tests. This has been known for a long time now. Indeed, as far back as in 1941, Berkson pointed out that in frequentist hypothesis tests, to obtain reasonable answers, the significance level $\alpha$ should depend on the sample size, $n$ (hence $\alpha = \alpha_n$) and moreover, $\alpha_n$ must decrease to 0 as $n$ increases.

(ii) Discrepancy similar to this exists between posterior probabilities and significance levels even when one considers tests of interval null hypotheses. This arises not only for the normal models and the normal priors but also for general classes of models and large classes of priors. Delampady (1989) discusses some of these issues.

(iii) There is literature on Bayesian approach to goodness of fit problems. Some of these articles are mentioned in Berger and Delampady (1987). Also, in the discussion of the above paper, Professor Bayarri mentions some other related articles such as Bayarri (1985).

## ADDITIONAL REFERENCES

Bayarri, M.J. (1985). A Bayesian test of goodness-of-fit. Technical Report, Departmento de Estadistica e Investigacion Operativa, University of Valencia.

Berger, J. and Delampady, M. (1987). Testing precise hypotheses (with discussion). *Statistical Science.* **2**, 317-348.

Berkson, J. (1942). Tests of significance considered as evidence. *Journal of American Statistical Association.* **37**, 325- 335.

Delampady, M. (1989). Lower bounds on Bayes factors for interval null hypotheses. *Journal of American Statistical Association.* **84**.

17

# Bayesian Approach to Some Problems in Life Testing and Reliability Estimation

Asit P. Basu [1]
Nader Ebrahimi[2]

**Abstract**

In this paper a survey of some inference problems relating to the exponential distribution is presented. Among the problems considered are: (1) estimation of parameters of exponential distribution using asymmetric loss function; and (2) reliability estimation of complex systems. Parametric empirical Bayes methods for the estimation of failure processes for repairable systems are also given.

*Key words and Phrases:* Reliability; Life testing; Asymmetric loss function; Linex loss; Empirical Bayes; Repairable systems; $k$-out-of-$p$ systems.

## 1. INTRODUCTION

The univariate exponential distribution is well known as an important model in reliability theory. Here the density function, conditional on a parameter $\theta$, given by

$$f(x|\theta) = \frac{1}{\theta}e^{-x/\theta},\ 0 < x < \infty,\ \theta > 0 \tag{1.1}$$

and the parameter $\theta$ has a given prior distribution $g(\theta)$. A prior distribution is chosen to reflect prior knowledge about $\theta$. See Berger (1985), DeGroot (1970) for some discussions to this end. For purpose of illustration, throughout this paper we shall assume that the prior distribution of $\theta$ is a gamma or the inverted gamma distribution with density

$$h(\theta) = \frac{\alpha^{\nu}}{\Gamma(\nu)}e^{-\alpha/\theta}(\frac{1}{\theta})^{\nu+1},\ \theta > 0,\ \nu > 0,\ \alpha > 0, \tag{1.2}$$

Here the parameters $\alpha$ and $\nu$ are chosen to reflect prior knowledge. We shall denote this by $\theta \sim IG(\alpha, \nu)$. Considerable literature exists for Bayesian

[1] University of Missouri - Columbia
[2] Northern Illinois University, DeKalb, Illinois

analysis of above model assuming various prior distributions for $\theta$ and using the squared error loss function. For a bibliography of relevant literature see Basu (1985), Basu and Tarmast (1987) and Johnson (1988). In this paper we present a survey of some recent work done involving the exponential distribution. In Section 2 we consider the problem of reliability estimation for complex systems using the squared error loss function. In Section 3 the case of asymmetric loss function is considered. Finally in Section 4 examples of parametric empirical Bayes estimation of failure processes for repairable systems are considered.

## 2. ESTIMATION FOR SQUARED ERROR LOSS FUNCTION

The reliability of a system can be defined in two different ways. Let $T$ denote the lifetime of a system with distribution function $F(t)$. Then the reliability function $R_1$ is the probability that the system will be in operating condition (and function satisfactorily) at mission time $t_0$. That is

$$R_1 = P(T > t_0). \tag{2.1}$$

We shall call this Model I.

A second definition of reliability function can be given as follows. Let $X$ and $Y$ be two random variables with cumulative distribution functions (*cdf*) $F(x)$ and $G(y)$ respectively. Suppose $Y$ is the strength of a component subject to a random stress $X$. Then the component fails if at any moment the applied stress (or load) is greater than its strength or resistance. The reliability of the component in this case is given by

$$R_2 = P(X < Y). \tag{2.2}$$

We shall call this model the stress-strength model, or Model II. The stress is a function of the environment to which the component is subjected. Strength depends on material properties, manufacturing procedures and so on. The above model has found an increasing number of applications in many different areas, especially in the structural and aircraft industries. As an example, consider the following. A solid propellant rocket engine is succesfully fired provided the chamber pressure ($X$) generated by ignition stays below the burst pressure ($Y$) of rocket chamber.

From practical considerations it is desirable to draw inference about $R_1, R_2$ and other similar measures. In many situations, the distribution of $X$ (or of both $X$ and $Y$) will be completely known except possibly for a few unknown parameters and it is desired to obtain parametric solutions.

Considerable literature exists dealing the problem of estimation of $R_1$ and $R_2$. For a bibliography on available results see Basu (1985), Basu and Tarmast (1987) and Johnson (1988). Most of these results, however, are based on the sampling theory approach. Basu and Tarmast (1987), Basu and Ebrahimi (1988), Bhattacharya (1967), Enis and Geisser (1971) and

Zacks (1977) have considered the problem from Bayesian point of view. Here we present some of these and some new results.

Consider a physical system. A system is called simple if it consists of a single component. Otherwise it is called a complex system. A complex system, consisting of a $p$ components, is called a $k$-out-of-$p$ system if it functions when at least $k$ of these $p$ components function successfully. Such a system occurs quite naturally in many physical and biomedical models. As an example of a 2-out-of-3 system, consider an airplane which can function satisfactorily if and only if at least two of its three engines are functioning. When $k = p$ (or $k = 1$) we obtain series (or parallel) systems as special cases of $k$-out-of-$p$ systems. We shall now consider reliability estimation for $k$-out-of-$p$ systems. Throughout we assume that the underlying distributions are of the form (1.1).

## 2.1 Model I

Let $(X_1, X_2, \cdots, X_n)$ be a random sample from the exponential distribution with density function (1.1), where the prior distribution of the random parameter $\theta$ is given by (1.2). In this case, the likelihood function, conditional on $\theta$, is given by

$$f(x_1, \cdots, x_n|\theta) = \frac{1}{\theta^n} e^{-\frac{t}{\theta}}, \tag{2.3}$$

where $t = \sum_{i=1}^n x_i$ is the value of the sufficient statistics $T$ for $\theta$. The posterior distribution of $\theta$, given the data $(x_1, x_2, \cdots, x_n)$ can be obtained by using Bayes theorem as

$$g(\theta|x_1, x_2, \cdots, x_n) = \frac{(\alpha+t)^{n+\nu}}{\Gamma(n+\nu)} \theta^{-(n+\nu+1)} e^{-(\frac{\alpha+t}{\theta})}, \tag{2.4}$$

which is $IG(\alpha + t, N + \nu)$. Using squared error loss function, the Bayes estimate of the mean life, $\theta$, is given by

$$\hat{\theta} = (\alpha + t)/(n + \nu - 1) \tag{2.5}$$

Similarly, using (2.4), the Bayes estimate of the reliability coefficient $R_1(t_0)$ $= \bar{F}(t_0) = 1 - F(t_0) = \exp(-t_0/\theta)$ is given by

$$\hat{R}_1(t_0) = \int_0^\infty e^{-t_0/\theta} g(\theta|x_1, x_2, \cdots, x_n) d\theta = (1 + \frac{t_0}{\alpha + t})^{-(n+\nu)}. \tag{2.6}$$

For the more complex model, we first consider, the distinct parameters case. Let $X_i$ denote the lifetime (failure time) of the $i$-th component with conditional density

$$f(x_i|\theta_i) = \frac{1}{\theta_i} e^{-x_i/\theta}, i = 1, 2, \cdots, p; x_i, \theta_i > 0 \tag{2.7}$$

Let $X$ denote the lifetime of a $k$-out-of-$p$ system $S$. Then the reliability of the system at time $t_0$ is

$$\begin{aligned}\hat{R}_3(t_0) &= P(X > t_0) \\ &= P(k \text{ or more components survive until time } t_0) \\ &= \sum_{j=k}^{p} \sum_{\alpha_i} \prod_{i=1}^{j} \bar{F}_{\alpha_i}(t_0) \prod_{i=j+1}^{p} F_{\alpha_i}(t_0). \end{aligned} \tag{2.8}$$

Here $\bar{F}_{\alpha_i}(t_0) = 1 - F_{\alpha_i}(t_0) = \exp -(t_0/\theta_{\alpha_i})$, is the reliability function of the $\alpha_i$-th component, and the sum $\sum_{\alpha_i}$ is over all $\binom{p}{j}$ distinct combinations of the integers $\{1, 2, \cdots, p\}$ taken $j$ at a time such that exactly $j$ of the $X_i$'s are greater than $t_0$ (and the remaining $X_i$'s are less than or equal to $t_0$).

Now we assume that $\theta_1, \theta_2, \cdots, \theta_p$ are independent apriori with $\theta_i \sim IG(\alpha_i, \nu_i), i = 1, 2, \cdots, p$. Let $(X_{ij}; j = 1, 2, \cdots, n_i)$ be an independent random sample of size $n_i$ from the $i$-th population with distribution given in (2.7); $i = 1, 2, \cdots, p$. Define

$$T_i = \sum_{j=1}^{n_i} X_{ij}.$$

Then $T_1, T_2, \cdots, T_p$ are sufficient statistics. Bayes estimators of $\theta_1, \theta_2, \cdots, \theta_p$ for squared error loss functions, based on $p$ independent random samples of sizes $n_1, n_2, \cdots, n_p$ respectively can be derived. Given $T_i = t_i, i = 1, 2, \cdots, p$ the parameters $\theta_1, \theta_2, \cdots, \theta_p$ are independent aposteriori, having inverted gamma posterior distributions $IG(\alpha + t_i, n_i + \nu_i), i = 1, 2, \cdots, p$. Thus the Bayes estimate of $R_3(t_0)$ is given by

$$E(R_3(t_0)|t_1, \cdots, t_p) = \hat{R}_3(t_0) = \sum_{j=k}^{p} \sum_{\alpha_i} \prod_{i=1}^{j} \hat{\bar{F}}_{\alpha_i}(t_0) \prod_{i=j+1}^{p} (1 - \hat{\bar{F}}_{\alpha_i}(t_0)), \tag{2.9}$$

where, $\hat{\bar{F}}_i(t_0) = E(\exp(-t_0/\theta_i)|\text{Data}) = (1 + t_0/(\alpha_i + t_i))^{-(n_i+\nu_i)}$, $i = 1, 2, \cdots, p$, as given in (2.6). From (2.9) we can now obtain, as special cases, the Bayes estimates of reliability for series and parallel systems. Thus taking $k = p$, the Bayes estimate of the reliability of a $p$-component series system is given by

$$\hat{R}_4(t_0) = \prod_{i=1}^{p} (1 + \frac{t_0}{\alpha_i + t_i})^{-(n_i+\nu_i)}. \tag{2.10}$$

And taking $k = 1$, the Bayes estimate of a reliability of a $p$-component parallel system is given by

$$\hat{R}_5(t_0) = 1 - \prod_{i=1}^{p} (1 - (1 + \frac{t_0}{\alpha_i + t_i})^{-(n_i+\nu_i)}). \tag{2.11}$$

In case the component lifetimes are independent and identically distributed with common density given in (1.1), (2.8) simplifies to

$$R_6(t_0) = \sum_{j=k}^{p} \binom{p}{j} (\bar{F}(t_0))^j (F(t_0))^{p-j}. \tag{2.12}$$

Here $\bar{F}(t_0) = e^{-t_0/\theta}$. Hence we can rewrite $R_6(t_0)$ as

$$R_6(t_0) = \sum_{j=k}^{p} \sum_{\alpha=0}^{p-j} (-1)^{\alpha} \binom{p}{j} \binom{p-j}{\alpha} e^{-\frac{(j+\alpha)t_0}{\theta}}. \tag{2.13}$$

As given before, let $(X_1, X_2, \cdots, X_n)$ be a random sample from (1.1). Then $T = \sum_1^n X_i$ is sufficient for $\theta$ and $g(\theta|t)$ is the posterior density of $\theta$ given the sufficient statistic $T = t$. Then using (2.6), the Bayes estimate of $R_6(t_0)$ is given by

$$\begin{aligned} \hat{R}_6(t_0) &= E(R_6(t_0)|t) = \int_0^{\infty} R_6(t_0) g(\theta|t) d\theta \\ &= \sum_{j=k}^{p} \sum_{\alpha=0}^{p-j} (-1)^{\alpha} \binom{p}{j} \binom{p-j}{\alpha} \left(1 + \frac{(j+\alpha)t_0}{\alpha + t}\right)^{-(n+\nu)}. \end{aligned} \tag{2.14}$$

## 2.2 Model II

Let us next consider the stress strength model. Let the random variable $X$ and $Y$ follow independent exponential distributions with parameters $\theta_1$ and $\theta_2$ respectively, where, as before, the density functions are

$$f(y|\theta_1) = \frac{1}{\theta_1} \exp(-\frac{x}{\theta_1}),$$

and

$$f(y|\theta_2) = \frac{1}{\theta_2} \exp(-\frac{y}{\theta_2}), \ (x, y, \theta_1, \theta_2 > 0).$$

Here

$$\begin{aligned} R_2 = P(X < Y) &= \theta_2/(\theta_1 + \theta_2). \\ &= \lambda/(1 + \lambda), \end{aligned} \tag{2.15}$$

where $\lambda = \theta_2/\theta_1$. Assume that the parameters $\theta_1, \theta_2$ are independently distributed, the prior distribution of $\theta_i$ being $IG(\alpha_i, \nu_i), i = 1, 2$. As before, the parameters $\alpha_i$ and $\nu_i$ are chosen to reflect prior information. Let $(X_1, X_2, \cdots, X_{n_1})$ and $(Y_1, Y_2, \cdots, Y_{n_2})$ be two random samples from the $X$ and $Y$ populations respectively. Then $T_1 = \sum_1^{n_1} X_i$ and $T_2 = \sum_1^{n_2} Y_i$ are sufficient for $\theta_1$ and $\theta_2$ respectively. Given $T_i = t_i, i = 1, 2$, let $\lambda = \theta_2/\theta_1$ and $U = (\alpha_2 + t_2)/(\alpha_1 + t_1)$. Basu and Tarmast (1987) obtain the following:

Lemma 2.1. The posterior density of $\lambda$ is given by

$$g(\lambda|t_1,t_2) = \frac{\Gamma(n_1+n_2+\nu_1+\nu_2)}{\Gamma(n_1+\nu_1)\Gamma(n_2+\nu_2)} \bullet \frac{u^{n_2+\nu_2}\,\lambda^{n_1+\nu_1-1}}{(\lambda+u)^{n_1+n_2+\nu_1+\nu_2}}. \tag{2.16}$$

Using (2.15) and (2.16), the Bayes estimate of $R_2$ can be obtained as

$$\begin{aligned} \hat{R}_2 &= E(R_2|t_1,t_2) \\ &= \int_0^\infty \frac{\lambda}{1+\lambda} g(\lambda|t_1,t_2). \end{aligned} \tag{2.17}$$

Basu and Tarmast (1987) have studied Bayes estimators of reliability for more complex systems and compared their bias and mean square errors with those of maximum likelihood estimators.

## 3. ESTIMATION FOR ASYMMETRIC LOSS FUNCTION

Recently Basu and Ebrahimi (1988) have considered estimates of reliability using asymmetric loss function. In many situations overestimate of reliability function or average failure time $\theta$ is usually much more serious than underestimate of reliability or mean failure time. For example, in a recent disaster of the U.S. space shuttle Challenger (See Feynmann (1987)) the management underestimated the failure rate and therefore overestimated the reliability of solid-fuel rocket booster.

As before, consider the exponential model (1.1) with inverted gamma prior distribution (1.2) for $\theta$. For other prior distributions see Basu and Ebrahimi (1988).

Suppose $\Delta = \hat{\theta}/\theta - 1$, where $\hat{\theta}$ is an estimate of $\theta$. Consider the following convex loss function

$$L(\Delta) = be^{a\Delta} - c\Delta - b, c \neq 0, b > 0. \tag{3.1}$$

Here $L(0) = 0$ and for the minimum to occur at 0 we must have $ab = c$. Note, first for $a = 1$, $L(\Delta)$ is quite asymmetric about zero with overestimation being more costly than underestimation. Second, for $a < 0$, $L(\Delta)$ rises exponentially when $\Delta < 0$ (underestimation) and almost linearly when $\Delta > 0$, (overestimation). This loss function is also known as the LINEX loss function. Finally, for small values of $|a|$,

$$L(\Delta) \doteq ba^2(\hat{\theta}-\theta)^2/\theta^2, \tag{3.2}$$

a symmetric function. Also note if we define $\Delta_1 = \hat{\theta} - \theta$, then $L(\Delta_1)$ is equivalent to Varian (1975) loss function. $L(\Delta_1)$ is also used by Zellnar (1986) to derive estimation for a scalar parameter and prediction of scalar random variable.

Using (1.1), (1.2) the optimal estimate of $\theta$ relative to $L(\Delta)$ is given by

$$\hat{\theta}_B = z_1(\alpha+t)/a \tag{3.3}$$

where $z_1 = 1 - \exp(-a/(n + \nu + 1))$. Similarly one can obtain an estimate of $R(t_0) = \exp(-t_0/\theta)$.

Basu and Ebrahimi (1988) have considered other prior distributions and computed corresponding estimates. Also these estimates have been compared with the estimates for squared error loss function.

## 4. PARAMETRIC EMPIRICAL BAYES METHODS

In this section we consider a parametric empirical Bayes procedure for the estimation of parameters of failure processes for repairable systems described in Basu and Rigdon (1986). Suppose successive failures of a physical system, e.g. computers, airplanes etc. follow a homogeneous Poisson process with rate $\lambda$. Then times between failures, after repairing, follow exponential distribution with mean time to failure given by $\theta = 1/\lambda$. We assume that the prior distribution of $\lambda$ is chosen to reflect our prior knowledge about $\boldsymbol{\lambda}$.

Suppose that $N$ similar systems are following $n$ independent Poisson processes with parameters $\lambda_1, \lambda_2, \cdots, \lambda_N$, and are subject to failure at random points in time. Repair time is assumed to be negligible. Assume that $(\lambda_1, \lambda_2, \cdots, \lambda_N)$, though unobservable, is a random sample from a gamma prior distribution with density

$$p(\lambda|\alpha, \beta) = \frac{\beta^\alpha}{\Gamma(\alpha)} \lambda^{\alpha-1} \exp(-\beta\lambda).$$

Let $t_{il} < t_{i2} < \cdots < t_{in(i)}$ be the observed failure times of system $i, i = 1, 2, \cdots, N$. Since the distribution of times between failures for each system are independent and exponentially distributed, and since the systems operate independently, the density of $\mathbf{t} = (t_{ij}; j = 1, 2, \cdots, n(i); i = 1, 2, \cdots, N)$ given $\boldsymbol{\lambda} = (\lambda_1, \lambda_2, \cdots, \lambda_N)$ is

$$p(\mathbf{t}|\boldsymbol{\lambda}) = \prod_{i=1}^{N} \lambda_i^{n(i)} \exp(-\lambda_i S_i), \tag{4.1}$$

where $S_i = \sum_{j=1}^{n(i)} t_{ij}$.

The density of $\boldsymbol{\lambda}$ given $\alpha$ and $\beta$ is then

$$p(\boldsymbol{\lambda}|\alpha, \beta) = (\frac{\beta^\alpha}{\Gamma(\alpha)})^N \prod_{i=1}^{N} \lambda_i^{\alpha-1} \exp(-\beta\lambda_i). \tag{4.2}$$

Hence

$$\begin{aligned} p(\mathbf{t}|\alpha, \beta) &= \int p(\mathbf{t}, \boldsymbol{\lambda}|\alpha, \beta) d\boldsymbol{\lambda} = \int p(\mathbf{t}|\boldsymbol{\lambda}) p(\boldsymbol{\lambda}|\alpha, \beta) d\boldsymbol{\lambda} \\ &= (\frac{\beta^\alpha}{\Gamma(\alpha)})^N \prod_{i=1}^{N} \int_0^\infty \lambda_i^{n(i)+\alpha-1} \exp[-(\theta + t_{i,n(i)})\lambda_i]\, d\lambda_i. \end{aligned} \tag{4.3}$$

Point estimates of $\boldsymbol{\lambda}$ can be obtained based on the posterior distribution of $\lambda_i$ given the estimated values of $\alpha$ and $\beta$. Details are given in Basu and Rigdon (1986).

## 5. CONCLUDING REMARKS

Exponential distribution has been considered as a model in a number of physical situations in addition to the ones described in previous sections. For example, Basu and Gupta (1974, 1977) have considered Bayesian classification problem using exponential distributions as models. Here we have only considered gamma (or inverted gamma) priors. One can carry out similar analyses for other priors. Also results can be extended to other physical models and to censored samples.

## ACKNOWLEDGMENTS

Research sponsored by the Air Force Office of Scientific Research, Air Force Systems Command, USAF, under grant number AFOSR-89-0406. The US Government is authoized to reproduce and distribute reprints for Governmental purposes notwithstanding any copyright notation thereon.

## REFERENCES

Barlow, R.E. and Proschan, F. (1986), Inference for the exponential life distribution. In *Proc. of the International School of Physics, Theory of Reliability*, eds: A. Serra and R.E. Barlow, 143-164, Amsterdam: North-Holland.

Basu, A.P. (1964), Estimates of reliability for some distributions useful in life testing, *Technometrics*, **6**, 215-219.

Basu, A.P. (1977a), Estimate of reliability in the stress-strength model. In *Proceedings of the 22nd conference of the design of experiments in army research development and testing*, 97-110.

Basu, A.P. (1977b), Reliability estimation for complex systems. In *Proceedings of the ARO workshop on reliability and probabilistic design, 85-99.*

Basu, A.P. (1977c), A Generalized Wilcoxon Mann-Whitney statistic with some applications in Reliability. In *The theory and applications of Reliability vol. 1* eds: C.P. Tsokos and I.N. Shimi, 131-149, New York: Academic Press.

Basu, A.P. (1981), The estimation of $P(X < Y)$ for distributions useful in life testing. *Naval research logistics quarterly*, **28**, 383-392.

Basu, A. P. (1985), Estimation of Reliability of Complex Systems-A Survey. In *The Frontiers of Modern Statistical Inference procedures* ed. E.J. Dudewicz, 271-287. Columbus: American Science Press.

Basu, A.P. and Ebrahimi, N. (1988), Bayesian approach to life testing and reliability estimation using asymmetric loss function. *Technical Report No. 144*, Dept. of Statistics, University of Missouri, Columbia

Basu, A.P. and El-Mawaziny, A.H. (1978), Estimates of reliability of $k$-out-of-$m$ structures in the independent exponential case. *Jour. Amer. Statist. Assoc.*, **73**, 850-854.

Basu, A.P. and Gupta, A.K. (1974), Classification rules for exponential populations. In *Proc. Conf. Reliability and Biometry*, 537-550, Philadelphia: SIAM Publ.

Basu, A.P. and Gupta, A.K. (1977), Classification rules for exponential populations: two parameter case. In *The Theory and Applications of Reliability* **Volume 1** eds: C.P. Tsokos and I.N. Shimi,.507-525, New York:Academic Press.

Basu, A.P. and Rigdon, S.E. (1986), Examples of parametric empirical Bayes methods for the estimation of failure processes for repairable systems. In *Reliability and Quality Control*, ed: A.P. Basu , 47-55, Amsterdam: North-Holland.

Basu, A.P. and Tarmast, G. (1987), Reliability of a complex system from Bayesian viewpoint. In *Probability and Baysian Statistics*, ed: R. Viertl, 31-38, Plenum.Publishers

Berger, James O. (1985), *Statistical Decision Theory and Bayesian Analysis* (2nd edition), New York:Springer-Verlag.

Bhattacharya, S.K. (1967), Bayesian approach to life testing and reliability estimation. *J. Amer. Stat. Assn.*, **62**, 48-62.

Bhattacharyya, G.K. and Johnson, R. (1974), Estimation of Reliability in a multicomponent stress-strength model. *J. Amer. Stat. Assn.*, 966-970.

DeGroot, M.H. (1970), *Optimal Statistical Decisions*, New York: McGraw-Hill.

El-Sayyad, G.M. (1967), Estimation of the parameter of an exponential distribution. *Jour. Royal Stat. Soc. Series B* **29, 525-535**.

Enis, P. and Geisser, S. (1971), Estimation of the probability that $Y < X$. *J. Amer. Stat. Soc. Assn.*, **66**, 162-168.

Feynman, R.P. (1987), Mr. Feynman goes to Washington. *Engineering and Science*, Fall 1987, California Institute of Technology, 6-22.

Johnson, R.A. (1988), Stress-Strength models for reliability *Handbook of Statistics*. **Volume 7**, eds: P.R. Krishnaiah and C.R. Rao, 27-54, Amsterdam: North-Holland.

Mann, N.R., Schafer, R.E. and Singpurwalla, N.D. (1974), *Methods for Statistical Analysis of Reliability and Life data*, New York: John Wiley and Sons.

Varian, Ha. R. (1975), A Bayesian approach to real estate assessment. In *Studies in Bayesian econometrics and Statistics in honor of L.J. Savage*,eds: S.E. Fienberg and A. Zellner, 195-208, Amsterdam: North-Holland.

Zacks, S. (1977), Bayes estimation of the reliability of series and parallel systems of independent exponential components. In *The theory and applications of Reliability Vol. II*, eds: C.P. Tsokos and I.N. Shimi, New York: Academic Press.

Zellner, A. (1986), Bayesian estimation and prediction using asymmetric loss functions. *J. Amer. Stat. Assn.*, 81, 446-451.

# 18

# When to Stop Testing Software? Some Exact Results


## S.R. Dalal [1]
## C.L. Mallows[2]



### Abstract

Developers of large software systems must decide how much to test a piece of software before it is released. We consider an explicit tradeoff between the costs of testing and releasing. The cost of testing may include the cost of a lost economic initiative because of continued testing and the cost of releasing may include the cost of customer dissatisfaction and cost of fixing an unknown number of bugs in the released version. The problem is formulated as a sequential Bayes problem in which information from past experience is also used. The structure of the optimal policy is determined.



*Key words and Phrases:* Software reliability; Software testing; Optimal policy; Tradeoff; Sequential rule.


## 1. INTRODUCTION AND SUMMARY

In a typical software development environment once a piece of software, usually referred to as a module, is created by a development group, it is sent to a testing group which must verify that the module meets the design specifications. A module can be as small as a program with a few hundred lines of code or as large as a system consisting of several submodules containing millions of lines of code. The testing group tests software by applying various inputs and tries to find "bugs" (defects in the software that cause failure to meet specifications).

Usually the testing activity is restricted to a maximum time, $t_\infty$, and a fixed staff level. Further, the number of possible paths through a program could be very large and consequently even for a moderate size program, it is not feasible to find every possible bug.

Given the inevitability of bugs, what should be the reasonable and verifiable goals of a software testing program? On a close examination it is apparent that there is a natural tradeoff between the decision to continue testing or to stop, namely:

---


[1] Bellcore, Morristown, N.J.

[2] AT&T Bell Laboratories, Murray Hill, N.J.

i) If testing stops too early, many bugs remain. Thus, we incur the cost of later bug-fixing and losses due to customers' dissatisfaction.

ii) If testing continues up to the maximum permissible time $t_\infty$, then there is the cost of testing effort and also a loss of market initiative.

In this paper, we take this tradeoff directly into account. Since testing is an ongoing activity, to capitalize on the past knowledge, we use a sequential Bayes approach.

Statistical underpinnings of our model follow a standard formulation in this context, and is presented in Section 2 along with the economic model. The optimal stopping rule is characterized in Section 3. The rule is obtained by solving a functional differential-difference equation iteratively, and may be difficult to compute.

Several results were stated without derivation in Dalal and Mallows (1988). Here the results are derived rigorously as this setting is one of the few instances where exact sequential Bayes rules are characterized and the proofs may extend to other more general cases. Some of intuitive reasoning behind the model, graphical aids and a real example are given in Dalal and Mallows (1990).

## 2. MODEL FORMULATION AND IMPLICATIONS

Our economic model includes costs and benefits derived from the natural tradeoff between further testing and stopping. Let $N$ be the total number of bugs in the module (unknown), $K(t)$ be the number of bugs observed up to time $t$, $a$ be the cost of fixing a bug when found while testing, $b$ be the cost of fixing a bug when found in the field, and $c = (b - a)$ be the net cost of fixing a bug after rather than before release.

Further, we assume that there is a known nonnegative monotone increasing function $f(t)$ reflecting the cost of testing plus the opportunity cost of not releasing the software up to time $t$. We assume the loss function to be:

$$L(t, K(t), N) = f(t) + aK(t) + b(N - K(t)) = f(t) - cK(t) + bN$$

Some comments on this formulation are in order. We assume that "time" is measured in some way (e.g. calendar time, cpu time, or work-hours) that makes sense of both $K(t)$ and $f(t)$. For a discussion of the definition of "time," see Musa (1975). The assumption that $f(t)$ is known for all $t$ implies that at each $t$, the only options available to the testing manager are to stop immediately, or to continue testing. A generalization, allowing for several classes of bugs with differing costs, is given in Dalal and Mallows(1988).

For our stochastic model we assume that (i) $N$, the unknown number of bugs in a given module, is random, and is distributed according to a Poisson distribution with unknown mean $\lambda$, (ii) $\lambda$ varies randomly across modules, and has a Gamma distribution $\Gamma(\alpha, \beta)$ where $\alpha$ and $\beta$ are known from past experience, (iii) given a bug, the amount of time it takes to find it during the testing is distributed with a known distribution $G$, and the failure times (i.e. the times at which individual bugs are observed) are independent. We assume $G$ has a density $g$.

As the number of bugs becomes large, one would expect that the specific values of $\alpha$ and $\beta$, and more generally the Gamma assumption, would be immaterial. This is supported by the connection of our asymptotic results with those of Starr (1974) and Vardi (1980) and by the asymptotic distribution theory developed in Dalal and Mallows (1988).

Models of this type have been proposed and used in the past with minor variations, see, e.g. Jelinski and Moranda (1972), Musa (1975), Goel and Okumoto (1979), and Langberg and Singpurwalla (1985). Currently we have extended our approach to allow dynamic updating of the scale parameter.

## 3. EXACT OPTIMAL POLICIES

### 3.1. Some Preliminary Results

Under the assumptions of Section 2, $K(t)$, the number of bugs observed up to time $t$, is a sufficient statistic, and thus, we limit our attention to decision policies based on $K(t)$ and $t$. Further, for setting up a dynamic programming argument, we deal with a reward function rather than a loss function. Clearly, the reward for stopping at time $t$ is $cK(t) - f(t) - bN$. The last term, $bN$, is irrelevant as far as stopping policies are concerned. Thus, the reward for stopping at time $t$ is taken to be:

$$R(K(t), t) = cK(t) - f(t) . \qquad (3.1)$$

Let $V(k, t) = V(k, t, f, G, \alpha, \beta)$ denote the expected *additional* reward, following the optimal policy, conditional on the fact that $K(t) = k$ bugs have been observed up to time $t$. Thus, the expected *total* reward, following the optimal policy from $(k, t)$ onward, is $cK(t) - f(t) + V(k, t)$. Clearly, $V(k, t) \geq 0$, and one should stop as soon as $V(k, t) = 0$. Hence, characterizing the stopping region is equivalent to characterizing the $t$'s for which $V(k, t)$ becomes zero. For this concept to be meaningful, we need to have monotonicity of the $V(\cdot)$ function in various arguments. In the following we deal with various aspects of monotonicity and derive sufficient conditions for it without restricting to the finite horizon case. Throughout this development, we assume that an optimal policy exists. This assumption could be avoided by working with $\epsilon$- optimal policies and letting $\epsilon \to 0$.

**Lemma 3.1** (a) $K(t) \mid \lambda$ is a Nonhomogeneous Poisson Process with increments $\lambda \cdot g(t)$;

(b) Conditional on $t$ and $K(t) = k$, (i) $K(t+s) - K(t) \mid \lambda$ is a nonhomogeneous Poisson Process with increments $\lambda \cdot g_t(s)$, where $g_t(s) =$

$d/ds\, G_t(s)$, $G_t(s) = (G(t+s) - G(t))/(1 - G(t))$, and (ii) $\lambda$ is distributed as $\Gamma(\alpha + k, (\beta + G(t))/(1 - G(t)))$.

The proof of this lemma is based on standard Bayesian and Poisson process manipulations. From this lemma it follows that having observed $(k, t)$, the optimal policy is optimal in a new game which is the same as the old game with changed parameters. Thus,

**Theorem 3.2** $V(k, t, f, g, \alpha, \beta) = V(0, 0, f_t, g_t, \alpha + k, (\beta + G(t))/(1 - G(t)))$, where, $f_t(\cdot) = f(t + \cdot) - f(t)$, $g_t(\cdot) = g(\cdot + t)/(1 - G(t))$.

**Corollary 3.3** For $g$ exponential ($g(t) = \mu e^{-\mu t}$) and $f$ linear ($f(t) = t f_1$), we have $g_t = g$, $f_t = f$, and

$$V(k, t, f, g, \alpha, \beta) = V(0, 0, f, g, \alpha + k, (\beta + G(t))/(1 - G(t))) .$$

Theorem 3.2 shows that monotonicity of $V$ in the arguments $k$ and $t$ is related to monotonicity of $V$ in the arguments $\alpha$ and $\beta$ for a problem with a different $f$ and $g$, and the corollary shows that in the linear-$f$, exponential-$g$ case the game remains the same except for the changes in $\alpha$ and $\beta$.

**Theorem 3.4** For arbitrary $f$ and $g$, $V(0,0,f,g,\alpha,\beta)$ is an increasing function of $\alpha$ and a decreasing function of $\beta$.

**Proof** First consider monotonicity in $\beta$. The proof will consist of showing that the function $V^0(\alpha, \beta) = V(0,0,f,g,\alpha,\beta)$ satisfies $V^0(\alpha, p\beta) \geq V^0(\alpha, \beta)$ for any $0 < p < 1$. Let $\beta_1 = p\beta$ and $\beta_2 = \beta$. Let $\tau^i$ be the optimal rule for the game $V(\alpha, \beta_i)$. Now let us sample a $\lambda$ from $\Gamma(\alpha_1, \beta_1)$ and let $K_1(t)$ be a sample path from a nonhomogeneous Poisson process with increments $\lambda_g(t)$. Construct a process $K_2(t)$ by $p$-thinning $K_1(t)$, i.e., by throwing away epochs of $K_1(t)$ with probability $1 - p$. Clearly $K_2(t)$ is a sample path of a $\Gamma(\alpha, \beta_2)$ process and $V^0(\alpha, \beta_1) = \{E\, cK_1(\tau_1) - f(\tau_1)\} \geq \{E\, cK_1(\tau_2(K_2)) - f(\tau_2(K_2))\} \geq \{E\, cK_2(\tau_2(K_2)) - f(\tau_2(K_2))\} = V^0(\alpha, \beta_2)$.

Here $\tau_2(K_2)$ refers to the stopping rule $\tau_2$ applied to $K_2$ derived from $K_1$. Since $K_2$ is obtained from $K_1$ by randomizing in a known manner, $\tau_2(K_2)$ is a genuine stopping rule for $K_1$. The first inequality and the last equality follows from the optimality of $\tau_i$ for $K_i$, and the second inequality follows from our thinning construction, since $K_1(t) \geq K_2(t)$ for all $t$.

Regarding monotonicity with respect to $\alpha$, let $\alpha_1 \geq \alpha_2$ and $\beta_1 = \beta_2 = \beta$. We use the same construction as before. However the thinning probability $p$ is chosen randomly from a Beta distribution with parameters $\alpha_2$ and $\alpha_1 - \alpha_2$. It is easily seen that $K_i(t)$, thus obtained are sample paths from the $(\alpha_i, \beta)$ processes, $i = 1, 2$, respectively. The rest of the proof now is similar to the $\beta$ case.

The proof of the following lemma is straightforward.

**Lemma 3.5** For any function $h$, $h(\cdot) \geq f(\cdot)$, $V(0, 0, f, g, \alpha, \beta) \geq V(0, 0, h, g, \alpha, \beta)$ .

Combining Theorem 3.2, Corollary 3.3, Theorem 3.4, and Lemma 3.5 gives:

**Theorem 3.6** For arbitrary $f$ and $g$, (i) $V(k, t, f, g, \alpha, \beta)$ is a monotonically increasing function of $k$; (ii) Let $h(y) = f\left(G^{-1}(1 - e^{-y})\right)$. Then a sufficient condition for $V(k, t, f, g, \alpha, \beta)$ to be monotonically decreasing in $t$ is that for each $s$, $h(y + s) - h(y)$ is an increasing function of $y$.

**Corollary 3.7** If $h$ is continuous nondecreasing and convex, the conclusions of Theorem 3.6 hold. If $G$ is exponential, then convexity of $f$ is enough for the conclusions.

**Proof of Theorem 3.6** Part (i) is direct from Theorem 3.2 and Theorem 3.4. For Part (ii), making the transformation $t \rightarrow G^{-1}(1 - e^{-t})$ is equivalent to rescaling time so that $G$ is exponential. Thus, it is enough to consider the case where $G$ is exponential and show that monotonicity follows if $f_t(\cdot) = f(t + \cdot) - f(t)$ is increasing in $t$. For this case use Theorem 3.4 to conclude the theorem.

It may be worth noting that besides convex functions, the above condition allows functions $f_t$ which are convex over a finite range and infinite outside. From now on we will assume that

**Assumption 4.** $f$ is monotone increasing, twice differentiable with continuous second derivatives in the range where it is finite. Further, $g$ is continuous, and $f\left(G^{-1}(1 - e^{-y})\right)$ is convex when finite.

### 3.2. Finite Horizon Case

First we consider a case where there is a finite horizon and obtain an explicit solution. The case with infinite horizon will be obtained as a limiting case. Let $t_\infty$ be the maximum permitted time on test.

**Assumption 5.** There exists a positive number $t_\infty$, $0 < t_\infty < \infty$ such that $f(t_\infty)$ is finite but $f(t_\infty +) = \infty$.

Under this assumption there will exist a minimal $\ell$ such that for $k \geq \ell$, $V(k, t) > 0$ for all $t < t_\infty$ (already $V(k, t_\infty) = 0$ for all $k$ as $f(t_\infty +) = \infty$). That is, at any time $t$, if one has seen at least $\ell$ bugs, then the optimal policy is a fixed time rule: continue all the way up to $t_\infty$. Having gotten this structure, which occurs only in the finite horizon case, it will be easy to obtain $\ell$ and start a backward induction argument.

**Theorem 3.8** Under assumptions 1-5 there exists an integer $\ell$, such that if $\ell$ or more bugs are observed up to some time $t < t_\infty$, then the optimal policy is to continue the testing up to time $t_\infty$. Further, for all $k \geq \ell$, $V(k, t) = V^*(k, t)$ where

$$V^*(k, t) = f(t) - f(t_\infty) + c(\alpha + k)\left\{((\beta + G(t_\infty))/(\beta + G(t))) - 1\right\} \quad (3.2.1)$$

and

$$\ell = [f'(t_\infty)(\beta + G(t_\infty))/cg(t_\infty) - \alpha], \tag{3.2.2}$$

where $[x]$ denotes the smallest integer not less than $x$.

**Proof** It is easily shown that $V^*(k,t)$ given by (3.2.1) is the expected future payoff at time $t$ of the policy which, having observed $K(t) = k$, continues till $t_\infty$. Now without the loss of generality (if necessary by rescaling $t$), assume that $G$ is exponential and $f$ is convex increasing on $[0, t_\infty]$. Then $d/dt\ V^*(k,t) = f'(t) - C(\alpha+k)(\beta+G(t_\infty))g(t)/(\beta+G(t))^2$ is strictly increasing in $t$ and thus is strictly convex. Also note that $V^*(k,t)$ is an increasing function $k$, $V^*(k,t_\infty) = 0$ and $V^*(k,t) \rightarrow \infty$ as $k \rightarrow \infty$ for all $t < t_\infty$. Thus, there exists a minimal $k$, say $\ell$, such that $V^*(\ell,t) > 0$ for all $t < t_\infty$ and it can be obtained by finding smallest $k$ such that $d/dt\ V^*(k,t_\infty) \leq 0$. Thus $\ell$ is as in (3.2.2). Let us now show that $V(k,t) = V^*(k,t)$ for $k \geq \ell$. Clearly $V(k,t) \geq V^*(k,t)$ as $V$ is the payoff of the optimal policy. Further, if $k \geq \ell$, then there is a policy (corresponding to $V^*(k,t)$) which gives positive payoff by continuing till $t_\infty$. Thus, the optimal policy has to continue till $t_\infty$. Further the optimal policy cannot continue beyond $t_\infty$. Thus, the optimal policy and the fixed time policy which continues till $t_\infty$ coincide if $k \geq \ell$. Hence $V(k,t) = V^*(k,t)$.

Note that $\ell \rightarrow \infty$ as $t_\infty \rightarrow \infty$. Now, given $V(k+1,t)$, we use backward induction to obtain $V(k\ ,\ t)$.

Let us suppose that we have $V(k+1,t)$ for $0 \leq t \leq t_\infty$, for some $k$. Then $V(k\ ,\ t)$, being the optimal additional reward, is obtained by comparing the strategy of terminating immediately with that of sampling for a small period of time, $dt$, and following the optimal policy from there on. In our problem, $K(t+dt) - K(t) \mid K(t)$ is a Poisson process; consequently we can ignore the possibility of finding more than one bug in $(t,\ t+dt)$. Thus

$$V(k\ ,\ t) \geq 0, \tag{3.2.3}$$

and when $V(k\ ,\ t) > 0$ it must satisfy

$$\begin{aligned} V(k\ ,\ t) = -f'(t)dt\ &+ Pr\{\text{ find a bug in } (t,t+dt) \mid k,t\} \\ &\times (c + V(k+1,t+dt))V(k,t+dt). \end{aligned}$$

By simple Bayes calculations and using the fact that $K(t)$ is a Poisson process,

$$Pr\{\text{ bug in } (t,t+dt) \mid k,t\} = E(\lambda dt \mid k,t) = g(t)dt(\alpha+k)/(\beta+G(t))$$

Substituting this and collecting terms, we get for $V(k\ ,\ t) > 0$:

$$\frac{d}{dt}V(k\,,\,t) = f'(t) - g(t)(\alpha + k)(c + V(k+1,t). \qquad (3.2.4)$$

It is easily shown that, for $k \geq \ell$, $V(k\,,\,t)$, given by (3.2.1) and (3.2.2), satisfies (3.2.3) and (3.2.4). To obtain $V(k\,,\,t)$ for $k \leq \ell - 1$, from (3.2.1), we use equation (3.2.4) for $k = \ell - 1$ in conjunction with (3.2.1) with $k + 1 = \ell$. Simplifying, we find that, for all $t$, such that $V(\ell - 1, t) > 0$,

$$\Delta(t)\frac{d}{dt}\frac{V(\ell-1,t)}{\Delta(t)} = \Delta(t)\frac{d}{dt}\frac{(V^*(\ell-1,t))}{\Delta(t)} \qquad (3.2.5)$$

where $\Delta(t) = (\beta + G(t))^{\alpha+\ell-1}$. Thus,

$$V(\ell - 1,t) = V^*(\ell - 1,t) + A_{\ell-1}\Delta(t) \qquad (3.2.6)$$

where, $A_{\ell-1}$ is a constant of integration which we shall obtain later on. Continuing this calculation iteratively, it can be shown that in general for all $k$, when $V(k\,,\,t) > 0$,

$$V(k\,,\,t) = V^*(k\,,\,t) + h_k(t), \qquad (3.2.7)$$

where

$$h_k(t) = (\beta + G(t))^{\alpha+k}\left\{A_k(\beta + G(s))^{\alpha+k+1}ds\right\} \text{ for all } k < \ell \qquad (3.2.8)$$

and 0 otherwise. Clearly, since $V \geq V^*$, we have $h_k(\cdot) \geq 0$. Thus, $A_k \geq 0$, $k = 1, 2, \ldots$. One can carry out the recurrence relation (3.2.9) explicitly to obtain a general formula for $h_k(t)$. It is given by:

$$h_k(t) = P_k(G(t))\,(\beta + G(t))^{\alpha+k} \qquad (3.2.9)$$

where $P_k(y)$ is an $(\ell - 1 - k)^{th}$ degree polynomial given by

$$P_k(y) = \sum_{j=0}^{\ell-k-1}(-1)^j A_{k+j}\frac{(\alpha + k + j - 1)!}{(\alpha + k - 1)!}\frac{y^j}{j!}. \qquad (3.2.10)$$

The polynomials $P_k(y)$ satisfy:

$$P'_k(y) = -(\alpha + k)\,P_{k+1}(y), \quad P_k(y) = 0 \quad k \geq \ell\,. \qquad (3.2.11)$$

The coefficients $A_k$ are to be determined using a side condition. Summarizing:

**Theorem 3.9** The optimal additional payoff function, $V(k\,,\,t)$, conditional on $t$ and $K(t) = k$, is given by (3.2.1), (3.2.2) and (3.2.7) - (3.2.10). The coefficients $A_k$ are nonnegative and the optimal policy is to stop as soon as $V(k\,,\,t)$ becomes 0.

Now we determine the constants $A_k$. Recall, by Theorem 3.6, that $V(k, t)$ is such that there exists $t_k$, $0 \leq t_k \leq t_\infty$ for which $V(k, t) > 0$ for $t < t_k$, and $V(k, t) = 0$ for $t \geq t_k$. Further, $0 \leq t_0 \leq \ldots \leq t_{\ell-1} \leq t_\ell = \ldots = t_\infty$ (the equality is by Theorem 3.8). Now we will give an iterative algorithm for computing the $t_k$'s and will prove various properties.

**Theorem 3.10** Given $t_{k+1}, \ldots, t_\ell = t_\infty$, let $B_k(\zeta) = -V^{**}(k, \zeta)/(\beta + G(\zeta))^{\alpha+k}$, where $V^{**}(k, t)$ is a particular solution of (3.2.4) with the side condition that $V^{**}(k, t_{k+1}) = 0$. Then $t_k$ is obtained as the $\zeta$ maximizing $B_k(\zeta)$. Further $t_k < t_{k+1}$, and if $V(k, 0) > 0$, then $t_k$ is a solution of $\frac{d}{d\zeta} V^{**}(k, \zeta)/(\beta + G(\zeta))^{\alpha+k} = 0$. Finally, $V(k, t)$ is such that $V(k, t_{k-}) > 0, V(k, t_k)0$ and $\frac{d}{dt} V(k, t) \mid_{t=t_k} = 0$.

**Proof** Let us determine $t_k$ given $t_{k+1}, \ldots, t_\ell = t_\infty$ for $k \leq \ell - 1$. Let the induction hypothesis be

$$\frac{f'(t_{k+1})(\beta + G(t_{k+1}))}{cg(t_k)} > \alpha + k, \tag{3.2.12}$$

which is clearly satisfied for $k = \ell - 1$ by (3.2.2). Now, given that one is at $(k, t)$, $t < t_{k+1}$, define a strategy $S_{k,t}(\zeta)$ as, stop at $\zeta$, $t \leq \zeta \leq t_{k+1}$ if there are no bugs found up to $\zeta$; otherwise follow the optimal strategy from there on. The latter is known to be $V(j, t)$ for all $j \geq k + 1$. Clearly, the optimal strategy at $(k, t)$ is within this class and corresponds to $\zeta = t_k$. Thus, one obtains $t_k$ by optimizing $S_{k,t}(\zeta)$. For this, we compute its value function $V(k, t; \zeta)$. Clearly,

$$\text{for } t \geq \zeta \quad V(k, t; \zeta) = 0 \tag{3.2.13}$$

By arguments similar to those used in deriving (3.2.4) and further simplifying, we get for $t < \zeta$

$$(\beta + G(t))^{\alpha+k} \frac{d}{dt} \frac{V(k, t; \zeta)}{(\beta + G(t))^{\alpha+k}} = \\ f'(t) - \frac{g(t)(\alpha + k)}{\beta + G(t)} \{c + V(k + 1, t)\}. \tag{3.2.14}$$

Thus, $V(k, t; \zeta)$ satisfies the same partial differential difference equation (PDDE) (3.2.4) as $V(k, t)$, but with side condition (3.2.13). Further, it need not be, and in general is not, non-negative. Let $V^{**}(k, t)$ denote a particular solution of (3.2.14), which vanishes at $t_{k+1}$. Thus, for $t < \zeta$,

$$V(k, t; \zeta) = V^{**}(k, t) + B_k(\zeta)\ (\beta + G(t))^{\alpha+k} \tag{3.2.15}$$

for some constant $B_k$ depending on $\zeta$. It may be helpful to recall that for $k = \ell - 1$, $V^{**}(\cdot) = V^*(\ell - 1, t)$. Since the side condition is $V(k, \zeta, \zeta) = 0$, we get

$$B_k(\zeta) \;=\; -V^{**}(k,\,\zeta)/(\beta\,+\,G(\zeta))^{\alpha+k}\,. \tag{3.2.16}$$

Thus, one obtains the optimal strategy for all $t$ by maximizing (3.2.15) with respect to $\zeta$ or equivalently (3.2.16). Now we show that $t_k < t_{k+1}$. This would happen if $\frac{d}{d\zeta}\, B_k(\zeta)\mid_{\zeta=t_{k+1}} < 0$. But as $V^{**}(k,\,t)$ satisfies PDDE (3.2.14), the sign of $\frac{d}{d\zeta} B_k(\zeta)$ at $\zeta = t_{k+1}$ equals the sign of

$$-(f'(t_{k+1}) - \frac{g(t_{k+1})(\alpha+k+1)}{\beta\,+\,G(t_{k+1})}(c\,+\,V(k+1,\,t_{k+1})),$$

which is negative by the induction hypothesis (3.2.12). Thus, $t_k < t_{k+1}$. Further, if $V(k,\,0) > 0$ (otherwise $t_0 = ... = t_k = 0$), the maximizing value, $t_k$, is in the interior. Consequently, at $\zeta \;=\; t_k$, $\frac{d}{d\zeta}\;\frac{V^{**}(k,\zeta)}{(\beta+G(\zeta))^{\alpha+k}} \;=\; 0$. As $V^{**}(\,\cdot\,)$ is also a solution of (3.2.15), this would imply that

$$f'(t_k) = \frac{g(t_k)(\alpha+k)}{\beta+G(t_k)}(c\,+\,V(k+1,\,t_k)) > \frac{cg(t_k)(\alpha+k-1)}{(\beta+G(t_k))},$$

i.e., the induction hypothesis (3.2.12) holds for $t_k$. Finally, as

$$V(k,\,t) \;=\; V^{**}(k,\,t) \;+\; B_k(t_k)(\beta+G(t))^{\alpha+k}\,, \tag{3.2.17}$$

we have

$$\frac{d}{dt}V(k,\,t)\mid_{t=t_k} = \; V^{**}(k,\,t_k)\frac{d}{dt}ln\,B_k(t)\mid_{t=t_k} = \frac{V^{**}(k,\,t_k)}{B_k(t_k)}\,\frac{d}{dt}B_k(t)\mid_{\{t=t_k\}},$$

thus $\frac{d}{dt}B_k(t)\mid_{\{t=t_k\}} = 0$ implies that

$$\frac{d}{dt}V(k,\,t)\mid_{t=t_k} \;=\; 0\,. \tag{3.2.18}$$

### 3.3. Infinite Horizon Case

We now show that even if $f(t)$, satisfying assumptions 1 and 2, is not restricted to $[0, t_\infty]$ for some $t_\infty < \infty$, there still exists a solution. For this, let us define a sequence of games with corresponding cost of testing being $f_\infty\,(t) \;=\; f(t),\; t \;\leq\; t_\infty$ and infinite otherwise. Thus, $f_{t_\infty}\,(t)$ is a decreasing function of $t_\infty$. Thus, by Lemma 3.5 $V(k,\,t,\,f_{t_\infty}) \;=\; V(k,\,t,\,f_{t_\infty},\,g,\,\alpha,\,\beta)$ obtained in Theorem 3.9 is an increasing function of $t_\infty$. Further, if $\tau\,(t_\infty)$ denotes the optimal rule corresponding to $f_{t_\infty}\,(\,\cdot\,)$, then
$V(k,t,t_\infty)$

$$\begin{aligned}
&= \;\; E\,\{c(K(\tau(t_\infty)) - K(t)) - (f_{t_\infty}(\tau(t_\infty)) - f_{t_\infty}(t)) \mid K(t) = k\}\\
&\leq \;\; c\;E\,\{(c(K(\tau(t_\infty)) \;-\; K(t)) \mid K(t) \;=\; k)\}\\
&\leq \;\; c\;E(N \;-\; K(t) \mid K(t) \;=\; k)\\
&= \;\; \frac{\alpha\,+\,k}{(\beta\,+\,G(t))}\,(1\,-\,G(t))\\
&\leq \;\; (\alpha\,+\,k)/\beta\,.
\end{aligned}$$

Thus, $V(k, t, t_\infty)$ is a bounded increasing function of $t_\infty$ and consequently converges to a function $V(k, t)$ as $t_\infty \to \infty$. The corresponding $t$'s such that $V(k, t-) > 0$ and $V(k, t) = 0$ would give the optimal stopping rule.

## REFERENCES


Dalal, S. R. and Mallows, C. L (1988). When should one stop testing software? *J. of Am. Statist. Assoc.*, **83**, 872-879

Dalal, S. R. and Mallows, C. L. (1990). Some graphical aids for deciding when to stop testing software. *IEEE Journal on selected areas in communications, Special issue on Telecommunications Software Quality and Productivity,* **8**, 169-175.

Goel, A. L. and Okumoto, K. (1979). Time dependent error detection rate model for software reliability and other performance measures. *IEEE Trans. Rel.,* **R-28**, 206-211.

Jelinski, Z. and Moranda, P. B. (1972). Software reliability research. In *Statistical Computer Performance Evaluation.* Ed. Freiberger, W., Academic Press, Inc., New York.

Langberg, N. and Singpurwalla, N. D. (1985). A unification of some software reliability models. *SIAM. J. Sci. Stat. Comput.* **6**, 781-790.

Musa, J. D. (1975). A theory of software reliability and its applications. *IEEE Trans. Software Eng.,* **SE-1**, 312-327.

Starr, Norman (1974). Optimal and adaptive stopping based on capture times. *J. Appl. Prob.* **11**, 194-302.

Vardi, Y. (1980). On a stopping time of Starr and its use in estimating the number of transmission sources. *J. Appl. Prob.* **17**, 235-242.

# 19

# Filtering, Smoothing, and Extrapolations in Dose-Response Experiments: Application to Data on Respiratory Tumors in Rats


## Nozer D. Singpurwalla & Jingxian Chen [1]



**Abstract**

A method for inference and extrapolations in certain dose-response, damage-assessments and accelerated life-testing studies as been proposed by Meinhold and Singpurwalla in 1986. The method is based on a use of the Kalman-filter algorithm and involves the double lognormal as the distributional assumption. In this paper, we discuss issues pertaining to a practical implementation of this methodology. This involves some insights based on a simulation study about the specification of prior parameters and an application of the proposed methodology to some published data on doses of bischloromethyl ether administered to rats.



*Key Words and Phrases:* Accelerated Testing, Bioassay, Kalman Filtering, Bayesian Methodology, Damage Assessment.


## 1. INTRODUCTION AND OVERVIEW

Let $x$ denote a dose or a stress that is applied to a biological or an engineering system, and suppose that $x$ takes values in $[0, \infty)$. Let $Y(x)$ be the response to $x$, and suppose that $0 \leq Y(x) \leq 1; Y(x)$ could be viewed as the propensity of each item to respond to $x$, or the extent of damage incurred by each item when subjected to stress $x$. Thus, for example, $Y(x) = 1(0)$ could imply total resistance (demolition) to (under) $x$. Often, it is true that we are able to test more than one item at any $x$, but that it may not be possible to repeat the test at any $x$, because doses and stresses cannot be exactly controlled. In what follows, we shall assume that the values of $x$ are known to an experimenter, but that the $Y(x)$ are unknown "states of nature" about which it is desired to make inferences. Such inferences are


[1]The George Washington University, Washington, DC

based on background knowledge about a particular scenario under consideration, an assumed relationship between $Y(x)$ and $x$, and tests conducted at several values of $x$. The tests conducted at the several values of $x$ give us information about $Y(x)$, but that $Y(x)$ cannot be directly observed.

It is often the case that $Y(x)$ responds to $x$ in a nonlinear fashion, and in view of this, plus considerations of the type cited in Section 1.1, we propose, as a model for relating $Y(x)$ and $x$, the relationship

$$Y(x) = \exp[-\alpha(x)x^{\beta(x)}], \alpha(x), \beta(x) > 0; \tag{1.1}$$

$\alpha(x)$ and $\beta(x)$ are unknown parameters which depend on $x$. In what follows, $E(Y(x))$ will denote expectation of the unknown quantity $Y(x)$.

### 1.1 Arguments Supporting Choice of the Relationship

As a special case of (1.1), suppose that

$$\alpha(x) = \alpha \text{ and } \beta(x) = \beta, \text{ for all } x \geq 0.$$

Then

$$Y(x) = \exp(-\alpha x^{\beta}), \alpha, \beta > 0; \tag{1.2}$$

the right hand side of the above is the survival function of a Weibull distribution. A virtue of (1.2) is flexibility in expressing a wide class of subjective opinions about the dose-response and stress-damage relationships. For example, it has been recommended for use in food safety assessment studies and clinical trials in the biological scenarios [cf. Final Report of the SCF-SC (1980)], and in studying the effects of underwater nuclear explosions on submarine miniatures in the engineering scenario [cf. McDonald (1989), Shaked and Singpurwalla (1990)]. The relationship (1.2) [and also (1.1)] can be linearized enabling one to employ standard filtering techniques, and when used in connection with data sets on human risk assessment studies, it gives an estimated risk at low doses which lies between the estimates for the "gamma multi-hit" and the "Armitage-Doll" models. Also, the relationship (1.2) has been reported to give good empirical fits to the data. In any particular application, it is suggested that the statistical analyst and the subject matter specialist examine plots of the Weibull dose-response curves for several combinations of values of $\alpha$ and $\beta$, and choose that curve or curves that best describe their judgments of the dose-response relationship. The above can be most effectively done on personal computers.

### 1.2 Statement of the Problem

Suppose that at $T$ distinct dose levels $x_1 > x_2 > \ldots > x_T$ an item or several items are tested and the corresponding observed responses $y(x_1), y(x_2), \ldots, y(x_T)$, recorded; note that $y(x_i)$ provides information about the unobservable $Y(x_i), i = 1, \ldots, T$. We are required to:

i) Make statements of uncertainty about the true responses $Y(x)$, at any $x$, including the $x_i$'s, $i = 1, \ldots, T$, at which the tests are conducted.

ii) Assess $Y(x_{T+1})$, the response at a very low dose $x_{T+1}$, where $x_{T+1} << x_T$

### 1.3 The Proposed Approach

Our approach for addressing the above issues is Bayesian, and involves a use of the technology of Kalman filtering and Kalman filter smoothing. The underpinnings of our approach are outlined in Meinhold and Singpurwalla (1987), where due to limitations of space and the need for an expository focus, applications of the methodology to practical scenarios was not undertaken. An aim of this paper is to fill the above gap and to attempt to demonstrate the potential usefulness of the Kalman filter approach to problems of the kind considered here. It is helpful to point out that Blackwell and Singpurwalla (1988) undertake an endeavor similar to the one discussed here, but focus attention on the case of exponentially distributed lifetimes; furthermore, they do not address the issue of smoothing that is relevant to problems of this type.

### 1.4 Overview of Paper

In Section 2 we motivate the Kalman filter model, state the distributional assumptions, and present the necessary smoothing and extrapolation formulae. In Section 3 we describe approaches for obtaining the starting values of the Kalman filter algorithm (i.e. specifying the prior parameters), and in Section 4 we describe an application of our approach to some data on respiratory tumor of rats subjected to doses of bischloromethyl ether. In Section 5 we offer some conclusions and suggest some direction for future research along the lines outlined here.

## 2. THE FILTERING AND SMOOTHING MODEL

Our review of the literature suggests that there is a dearth of dose-response relationships based on pharmacokinetic, oncological or engineering considerations. Thus one's choice of (1.2), the Weibull survival function, is at best a reasonable approximation, and hence one should incorporate into one's analyses some measure of uncertainty about such approximations. Furthermore, there is no reason to sacrifice flexibility by making the Weibull scale parameter $\alpha$, and the shape parameter $\beta$, invariant with the dose $x$. On the contrary, there is evidence in the engineering sciences, that high stresses cause a change in the basic failure mechanisms, making it reasonable to assume that $\alpha$ and $\beta$ depend on $x$ - thus our choice of the relationship (1.1). The dynamic nature of $\alpha$ and $\beta$ introduces the novelty of our approach over the currently used ones. Once the above is undertaken, our set-up fits into the general form of a Kalman-filter model - provided that some details which facilitate an iterative computational scheme are attended to.

### 2.1 The Kalman-Filter Model and Distributional Considerations

Assume that $E(y(x)) = \exp(-\alpha(x)x^{\beta(x)})$, and suppose that we require $y^*(x) \stackrel{\text{def}}{=} \log\{-\log y(x)\}$, to be such that

$$y^*(x) \sim \mathcal{N}(\mu(x), \sigma^2(x)),$$

where "$\sim \mathcal{N}(\mu, \sigma^2)$" denotes "normally distributed with mean (variance) $\mu(x)(\sigma^2(x))$". Then, it can be seen that $y(x)$ must have a "double lognormal distribution" [cf. Meinhold and Singpurwalla (1987)] with parameters $\mu(x)$ and $\sigma^2(x)$, where $\mu(x)$ is the median of the distribution of $y(x)$. The distribution cited above, has a density which is flexible enough to express a wide variety of subjective opinions about $y(x)$, and is an attractive alternative to the beta density for modelling data on proportions - see Ahsanullah and Holland (1989). Motivated by the fact that when $\sigma^2(x)$ is small, $E(y(x)) \approx \exp\{-\exp(\mu(x))\}$, we propose, as the *observation equation* of the Kalman filter

$$y^*(x) = (1, \log x)\begin{bmatrix}\gamma \\ \beta\end{bmatrix}_x + \nu(x), \tag{2.1}$$

where $\nu(x) \sim \mathcal{N}(0, \boldsymbol{V}(x))$, $\boldsymbol{V}(x)$ is the conditional variance of $y^*(x)$, $(\gamma, \beta)'_x = (\gamma(x), \beta(x))'$, and $\gamma(x) = \log \alpha(x)$.

The set-up (2.1) also implies that $y(x) = (Y(x))^{\lambda(x)}$, where $\log \lambda(x) = \nu(x)$, and $y^*(x) = \log\{-\log y(x)\}$; the innovation $\lambda(x)$ has a lognormal distribution with parameters 0 and $\sigma^2(x)$. For the *system equation* of the Kalman filter, we propose the "steady model" [cf. Meinhold and Singpurwalla (1983)]

$$\begin{bmatrix}\gamma \\ \beta\end{bmatrix}_{(x)} = \begin{bmatrix}\gamma \\ \beta\end{bmatrix}_{(x-1)} + \boldsymbol{w}(x), \tag{2.2}$$

where $\boldsymbol{w}(x) \sim \mathcal{N}(\boldsymbol{0}, \boldsymbol{W}(x))$, and $\boldsymbol{W}(x)$ is the variance-covariance matrix of $\gamma(x)$ and $\beta(x)$ conditional on $\gamma(x-1)$ and $\beta(x-1)$; $\nu(x)$ is assumed independent of $\boldsymbol{w}(x)$; $(x-1)$ is the dose previous to dose $x$. The values $\boldsymbol{V}(x)$ and the four entries of $\boldsymbol{W}(x)$ must be specified by a user. Also to be specified by the user are the "starting values" of the Kalman filter algorithm; these are $\hat{\gamma}(0)$, $\hat{\beta}(0)$, $\boldsymbol{V}(x_0)$ and $\boldsymbol{W}(x_0)$. A strategy for specifying these is outlined in Section 3.

## 2.2 Results from Filtering and Smoothing

For notational convenience, we let $y^\star(x_j) = y_j^\star$, $[\gamma, \beta]'_{x_j} = \boldsymbol{\theta}_j$, $[1, \log x_j] = \boldsymbol{F}_j$, $\boldsymbol{V}(x_j) = \boldsymbol{V}_j$, $\boldsymbol{W}(x_j) = \boldsymbol{W}_j$, $\nu(x_j) = \nu_j$ and $\boldsymbol{w}(x_j) = \boldsymbol{w}_j$. Then (2.1) and (2.2) can be written as

$$y_j^* = \boldsymbol{F}_j \boldsymbol{\theta}_j + \nu_j, \text{ and}$$

$$\boldsymbol{\theta}_j = \boldsymbol{G}_j \boldsymbol{\theta}_{j-1} + \boldsymbol{w}_j,$$

where $\boldsymbol{G}_j$ is a $2 \times 2$ identity matrix. If we assume that $\boldsymbol{\theta}_{j-1} \sim \mathcal{N}(\hat{\boldsymbol{\theta}}_{j-1}, \boldsymbol{\Sigma}_{j-1})$, then, upon observing $y_j^*$, $\boldsymbol{\theta}_j \sim \mathcal{N}(\hat{\boldsymbol{\theta}}_j, \boldsymbol{\Sigma}_j)$, and

$$\hat{\boldsymbol{\theta}}_j = \boldsymbol{G}_j \hat{\boldsymbol{\theta}}_{j-1} + \boldsymbol{K}_j(y_j^* - \boldsymbol{F}_j \boldsymbol{G}_j \hat{\boldsymbol{\theta}}_{j-1}), \boldsymbol{\Sigma}_j = (\boldsymbol{I} - \boldsymbol{K}_j \boldsymbol{F}_j)\boldsymbol{R}_j, \tag{2.3}$$

where

$$\boldsymbol{R}_j = \boldsymbol{G}_j \boldsymbol{\Sigma}_{j-1} \boldsymbol{G}'_j + \boldsymbol{W}_j, \boldsymbol{K}_j = \boldsymbol{R}_j \boldsymbol{F}'_j (\boldsymbol{F}_j \boldsymbol{R}_j \boldsymbol{F}'_j + \boldsymbol{V}_j)^{-1},$$

and $\boldsymbol{I}$ is an identity matrix. The relationships in (2.3) are referred to as the *forward-recurrence equations* of the Kalman filter. Note that, in the above scheme, inference for $\boldsymbol{\theta}_j$ is based upon the previous data, $y_j^*, y_{j-1}^*, \ldots, y_1^*$, $j = 1, 2, \ldots, T$, only. Thus, with the exception of $\boldsymbol{\theta}_T$, inference for the other $\boldsymbol{\theta}_j$'s is not based on all the data. Should we desire inference for $\boldsymbol{\theta}_j$ based on <u>all</u> the data $y_1^*, \ldots, y_T^*$, then we need to smooth using the *backward-recurrence equations* [see Appendix A of Meinhold and Singpurwalla (1987)], whereby

$$\begin{cases} \hat{\boldsymbol{\theta}}_j(T) = \hat{\boldsymbol{\theta}}_j(j) + \boldsymbol{J}_{j+1}[\hat{\boldsymbol{\theta}}_{j+1}(T) - \boldsymbol{G}_{j+1}\hat{\boldsymbol{\theta}}_j(j)], \\ \boldsymbol{\Sigma}_j(T) = \boldsymbol{\Sigma}_j(j) - \boldsymbol{J}_{j+1}[\boldsymbol{\Sigma}_{j+1}(T) - \boldsymbol{R}_{j+1}]\boldsymbol{J}'_{j+1}, \end{cases} \tag{2.4}$$

where $\boldsymbol{J}_j = \boldsymbol{\Sigma}_{j-1}(j-1)\, \boldsymbol{G}'_j\, \boldsymbol{R}_j^{-1}$ and $\hat{\boldsymbol{\theta}}_j(k)[\boldsymbol{\Sigma}_j(k)]$ is the mean [covariance] of the normal distribution of $\boldsymbol{\theta}_j$ based on $y_1^*, \ldots, y_k^*$. The standard Kalman-filter algorithm enables us to produce $\hat{\boldsymbol{\theta}}_1(1), \hat{\boldsymbol{\theta}}_2(2), \ldots, \hat{\boldsymbol{\theta}}_T(T)$ in an efficient manner; see (2.3).

Once $\hat{\boldsymbol{\theta}}_T(T)$ is obtained, inference about $Y(x_{T+1})$ follows from the fact that $\log(-\log(Y(x_{T+1}))) \sim \mathcal{N}(\boldsymbol{F}_{T+1}\, \hat{\boldsymbol{\theta}}_T(T),\, \boldsymbol{F}_{T+1}\boldsymbol{\Sigma}_T(T)\, \boldsymbol{F}'_{T+1})$, and so

$$E(Y(x_{T+1})) \approx \exp(-\exp(\boldsymbol{F}_{T+1}\hat{\boldsymbol{\theta}}_T(T))). \tag{2.5}$$

# 3. SPECIFICATION OF INPUTS FOR THE KALMAN FILTER

Let $x_0$ and $x_{-1}$, $x_{-1} > x_0 > x_1$, be two dose levels at which the subject matter specialist has the most knowledge about the responses, $Y(x_0)$ and $Y(x_{-1})$ respectively. Typically, these would be very large dose levels; also, we require that $x_0$ and $x_{-1}$ be close to each other. Suppose that, in the opinion of the specialist, the *most likely* values of $Y(x_0)$ and $Y(x_{-1})$ are $\tilde{Y}(x_0) = 1-\delta_0$ and $\tilde{Y}(x_{-1}) = 1-\delta_1$, where $\delta_0$ and $\delta_1$ are large with $\delta_0 < \delta_1$; what we have in mind is $\delta_0 = .995$ and $\delta_1 = .999$. Then, for $x_0$ close to $x_{-1}$, we may write

$$\log\{-\log(1-\delta_0)\} \approx \hat{\gamma}(x_0) + \hat{\beta}(x_0)\log x_0,$$
$$\log\{-\log(1-\delta_1)\} \approx \hat{\gamma}(x_0) + \hat{\beta}(x_0)\log x_{-1}, \tag{3.1}$$

from which it follows that

$$\hat{\beta}(x_0) = \frac{\{\log(-\log(1-\delta_0)) - \log(-\log(1-\delta_1))\}}{\{\log x_0 - \log x_{-1}\}},$$

and

$$\hat{\gamma}(x_0) = \log(-\log(1-\delta_0)) - \hat{\beta}(x_0)\log x_0.$$

The values $\hat{\beta}(x_0)$ and $\hat{\gamma}(x_0)$ thus computed will become the starting values for the Kalman-filter algorithm.

Our next task is to pin down $\boldsymbol{V}(x_0), \ldots, \boldsymbol{V}(x_T)$, the variances of the innovations $\nu(x_i), i = 0, \ldots, T$. In specifying the above, two considerations must be borne in mind. The first, is that since $y(x)$ is between 0 and 1, the variance of $\lambda(x)$ must be the smallest when $Y(x)$ is either 0 or 1, and the largest when $Y(x) = .5$. Second, under binomial testing, the variance of $y(x)$ is approximately of the form $m(x)\ (1-m(x))/n(x)$, where $m(x)$ is the median of the distribution of $y(x)$ and $n(x)$ is the number of units tested at dose $x$. Following a line of reasoning given in Appendix A, a first order approximation to $\boldsymbol{V}(x) = \sigma^2(x)$ is of the form

$$\sigma^2(x) \approx \log\left\{\frac{1}{2}(1 + [1 + 4(1-m(x))/n(x)m(x)(\log(m(x)))^2]^{\frac{1}{2}})\right\}. \tag{3.2}$$

Note that in the above expression, $\sigma^2(x) \to 0$ as $m(x) \to 0[1]$ provided that $n(x) > (m(x))^{-1}\ [(1-m(x))^{-1}]$, and so as is commonly done by engineers, we may set $\boldsymbol{V}(x_0) = 0$. In (3.2) above, $m(x) = \exp\{-\exp(\mu(x))\}$, with $\mu(x)$ replaced by its predicted value $\hat{\mu}(x)$ obtained from the Kalman-filter algorithm. Specifically,

$$\hat{\mu}(x_1) = \hat{\gamma}(x_0) + \hat{\beta}(x_0)\log x_1, \text{ and}$$
$$\hat{\mu}(x_i) = \boldsymbol{F}_i\hat{\boldsymbol{\theta}}_i, \quad i = 1, \ldots, T,$$

where $\hat{\boldsymbol{\theta}}_i$ is given by (2.3), the forward mechanism of the Kalman filter.

Our next step is to pin down $\boldsymbol{W}(x_0)$, the variance covariance matrix of the starting values $\hat{\gamma}(x_0)$ and $\hat{\beta}(x_0)$. For this, we require that the subject matter specialist specify, in addition to $\tilde{Y}(x_0)$ and $\tilde{Y}(x_{-1}), \nu_1, \nu_2$ and $\rho$, the variances of $Y(x_0), Y(x_{-1})$ and the correlation between $Y^*_{(x_0)}$ and $Y^*_{(x_{-1})}$, as perceived by the specialist. The variances $\nu_1$ and $\nu_2$ will typically be very small, something like $(.001)^2$ and $(.0003)^2$, respectively, and $\rho$ should be nonnegative, say something of the order of .7 or .8. Once the above are done, the relationship (3.1) can be used to show that if $\boldsymbol{b} = [\nu_1, \nu_2, \rho]'$, and if $\boldsymbol{Z} = [\text{Var}(\hat{\gamma}(x_0)), \text{Var}(\hat{\beta}(x_0)), \text{Cov}(\hat{\gamma}(x_0), \hat{\beta}(x_0))]'$, then $\boldsymbol{Z} = \boldsymbol{A}^{-1}\boldsymbol{b}$, where the matrix $\boldsymbol{A}$ is of the form

$$\boldsymbol{A} = \begin{bmatrix} 1 & (\log x_0)^2 & 2\log x_0 \\ 1 & (\log x_{-1})^2 & 2\log x_{-1} \\ \frac{4\rho}{\nu_1+\nu_2+2\rho\sqrt{\nu_1\nu_2}} & \frac{\rho(\log x_0+\log x_{-1})^2}{\nu_1+\nu_2+2\rho\sqrt{\nu_1\nu_2}} & \frac{4\rho(\log x_0+\log x_{-1})}{\nu_1+\nu_2+2\rho\sqrt{\nu_1\nu_2}} \end{bmatrix}.$$

Once $\boldsymbol{Z}$ is known, the matrix $\boldsymbol{\Sigma}(x_0)$ can be constructed, and this enables us to undertake the first iteration of the Kalman-filter algorithm. Subsequent iterations of the algorithm require that we specify $\boldsymbol{W}(x_j)$, $j = 1, \ldots, T$. For this, we propose, based on some simulation studies, that

$$\boldsymbol{W}(x_j) = .5(1.5)^j(x_{j-1} - x_j)\boldsymbol{\Sigma}(x_0). \tag{3.3}$$

The multipliers of $\boldsymbol{\Sigma}(x_0)$ given above, reflect the following considerations:

i) The factor .5 (or any other number less than 1) reflects the fact that our uncertainty about the parameters, subsequent to observing data, should be less than our uncertainty about them prior to the data.

ii) The factor $(1.5)^j$ (or for that matter any other number greater than 1) reflects the fact that our uncertainty about the parameters should increase as we get closer to the low dose levels.

iii) The factor $(x_{j-1} - x_j)$ reflects the fact that if the separation between two consecutive doses is large then our uncertainty about the model parameters should increase.

It should be clear from the above, that the scheme proposed here could be automated once $\tilde{Y}(x_0)$, $\nu_1$, $\tilde{Y}(x_{-1})$, $\nu_2$ and $\rho$ are specified.

## 4. APPLICATION TO DATA ON RESPIRATORY TUMORS IN RATS

For purposes of illustration, we consider some data on doses (the number of six hour exposures by inhalation of 100 parts per billion) of bis-chloromethyl ether administered to rats. This data has been abstracted from the Final Report of the SCFSC (1980), wherein its analysis using the Weibull survival model has been advocated and undertaken. In Table 4.1 below, we present our data; the response is 0 if a rat develops respiratory tumor and 1 otherwise - thus 43/46 denotes the fact that 43 out of 46 rats have not developed tumors. Our aim is to predict the responses at low doses, say 7, 5, 2 and 1.

Table 4.1. Dose-Response Data on Respiratory Tumors in Rats

| Dose: $x_i$ | 10 | 20 | 40 | 60 | 80 | 100 |
|---|---|---|---|---|---|---|
| Observed Response | 40/41 | 43/46 | 14/18 | 14/18 | 19/34 | 8/20 |

We let $x_1 = 100, x_2 = 80, \ldots, x_6 = 10$, and $y(x_1) = 40/41, y(x_2) = 43/46, \ldots, y(x_6) = 8/20$. For $x_0$ and $x_{-1}$ we choose the doses 250 and 280 respectively, and for $\delta_0$ and $\delta_1$ we choose .995 and .999, respectively. Our other choices follow the recommendations of Section 3; specifically, $\nu_1 = (.001)^2, \nu_2 = (.0003)^2$ and $\rho = .7$. Substituting the above in the formulae for $\hat{\beta}(x_0), \hat{\gamma}(x_0)$ and $\boldsymbol{A}$, we obtain, as starting values for the Kalman filter, the following:

$$\hat{\gamma}(x_0) = -11.256, \hat{\beta}(x_0) = 2.3405, \mathrm{Var}(\hat{\gamma}(x_0)) = 1.668 \times 10^{-3},$$

$$\mathrm{Var}(\hat{\beta}(x_0)) = 5.217 \times 10^{-5} \text{ and Cov } (\hat{\gamma}(x_0), \hat{\beta}(x_0)) = -2.95 \times 10^{-4}.$$

In Table 4.2 below, we give filtered and smoothed estimates of the parameters $\gamma(x_i)$ and $\beta(x_i), i = 1, \ldots, 6$. The filtered estimates are obtained via the forward recurrence equations (2.3) whereas the smoothed estimates are obtained via the backward recurrence equations (2.4). The entries in Table 4.2 indicate that smoothing does have an effect on the filtered estimates and that the estimates of $\gamma(x_i)$ and $\beta(x_i)$ do change with $x_i$. Note that when $i = 6$, that is, for $x_6 = 10$, the smoothed estimate is indeed the filtered estimate.

Table 4.2. Filtered and Smoothed Estimates of $\gamma(x_1)$ and $\beta(x_i)$

| Dose: $x_i$ | 100 | 80 | 60 | 40 | 20 | 10 |
|---|---|---|---|---|---|---|
| Filtered Est. of $\gamma(x_i)$ | -10.652 | -9.563 | -10.079 | -7.702 | -7.965 | -7.604 |
| Smoothed Est. of $\gamma(x_i)$ | -9.801 | -9.780 | -9.651 | -7.704 | -7.965 | -7.604 |
| Filtered Est. of $\beta(x_i)$ | 2.234 | 2.041 | 2.132 | 1.712 | 1.759 | 1.695 |
| Smoothed Est. of $\beta(x_i)$ | 2.083 | 2.080 | 2.057 | 1.713 | 1.759 | 1.695 |

In Table 4.3 we give the observed values $y(x_i)$ and the predicted, filtered and smoothed values of $Y(x_i), i = 1, \ldots, 6$. Note that

i) $E(Y(x_i)|y(x_{i-1}), \ldots, y(x_1))$, is the predicted value of $Y(x_i)$, where $y(x_0) = Y(x_0)$,

ii) $E(Y(x_i)|y(x_i), \ldots, y(x_0))$ is the filtered value of $Y(x_i)$, and

iii) $E(Y(x_i)|y(x_6),\ldots,y(x_0))$ is the smoothed value of $Y(x_i)$.

The above conditional expectations can be obtained via formulae analogous to (2.5).

Table 4.3. Predicted, Filtered, Smoothed and Observed Values of $Y(x_i)$

| Dose: $x_i$ | 100 | 80 | 60 | 40 | 20 | 10 |
|---|---|---|---|---|---|---|
| Observed Val. of $y(x_i)$ | .4 | .5588 | .7778 | .7778 | .9348 | .9756 |
| Predicted Val. of $Y(x_i)$ | .5377 | .6559 | .7411 | .8963 | .9265 | .9756 |
| Filtered Val. of $Y(x_i)$ | .4995 | .5834 | .7712 | .7788 | .9348 | .9756 |
| Smoothed Val. of $Y(x_i)$ | .4439 | .5986 | .7465 | .7789 | .9348 | .9756 |

We note from Table 4.3, that smoothing has a tendency to lower the filtered values of $Y(x_i)$ and that the predicted values have a tendency to be larger than the observed values. Finally, in Table 4.4, we give predicted values of $Y(x_i)$, for $x_i = 7$, 5, 2 and 1, low doses, at which no testing was done and at which inference is most crucial. Note that here the predicted values $\hat{y}(x_i)$ are given by $E(Y(x_i)|y(x_6),\ldots,y(x_0))$, and that this quantity is obtained via (2.5). Also given in Table 4.4 are the 90% Probability of Coverage Intervals $(L_i, U_i)$ for the above $Y(x_i)$'s.

Table 4.4. Predicted Values of $Y(x_i)$ at Low Doses

| Dose: $x_i$ | | 7 | 5 | 2 | 1 |
|---|---|---|---|---|---|
| $\hat{y}(x_i)$ | | .9866 | .9924 | .9984 | .9995 |
| 90% PCI | L | .98658 | .99239 | .9984 | .9995 |
| for $Y(x_i)$ | U | .98662 | .99242 | .9984 | .9995 |

### 4.1. Comparison with Maximum Likelihood Approach

It is of interest to compare the results produced by our approach with those produced via the conventional approach, in which the relationship (1.2) is assumed, and $\alpha$ and $\beta$ estimated by the method of maximum likelihood. It can be easily verified that $\hat{\alpha}$ and $\hat{\beta}$, the maximum likelihood estimates of $\alpha$ and $\beta$, respectively, are $\hat{\alpha} = 7.496 \times 10^{-4}$, and $\hat{\beta} = 1.513$. Replacing $\alpha$ and $\beta$ by $\hat{\alpha}$ and $\hat{\beta}$ respectively, in (1.1), we obtain the maximum likelihood estimates of $Y(x_i)$. These are shown in Table 4.5; also shown there, for purposes of comparison, are the smoothed values of $Y(x_i)$ and the observed values $y(x_i)$. We note from Table 4.5, that whereas the differences between the maximum likelihood estimates of $Y(x_i)$ and the smoothed values of $Y(x_i)$ are significant at the higher doses, the differences at the lower doses, particularly the doses at which predictions are sought, are negligible if not nonexistent. Thus it appears that a use of our approach does not lead one to conclusions that would be significantly different from those obtained via a more conventional approach, except that should the situation so demand, our approach would provide a greater flexibility than the conventional one. Finally, we also note that the smoothed values of $Y(x_i)$ tend to be larger than those of the maximum likelihood estimates.

Table 4.5. A Comparison of the Observed and Smoothed Values and Maximum Likelihood Estimates of $Y(x_i)$

| Dose: $x_i$ | Observed Value of $y(x_i)$ | Smoothed Value of $Y(x_i)$ | Max. Likelihood Est. of $Y(x_i)$ |
|---|---|---|---|
| 100 | .4000 | .4439 | .4515 |
| 80 | .5588 | .5986 | .5667 |
| 60 | .7778 | .7465 | .6926 |
| 40 | .7778 | .7789 | .8197 |
| 20 | .9348 | .9348 | .9327 |
| 10 | .9756 | .9756 | .9759 |
| 7 | | .9866 | .9859 |
| 5 | | .9924 | .9915 |
| 2 | | .9984 | .9979 |
| 1 | | .9995 | .9993 |

## 5. SUMMARY AND CONCLUSIONS

It appears to us that using the dynamic linear model set-up of Kalman filtering is a potentially useful approach for making inference under dose-response experiments. Its chief virtues are flexibility in modelling - specifically, making the parameters dose dependent - and the ease of undertaking inference. Its chief disadvantage, especially to a non-Bayesian, is the need to specify the starting values. However, as indicated in Section 3, some general guidelines can be followed, and once this is done the procedure is almost automatic. A computer code which facilitates the required computations are given by Chen and Campodonico (1989). An issue that remains to be addressed, and one that we have not been able to satisfactorily undertake, is that pertaining to the determination of a "safe dose". Specifically, what is needed is inference about $x_s$ when $Y(x_s)$ is specified; $Y(x_s)$ is typically, a number close to 1, say .999 or .9999. Finally, regarding further research along the above lines, it would be desirable to develop an inference mechanism which does not rely, as heavily as we have, on a use of the Gaussian distribution.

## APPENDIX A

### THE SPECIFICATION OF $\sigma^2(x)$

Let $\boldsymbol{Z}(x) = -\log y(x) = (-\log Y(x))\lambda(x)$, then $\boldsymbol{Z}(x)$ has a lognormal distribution with parameters $\log(-\log Y(x))$ and $\sigma^2(x)$.

Since $y(x) = m(x) \cdot \exp\{-(\boldsymbol{Z}(x) + \log m(x))\}$, we have by a Taylor's series expansion

$$\begin{aligned} y(x) &= m(x) \sum_{k=0}^{\infty} (-1)^k (\boldsymbol{Z}(x) + \log m(x))^k / k! \\ &\approx m(x) \left(1 - (\boldsymbol{Z}(x) + \log m(x))\right), \end{aligned}$$

from which it follows that

$$\begin{aligned} \text{Var}\ (y(x)) &\approx m^2(x) \cdot \text{Var}\ (\boldsymbol{Z}(x)) \\ &= m^2(x)(\log Y(x))^2 e^{\sigma^2(x)}(e^{\sigma^2(x)} - 1). \end{aligned}$$

However, $Y(x) = m(x)$, since

$$\begin{aligned} .5 &= P(y(x) \leq m(x)) \\ &= P(Y(x)^{\lambda(x)} \leq m(x)) \\ &= P(\lambda(x) \geq \log m(x) / \log Y(x)). \end{aligned}$$

Therefore, $\log m(x)/\log Y(x)$ equals to the median of $\lambda(x)$, which is equal to 1. Now (3.2) follows from simple algebra and the fact that $\text{Var}(y(x)) = m(x)(1 - m(x))/n(x)$.

## ACKNOWLEDGMENT

Our thanks to Professor William DuMouchel for his excellent and penetrating comments on the proposed approach and to Professor Robert S-mythe, who provided us with a copy of the Final Report of the SCFSC (1980). Research supported by Office of Naval Research Contract N00014-85-K-0202, Project NR 042-372, U.S. Army Research Office Grant DAAL 03-87-K-0056 and The Air Force Office of Scientific Research Grant AFOSR-89-0381.

## REFERENCES

Ahsanullah, M. and Holland, B. (1989). Further Results on a Distribution of Meinhold and Singpurwalla. *The American Statistician*, **43**, 216-219.

Blackwell, L. M. and Singpurwalla, N. D. (1988). Inference from Accelerated Life Tests Using Filtering in Coloured Noise. *Journal of the Royal Statistical Society*, Ser. B, **50**, 281-292.

Chen, J. and Campodonico, S. (1989). A Computer Program for "Filtering, Smoothing, and Extrapolations in Dose-Response Experiments with Application to Data on Respiratory Tumor of Rats." The George Washington University, Technical Report GWU/IRRA/Serial TR-89/6.

McDonald, W. W. (1989). A Statistical Approach to the Prediction of Dynamic Structure Failure. ASME, 89-OCN-14.

Meinhold, R. J. and Singpurwalla, N. D. (1987). A Kalman Filter Smoothing Approach for Extrapolations in Certain Dose-Response, Damage-Assessment, and Accelerated-Life-Testing Studies. *The American Statistician*, **41**, 101-106.

Meinhold, R. J. and Singpurwalla, N. D. (1983). Understanding the Kalman Filter. *The American Statistician*, **37**, 123-127.

Scientific Committee of the Food Safety Council. (1980). Proposed System for Food Safety Assessment: Final Report, Food Safety Council, 1725 K Street N.W. Washington, D.C.

Shaked, M. and Singpurwalla, N. D. (1990). A Bayesian Approach for Quantile and Response Probability Estimation with Applications to Reliability. *Annals of the Institute of Statistical Mathematics*, **42**, 1-19. (1980).

# 20

# Bayesian Perturbation Diagnostics and Robustness


## Seymour Geisser [1]



**Abstract**

A Bayesian analysis may depend critically on the modeling assumptions which include prior, likelihood and loss function. A model that has been judged adequate in previous more or less similar situations may be assumed to be the standard. However one ought to consider the effect of perturbing the standard model in potentially conceivable directions especially if graphical procedures indicate the standard may only be marginally adequate. We discuss a variety of perturbation models and Bayesian diagnostics that can be helpful in a local or a more global analysis of the robustness of the sample.



*Key words and Phrases:* Perturbation models ; Robustness; Bayesian diagnostics; Local analysis, Global Analysis, Predictive Analysis.


## 1. INTRODUCTION

To paraphrase the opening remarks of G.E.P. Box (1980) "No [Bayesian] analysis can safely be assumed to be adequate. Perspicacious criticism employing diagnostic checks must therefore be applied."

A Bayesian analysis may depend critically on the modeling assumptions which include prior, likelihood and loss function. While a loss function is presumably a choice made in the context of particular situations, there is no harm and potentially some gain in investigating the effect on an analysis using alternative loss functions. The likelihood is supposed to represent to some approximation the physical process generating the data while the prior reflects subjective views about some of the assumed constructs of this process. Now a likelihood model that has been judged adequate in previous situations similar to a current one is certainly a prime candidate for modeling. However even in such situations the statistician is still obliged to investigate its present adequacy. A way of addressing this problem is to perturb the "standard" model to a greater or lesser degree in potentially conceivable directions to determine the effect of such alterations on the


[1] University of Minnesota

analysis. While for the strict Bayesian the prior is subjective, it is common knowledge how difficult it often is to subject an investigator or even a statistician to an elicitation procedure that convincingly yields an appropriately subjective prior. Hence to perturb an investigator's prior or some standard one that appears appropriate, is also sensible.

## 2. TYPES OF PERTURBATION

Even when a standard statistical model has proven adequate in data sets similar to a current one at hand, one is obliged to consider the effect of perturbing the standard model in one way or another on the analysis especially if graphical procedures indicate the possibility that the standard model may only be marginally adequate.

There are a large number of possible perturbation schemata. A typically useful one is where $\omega \in \Omega$ an index governing a perturbation schema is a set of hyperparameters. For $X^{(N)} = (X_1, \ldots, X_N)$, a set of observables, a rather simple example is

$$f(x^{(N)}|\theta,\omega) \propto \prod_{j=1}^{N} \left\{1 + \frac{(x_j - \mu)^2}{\omega\sigma^2}\right\}^{-(\omega+1)/2} , \quad \omega \geq 1$$

where the standard is $\omega \to \infty$, the normal distribution and the most deviant $\omega = 1$, the Cauchy distribution.

A second set is exemplified by a mixture e.g.

$$f(x^{(N)}|\theta,\omega) \propto \prod_{j=1}^{N} [\omega f_1(x_j|\alpha) + (1-\omega) f_2(x_j|\beta)], \quad 0 \leq \omega \leq 1$$

where say $\omega = 1$ is the standard and $\alpha$ and $\beta$ are subsets of $\theta$.

Use of $\omega$ as an indicator is relevant to situations where $\omega$ changes the model distribution to varyingly different but known distributional forms not necessarily in the same family. Although this can often be regarded as a special case of either of the first two methods it is best to consider it separately.

A fourth possibility is the use of $\omega$ as an exclusion indicator i.e. $X^{(N)} = (X_1, \ldots, X_N)$ has some standard distributional form under $\omega_o$ but for $\omega \neq \omega_o$ one or more of the $X_i$'s have either another distributional form or a completely unspecifiable distribution. In the former case this could mean for example that an observation's variance differs from the others or more generally that a parameter set not under scrutiny differs for a few of the observations. The latter situation is typically reflected in problems with outliers and aberrant observations that defy satisfactory alternative modeling.

A fifth possibility has to do with what one may term periparametric models. Here $\omega = \omega_o$ specifies a standard density while $\omega \neq \omega_o$ specifies all model densities $f(x^{(N)}|\omega)$ that are within a given neighborhood of $f(x^{(N)}|\omega_o)$ determined by varying $\omega$.

A sixth may have to do with possibly inaccurate measurement of the covariates under $\omega \neq \omega_o$ or even the actual responses themselves. All of the above have to do essentially with perturbation of the likelihood. Similar remarks may be made regarding the prior $g(\theta|\omega)$ and combinations of both likelihood and prior. As a typical example the prior could be a mixture e.g.

$$g(\theta|\omega) = \omega g_1(\theta) + (1-\omega)g_2(\theta)$$

with $\omega = 1$ resulting on the standard $g_1(\theta)$ based on previous information while $g_2(\theta)$ expresses the possibility of another view of the situation. This, for whatever it is worth, results in simpler calculations than having to deal with a likelihood mixture. In particular the use of periparametric perturbation models for additional uncertainty about a "standard" prior seems to be a promising approach especially when combined with a standard likelihood. Here one can examine the extent to which bounds on the "standard" prior can be expanded and still yield moderate sample size robustness, e.g. Lavine (1988).

## 3. FORMAL PARAMETRIC ANALYSES

A formal Bayesian framework for a perturbation analysis either for a "relevant" parameter or future observables can be delineated.

For the relevant parameter say $\theta$, we can consider the modeling is such that for a given perturbation index $\omega$, the posterior probability function for $\theta$ is specified as

$$\mathcal{P}(\theta|x^{(N)},\omega) \propto f(x^{(N)}|\theta,\omega)g(\theta|\omega)$$

where $g(\theta|\omega)$ is an assumed prior density for $\theta$ conditional on $\omega \in \Omega$ where $\omega = \omega_o$ is the standard.

A loss function $L(a,\theta)$ for taking action $a(X^{(N)}) \in \mathbf{A}$ upon observing $X^{(N)}$, given $\theta$ is the true value, is assumed (the loss function itself may also be perturbed but we shall not consider this possiblity in what follows). The average loss

$$\bar{L}_\omega(a) = \int L(a,\theta)\mathcal{P}(\theta|x^{(N)},\omega)d\theta,$$

which depends on $\omega$, is now minimized

$$\min_a \bar{L}_\omega(a) = \bar{L}_\omega(a^*_\omega)$$

yielding optimal action $a^*_\omega$ when $\omega$ is "true". We then consider the difference in the loss when taking action $a^*_{\omega_0} = a^*$, the optimal action under the standard and when $\omega \neq \omega_o$ is true. We define the differential loss as

$$d(\omega) = \bar{L}_\omega(a^*) - \bar{L}_\omega(a^*_\omega) \geq 0.$$

One then can examine this loss over a possible range of $\omega$ to assess its importance with regard to the action taken under $\omega_o$ and in particular $d^* = \max_{\omega \in \Omega} d(\omega)$. We could also assess its local significance by examining $d(\omega)$ in a neighborhood about $\omega_o$. In fact if $\omega$ is a scalar and the second derivative of $d(\omega)$ exists and is continuous the calculation of the curvature at $\omega = \omega_o$ i.e. $d''(\omega_o)$, since $d'(\omega_o) = 0$, could be rather informative regarding local perturbations. For example a large curvature would indicate that the actions taken could be highly sensitive to a slight variation in the standard model. For a vector $\omega$, the matrix of second derivatives will govern the local curvature and one could assess the maximum curvature i.e. in the direction of the normed vector associated with the largest root of the matrix of second derivatives evaluated at the standard $\omega = \omega_o$. Cook (1986) has proposed probing local curvature with regard to the displacement of maximized log-likelihoods.

## 4. PREDICTIVE ANALYSIS

We now outline the situation for prediction. The model considered is the joint probability function

$$f(x^{(N)}, x_{(M)}, \theta|\omega) = f(x_{(M)}|x^{(N)}, \theta, \omega) f(x^{(N)}|\theta, \omega) g(\theta|\omega)$$

whence we obtain

$$f(x_{(M)}|x^{(N)}, \omega) = \frac{\int f(x^{(N)}, x_{(M)}, \theta|\omega)\, d\theta}{\int\int f(x^{(N)}, x_{(M)}, \theta|\omega)\, d\theta\, dx_{(M)}}.$$

Now assume that $L(a, x_{(M)})$ is the loss incurred in taking action $a$ when observing $x^{(N)}$ given a future realization $x_{(M)}$. The average predictive loss

$$\bar{L}_\omega(a) = \int L(a, x_{(M)}) f(x_{(M)}|x^{(N)}, \omega)\, dx_{(M)}$$

is then minimized

$$\min_a \bar{L}_\omega(a) = \bar{L}_\omega(a^*_\omega)$$

where $a^*_\omega$ is the optimal action. As before letting $a^*_{\omega_o} = a^*$, we define the differential loss as $d(\omega) = \bar{L}_\omega(a^*) - \bar{L}_\omega(a^*_\omega)$ and examine globally $\max_w d(w)$ to determine the possible extent of the maximum effect of the perturbations.

Further in regular cases one can again study locally the maximum curvature which occurs in the direction of the normal vector associated with the largest root of the Hessian matrix, say $d''(\omega_o)$. If local curvature is appreciable it would appear that the sample is not even robust locally and a review of the standard model is in order. Of course if the perturbed $\omega$ model is deemed reasonable one possibility is to define a prior distribution for $\omega$ and then integrate it out to obtain

$$f(x_{(M)}|x^{(N)}) = \int g(\omega) f(x_{(M)}|x^{(N)}, \omega)\, d\omega.$$

## 5. OTHER PERTURBATION DIAGNOSTICS

Often, we are not in a position to discuss decisions or actions which would necessarily flow from a data set and consequently report either the posterior or predictive distribution itself or some high probability density region for $\theta$ or $X_{(M)}$. For reporting the entire posterior distribution the Kullback-Leibler estimative divergence,

$$K(\mathcal{P}_\omega, \mathcal{P}_{\omega_o}) = E[\log \mathcal{P}_\omega - \log \mathcal{P}_{\omega_o}],$$

where $\mathcal{P}_\omega = \mathcal{P}(\theta|y^{(N)}, \omega)$, is a reasonable diagnostic to consider when it exists and is finite, [see, e.g., Geisser (1985), Johnson and Geisser (1985), McCulloch (1986)], and can be investigated in a variety of paradigms. Similarly for predictive distributions, a predictive divergence

$$K(\omega, \omega_o) = E[\log f_\omega - \log f_{\omega_o}]$$

where $f_\omega = f(x_{(M)}|x^{(N)}, \omega)$, will serve as a reasonble diagnostic. Divergences of this sort were introduced by Johnson and Geisser (1982,1983) for determining influential observations, one of the particular types of perturbation previously mentioned, and were termed predictive influence functions (PIF).

Both, estimative and predictive diagnostics, are most useful in indicating the relative effect of various perturbations. There may be, however, some difficulty in adequately interpreting globally

$$\max_{\omega \in \Omega} K(\mathcal{P}_\omega, \mathcal{P}_{\omega_o}), \text{ or } \max_{\omega \in \Omega} K(\omega, \omega_o)$$

for some of these paradigms.

Another use is to find the direction in which local perturbations have the greatest effect in terms of normal curvature. It can be shown that under suitable regularity conditions that the matrix of second derivatives of $K(\mathcal{P}_\omega, \mathcal{P}_{\omega_o})$ or $K(\omega, \omega_o)$ for $\omega$ a vector of perturbations, say

$$K''_{\omega=\omega_o} = I(\omega_o)$$

where $I(\omega_o)$ is the Fisher Information matrix for either the posterior or predictive distribution at $\omega = \omega_o$, Kullback (1959). The curvature in direction $z$ where $\omega(t) = \omega_o + tz$ and $z'z = 1$ is

$$C_z = z'I(\omega_o)z$$

so that the maximum curvature $C^*$ is in the direction $z^*$, the vector associated with the maximum root of $I(\omega_o)$, where

$$C^* = z^{*'}I(\omega_o)z^*$$

An examination of the components of $z^*$ will indicate which ones, namely the larger ones, are those perturbations which relatively most alter the posterior or predictive distribution in terms of the divergence.

Once potentially significant directions are identified, an analysis involving these directions is in order to ascertain whether local departures for them are important enough to vitiate the standard analysis.

The $L^1$ norm between two densities $f$ and $g$, favored by Devroye (1987), or the $L^2$ norm between $\sqrt{f}$ and $\sqrt{g}$ favored by Pitman (1979) as measures of distance between densities can also be used here as diagnostics for the posterior distribution. More generally the Hellinger distance between densities raised to the $n$-th power

$$H^n = \int |f^{1/n} - g^{1/n}|^n dx$$

yields these as special cases. For the case here with $n = 2$, we have

$$H^2(\mathcal{P}_\omega, \mathcal{P}_{\omega_o}) = \int (\mathcal{P}_\omega^{1/2} - \mathcal{P}_{\omega_o}^{1/2})^2 \quad d\theta$$

for posterior densities and for predictive densities,

$$H^2(\omega, \omega_o) = \int (f_\omega^{1/2} - f_{\omega_o}^{1/2})^2 \quad dx_{(M)},$$

which accords with the norm favored by Pitman. Under suitable smoothness conditions, twice the matrix of second derivatives evaluated at $\omega = \omega_o$ when $\omega$ is a vector, is

$$2H^{2''}_{\omega=\omega_o} = I(\omega_o)$$

again Fisher's Information matrix. More generally, there is a whole class of such distance functions that yield the same Hessian matrix at $\omega = \omega_o$, Burbea and Rao (1984), indicating the wide- ranging usefulness of $I(\omega_o)$ for local perturbation analysis.

The $L^1$ norm may also be used. While it is unaffected by any one-to-one transformation as is the divergence and $H^2$, it is analytically awkward and does not discriminate between differences of the two densities when the smaller of the two is large or small, as does $H^2$ and the divergence.

While the divergence and $H^2$ are as sensible as any measure of how densities differ overall it is difficult to establish a reasonable calibration that different values of the divergence or $H^2$ entail except in a relative sense. Methods for a more suitably direct interpretation that a statistician, and more to the point an investigator, can readily understand can also be defined but they involve rather specific situations. We now present some of these ways of assessing the robustness in terms of posterior or predictive regions for $\theta$ or $X_{(M)}$. One could restrict oneself to perturbations that could matter as determined locally but we shall retain the same notation as before for two reasons. First for convenience in that it is possible that the entire $\omega$ set may matter and secondly in certain instances one may not be specifically interested in a local determination. The potential value of the local analysis is the possibility of restricting the dimension of the vector of perturbations to a small set that can more easily be managed by the assessments we now shall propose.

The first method is to assess the robustness of a $1-\alpha$ highest probability density region based on the standard $\omega_o$. Suppose this region denoted by $R_{1-\alpha}(\omega_o)$ has volume $V(\omega_o)$ and when perturbed the highest probability density region $R_{1-\alpha}(\omega)$ has volume $V(\omega)$. Let $\nu(\omega)$ be the volume of the intersection of $R_{1-\alpha}(\omega)$ and $R_{1-\alpha}(\omega_o)$ as a function of $\omega$,

$$\nu(\omega) = \text{ volume } [R_{1-\alpha}(\omega) \cap R_{1-\alpha}(\omega_o)]$$

and let

$$\Gamma_\omega = \frac{\nu(\omega)}{M(\omega)},$$

where $M(\omega) = \max\,(V(\omega), V(\omega_o))$, be the ratio of the volume of the intersection to whichever is larger the standard or the perturbed for the given $\omega$. Then calculate

$$\min_{\omega \in \Omega} \Gamma_\omega = \Gamma_{\omega^*}$$

which now yields the proportion of the region for the "worst" possible case at a given probability $1-\alpha$. Hence one has an easily interpretable value for assessing the robustness of the data set in terms of a standard analysis involving a $1-\alpha$ region in the presence of presumably anticipated perturbations. A second method focuses on the use of the standard region's $R_{1-\alpha}(\omega_o)$ perturbed probability when $\omega \neq \omega_o$. Here we use either

$$Pr[\theta \in R_{1-\alpha}(\omega_o)|\omega] = \int_{R_{1-\alpha}(\omega_o)} \mathcal{P}(\theta|y^{(N)}, \omega)d\theta = 1 - \alpha_\omega$$

or

$$Pr[X_{(M)} \in R_{1-\alpha(\omega_o)}|\omega] = \int_{R_{1-\alpha(\omega_o)}} f(x_{(M)}|x^{(N)}, \omega)\; dx_{(M)} = 1 - \alpha_\omega$$

and either

$$\max_{\omega \in \Omega} |1 - \alpha - (1 - \alpha_\omega)| = \max_{\omega \in \Omega} |\alpha_\omega - \alpha|$$

or

$$\max_{\omega \in \Omega} \frac{|\alpha_\omega - \alpha|}{1 - \alpha_\omega}$$

as easily interpretable values. This second method is most compelling when some specified region is critical to an analysis, e.g. the effect of the perturbation on the calculation of the probability of an observable exceeding some threshold.

In fact as a very simple illustration of this consider $X^{(N)} = (X_1, \ldots, X_N)$ a random sample from

$$f(x|\theta,\omega) = \theta e^{-\theta(x-\omega)} \qquad x \geq \omega \geq 0$$

and noninformative prior

$$g(\theta) \propto \theta^{-1}.$$

Suppose $x_1, \ldots, x_d$ are fully observed realizations and $X_{d+1}, \ldots, X_N$ are independently censored at values $x_{d+1}, \ldots, x_N$. We further suppose, as is almost always the case, that

$$x_{(1)} = \min(x_1, \ldots, x_d) \leq \min(x_{d+1}, \ldots, x_N)$$

The predictive distribution function, Geisser (1982), is then easily calculated to be

$$Pr[X_{N+1} \leq x|x^{(N)},\omega] = 1 - \left(1 + \frac{x-\omega}{N(\bar{x}-\omega)}\right)^{-d} \quad 0 \leq \omega \leq \min(x, x_{(1)}).$$

Here it is of interest to calculate the probability of a survival threshold say $y$

$$Pr[X_{N+1} > y|x^{(N)},\omega] = \left(1 + \frac{y-\omega}{N(\bar{x}-\omega)}\right)^{-d}$$

where the standard say is $\omega = 0$. Of course the divergence and the $H^n$ distances are largely irrelevant for this case but we can easily calculate

$$\max_{0 \leq \omega \leq x_{(1)}} |\alpha_\omega(y) - \alpha(y)| = \left(1 + \frac{y - x_{(1)}}{N(\bar{x} - x_{(1)})}\right)^{-d} - \left(1 + \frac{y}{N\bar{x}}\right)^{-d}$$

for a fixed $y$ or conversely for those values of $y$ such that the quantity on the right is no larger that a given value considered negligible with respect to stating a probability for surviving the threshold.

As an example consider the following data reported in Gnedenko et al (1969, p. 176) consisting of a sample of N=100 items tested and time to failure recorded for each item until 500 standard time units have elapsed. The recorded failure times for 11 items were: 31, 49, 90, 135, 161, 249, 323, 353, 383, 436, 477. The remaining 89 items survived the test termination time. If interest is focused on the probability of a future item surviving 500 time units then

$$\begin{aligned} Pr[X_{N+1} > 500|\omega] &= \left(1 + \frac{500-\omega}{47,187 - 100\omega}\right)^{-11} \\ &= .891 \quad \text{for } \omega = 0 \\ &= .890 \quad \text{for } \omega = x_{(1)} = 31 \end{aligned}$$

Hence

$$\max_{0\leq\omega\leq 31} |\alpha_{31}(500) - \alpha(500)| = .001$$

On the other hand one might be interested in that value $y$ such that

$$Pr[X_{N+1} > y|\omega] = .5$$

Here for $\omega = 0, y = 3069$ and for $\omega = 31, y = 2898$ yielding a maximum relative difference of 5.6%. In passing we also point out here that the maxima for the divergence and the two norms are $K = \infty, H^2 \doteq H^1 = .01$ and are not particularly informative. The divergence indicates only a difference in support while the norms are approximately and exactly twice the probability assigned to the largest interval over which only one of the densities is supported.

More generally, implementation of these methods in other cases could involve the algebraic or numerical calculation of the intersection of two n-dimensional hyperellipsoids which could be quite burdensome for $n > 3$.

Even more complex situations arise where the highest probability density regions are disconnected. Here one may also want to take into consideration the distance from the standard a perturbed and disconnected region is in ordering the diagnostics discussed above, i.e. not only the size of the non-intersecting disconnected region but its distance in some sense from the standard.

## ACKNOWLEDGMENT

This work was sponsored in part by an NIH grant GMS-25271 and NSF grant DMS-8601314.

## REFERENCES

Box, G.E.P.(1980). Sampling and Bayes' inference in scientific modelling and robustness. *Jour. of the Royal Statistical Society* A, **143**, 383-430.

Burbea, J. and Rao, C.R. (1982). Entropy differential metric, distance and divergence measures in probability spaces: a unified approach. *Jour. of Multivariate Analysis,* **12**, 575-596.

Cook, R.D.(1986). Assessment of local influence (with discussion). *Jour. of the Royal Statistical Society* **B, 48**, 2, 133-169.

Devroye, L. (1987). *A Course in Density Estimation.* Birkhauser.

Geisser, S. (1982) Aspects of predictive and estimative approaches in the determination of probabilities, Biometrics Supplement: *Current Topics in Biostatistics and Epidemiology* **38**, 1, March, 75-85.

Geisser, S. (1985) On the predicting of observables: a selective update, in: Bernardo, J.M. et al. (Ed.) *Bayesian Statistics* **2**, (with discussion) 203- 230. Amsterdam, North-Holland.

Gnedenko, B.B., Belyayev, Y.K., and Solovyev, A.D. (1969). *Mathematical Methods of Reliability Theory.* New York and London: Academic Press.

Johnson, W. and Geisser S. (1982) Assessing the predictive influence of observations, in: G. Kallianpur, P.R. Krishnaiah & J.K. Ghosh (Eds) *Statistics and Probability Essays in Honor of C.R. Rao*, 343-358. Amsterdam, North-Holland.

Johnson, W. and Geisser, S. (1983) A predictive view of the detection and characterization of influential observations in regression analysis, *Journal of American Statistical Association* **78**, 137-144.

Johnson, W. and Geisser, S. (1985) Estimative influence measures for the multivariate general linear model, *Journal of Statistical Planning and Inference* **11**, 33-56.

Kullback, S. (1959). *Information Theory and Statistics.* New York, John Wiley and Sons.

Lavine, M. (1988). Prior influence in Bayesian Statistics. University of Minnesota Technical Report No. 504

McCulloch, R. (1986). Local prior influence. University of Minnesota Technical Report No 477.

Pitman, E.J.G. (1979). *Some Basic Theory for Statistical Influence.* London, Chapman and Hall.

## COMMENTS

Joseph B. Kadane
Carnegie-Mellon University

Seymour Geisser has given us an interesting think-piece on Bayesian robustness. The first thing to say about Bayesian robustness is that the Bayesian paradigm is inherently robust. The user is asked to specify all the possibilities he or she is prepared to entertain, asked to weight them with the probabilities of prior plausibility and then, using Bayes' Theorem, allows the data to influence how those possibilities are to be weighted in the posterior. Thus, a Bayesian would have a prior on Seymour's "$\omega$", thus closing the model and making "robustness" a special case, once again, of the usual Bayesian paradigm.

Second, there is the question of one's attitude toward the ingredients of the Bayesian method: prior, Likelihood and loss. While Geisser takes the first and third to be subjective, not so the second, the likelihood. In fact, he writes, "the likelihood is supposed to represent to some approximation the physical process generating the data." This one is to investigate the likelihood's "adequacy," as though this were not a matter of subjective opinion. With this I disagree. Herman Rubin has been telling us for years that only the product of likelihood and prior matter. Bayarri, DeGroot, and Kadane (1988) use this idea to show that the likelihood itself is not well- defined: factors involving the parameter may be shifted arbitrarily one way or the other. The parameter space itself is subjective. Hence the whole weight of modern Bayesian thought lies on the side that all the ingredients to a Bayesian analysis are subjective (except possibly the data themselves?).

To make these two points is not to take the view that robustness is unimportant in the Bayesian perspective. On the contrary it is very important, but its role needs to be understood in a properly subjective, non-mechanical way.

The importance of a robustness study in a Bayesian analysis is that it can reveal the sensitivity of the analysis to particular features of the subjective specification that may have been assumed too quickly and without appropriate reflection. As such it is part of the elicitation process. The results of a robustness study, then, are fundamentally an appeal to the conscience. Are you aware that the following (seemingly minor) change in your specifications can lead to the following (seemingly major) change in your conclusions? Does this cause you to think again about the loss, prior, and likelihood you tentatively wrote down to begin the analysis? Do you think that your write-up should justify these choices, pointing out the sensitivity so that readers can make their own informal determinations of what assumptions they believe, and what conclusions they embrace?

Geisser also envisions situations in which it is difficult to write down a loss function. However, his suggestion of Kullback-Liebler distance in this context strikes me as the introduction of a loss function nonetheless.

I also found myself wondering about Geisser's use of maximum curvature to indicate directions of greatest sensitivity to prior specifications. But if I am more sure of my prior specifications in some directions than others, the directions of maximum curvature may not be the most important direction of sensitivity to me.

Viewed in this light, Geisser has presented us with a good list of possible ways to think about "close" models. These all amount to particular metrics or topologies, especially with respect to likelihoods, but also with respect to priors and losses.

Some non-Minnesota references in this general area include much of the recent work of Jim Berger (see, for example, his review paper (1984)), the early paper of Edwards, Lindman, and Savage (1963), the book of Box and Tiao (1973), and a series of papers by Hill (see Berliner and Hill (1988) and the references cited there). Also, much recent work on capacities is motivated by Bayesian robustness considerations. Finally, I would mention a recent paper of mine, joint with Tierney and Kass (1989), in which we show how to use Laplacean asymptotics to compute sufficiently the effects of perturbations on posterior expectation.

I welcome Geisser's paper as a summary of his view of this important area.

## REJOINDER

J.B. Kadane makes several points. The first is always to assume a prior for the hyperparameter set $\omega$ and then integrate it out and so be done with the problem. Under certain circumstances, as I indicated, this was recommended. In general, it is difficult enough to imagine and construct a prior for the set $\theta$, but now we are required to conjure up one for the hyperparameter set $\omega$ which will depend on a hyper-hyperparameter set $\tau$. And so it goes on and on never getting off the Kadanian infinite regress of assuming probabilities for entities that are mostly artifacts, intrinsically unverifiable and certainly unobservable.

Undoubtedly the Bayesian model consists of the likelihood and the prior. But in so far as the likelihood depends on observables it has a reasonable degree of objectivity in that very often validation is possible. Kadane, however, prefers Rubinesque Herman-eutics which, by the way, also includes the loss function along with the specification of all possible actions to be taken - all in one tidy package. This approach has never gained credence with users of statistics because of the great burden of information and misinformation required on parameters inarguably incapable of verification which often drives the Bayesian view to the point of absurdity. It would appear that users find this cost too prohibitive. Only the prediction of observables compared with new observations generated from the process can fully be relied on to support the adequacy of a model - although certain sample reuse approaches may also be informative in this regard. The Bayesian game is basically a "what if" game, i.e., we say, "what if this is assumed" then we expect the observations to exhibit certain values as compared to "what if that were assumed" and decide which assumptions are more useful for the purposes at hand.

I have not denied the fact that the divergence is a loss function. I have suggested that when investigator and statistician are at a loss (no pun intended) in that no particular loss function surfaced from the depths of their collective psyches, they consider the divergence as a possibility along with Hellinger distances. And that the $L^2$ norm, the divergence and a variety of other distance functions are locally equivalent.

The use of maximum curvature is merely to disclose the worst possible direction in a multiparameter perturbation case. The investigator may or may not want to be protected against this possibility but should be aware of it. At any rate any direction can be explored.

Lastly, Kadane misinterprets the intent of this short paper. It was not intended to be a review of Bayesian robustness. Here the focus is on a specific parametric model that is probed in various ways to ascertain whether an analysis is robust against particular anticipated expansions of the model. Hence many Bayesian robustness references especially of the nonparametric variety and of course those which were unavailable when this paper was written, were not included. The reader and I are surely grateful to Kadane for remedying this lacuna.

## ADDITIONAL REFERENCES

Bayarri, S.J., M. H. DeGroot and Kadane, J.B. (1988), What is the Likelihood Function? (with discussion), in *Proceedings of the Fourth Purdue Symposium on Decision Theory and Related Topics,* S. Gupta and J. Berger, eds. Springer-Verlag, New York, 3-27.

Berliner, L.M. and B. Hill (1988), Bayesian Nonparametric Survival Analysis, (with discussion), *Journal of the American Statistical Association,* **83**, 772-784.

Berger, J. (1988), The Robust Bayesian Viewpoint, in *Robustness of Bayesian Analysis,* J. Kadane, ed., North-Holland Publishing Company, Amsterdam, 63-144.

Box, G.E.P. and Tiao, G. (1973), *Bayesian Inference in Statistical Analysis,* Addison-Wesley, MA.

Edwards, W., Lindman, H. and Savage, L.J. (1963), Bayesian Statistical Inference, *Psychological Research.*

Tierney, L.J., Kass, R. and Kadane, J.B. (1989), Approximate Methods for Assessing Influence and Sensitivity in Bayesian Analysis, Technical Report No. 430, Department of Statistics. Carnegie-Mellon University, *Biometrika.* To appear.

# 21

# Forecasting Similar Time Series with Bayesian Pooling Methods: Application to Forecasting European Output Growth

## Richard A. Highfield[1]

**Abstract**

It is shown how simple, easily motivated, pooling methods can be used to increase the accuracy of forecasts of similar time series, and how the use of parametrized prior distributions on the linear parameters of the models can provide large further improvements in accuracy. Two principle approaches are discussed: those which pool the data to arrive at similar models for all of the series and those which shrink the forecasts themselves toward a common value. All methods are explicated within the context of a simple Bayesian state-space structure, and issues regarding parameter estimation are discussed. The techniques are applied to forecasting the real output growth of several countries.

*Key Words and Phrases:* Forecasting; Autoregressive models; Leading indicator variables; Structural models, Macroeconomic models; Bayesian shrinkage; Comparison with OECD Forecasts.

### 1. INTRODUCTION

In earlier work Garcia-Ferrer, Highfield, Palm and Zellner (1987) (hereafter referred to as GHPZ) demonstrated that the interdependence among nine industrial economies could be econometrically exploited to produce forecasts of output growth that are superior to those obtained from separate models for the countries. Two basic pooling concepts were examined: ones in which the parameters of the models for the separate countries were related (shrunk toward common values) and one in which the forecasts themselves were averaged. More recently Zellner and Hong (1988) (hereafter called ZH) extended this work on international pooling models to include nine additional countries, more recent data, and more complex

---

[1]Cornell University

models. ZH have shown the basic pooling concept to be quite robust in this way.

In separate work on the U.S. macroeconomic data, Litterman (1980, 1984,1986) and Highfield (1986) have shown that Bayesian approaches to vector autoregression (BVAR's) can result in forecasts from simple models that rival and often surpass the accuracy of the published forecasts from complicated (and judgementally adjusted) econometric models. The improvement in forecast accuracy in BVAR's stems from the imposition of stochastic restrictions on the linear parameters of the model through the use of a statistically motivated prior distribution on these parameters.

One purpose to this paper is to combine these two approaches to forecast accuracy improvement. To do this we describe a simple and flexible framework for the pooling of information in linear models for several similar time series. Typically, the degree to which information is pooled across equations is controlled by a small number of hyperparameters. We provide a simple method for choosing the values of these hyperparameters based on the data available. In a similar way the Bayesian prior distributions also depend on a small number of hyperparameters, and these are also be chosen on the basis of the data. This approach provides a basis for the complete specification of the pooling model prior to forecasting, and thus represents an extension of the results in GHPZ and ZH. We test the approach in a forecasting example. A second purpose to this work is to investigate the sensitivity of our forecasting results to various simplifying assumptions regarding the contemporaneous variance-covariance matrix of the pooling model errors.

The paper proceeds as follows. In section 2 we outline a general multivariate dynamic linear model that can be specialized to include a wide variety of pooling models and discuss the estimation of the hyperparameters on which these models depend. In section 3 four possible pooling models that fit the general framework will be set out. These models are applied to the problem of forecasting output growth in seven European countries in section 4. We show that the application of these simple Bayesian methods leads to substantial improvements in the accuracy of the forecasts. Finally, section 5 contains a summary and concluding remarks.

## 2. GENERAL FRAMEWORK

As a general framework for the pooling models presented here we adopt the well-known form of multivariate linear state-space models:

$$\mathbf{y}_t = \mathbf{Z}_t \boldsymbol{\delta}_t + \mathbf{F}\ \mathbf{u}_t \tag{2.1}$$

$$\boldsymbol{\delta}_t = \mathbf{H}\ \boldsymbol{\delta}_{t-1} + \mathbf{G}\ \mathbf{w}_t \tag{2.2}$$

where $\mathbf{y}_t$ is an $m \times 1$ vector of dependent variables to be forecast. In this pooling model context, the elements of $\mathbf{y}_t$ are measurements of the same conceptual variable across different regimes; growth of GNP in various countries for example. The matrix $\mathbf{Z}_t$ is $m \times s$ and contains predetermined explanatory variables and $\boldsymbol{\delta}_t$ is the $s \times 1$ vector of states. The vectors $\mathbf{u}_t$ and $\mathbf{w}_t$ are disturbance vectors of dimension $m \times 1$ and $q \times 1$, and $\mathbf{H}, \mathbf{F}$ and $\mathbf{G}$ are conformable matrices. The stochastic specification for the disturbance vectors is:

$$\begin{pmatrix} \mathbf{u}_t \\ \mathbf{w}_t \end{pmatrix} \sim \text{NID}\left(\begin{pmatrix} \mathbf{0} \\ \mathbf{0} \end{pmatrix}, \begin{bmatrix} \boldsymbol{\Sigma} & \mathbf{0} \\ \mathbf{0} & \mathbf{W} \end{bmatrix}\right) \tag{2.3}$$

so that the disturbance vectors are not serially correlated and are independent of one another. To complete the stochastic specification we require the distribution of the state vector at time zero:

$$\boldsymbol{\delta}_0 \sim N(\boldsymbol{\delta}^*{}_{0|0}, \Delta_{0|0}) \tag{2.4}$$

where we adopt the notation that $\boldsymbol{\delta}^*{}_{t|s}$ represents the conditional mean of $\boldsymbol{\delta}_t$ given observations on the dependent variable through time period $s$, and $\boldsymbol{\Delta}_{t|s}$ represents the corresponding conditional variance-covariance matrix.

In section 3 we outline several simple pooling models within this general framework. The state vector $\boldsymbol{\delta}_t$ contains the linear parameters that describe the dependence of the $y$'s on the explanatory variables. Typically these parameters have interpretations as regression coefficients. All of the matrices $\boldsymbol{Z}_t, \mathbf{F}, \mathbf{H}, \mathbf{G}, \mathbf{W}$ and $\boldsymbol{\Delta}_{0|0}$ have a known structure but may contain parameters that enter the model in a non-linear way. For this reason any parameters appearing in these matrices will be referred to as hyperparameters. A framework this general is not required for all models implemented here, but it does provide a common basis for thinking about these models and other models contemplated for future work.

Given known hyperparameters, the states can be obtained via the Kalman filter. These well-known recursions are:

$$\boldsymbol{\delta}^*{}_{t|t-1} = \boldsymbol{H}\boldsymbol{\delta}^*{}_{t-1|t-1} \tag{2.5}$$

$$\boldsymbol{\Delta}_{t|t-1} = \mathbf{H}\boldsymbol{\Delta}_{t-1|t-1}\boldsymbol{H}' + \mathbf{G}\mathbf{W}\mathbf{G}' \tag{2.6}$$

$$\boldsymbol{y}^*{}_{t|t-1} = \boldsymbol{Z}_t\boldsymbol{\delta}^*{}_{t|t-1} \tag{2.7}$$

$$\boldsymbol{P}_{t|t-1} = \boldsymbol{Z}_t\boldsymbol{\Delta}_{t|t-1}\boldsymbol{Z}_t' + \mathbf{F}\boldsymbol{\Sigma}\mathbf{F}' \tag{2.8}$$

$$\boldsymbol{\delta}^*{}_{t|t} = \boldsymbol{\delta}^*{}_{t|t-1} + \mathbf{K}_t\mathbf{e}_t \tag{2.9}$$

where $\boldsymbol{e}_t = \boldsymbol{y}_t - \boldsymbol{y}^*{}_{t|t-1}$ and $\mathbf{K}_t = \boldsymbol{\Delta}_{t|t-1}\boldsymbol{Z}_t'\boldsymbol{P}_{t|t-1}^{-1}$

$$\boldsymbol{\Delta}_{t|t} = (\mathbf{I} - \mathbf{K}_t\boldsymbol{Z}_t)\boldsymbol{\Delta}_{t|t-1} \tag{2.10}$$

Equations (2.7) and (2.8) generate the mean and variance-covariance matrix of the predictive distribution for $\boldsymbol{y}_t$ given observations through time period $t-1$, known hyperparameters and known $\boldsymbol{\Sigma}$. This distribution is multivariate normal and not dependent on the states $\boldsymbol{\delta}_t$. From this we see that the one-step-ahead forecast error $\boldsymbol{e}_t$ (the innovation) is normally distributed with zero mean and variance-covariance matrix $\boldsymbol{P}_{t|t-1}$, and the innovation property of the Kalman filter [see Anderson and Moore (1979)] gives the result that $\boldsymbol{e}_t$ is independent of $\boldsymbol{e}_s$ for all $s < t$. Thus the product of these conditional normal distributions for the innovations for $t = 1, ..., T$ can be viewed as the likelihood function for all unknown hyperparameters and $\boldsymbol{\Sigma}$. The logarithm of this likelihood function, after subtracting a constant and multiplying by a factor of 2, has the form:

$$ll\ f(\boldsymbol{\pi}, \boldsymbol{\Sigma}|\boldsymbol{y}_T, \ldots, \boldsymbol{y}_1) = -\left(\sum_{t=1}^{T} ln|\boldsymbol{P}_{t|t-1}| + \sum_{t=1}^{T} \boldsymbol{e}_t'\ \boldsymbol{P}_{t|t-1}^{-1}\ \boldsymbol{e}_t\right) \quad (2.11)$$

where $\boldsymbol{\pi}$ represents a vector of unknown hyperparameters. In principle (2.11) can be maximized to yield estimates of the elements of $\boldsymbol{\pi}$ and $\boldsymbol{\Sigma}$. In practice, however, this can be very difficult in pooling models since there are $m(m+1)/2$ unknown elements in $\boldsymbol{\Sigma}$, and $m$ can be quite large.

In section 3 we develop simple pooling models in which the number of elements in $\boldsymbol{\pi}$ is at most two. These hyperparameters control both the prior distribution on the states and the nature of the pooling of information across equations. To avoid non-linear estimation of the elements of $\boldsymbol{\Sigma}$ either some simplifying assumptions must be made about the structure of this matrix or an alternative estimation method must be used. If the structure of the matrices $\boldsymbol{Z}_t, \mathbf{F}, \mathbf{H}, \mathbf{G}, \mathbf{W}$ and $\boldsymbol{\Delta}_{0|0}$ is such that the innovation covariance matrix $\boldsymbol{P}_{t|t-1}$ has the form $\boldsymbol{P}_{t|t-1} = p_{t|t-1}\boldsymbol{\Sigma}$ for all $t$, where $p_{t|t-1}$ is a scalar, then $\boldsymbol{\Sigma}$ is easily concentrated out of the likelihood function. This will not be the case in the models considered here, because, among other things this would require the explanatory variables in each equation to be identical (as they are, for example, in a VAR).

If we assume that $\boldsymbol{\Sigma} = \sigma^2\mathbf{I}_m$, however, then $\boldsymbol{P}_{t|t-1}$ can be written $\boldsymbol{P}_{t|t-1} = \sigma^2\boldsymbol{P}^*_{t|t-1}$. If we assume a standard diffuse prior distribution on $\sigma, p(\sigma) \propto 1/\sigma$, we can integrate $\sigma$ out of the likelihood function (2.11) to yield:

$$ll\ f(\boldsymbol{\pi}|\boldsymbol{y}_T, \ldots, \boldsymbol{y}_1) = -\left(\sum_{t=1}^{T} ln\ |\boldsymbol{P}^*_{t|t-1}| + T\ m\ ln\left(\sum_{t=1}^{T} \boldsymbol{e}_t'\ \boldsymbol{P}^*_{t|t-1}\boldsymbol{e}_t\right)\right) \quad (2.12)$$

Finally, if we are willing to assume that $\boldsymbol{\Sigma}$ is a diagonal matrix, and if the structure of $\boldsymbol{Z}_t, \mathbf{F}, \mathbf{H}, \mathbf{G}, \mathbf{W}$ and $\boldsymbol{\Delta}_{0|0}$ is such that the $m$ innovation distributions are related only through the common hyperparameters $\boldsymbol{\pi}$, i.e., $\boldsymbol{P}_{t|t-1} = \text{diag}\ \{\sigma_1^2 p_{1t|t-1}, ..., \sigma_m^2 \boldsymbol{P}_{mt|t-1}\}$, then the $m$ equations are stochastically independent and the log likelihood function for the system is simply the sum of the $m$ log likelihood functions for the individual equations:

$$ll\ f_i(\boldsymbol{\pi}|\boldsymbol{y}_T,\ldots,\boldsymbol{y}_1) = -\left(\sum_{t=1}^{T} ln\ p_{it|t-1} + T\ ln\left(\sum_{t=1}^{T}\frac{e_{it}^2}{p_{it|t-1}}\right)\right) \tag{2.13}$$

for $i = 1, ..., m$. It may not always be reasonable to assume that the disturbances in each equation have identical variance or are uncorrelated with one another. If this is the case, one must either estimate the elements of $\boldsymbol{\Sigma}$ jointly with the elements of $\boldsymbol{\pi}$ using (2.11), or appeal to some conditional procedure. A practical procedure will now be described.

As will be discussed in Section 3, the seemingly unrelated regression (SUR) model of Zellner (1962) is one special case of the model described in (2.1) and (2.2). In the standard implementation of this model a two-stage procedure is used. In the first stage each model in the system is estimated via ordinary least squares, and the residuals are used to form an estimate of $\boldsymbol{\Sigma}$ that is used in the second stage. We could proceed in a similar manner here. In the first stage, the Kalman filter recursions are computed for a given $\boldsymbol{\pi}$ vector and under the assumption that $\boldsymbol{\Sigma} = \sigma^2\mathbf{I}_m$. Estimates of the elements of $\boldsymbol{\Sigma}$ are then formed from the estimated residuals obtained with the linear parameters are set equal to their smoothed estimates. If $\boldsymbol{\Sigma}$ is assumed diagonal then the diagonals are estimated from these residuals. At the second stage, (2.11) is computed under the assumption that $\boldsymbol{\Sigma}$ is equal to $\boldsymbol{\Sigma}^*$ and becomes the likelihood function for $\boldsymbol{\pi}$ with known $\boldsymbol{\Sigma}$, i.e.,

$$ll\ f(\boldsymbol{\pi}|\boldsymbol{y}_T,\ldots,\boldsymbol{y}_1,\boldsymbol{\Sigma}=\boldsymbol{\Sigma}^*) = -\left(\sum_{t=1}^{T} ln|P_{t|t-1}| + \sum_{t=1}^{T} e_t'\ P_{t|t-1}^{-1}e_t\right) \tag{2.14}$$

and where

$$\boldsymbol{P}_{t|t-1} = \boldsymbol{Z}_t\boldsymbol{\Delta}_{t|t-1}\boldsymbol{Z}_t' + \mathbf{F}\boldsymbol{\Sigma}^*\mathbf{F}' \tag{2.15}$$

replaces (2.8) in the filter recursions above.

Although only results obtained from the joint estimation of $\boldsymbol{\Sigma}$ and $\boldsymbol{\pi}$ are presented in Section 4 below, some experimentation with the conditional procedure just described has also produced good results.

## 3. POOLING MODELS

### SUR Model

Many different pooling models can be defined within the general framework set out above in (2.1) - (2.4). The first that we discuss is the seemingly unrelated regression model of Zellner (1962). In the SUR model each equation has the form:

$$y_{it} = \mathbf{x}_{it}'\boldsymbol{\delta}_{it} + u_{it} \text{ for } i = 1, ..., m \tag{3.1}$$

and where $\mathbf{x}_{it}$ is a $k \times 1$ vector of explanatory variables (not identical across equations). The SUR model can be written as a system as in equations (2.1) and (2.2) if we make the following definitions:

$$y_t \equiv (y_{1t}, \ldots, y_{mt})'; \mathbf{u}_t \equiv (u_{1t}, \ldots, u_{mt})'; \boldsymbol{\delta}_t \equiv (\boldsymbol{\delta}'_{1t}, \ldots, \boldsymbol{\delta}'_{mt})';$$

$$\boldsymbol{H} \equiv \mathbf{I}_{mk}; \mathbf{F} \equiv \mathbf{I}_m; \mathbf{W} \equiv \mathbf{0}_{mk} \tag{3.2}$$

and

$$\boldsymbol{Z}_t \equiv \mathbf{X}_t \equiv \begin{bmatrix} \mathbf{x}'_{1t} & \mathbf{0}' & \cdot & \cdot & \mathbf{0}' \\ \mathbf{0}' & \mathbf{x}'_{2t} & \mathbf{0}' & \cdot & \cdot \\ \cdot & \mathbf{0}' & \cdot & \cdot & \cdot \\ \cdot & \cdot & \cdot & \cdot & \cdot \\ \cdot & \cdot & \cdot & \cdot & \mathbf{0}' \\ \mathbf{0}' & \cdot & \cdot & \mathbf{0}' & \mathbf{x}'_{mt} \end{bmatrix} \tag{3.3}$$

For the stochastic specification we have:

$$\mathbf{u}_t \sim \text{NID}\ (\mathbf{0}, \boldsymbol{\Sigma}) \tag{3.4}$$

as in (2.3) and we define the variance-covariance matrix of the initial state distribution to be:

$$\boldsymbol{\Delta}_{0|0} \equiv \boldsymbol{\Sigma} \otimes \boldsymbol{A} \tag{3.5}$$

As we have defined this model, the only place for hyperparameters to appear is the $k \times k$ matrix $\boldsymbol{A}$ in (3.5), i.e., they will be parameters of the variance-covariance matrix of the distribution of the initial states. To keep things simple, we will consider only the following minimal parameterization of $\boldsymbol{A}$:

$$\boldsymbol{A} \equiv \tau \mathbf{I}_k; \tau \geq 0 \tag{3.6}$$

Thus, in addition to the elements of $\boldsymbol{\Sigma}$, there is only a single hyperparameter, $\tau$, and it controls the tightness of the prior distribution on the linear parameters. As it stands, this is simply a Bayesian formulation of the standard SUR model, and as $\tau \to \infty$ it becomes the usual diffuse prior SUR model. Since there are no cross-equation restrictions on the linear parameters in the individual equations (except that they have virtually exchangeable prior distributions), the only pooling of information is through the common prior, and if $\boldsymbol{\Sigma}$ is known or can be estimated, in the efficient estimation of the individual $\boldsymbol{\delta}_{it}$'s. Of course, if $\boldsymbol{\Sigma}$ is diagonal, this efficiency gain in the estimation of the states is lost, and this model retains its pooling nature only in the common elements of the prior. Had $\mathbf{W}$ in (2.3) been defined with positive elements on the diagonal, and perhaps some hyperparameters, we would have an SUR model with time-varying parameters.

Estimation of the hyperparameter $\tau$ can be accomplished using (2.13) if $\boldsymbol{\Sigma}$ is diagonal and using (2.12) if $\boldsymbol{\Sigma} = \sigma^2\mathbf{I}$. If $\boldsymbol{\Sigma}$ has nonzero off-diagonal elements, however, either a two-step conditional procedure is required and the conditional log-likelihood given in (2.14) must be used, or the elements of $\boldsymbol{\Sigma}$ must be estimated jointly with the hyperparameters using (2.11).

**Lindley-Smith Model**

The second pooling model is an extension of the first, and is a modified version of the Bayesian shrinkage model due to Lindley and Smith (1972). This model allows for shrinkage in the coefficient estimates in all equations toward a common coefficent vector. Each equation in this model can be written :

$$y_{it} = \mathbf{x}'_{it}(\boldsymbol{\delta}_{it} + \boldsymbol{\theta}_t) + u_{it}; \quad i = 1, \ldots, m \tag{3.7}$$

where $\boldsymbol{\theta}_t$ is a $k \times 1$ coefficient vector common to all equations. We can write this model in the form of (2.1) - (2.4) with the following definitions:

$$\boldsymbol{\delta}_t \equiv (\boldsymbol{\delta}'_{1t}, \ldots, \boldsymbol{\delta}'_{mt}, \boldsymbol{\theta}'_t)' \tag{3.8}$$

$$\boldsymbol{Z}_t \equiv [\mathbf{X}_t \ \mathbf{X}_t \ \mathbf{J}] \ ; \mathbf{J} \equiv [\mathbf{I}_k \ldots \mathbf{I}_k]'; \boldsymbol{H} \equiv \mathbf{I}_{mk+k}$$

$$\mathbf{G} \equiv \mathbf{I}_{mk+k} \ ; \mathbf{F} \equiv \mathbf{I}_m \ ; \mathbf{W} \equiv \mathbf{0}_{mk+k} \tag{3.9}$$

so that $\boldsymbol{Z}_t$ is an $m \times (mk + k)$ matrix, $\mathbf{J}$ is $mk \times k$, and $\mathbf{X}_t$ is the $m \times mk$ matrix of explanatory variables defined in (3.3). The stochastic specification of $\mathbf{u}_t$ is as in (3.4) and the variance-covariance matrix of the initial state distribution is:

$$\boldsymbol{\Delta}_{0|0} \equiv \begin{bmatrix} \boldsymbol{\Sigma} \otimes \mathbf{A}_1 & \mathbf{0} \\ \mathbf{0} & \mathbf{A}_2 \end{bmatrix} \tag{3.10}$$

and again we make a simple parameterization of the $\mathbf{A}$ matrices:

$$\mathbf{A}_1 \equiv \tau_1 \mathbf{I}_{mk} \ ; \mathbf{A}_2 \equiv \tau_2 \mathbf{I}_k. \tag{3.11}$$

The Lindley-Smith model is closely related to the SUR model, and has some interesting features. The first is that it reduces to SUR when $\tau_2 = 0$ and the prior mean on the common coefficients $\boldsymbol{\theta}_0$ is zero. This is true regardless of any positive value for $\tau_1$. If the prior mean on $\boldsymbol{\theta}_0$ is nonzero, this prior mean becomes a fixed shrinkage point. As $\tau_1 \to \infty$ the model reduces to the diffuse prior SUR model regardless of the prior on $\boldsymbol{\theta}_0$. If, on the other hand, $\tau_1 = 0$ and the prior mean on the individual equation coefficients is zero, this model reduces to one big pooled regression in which the coefficients in all equations are constrained to be equal. Thus this model nests both the SUR model and the "big regression" model that GHPZ and found to yield good forecasting results. The explicit pooling of information across equations in this model, is therefore completely determined by the prior distributions on the states, although information in all equations is also used in the estimation of the common hyperparameters $\tau_1$ and $\tau_2$.

Like the SUR model, if we assume $\boldsymbol{\Sigma} = \sigma^2\mathbf{I}$ we can use equation (2.12) for the estimation of the hyperparameters. If $\boldsymbol{\Sigma}$ has nonzero off-diagonal elements, however, it does not conveniently factor out of the likelihood function, and a conditional method of estimation or full ML estimation of $\boldsymbol{\Sigma}$ is required.

## Forecast Averaging Model

Our third model is designed to capture the idea that, quite apart from the values of the coefficients across models, the best forecasts are likely to be very similar in magnitude. It is in the same spirit as the $\eta$-shrinkage model employed with great sucess by GHPZ and ZH. Again, it can be viewed as a modification of the SUR model. Each equation in this model has the form:

$$y_{it} = \eta y_t^\dagger + (1-\eta)\mathbf{x}_{it}{}'\boldsymbol{\delta}_{it} + u_{it} \ ; \ 0 \le \eta < 1 \ ; \ i = 1, ..., m \tag{3.12}$$

where

$$y_t^\dagger = \frac{\sum_{i=1}^m y_{it}}{m}. \tag{3.13}$$

In this system, the mean of the $y$'s represents a common variable in all of the equations and might be regarded as a "world" measure of $y_t$. Viewed in this way (3.12) is a variant of a model given in ZH in which the median of the $y_{it}$'s was included on the right-hand side of all equations and an additional equation was included to provide the required forecast of the median. Here we eliminate the need for the additional equation by writing the system (3.12) in its reduced form:

$$\boldsymbol{y}_t = (1-\eta)\mathbf{R}\mathbf{X}_t\boldsymbol{\delta}_t + \mathbf{R}\mathbf{u}_t \tag{3.14}$$

where

$$\mathbf{R} = \left[\ \mathbf{I}_m - \tfrac{\eta}{m}\boldsymbol{\Gamma}\ \right]^{-1} \tag{3.15}$$

and $\boldsymbol{\Gamma}$ is an $m \times m$ matrix of ones. This model is put into the form of (2.1) - (2.4) by defining all relevant matrices the same as in the SUR model, with the following exceptions:

$$\boldsymbol{Z}_t \equiv (1-\eta)\mathbf{R}\mathbf{X}_t \ ; \ \mathbf{F} \equiv \mathbf{R} \tag{3.16}$$

Thus $\eta$ controls the correspondence between forecasts of $y_{it}$ and forecasts of $\bar{y}_t$. When $\eta = 0$, we again have the SUR model. The mean of the predictive distribution for each equation from equation (2.7) above has the following form:

$$y^*_{it|t-1} = (1-\eta)\mathbf{x}'_{it}\boldsymbol{\delta}^*_{it|t-1} + \frac{\eta}{m}\sum_{j=1}^m \mathbf{x}'_{jt}\boldsymbol{\delta}^*_{jt|t-1} \tag{3.17}$$

Thus each forecast is an average of information in its own equation and information in the others. As $\eta$ tends to one, we get the same forecast for all equations.

The hyperparameters in this model are thus $\eta$ and the common tightness parameter for the prior, $\tau$. As in the Lindley-Smith model, either a two-stage procedure or joint estimation of $\boldsymbol{\Sigma}$ and the hyperparameters using (2.11) must be employed when $\boldsymbol{\Sigma} \neq \sigma^2\mathbf{I}$.

## Multivariate Regression Model

Our final model is the familiar multivariate regression model in which several different dependent variables are regressed on the same set of regressors. Let the vector of dependent variables be the $m \times 1$ vector

$$\boldsymbol{y}_t \equiv (y_{1t}, ..., y_{mt})' \tag{3.18}$$

where, as above, each $y_{it}$ represents a measurement on some economic quantity for regime $i$. Let the vector of common explanatory variables be the $(mk+1) \times 1$ vector

$$\mathbf{z}_t \equiv (1, \mathbf{x}'_{1t}, ..., \mathbf{x}'_{mt})' \tag{3.19}$$

where $\mathbf{x}_{it}$ is a $k \times 1$ vector of explanatory variables from regime $i$, perhaps including lags of $y_{it}$. In this way we relate each dependent variable $y_{it}$ to a vector of "own" variables $\mathbf{x}_{it}$ and $m-1$ vectors of "foreign" variables, $\mathbf{x}_{jt}$ where $j \neq i$. Of course, if all of the explanatory variables are lags of the dependent variables, this system is simply a vector autoregression. The system can be put into the framework of (2.1) to (2.4) by defining $\mathbf{u}_t$, $\mathbf{H}, \mathbf{F}, \mathbf{G}$ and $\mathbf{W}$ exactly as is done in (3.3) for the SUR Model, but with the following definitions for the data matrix $\boldsymbol{Z}_t$ and the variance-covariance matrix for the initial state vector. For $\boldsymbol{Z}_t$ we define:

$$\boldsymbol{Z}_t \equiv \mathbf{I}_m \otimes \mathbf{z}'_t \tag{3.20}$$

where $\mathbf{z}_t$ is defined in (3.19). For $\boldsymbol{\Delta}_{0|0}$ we will use a specification that allows for different prior tightness on the coefficients of "own" variables and "foreign" variables:

$$\boldsymbol{\Delta}_{0|0} \equiv \text{diag}\{\sigma_1^2 \mathbf{A}_1, ..., \sigma_m^2 \mathbf{A}_m\} \tag{3.21}$$

where $\sigma_i^2$ is the $i$'th diagonal element of $\boldsymbol{\Sigma}$ and we adopt the following simple parameterization of the $\mathbf{A}$ matrices:

$$\mathbf{A}_j \equiv \tau \cdot \text{diag}\{1, \alpha \cdot \mathbf{1}'_{k(j-1)}, \mathbf{1}'_k, \alpha \cdot \mathbf{1}'_{k(m-j)}\} \; ; \; j = 1, ..., m \; ; \; \tau \geq 0 \; ; \; \alpha \geq 0 \tag{3.22}$$

where $\mathbf{1}_q$ represents a $q \times 1$ unit vector. This parameterization is quite similar to that recommended by Litterman(1980) for vector autoregressions. The first hyperparameter, $\tau$, controls the overall tightness of the prior on $\boldsymbol{\delta}_0$ and the second, $\alpha$, controls the relative tightness of the prior on the elements of $\boldsymbol{\delta}_0$ that relate to the influence of "foreign" variables in each equation. If $\alpha =$ 0 and the prior means on the elements of $\boldsymbol{\delta}_0$ that relate foreign variables are all zero, this model reduces to an equation by equation Bayesian regression of each dependent variable on its "own" explanatory variables.

As it has just been defined, the Multivariate Regression model is not adequate for handling the case of a full $\boldsymbol{\Sigma}$ matrix. The hope is that the

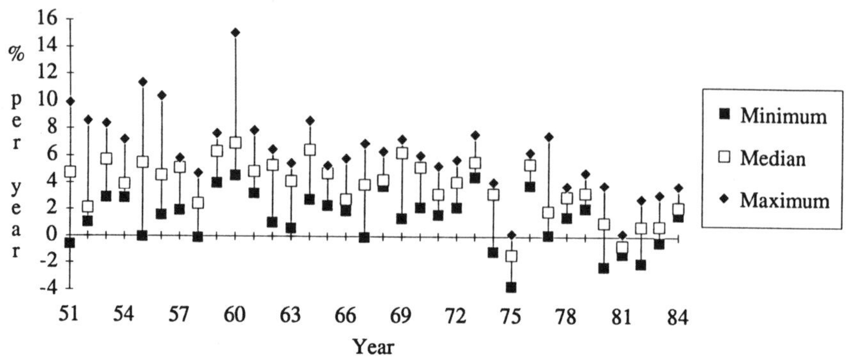


Figure 1: Output Growth in Seven European Countries

inclusion of the "foreign" variables in each equation will account for all common movements among the elements of $\boldsymbol{y}_t$. When estimating the hyperparameters $\tau$ and $\alpha$, the log-likelihood given in (2.12) is employed if we assume $\boldsymbol{\Sigma} = \sigma^2 \mathbf{I}_m$, and (2.13) is used if it is assumed that $\boldsymbol{\Sigma}$ is diagonal.

## 4. FORECASTING RESULTS

In this section we apply the four pooling methods just described to the problem of forecasting annual real output growth rates in seven European countries: Belgium, Denmark, France, Germany, Italy, The Netherlands and the United Kingdom. The data on output and the other variables discussed below are available, in the main, in the International Monetary Fund's International Financial Statistics data base (IMF/IFS), and are the same data used by Zellner and Hong (1988) and were provided by them. The time period covered is 1951 - 1984. Real output growth is defined as the year to year differences in the natural logarithm of GNP (or GDP) for each country, multiplied by 100. The minimum, median and maximum output growth rates are plotted for the years in our sample in Figure 1. As can be seen, the data series do indeed seem similar in nature, reflecting the interdependence of these economies.

Our approach is to use the techniques discussed in Section 2 to estimate the hyperparameters of the various pooling models, and elements of $\boldsymbol{\Sigma}$ when necessary, using a subset of the data (1951 - 1979). We then proceed to forecast the remaining five years of data one year at a time, updating the linear parameters of the model each year via the Kalman filter with the hyperparameters, and $\boldsymbol{\Sigma}$ elements when necessary, set equal to their estimates. In this way we are able to evaluate the out-of-sample performance of the models that would have been achieved had our procedure been applied in real time. This represents an extension of the work of GHPZ and ZH, in which the pooling parameters were not chosen ex ante.

All of the pooling models employed in this section are based on an output growth forecasting model found to provide good accuracy for nine countries by GHPZ. In this model the explanatory variables for each country are a constant term, three lags of the output growth rate being forecast (g), two lags of the growth rate of that country's share price index in real terms (SR), one lag of that country's money supply growth in real terms (GM), and one lag of the median real share price growth of the seven countries (WR). [2]

This model, called the AR(3)LI model by ZH, thus includes leading indicator variables to increase the accuracy of the output growth forecasts, and the median share price growth variable helps pick up common effects across countries. In the notation of section 3, at each time period the vector of explanatory variables for each country is:

$$\mathbf{x}'_{it} \equiv (1, g_{it-1}, g_{it-2}, g_{it-3}, SR_{it-1}, SR_{it-2}, GM_{it-1}, WR_{t-1}) \ ; \ i = 1, ..., m \tag{4.1}$$

and the dependent variable is:

$$y_{it} = g_{it} \ ; \ i = 1, ..., m \tag{4.2}$$

This specification is maintained in three of the four pooling models, so that the focus is on the pooling results. In the multivariate regression model, however, we must specify the same independent variables in all equations. In all equations the independent variables are three lags of real output growth for all seven countries, two lags of the real share price growth for all countries and one lag of real money supply growth for all countries, plus a constant term. Thus we have forty-three right-hand-side variables in each equation. The lagged median real share price growth of the seven countries is dropped as an explanatory variable, since all equations include all lagged share price growth measurements on the right hand side.

The pooling models discussed in Section 3 were estimated under both diffuse and proper priors on the distribution of the initial states $\boldsymbol{\delta}_0$ and under several alternative assumptions regarding the variance-covariance matrix $\boldsymbol{\Sigma}$. This is to evaluate the benefits from pooling that may exist apart from the use of Bayesian prior distributions on the linear parameters as well as the effect of possible assumptions regarding $\boldsymbol{\Sigma}$. Our five alternative assumptions regarding $\boldsymbol{\Sigma}$, in increasing order of the number of parameters introduced, are:

---

[2] The nominal share price indices and money supply figures were deflated by the individual countries' price indices. SR and WR are the differences of the logarithms of these deflated series multiplied by 100.

**CZ:** common variance, zero correlation, i.e., $\boldsymbol{\Sigma} = \sigma^2 \, \mathbf{I}_m$.

**CC:** common variance, common correlation, i.e., $\boldsymbol{\Sigma} = \sigma^2 \Psi$, where

$$\Psi = (1 - \rho)\mathbf{I}_m + \rho \mathbf{J}_{mxm}, \tag{4.3}$$

with $\mathbf{J}$ being a matrix of all 1's.

**SZ:** separate variances, zero correlation, i.e., $\boldsymbol{\Sigma}$ is diagonal.

**SC:** separate variances, common correlation, i.e., $\boldsymbol{\Sigma} = \mathbf{S}^{1/2}\Psi\mathbf{S}^{1/2}$ where $\mathbf{S}$ is diagonal and $\Psi$ is defined as in (4.3).

**SS:** separate variances, separate correlations, i.e., $\boldsymbol{\Sigma}$ is a full pds matrix with no further restriction.

The out-of-sample forecasting results obtained under diffuse priors for the elements of $\boldsymbol{\delta}_0$ are given in Table 1. Diffuse priors in this context simply involves setting the $\tau$ parameters of the models equal to infinity. [3] The SUR model reduces to equation by equation ordinary least squares under diffuse priors and $\boldsymbol{\Sigma}$ assumptions CZ or SZ, as does the Lindley-Smith model. For contrast, therefore, the Lindley-Smith model was computed for $\tau_1 = 0, \tau_2 = \infty$. In this way the Lindley-Smith model is "one big regression" in which the linear parameters in all equations are constrained to be equal. Thus the only hyperparameters to be estimated are $\alpha$ in the Multivariate Regression Model and $\eta$ in the Forecast Averaging Model. Parameter estimates for all models are provided in Table 2. The parameters and, when required, elements of $\boldsymbol{\Sigma}$ were estimated using the 1951-1979 data by maximizing the relevant likelihood function via a Quasi-Newton method, and can thus be interpreted as posterior mode estimates under diffuse priors. After estimating the values for the hyperparameters and $\boldsymbol{\Sigma}$ elements, the 1980-1984 period was forecast recursively, one-year-ahead, and the root mean squared errors of these out-of-sample forecasts are given in Table 1.

For comparison, Table 1 also includes the RMSE achieved by three "naive" forecasting models; In Naive Model I output growth is always forecast to be zero. This forecast is, of course, the one implied by the prior mean specification in all four of the pooling models attempted (i.e., $\boldsymbol{\delta}^*_{0|0} = 0$). Naive Model II forecasts next year's growth to be the same as this year, and Naive Model III forecasts next year's growth to be equal to the historical average (to date) for each country. Finally, Table 1 also includes the comparable forecasts published for these countries by the OECD in

---

[3]The exception is the multivariate regression model in which $\tau$ is set equal to a large finite number (one million). This is to make this model computable given the large number of parameters and small number of observations, and to identify $\alpha$.

*OECD Economic Outlook.* Thus we can compare the results from our models, which are relatively simple and mechanical, with those obtained from presumeably more complex and judgemental analysis and a much larger information set.

In Part 1 of Table 1 we examine our for models under the CZ assumption, i.e., $\boldsymbol{\Sigma} = \sigma^2\mathbf{I}$. The benefits of pooling can be seen in relation to the ordinary least squares result provided by the SUR model. The Lindley-Smith model with $\tau_1 = 0$ resulted in lower RMSEs in six of the seven countries, and often by a large margin. On average the reduction in RMSE is about 18 percent and the reduction in median RMSE is even larger. Further, the Lindley- Smith Model resulted in a tighter distribution of RMSEs across countries. These results are completely consistent with the GHPZ and ZH results in which the "big regression model" was shown to produce significant forecast accuracy improvements ex post. The Forecast Averaging model shows similar improvement over the SUR model. The optimal value for $\eta$ is about 0.4 which results in forecasts that are more heavily weighted toward "own-equation" information. Our fourth model, the Multivariate Regression Model with a estimated to about 0.3, does particularly poorly. This is not a surprising result when you consider that there were only twenty-six time periods in the hyperparameter estimation sample, and each equation in this model has forty- three coefficients. Indeed, it is only the use of a spread-out proper prior that allowed estimation of this model. Finally, it must be admitted that none of the models achieve average results as good as the Naive I forecast or the OECD results, and the discrepancy is quite large.

The remaining parts of Table 1 give the forecasting results achieved under the other four $\boldsymbol{\Sigma}$ assumptions. For the SUR model we see the result that forecast accuracy generally degrades as more $\boldsymbol{\Sigma}$ elements must be estimated under the CC, SC and SS assumptions. Under diffuse priors, therefore, it seems that the simplifying assumptions embodied in CZ and SZ are useful in achieving greater precision. The results for the Lindley-Smith model, however, seem less sensitive to the $\boldsymbol{\Sigma}$ assumption. Indeed, the lowest mean and median RMSEs (although only by about 5%) are obtained under the SS assumption which required the estimation of the entire $\boldsymbol{\Sigma}$ matrix. This might follow from the fact that when $\tau_1 = 0$, the Lindley-Smith model has far fewer linear parameters than any of the other models. Finally, the Forecast Averaging Model also achieves some improvement in accuracy, again not large, under all of the less restrictive $\boldsymbol{\Sigma}$ assumptions other than SS, and approaches the accuracy of Naive Model I. Under assumptions CC, SZ, and SC we also see higher estimated values for $\eta$, thus weighting the forecasts from this model more heavily toward the grand mean forecast than is the case under the CZ assumption. When the entire $\boldsymbol{\Sigma}$ matrix must be estimated, however, accuracy suffers. As in the SUR model, this is probably due to overparameterization. In summary, none of the models yet compare favorably with the OECD forecasts, although the Lindley-Smith and Forecast Averaging methods represent a move in the right direction.

In Table 3 we present the forecast accuracy obtained when all of the $\tau$ parameters are also estimated, resulting in conditional proper prior distributions on the linear parameters of the models. [4] In all cases the priors implied by the optimal $\tau$ values are quite "tight", with the largest estimated value in any of the models being less than .5, and most much smaller, often by several orders of magnitude. Looking at Part 1 of the Table, for the results under the simple CZ assumption, we see that the effect of the tight priors is quite dramatic. Under this $\boldsymbol{\Sigma}$ assumption the median RMSEs achieved by the SUR, Forecast Averaging and Multivariate Regression models is virtually halved in comparison to those achieved under diffuse priors. Indeed, even the model most profligate with linear parameters, the Multivariate Regression Model, achieved mean and median RMSE results better (albeit very slightly) than the best results in Table 1. All four models are competitive with or superior to Naive Model I, and the Forecast Averaging model compares quite favorably with the OECD forecast – having lower median RMSE, but more variability across countries. In the other parts of Table 3 we see, unlike the diffuse prior case, that the accuracy of the models either improves (SUR and Lindley-Smith), or degrades little (Forecast Averaging) as we move to less restrictive assumptions regarding $\boldsymbol{\Sigma}$. We also see a reversal of one of the results of Table 1. Here the SUR model does not seem overparameterized and inferior in accuracy. In Table 3, the SUR model is often more accurate than Lindley-Smith, and is often not far from the performance of the Forecast Averaging Model.

Although the results in Table 3 suggest that there is little to be lost and, for some models, much to be gained from the use of less restrictive $\boldsymbol{\Sigma}$ assumptions, it must be noted that this statement ignores computational ease and cost. While our typical proper prior model with the CZ or CC assumption required on the order of 10 CPU minutes to estimate on a VAX 8530, the typical full $\boldsymbol{\Sigma}$ model required 2.5 to 3.5 CPU hours. In contrast, however, the proper prior models generally required no more than one half to one third of the time required to estimate the corresponding diffuse prior model. Thus, with the Bayesian approach, we gain advantages in both accuracy and cost.

As a final comparison, we include in Table 4 the results obtained for the simplest versions of our models (i.e. under the CZ assumption) and those obtained by ZH for the forecast period 1974-84. [5] The Zellner-Hong (1987) results are excerpted from their tables in which they present results for eighteen countries of which our seven are a subset. The ZH results,

---

[4] In all models were assume a zero prior mean vector for the linear parameters, i.e., we assume $\boldsymbol{\Delta}^*0|0 = \mathbf{0}$.

[5] We applied the identical procedures for our models discussed above, but our parameter estimation period was 1951-1973, and our forecast evaluation period 1974-1984.

therefore, represent forecasts from models which pool eighteen countries, much more information than we employ. Further, for their $\eta$-forecasts and their $\gamma$-forecasts [6] , we have excerpted the lowest RMSEs for each of the seven countries, essentially choosing the optimal parameter values ex post and "stacking the deck" in favor of the ZH models. What we see in Table 4 is that all four of our models outperform the ZH $\eta$- and $\gamma$-forecasting models in terms of average and median RMSE even though our parameters were estimated ex ante. The advantage that our models gain, therefore, must be attributed to the use of our parameterized prior distributions. The ZH world growth rate model, which is non-Bayesian in nature, does quite well and achieves results very similar to those achieved by our Forecast Averaging Model. As noted above, our model can be seen as a reduced form version of the ZH world growth rate in which the average growth rate, rather than the median used by ZH, is included in each country's equation as an explanatory variable. Whether the success of the ZH model is due to the additional countries employed or to the use of a nonparametric measure of the world growth rate, is a matter for future research.

## 5. SUMMARY AND CONCLUDING REMARKS

We have developed a framework for a wide class of pooling models that depend on a small number of hyperparameters and have outlined a straightforward method for choosing the values of those hyperparameters. We have shown that some common pooling models, such as SUR, Lindley-Smith and "one big regression" fit into this framework and have provided four specific example models that use the idea of information pooling to estimate hyperparameters, to compute models with cross-equation parameter restrictions and/or forecast averaging restrictions, and that employ Bayesian prior distributions on the linear parameters of the models. These models and methods have been applied to the forecasting of output growth rates in seven European countries in a simulate "real world" example in which all hyperparameters of the models were estimated using an estimation sample prior to generating and evaluating out-of-sample predictions. The results are instructive, but not (to this Bayesian) surprising. Although classical (diffuse prior) pooling models are shown to provide increases in forecast accuracy relative to a simple single equation approach, the improvements were greatly increased when simple Bayesian prior distributions are employed.

Extensions of this work will include investigation of time-varying parameters (which naturally fit into the framework above when the matrix $\mathbf{W}$ in equation (2.3) is non-zero) and pooling in non-linear time series models such as ARIMA models, which also fit the framework quite conveniently.

---

[6] See the notes to Table 4 for a brief description of the ZH models.

TABLE 1
RMSEs of One-Year-Ahead Forecasts, 1980-1984
with diffuse priors on coefficients

| MODEL | BEL | DEN | FRN | GRM | ITL | NET | UK | AVG | MED | RANGE |
|---|---|---|---|---|---|---|---|---|---|---|
| Naive I[1] | 1.75 | 2.35 | 1.28 | 1.62 | 2.08 | 1.30 | 2.12 | 1.79 | 1.75 | 1.28 - 2.35 |
| Naive II[2] | 2.28 | 2.56 | 1.51 | 1.80 | 2.17 | 1.73 | 2.59 | 2.09 | 2.17 | 1.51 - 2.59 |
| Naive III[3] | 2.97 | 2.67 | 3.50 | 4.33 | 4.00 | 4.34 | 2.82 | 3.52 | 3.50 | 2.67 - 4.34 |
| OECD f'casts | 1.19 | 2.05 | 0.96 | 1.28 | 1.25 | 1.04 | 1.26 | 1.29 | 1.25 | 0.96 - 2.05 |
| **1. Common Variance, Zero Correlation (CZ)** | | | | | | | | | | |
| SUR | 1.04 | 2.69 | 2.23 | 2.72 | 2.85 | 2.71 | 3.03 | 2.47 | 2.71 | 1.04 - 3.03 |
| Lindley-Smith | 1.32 | 2.07 | 1.93 | 1.91 | 1.59 | 2.64 | 2.71 | 2.02 | 1.93 | 1.32 - 2.71 |
| Forecast Ave. | 0.70 | 2.46 | 1.80 | 2.15 | 2.51 | 2.36 | 2.53 | 2.07 | 2.36 | 0.70 - 2.53 |
| Multivariate | 2.93 | 2.94 | 2.00 | 5.85 | 4.46 | 4.53 | 2.15 | 3.55 | 2.94 | 2.00 - 5.85 |
| **2. Common Variance, Common Correlation (CC)** | | | | | | | | | | |
| SUR | 1.41 | 2.75 | 2.75 | 2.87 | 3.65 | 2.84 | 3.04 | 2.76 | 2.84 | 1.41 - 3.65 |
| Lindley-Smith | 1.50 | 1.96 | 1.89 | 1.99 | 1.86 | 2.71 | 2.72 | 2.09 | 1.96 | 1.50 - 2.72 |
| Forecast Ave. | 0.67 | 2.30 | 1.60 | 2.03 | 2.30 | 2.30 | 2.42 | 1.94 | 2.30 | 0.67 - 2.42 |
| **3. Separate Variances, Zero Correlation (SZ)** | | | | | | | | | | |
| SUR | 1.04 | 2.69 | 2.23 | 2.72 | 2.85 | 2.71 | 3.03 | 2.47 | 2.71 | 1.04 - 3.03 |
| Lindley-Smith | 1.30 | 2.06 | 1.92 | 1.90 | 1.58 | 2.60 | 2.69 | 2.01 | 1.92 | 1.30 - 2.69 |
| Forecast Ave. | 0.62 | 2.40 | 1.72 | 1.97 | 2.17 | 2.22 | 2.33 | 1.92 | 2.17 | 0.62 - 2.40 |
| Multivariate | 2.97 | 2.95 | 1.98 | 5.85 | 4.44 | 4.53 | 2.16 | 3.55 | 2.97 | 1.98 - 5.85 |
| **4. Separate Variances, Common Correlation (SC)** | | | | | | | | | | |
| SUR | 1.49 | 2.78 | 2.61 | 2.94 | 3.93 | 2.95 | 3.04 | 2.82 | 2.94 | 1.49 - 3.93 |
| Lindley-Smith | 1.54 | 1.98 | 1.87 | 2.01 | 1.89 | 2.64 | 2.73 | 2.09 | 1.98 | 1.54 - 2.73 |
| Forecast Ave. | 0.70 | 2.00 | 1.55 | 1.88 | 1.88 | 2.16 | 2.21 | 1.77 | 1.88 | 0.70 - 2.21 |
| **5. Separate Variances, Separate Correlations (SS)** | | | | | | | | | | |
| SUR | 3.68 | 3.77 | 3.60 | 3.70 | 4.30 | 3.90 | 3.33 | 3.75 | 3.70 | 3.33 - 4.70 |
| Lindley-Smith | 1.28 | 1.86 | 1.70 | 1.80 | 1.64 | 2.50 | 2.53 | 1.90 | 1.80 | 1.28 - 2.53 |
| Forecast Ave. | 1.31 | 2.85 | 2.54 | 2.56 | 3.04 | 3.53 | 2.86 | 2.67 | 2.85 | 1.31 - 3.53 |

[1] Zero Growth [2] Same Growth [3] Past Average

TABLE 2
Values of Estimated Hyperparameters
Estimation Period: 1951 - 1979

| MODEL | Diffuse Priors | Proper Priors |
|---|---|---|
| **1. Common Variance, Zero Correlation (CZ)** | | |
| SUR | $\tau = \infty$ | $\tau = 0.01269$ |
| Lindley-Smith | $\tau_1 = 0, \tau_2 = \infty$ | $\tau_1 = .0001732$ , $\tau_2 = .1449$ |
| Forecast Ave. | $\tau = \infty$ , $\eta = .4107$ | $\tau = .02825$ , $\eta = .6806$ |
| Multivariate | $\tau = 10^6$ , $\alpha = .3107$ | $\tau = .01574$ , $\alpha = .009527$ |
| **2. Common Variance, Common Correlation (CC)** | | |
| SUR | $\tau = \infty$ | $\tau = .0004862$ |
| Lindley-Smith | $\tau_1 = 0$ , $\tau_2 = \infty$ | $\tau_1 = .0001509$ , $\tau_2 = .2701$ |
| Forecast Ave. | $\tau = \infty$ , $\eta = .4753$ | $\tau = .02942$ , $\eta = .6849$ |
| **3. Separate Variances, Zero Correlation (SZ)** | | |
| SUR | $\tau = \infty$ | $\tau = .01384$ |
| Lindley-Smith | $\tau_1 = 0$ , $\tau_2 = \infty$ | $\tau_1 = .0001678$ , $\tau_2 = .4169$ |
| Forecast Ave. | $\tau = \infty$ , $\eta = .5269$ | $\tau = .03880$ , $\eta = .7002$ |
| Multivariate | $\tau = 10^6$ , $\alpha = .2781$ | $\tau = .01781$ , $\alpha = .009684$ |
| **4. Separate Variances, Common Correlation (SC)** | | |
| SUR | $\tau = \infty$ | $\tau = .0002322$ |
| Lindley-Smith | $\tau_1 = 0$ , $\tau_2 = \infty$ | $\tau_1 = .001304$ , $\tau_2 = .05340$ |
| Forecast Ave. | $\tau = \infty$ , $\eta = .6080$ | $\tau = .07421$ , $\eta = .7616$ |
| **5. Separate Variances, Separate Correlations (SS)** | | |
| SUR | $\tau = \infty$ | $\tau = .0002855$ |
| Lindley-Smith | $\tau_1 = 0$ , $\tau_2 = \infty$ | $\tau_1 = .00009383$ , $\tau_2 = .04290$ |
| Forecast Ave. | $\tau = \infty$ , $\eta = .4713$ | $\tau = .05855$ , $\eta = .8119$ |

TABLE 3
RMSEs of One-Year-Ahead Forecasts, 1980-1984
with proper priors on coefficients

| MODEL | BEL | DEN | FRN | GRM | ITL | NET | UK | AVG | MED | RANGE |
|---|---|---|---|---|---|---|---|---|---|---|
| Naive I[1] | 1.75 | 2.35 | 1.28 | 1.62 | 2.08 | 1.30 | 2.12 | 1.79 | 1.75 | 1.28 - 2.35 |
| Naive II[2] | 2.28 | 2.56 | 1.51 | 1.80 | 2.17 | 1.73 | 2.59 | 2.09 | 2.17 | 1.51 - 2.59 |
| Naive III[3] | 2.97 | 2.67 | 3.50 | 4.33 | 4.00 | 4.34 | 2.82 | 3.52 | 3.50 | 2.67 - 4.34 |
| OECD f'casts | 1.19 | 2.05 | 0.96 | 1.28 | 1.25 | 1.04 | 1.26 | 1.29 | 1.25 | 0.96 - 2.05 |
| **1. Common Variance, Zero Correlation (CZ)** | | | | | | | | | | |
| SUR | 1.07 | 2.81 | 1.13 | 1.37 | 0.56 | 2.19 | 2.14 | 1.61 | 1.37 | 0.56 - 2.81 |
| Lindley-Smith | 1.18 | 2.43 | 1.59 | 1.78 | 1.36 | 2.60 | 2.39 | 1.90 | 1.78 | 1.18 - 2.60 |
| Forecast Ave. | 1.05 | 2.42 | 0.89 | 1.14 | 0.61 | 1.68 | 2.00 | 1.40 | 1.14 | 0.61 - 2.42 |
| Multivariate | 1.45 | 2.45 | 1.30 | 1.30 | 1.39 | 2.61 | 1.69 | 1.74 | 1.45 | 1.30 - 2.61 |
| **2. Common Variance, Common Correlation (CC)** | | | | | | | | | | |
| SUR | 1.39 | 2.46 | 0.65 | 0.95 | 1.11 | 1.08 | 1.86 | 1.38 | 1.11 | 0.65 - 2.46 |
| Lindley-Smith | 1.16 | 2.14 | 1.29 | 1.63 | 1.46 | 2.33 | 2.30 | 1.76 | 1.63 | 1.16 - 2.33 |
| Forecast Ave. | 1.04 | 2.40 | 0.90 | 1.14 | 0.61 | 1.69 | 2.00 | 1.40 | 1.14 | 0.61 - 2.40 |
| **3. Separate Variances, Zero Correlation (SZ)** | | | | | | | | | | |
| SUR | 1.06 | 2.81 | 1.15 | 1.39 | 0.57 | 2.21 | 2.15 | 1.62 | 1.39 | 0.57 - 2.81 |
| Lindley-Smith | 1.16 | 2.44 | 1.54 | 1.74 | 1.36 | 2.53 | 2.40 | 1.88 | 1.74 | 1.16 - 2.53 |
| Forecast Ave. | 1.09 | 2.36 | 1.00 | 1.24 | 0.60 | 1.72 | 2.02 | 1.43 | 1.24 | 0.60 - 2.36 |
| Multivariate | 1.47 | 2.45 | 1.31 | 1.32 | 1.45 | 2.65 | 1.66 | 1.76 | 1.47 | 1.31 - 2.65 |
| **4. Separate Variances, Common Correlation (SC)** | | | | | | | | | | |
| SUR | 1.49 | 2.31 | 0.84 | 1.01 | 1.31 | 0.91 | 1.84 | 1.39 | 1.31 | 0.84 - 2.31 |
| Lindley-Smith | 1.06 | 2.19 | 0.94 | 1.32 | 1.37 | 1.78 | 2.31 | 1.57 | 1.37 | 0.94 - 2.31 |
| Forecast Ave. | 1.06 | 2.06 | 1.12 | 1.31 | 0.68 | 1.81 | 2.01 | 1.44 | 1.31 | 0.68 - 2.06 |
| **5. Separate Variances, Separate Correlations (SS)** | | | | | | | | | | |
| SUR | 1.50 | 2.57 | 0.99 | 1.04 | 1.42 | 1.00 | 1.90 | 1.49 | 1.42 | 0.99 - 2.57 |
| Lindley-Smith | 1.07 | 2.06 | 0.87 | 1.24 | 1.20 | 1.74 | 2.24 | 1.49 | 1.24 | 0.87 - 2.24 |
| Forecast Ave. | 0.96 | 2.03 | 1.11 | 1.11 | 0.80 | 2.05 | 2.11 | 1.45 | 1.11 | 0.80 - 2.11 |

[1] Zero Growth [2] Same Growth [3] Past Average

TABLE 4
RMSEs of One-Year-Ahead Forecasts, 1974-1984
with proper priors on coefficients
Comparison with Results of Zellner and Hong (1988)

| Model | BEL | DEN | FRN | GRM | ITL | NET | UK | AVE | MED | RANGE |
|---|---|---|---|---|---|---|---|---|---|---|
| ZH $\eta$-forecast[1] | 1.70 | 2.37 | 1.77 | 1.94 | 2.27 | 2.41 | 2.32 | 2.11 | 2.27 | 1.70 - 2.41 |
| ZH $\gamma$-forecast[2] | 1.60 | 2.16 | 1.66 | 1.80 | 2.45 | 2.34 | 2.18 | 2.02 | 2.16 | 1.60 - 2.45 |
| ZH wld grth[3] | 1.79 | 2.04 | 1.36 | 1.35 | 1.82 | 2.27 | 2.21 | 1.83 | 1.82 | 1.35 - 2.27 |
| SUR[4] | 1.43 | 2.27 | 1.83 | 1.66 | 2.00 | 2.28 | 1.89 | 1.91 | 1.89 | 1.43 - 2.28 |
| Lindley-Smith[5] | 1.37 | 2.06 | 1.50 | 1.75 | 1.98 | 2.51 | 2.07 | 1.89 | 1.98 | 1.37 - 2.51 |
| Forecast Ave.[6] | 1.61 | 1.73 | 1.47 | 1.47 | 2.09 | 2.52 | 1.73 | 1.80 | 1.73 | 1.47 - 2.52 |
| Multivariate[7] | 1.78 | 2.02 | 1.90 | 1.65 | 2.39 | 2.37 | 1.74 | 1.98 | 1.90 | 1.65 - 2.39 |

---

[1] 18 countries, of which our 7 countries are a subset. Each country's model is specified as in (4.1) and (4.2) above and estimated by least squares. The $\eta$-forecast is $y_{it}^{**} = \eta_i \mu_t + (1-\eta_i) y_{it}^{*}$, where $\mu_t$ is the mean of the 18 countries least squares forecasts, $y_{it}^{*}$ is the i'th country's least squares forecast and $\eta_i$ is an optimal value chosen ex post. RMSEs for the 7 countries were taken from ZH Table 4.4.

[2] The $\gamma$-forecasts are those obtained from a model similar to the Lindley-Smith model in which the coefficients of the individual country models are shrunk toward the coefficients of "one big regression" for the 18 countries, and the parameter $\gamma$ controls the amount of shrinkage. RMSEs above are taken from ZH Table 4.5 and represent the results obtained if a separate $\gamma_i$ is chosen optimally ex post for each country.

[3] In this model the contemporaneous world median growth rate (i.e., the median of the 18 countries) is included as an additional regressor in each country's equation, and an additional equation was added to the system to predict the world median growth rate. RMSEs for the 7 countries are taken from ZH Table 4.7, column 3.

[4] $\Sigma = \sigma^2 I$, $\tau = .01292$

[5] $\Sigma = \sigma^2 I$, $\tau_1 = 0.0$, $\tau_2 = .07261$

[6] $\Sigma = \sigma^2 I$, $\tau = .02600$, $\eta = .7111$

[7] $\Sigma = \sigma^2 I$, $\tau = .01606$, $\alpha = .002766$

## REFERENCES

Anderson, B.D.O. and J.B. Moore (1979): *Optimal Filtering,* Englewood Cliffs, New Jersey: Prentice-Hall.

Garcia-Ferrer, A., R.A. Highfield, F. Palm, and A. Zellner (1987): Macroeconomic Forecasting Using Pooled International Data, *Jour. of Business and Economic Statistics,* **5**, 53-67.

Highfield, R.A. (1986): *Forecasting with Bayesian State Space Models,* unpublished doctoral dissertation, Graduate School of Business, University of Chicago.

Lindley, D.V. and A.F.M. Smith (1972): Bayes' Estimates for the Linear Model, *Journal of the Royal Statistical Society,* **Series B, 34**, 1-41.

Litterman, R.B. (1980): *Techniques for Forecasting Using Vector Autoregressions,* Unpublished doctoral dissertation, Dept. of Economics, University of Minnesota.

Litterman, R.B. (1984): *Specifying Vector Autoregressions for Macroeconomic Forecasting,* Staff Report 92, Federal Reserve Bank of Minneapolis Research Department (See also *Bayesian Inference and Decision Techniques With Applications: Essays in Honor of Bruno de Finetti,* ed. P.K. Goel and A. Zellner, Amsterdam: North Holland).

Litterman, R.B. (1986): A Statistical Approach to Economic Forecasting, *Journal of Business and Economic Statistics,* **4**, 1-4.

Zellner, A. (1962): An Efficient Method of Estimating Seemingly Unrelated Regressions and Tests for Aggregation Bias, *Journal of the American Statistical Association*, **57**, 348-368.

Zellner, A. and C. Hong (1988): Forecasting International Growth Rates Using Bayesian Shrinkage and Other Procedures, *Journal of Econometrics,* **40**, 183-202. [see Chapter 22 in this volume for a reprint with Comments].

## COMMENTS

Arnold Zellner
University of Chicago

This is an interesting paper which illustrates well the flexibility and usefulness of the Bayesian approach in the analysis of an important problem, forecasting of countries' annual output growth rates. The finding that the relatively simple autoregressive-leading indicator (ARLI) model, employed in our previous work–see references in the paper, performs well in forecasting in his current forecasting experiments is indeed satisfying. That he is able to estimate certain hyperparameters is also a useful contribution.

As regards the forecasting performance of our relatively simple ARLI model and that of more complex models, it appears from Highfield's results that the more complicated VAR models using various prior assumptions did not perform nearly as well as the simpler ARLI model. These results add further support to Jeffreys' and others' view that sophisticatedly simple models will perform better than complicated models. Note that in the present instance, the complicated models involved the use of 43 variables in the forecasting equation for each country and a large number of parameters. With so many parameters, it is difficult to obtain precise forecasts. The old advice, "Make your models as big as an elephant" is certainly misguided as shown by the present study and others–see McNees (1986). Also, as regards Litterman's Bayesian vector autoregressive models, cited by Highfield, it should be recognized that they also contain a very large number of parameters and while their performance in forecasting several real variables has been good, their performance in forecasting financial variables has not been as good–see, e.g., McNees (1986).

As regards the time-varying parameter (TVP) model described in equations (2.1)-(2.3) with recursions given in (2.5)-(2.10), there are many matrices in this model which are difficult to estimate. Note that this general TVP model is not used in the present paper. In our past work, Garcia-Ferrer et al. (1987), we found that very simplified versions of (2.1)-(2.3) worked reasonably well in forecasting and suggested that forecasting results might be improved by use of pooling techniques. See below for some recently obtained forecasting results using this approach. Also, we found that use of an unrestricted covariance matrix, $\boldsymbol{\Sigma}$, led to poor forecasting results, also encountered in Highfield's present study. Again the problem probably arises because of the many free parameters in the large matrix $\boldsymbol{\Sigma}$. Earlier, we used a "world return" or common effect variable which reduced contemporaneous disturbance terms' correlations considerably. It thus became possible to employ a diagonal disturbance term covariance matrix. Similar assumptions appear in the present paper.

Further, it has to be realized that the RMSEs reported in the present study are based on just five out-of-sample forecasts for the years 1980-84. With so few forecasts, the RMSEs are probably not very precise. It would be desirable to provide some measures, Bayesian or other, to indicate the precision of the RMSEs.

On "choosing" values for the hyperparameters, it would be useful to provide more details and to indicate in what sense, if any, the choices are optimal. Also, since the choices are based on relatively small samples of data, it would be interesting to learn how sensitive forecasting results are to variation in these values.

As regards the comparisons with results given in our past work, we found the performance of our "ZH world growth rate model" to be best among the fixed parameter models considered–[see Zellner and Hong (1988), reprinted in this volume]. Without going into details, it provides an economic interpretation for some of the mechanical shrinkage techniques with which we experimented. It's not clear why Highfield refers to this model as "non-Bayesian." We employed an approximate Bayesian forecasting procedure to generate point forecasts quite simply. It's gratifying to see that our model performs so well in the present set of experiments. Also, as mentioned above, in recent work, we have developed and applied a pooled TVP version of the ZH world growth rate model and used it to forecast growth rates year-by-year for the period 1974-86, thirteen years for 18 countries. The results for the countries considered in Highfield's paper are as follows:

| Country | RMSE | Country | RMSE |
|---|---|---|---|
| Belgium | 1.93 | Netherlands | 1.32 |
| Denmark | 1.90 | U. K. | 1.78 |
| France | 1.37 | Minimum | 1.20 |
| Germany | 1.20 | Median | 1.62 |
| Italy | 1.62 | Maximum | 1.93 |

From these results it is seen that the median RMSE of 1.62 is somewhat below all of those reported in Table 4. Also, the RMSEs for several of the countries are much lower than those reported in Table 4. However Belgium's RMSE, 1.93 is somewhat higher than those for Belgium reported in Table 4.

The results of the above experiment, and others that we are carrying forward using exponentially weighted autoregressive techniques seem to indicate that allowing for parameter variation will improve forecasting performance.

In summary, Highfield is to be congratulated for his fine analysis indicating the relationships connecting various models and shrinkage approaches and for providing methods for estimating hyperparameters' values. Also, his empirical results are a welcome addition to those previously obtained. It is through such research that better forecasts can be obtained which will be of great value to private and public decision-makers.

## REJOINDER

I am grateful to Arnold Zellner for his kind and constructive comments. He quite correctly points to the value of "sophisticatedly simple" approaches to economic modeling. His comments on this subject, both here and in the past, apply to Bayesians and non-Bayesians alike. What I have illustrated here, and what Robert Litterman has demonstrated in the VAR context, is that a sophisticatedly complex model can provide a high accuracy benchmark that is attainable at very low cost. My intention is not to deny the virtue of simplicity. Rather it is to extol the virtue of a Bayesian kind of sophistication.

Another issue raised by Professor Zellner is the matter of estimation of hyperparameters and the sensitivity of forecast accuracy to these estimates. In the current work the hyperparameter estimates have an interpretation as posterior modes under diffuse priors, or alternatively as Type II maximum likelihood estimates in the sense of Good (1965). All standard Bayesian analysis of the posterior distributions of these parameters is possible, however, via numerical methods. Whether any increase in forecasting accuracy obtained is worth the time cost of these methods is an open issue. My experience with this data is that the results are relatively insensitive to choice of parameters in the prior (i.e., there is little change in the results if $t$ is chosen from a fairly large neighborhood around the "optimal" value in the sense just described) but more sensitive to the choice of other model parameters such as $\alpha$ and $\eta$. This suggests that the method might be simplified with little loss of accuracy by estimating the parameters of the prior over a discrete grid.

The results reported on the TVP version of the ZH world growth rate model are quite interesting. This type of model continues to show great promise and is a hard one to beat. The results reported by Professor Zellner are not strictly comparable to those in my Table 4 as the years 1985 and 1986 are included in the numbers. Nevertheless I share his view that simple TVP techniques are likely to further improve forecast performance.

Finally, Professor Zellner takes me to task for referring to fixed parameter version of the ZH world growth rate model as "non-Bayesian" – and justly so. I fell into the trap of treating a Bayesian model with diffuse priors as if it were a non-Bayesian counterpart. Even if I didn't know the difference, however, I should have recognized the folly of a sentence suggesting that Professor Zellner advocated something non-Bayesian.

## ADDITIONAL REFERENCES

Good, I.J. (1965) Estimation of Probabilities: An Essay on Modern Bayesian Methods, Cambridge: The M.I.T. Press

McNees, S.K. (1986), "Forecasting Accuracy of Alternative Techniques: A Comparison of U.S. Macroeconomic Forecasts," (with discussion) *J. of Business and Economic Statistics,* **4**, 5-15 .

# 22

# Forecasting International Growth Rates Using Bayesian Shrinkage and Other Procedures


## Arnold Zellner & Chansik Hong [1]



[Reprinted from Journal of Econometrics **40** (1989) 183-202 with permission from Elsevier Science Publishers B.V. (North Holland)]



**Abstract**

Forecasts of eighteen countries' annual output growth rates, 1974-1984, are computed using an autoregressive model containing leading indicator variables and various forecasting procedures. It is found that Bayesian shrinkage forecasting techniques produced improved forecasts in terms of an out-of-sample root-mean-square criterion relative to those provided by the naive models and by autoregressive models with and without leading indicator variables. The precision of our forecasts compares favorably with that of OECD forecasts derived from complicated econometric 'structural' macroeconometric models and subjected to judgmental adjustments. Work to rationalize our forecasting models in terms of macroeconomic theory is in progress.



*Key words and Phrases:* Forecasting; Autoregressive models; Leading indicator variables; Comparison with OECD Forecasts; Structural models, Macroeconomic models; Bayesian shrinkage.


## 1. INTRODUCTION

In our past work, Garcia-Ferrar, Highfield, Palm and Zellner (1987), we employed several methods to forecast growth rates of real output (GNP or GDP ) for eight European Economic Community countries and the U.S. year by year for the period 1974-81. It was found that diffuse prior or least squares forecasts based on an autoregressive model of order 3 including leading indicator variables, denoted by AR(3)LI, were reasonably good in terms of forecast root mean squared error (RMSE) relative to those of three naive models and of AR(3) models without leading indicator variables. Also it was found that certain shrinkage forecasting techniques produced


[1] University of Chicago

improved forecasting results for many countries and that our simple mechanical forecasts compared favorably with OECD annual forecasts which were constructed using elaborate models and judgmental adjustments.

In the present paper our main objectives are to extend our earlier work by (1) providing further analysis of shrinkage forecasting techniques, (2) providing forecasting results for an extended time period, 1974-84, for our past sample of nine countries, (3) applying our forecasting techniques to data relating to nine additional countries, and (4) reporting results of forecasting experiments using a simple modification of our AR(3)LI model.

The importance of checking the forecasting performance of our techniques using new data is reflected in objectives (2) and (3) above. The modification of our AR(3)LI model, mentioned in (4), was motivated by macroeconomic considerations embedded in structural models currently being formulated, which yield reduced form equations similar in form to our AR(3)LI forecasting model and reduced form equations for other variables (e.g., the rate of inflation, employment growth, etc.) that will be subjected to forecasting tests in the future.

The plan of the paper is as follows. In Section 2 our AR(3)LI model is explained and analysis yielding several shrinkage forecasts is presented. Also, an extended version of our AR(3)LI model is specified and it is shown how it can be employed to yield forecasts. Section 3 is devoted to a presentation of our data, and in Section 4 previous forecasting results are compared with those relating to our broader data set and with those yielded by our extended AR(3)LI model. Finally, we provide a summary of results and some concluding remarks in Section 5.

## 2. MODEL DESCRIPTION AND FORECASTING

In this section, we shall describe the autoregressive-leading indicator (ARLI) model employed in our past work as well as some possible extensions of it. Then we shall consider various forecasting procedures of our ARLI models.

### 2.1 Model Description

In Garcia-Ferrer et al. (1987), the following AR(3)LI model was employed to generate one-year-ahead forecasts of the growth rate of real output, $y_{it}$, for eight years, 1974-81, for nine countries:

$$\begin{aligned} y_{it} &= \beta_{0i} + \beta_{1i} y_{it-1} + \beta_{2i} y_{it-2} + \beta_{3i} y_{it-3} + \beta_{4i} SR_{it-1} + \beta_{5i} SR_{it-2} \\ &\quad + \beta_{6i} GM_{it-1} + \beta_{7i} WR_{t-1} + u_{it}, \quad i = 1, 2, \ldots, 9, \; t = 1, 2, \ldots, T, \end{aligned} \tag{1}$$

or,

$$\boldsymbol{y}_i = \boldsymbol{X}_i \boldsymbol{\beta}_i + \boldsymbol{u}_i, \quad i = 1, 2, \ldots, 9, \tag{1a}$$

where, with $L$ being the lag operator and subscript $i, t$ denoting the value of a variable for the $i$th country in the $t$th time period,

$y_{it}$: rate of growth of output = $(1 - L)$ log $O_{it}$ with $O_{it}$ real output,

$SR_{it}$: real stock return = $(1 - L)$ log $(SP_{it}/P_{it})$ with $SP_{it}$ a stock price index and $P_{it}$ a general price index,

$GM_{it}$: growth rate of real money supply = $(1 - L)$ $\log(M_{it}/P_{it})$ with $M_{it}$ = nominal money supply,

$WR_t$: world return in period $t$ = median of countries' real stock returns $SR_{it}$ ,

$\beta_{ji}$: parameters for $i$th country, j=0,1,...,7,

$u_{it}$: disturbance term.

In (1$a$) the model for each country is expressed in matrix notation with $y_{it}$ a typical element of $\boldsymbol{y}_i$, $(1, y_{it-1}, y_{it-2}, y_{it-3}, SR_{it-1}, SR_{it-2}, GM_{it-1}, WR_{t-1})$ a typical row of $\boldsymbol{X}_i$, and $u_{it}$ a typical element of $\boldsymbol{u}_i$. Some comments regarding the model in (1) follow:

1) An autoregression of order 3 was chosen to permit the possibility of having two complex roots associated with a cycle and a real root associated with a trend. Past calculations indicated that estimated roots had these properties for eight of nine countries. Also, use of just an AR(3) process without leading indicator variables did not perform well in actual forecasting. Use of leading indicator variables led to improved forecasts in most cases as measured by RMSEs of forecast.

2) The disturbance terms in (1) were found to be practically serially uncorrelated for most countries and not highly correlated across countries, results based on least squares analyses of (1) using initial annual data of 1951-73 for estimation. The introduction of the 'common effect' variable, $WR_t$, reduced contemporaneous disturbance terms' correlations considerably.

3) The leading indicator stock return variables and money growth rate variable apparently caught the effects of oil price shocks, policy changes, etc. in the period of fit 1951-73 and in our previous forecast period 1974-81. Here we are employing market variables to take rough account of expectational and other effects influencing countries' output growth rates.

4) Macroeconomic considerations suggest that a measure of world output growth and changes in countries' real exchange rates affect countries' exports and these should be included in our ARLI model. Since these variables are close to being white noise, they may be buried in the disturbance terms of (1). Below, we shall report some results using an ARLI model including a measure of world output growth.

5) In our past work, forecasts from (1), using least squares and some shrinkage forecast procedures, were reported. Also, some forecasts yielded by a time-varying-parameter version of (1) were reported. Some of these

results will be presented below and compared with more recently obtained results.

In our forecasting experiments, we employ annual data, usually 1954-73, twenty observations with data from 1951-53 used for initial lagged values to fit our models. [2] Then the fitted models are employed to forecast outcomes for 1974 and subsequent years with the models re-estimated year by year. Multi-year-ahead forecasts have not as yet been calculated. For the forecast period 1974-81 (eight years), least squares forecasts using (1) have yielded forecast RMSEs ranging from 1.47 to 2.92 with a median of 2.23 percentage points for eight EEC countries and the U.S. - see Table 2, line F of Garcia-Ferrer et al. (1987). Our '$\eta$-shrinkage' forecasts described below, yielded forecast RMSEs ranging from 1.25 to 2.52 percentage points with a median of 1.78 percentage points - see Table 4, line G3 of Garcia-Ferrer et al. (1987). Similar results for an extended time period and for nine additional countries are presented below.

## 2.2 Derivation and Description of Shrinkage Forecasts

In this subsection, we provide derivations of several shrinkage forecasts, including the '$\eta$-forecast' and the '$\gamma$-forecast'. The performance of these forecasts will be compared with those of naive models and diffuse prior forecasts or least squares forecasts derived from the ARLI model, or variants of it, shown in (1).

The $\eta$-forecast involves averaging a forecast from (1), say a diffuse prior or least squares forecast for a particular country, $\hat{y}_{if}$, with the mean of all the $N$ countries' forecasts, $\bar{\hat{y}}_f = \sum_{i=1}^N \hat{y}_{if}/N$, as follows:

$$\begin{aligned} \tilde{y}^*_{if} &= \eta\bar{\hat{y}}_f + (1-\eta)\hat{y}_{if} \\ &= \bar{\hat{y}}_f + (1-\eta)(\hat{y}_{if} - \bar{\hat{y}}_f). \end{aligned} \tag{2}$$

From the second line of (2), it is seen that for $0 < \eta < 1$, a country's forecast, $\hat{y}_{if}$, is shrunk toward the average forecast $\bar{\hat{y}}_f$ for all countries.

One way to obtain an optimal forecast in the form of (2) is to employ the following predictive loss function:

$$L = (y_{if} - \tilde{y}_{if})^2 + c\left(\frac{1}{N}\sum_{i=1}^N y_{if} - \tilde{y}_{if}\right)^2, \tag{3}$$

where $c > 0$ is a given constant, the $y_{if}$'s are the future unknown values, $i = 1, 2, \ldots, N$, and $\tilde{y}_{if}$ is some forecast. Note that the loss function in (3) incorporates an element of loss associated with being away from the mean outcome in its second term. Under the assumption that the $y_{if}$'s are

---

[2] The 1954-73 period was used for all countries except Australia 1960-73, Canada 1959-73, Japan 1956-73, and Spain 1958-73.

independent [common influences have been represented by input variables in (1)] and have predictive probability density functions (pdfs) with mean $m_i$ and variance $v_i, i = 1, 2, \ldots, N$. The predictive expectation of the loss function in (3) is

$$EL = v_i + (m_i - \tilde{y}_{if})^2 + c[E(\bar{y}_f - E\bar{y}_f)^2 + (E\bar{y}_f - \tilde{y}_{if})^2], \tag{4}$$

where $\bar{y}_f = \sum_{i=1}^{N} y_{if}/N$. On minimizing (4) with respect to $\tilde{y}_{if}$, the result is

$$\tilde{y}_{if}^* = \eta \frac{1}{N} \sum_{i=1}^{N} m_i + (1 - \eta) m_i, \tag{5}$$

where $\eta = c/(1 + c)$. If diffuse priors for the $\boldsymbol{\beta}_i$'s in (1) are employed, the means of the predictive pdfs are $m_i = \boldsymbol{x}_{if}^{'} \hat{\boldsymbol{\beta}}_i$, $i = 1, 2, \ldots, N$, where $\boldsymbol{x}_{if}^{'}$ is a vector of observed inputs for the first future period and $\hat{\boldsymbol{\beta}}_i = (\boldsymbol{X}_i^{'} \boldsymbol{X}_i)^{-1} \boldsymbol{X}_i^{'} \boldsymbol{y}_i$, the least squares estimate for country $i$. Under these conditions (5) takes the form of (2) with $\hat{y}_{if} = \boldsymbol{x}_{if}^{'} \hat{\boldsymbol{\beta}}_i$. This is the 'diffuse prior $\eta$-forecast'.

Another approach for obtaining relatively simple shrinkage forecasts is a slightly modified form of the Lindley-Smith (1972) procedure in which the coefficient vectors are assumed generated by

$$\boldsymbol{\beta}_i = \boldsymbol{\theta} + \boldsymbol{\delta}_i, \quad i = 1, 2, \ldots, N, \tag{6}$$

with the $\boldsymbol{\delta}_i$'s assumed independently distributed, each having a $N(\boldsymbol{0}, \phi^{-1} \sigma_u^2 \boldsymbol{I}_k)$ distribution where $0 < \phi < \infty, \sigma_u^2$ is a common variance of $u_{it}$ for all $i$ and $t$, and $\boldsymbol{\theta}$ is a $k \times 1$ mean vector. If the $u_{it}$'s are assumed normally and independently distributed, each with zero mean and common variance $\sigma_u^2$, then a conditional point estimate for $\boldsymbol{\beta}^{'} = (\boldsymbol{\beta}_1^{'}, \boldsymbol{\beta}_2^{'}, \ldots, \boldsymbol{\beta}_N^{'})$, denoted by $\tilde{\boldsymbol{\beta}}_a$, an $Nk \times 1$ vector, is given by

$$\tilde{\boldsymbol{\beta}}_a = (\boldsymbol{Z}^{'} \boldsymbol{Z} + \phi \boldsymbol{I}_{NK})^{-1} (\boldsymbol{Z}^{'} \boldsymbol{Z} \hat{\boldsymbol{\beta}} + \phi \boldsymbol{J} \tilde{\boldsymbol{\beta}}), \tag{7}$$

where $\boldsymbol{Z}$ is a block-diagonal matrix with $\boldsymbol{X}_1, \boldsymbol{X}_2, \ldots, \boldsymbol{X}_N$ on the main diagonal, $\hat{\boldsymbol{\beta}} = (\boldsymbol{Z}^{'} \boldsymbol{Z})^{-1} \boldsymbol{Z}^{'} \boldsymbol{y}$, where $\boldsymbol{y}^{'} = (\boldsymbol{y}_1^{'}, \boldsymbol{y}_2^{'}, \ldots, \boldsymbol{y}_N^{'}), \hat{\boldsymbol{\beta}}^{'} = (\hat{\boldsymbol{\beta}}_1^{'}, \hat{\boldsymbol{\beta}}_2^{'}, \ldots, \hat{\boldsymbol{\beta}}_N^{'})$ with $\hat{\boldsymbol{\beta}}_i = (\boldsymbol{X}_i^{'} \boldsymbol{X}_i)^{-1} \boldsymbol{X}_i^{'} \boldsymbol{y}_i$, $\boldsymbol{J}^{'} = (\boldsymbol{I}_k, \boldsymbol{I}_k, \ldots, \boldsymbol{I}_k)$, and

$$\tilde{\boldsymbol{\beta}} = \left( \sum_{i=1}^{N} \boldsymbol{X}_i^{'} \boldsymbol{X}_i \right)^{-1} \sum_{i=1}^{N} \boldsymbol{X}_i^{'} \boldsymbol{X}_i \hat{\boldsymbol{\beta}}_i, \tag{8}$$

a matrix-weighted average of the least squares estimates, the $\hat{\boldsymbol{\beta}}_i$'s, which replaces $\boldsymbol{\theta}$ in (7). Also, $\tilde{\boldsymbol{\beta}}$ in (8) can be obtained by regressing $\boldsymbol{y}^{'} = (\boldsymbol{y}_1^{'}, \boldsymbol{y}_2^{'}, \ldots, \boldsymbol{y}_N^{'})$ on $\boldsymbol{X}$, where $\boldsymbol{X}^{'} = (\boldsymbol{X}_1^{'}, \boldsymbol{X}_2^{'}, \ldots, \boldsymbol{X}_N^{'})$, that is, from one big regression in which it is assumed that the $\boldsymbol{\beta}_i$'s are equal. Point forecasts can be obtained using the coefficient estimate $\tilde{\boldsymbol{\beta}}_a$ for various selected values of $\phi$. When $\phi$ is very large, (7) reduces approximately to (8). Forecasts based on the estimate in (8) have been reported earlier in Garcia-Ferrer et al. (1987). It should be recognized that, while the $\boldsymbol{\beta}$'s are probably not all the same, the bias introduced by assuming them to be the same may be more than offset in a MSE error sense by a reduction in variance.

As an alternative to the assumptions used in connection with (6), following the '$g$-prior' approach of Zellner (1983, 1986), we assume that the $\boldsymbol{\delta}_i$'s in (6) are independently distributed with normal distributions $N[\mathbf{0}, \sigma_\delta^2 (\boldsymbol{X}_i'\boldsymbol{X}_i)^{-1}]$. With this assumption and the earlier assumption made about the $u_{it}$'s, the joint pdf for $\boldsymbol{y}' = (\boldsymbol{y}_1', \boldsymbol{y}_2', \ldots, \boldsymbol{y}_N')$ and $\boldsymbol{\beta}' = (\boldsymbol{\beta}_1', \boldsymbol{\beta}_2', \ldots, \boldsymbol{\beta}_N')$ is proportional to

$$exp\{-[(\boldsymbol{y} - \boldsymbol{Z}\boldsymbol{\beta})'(\boldsymbol{y} - \boldsymbol{Z}\boldsymbol{\beta}) + \gamma(\boldsymbol{\beta} - \boldsymbol{J}\boldsymbol{\theta})'\boldsymbol{Z}'\boldsymbol{Z}(\boldsymbol{\beta} - \boldsymbol{J}\boldsymbol{\theta})]/2\sigma_u^2\}, \quad (9)$$

where $\gamma = \sigma_u^2/\sigma_\delta^2$ and other quantities have been defined in connection with (6)-(7). On completing the square on $\boldsymbol{\beta}$ in the exponential terms of (9), the mean of $\boldsymbol{\beta}$ given $\boldsymbol{y}, \gamma$ and $\boldsymbol{\theta}$ is

$$\bar{\boldsymbol{\beta}} = [(\boldsymbol{Z}'\boldsymbol{Z})^{-1}\boldsymbol{Z}'\boldsymbol{y} + \gamma\boldsymbol{J}\boldsymbol{\theta}]/(1+\gamma), \quad (10)$$

with the $i$-th subvector of $\bar{\boldsymbol{\beta}}$ given by

$$\bar{\boldsymbol{\beta}}_i = (\hat{\boldsymbol{\beta}}_i + \gamma\boldsymbol{\theta})/(1+\gamma), \quad (10b)$$

where $\hat{\boldsymbol{\beta}}_i = (\boldsymbol{X}_i'\boldsymbol{X}_i)^{-1}\boldsymbol{X}_i'\boldsymbol{y}_i$, the least squares estimate for the $i$th country's data. Thus (10b) is a simple average of $\hat{\boldsymbol{\beta}}_i$ and $\boldsymbol{\theta}$ with $\gamma = \sigma_u^2/\sigma_\delta^2$ involved in the weights. When a diffuse prior pdf for $\boldsymbol{\theta}, p(\boldsymbol{\theta}) \propto$ const., is empoyed, the posterior pdf for $\boldsymbol{\theta}$ can be derived from (9) and employed to average the expression in (10) to obtain the marginal mean of $\boldsymbol{\beta}_i$ given $\gamma$ and the data, namely,

$$\bar{\boldsymbol{\beta}}_i^m = (\hat{\boldsymbol{\beta}}_i + \gamma\tilde{\boldsymbol{\beta}})/(1+\gamma), \quad (11)$$

with $\tilde{\boldsymbol{\beta}}$ given in (8), the estimate resulting from a big regression in which the $\boldsymbol{\beta}_i$'s are assumed equal. As $\gamma = \sigma_u^2/\sigma_\delta^2$ grows in value, $\bar{\boldsymbol{\beta}}_i^m \to \tilde{\boldsymbol{\beta}}$ while as $\gamma = \sigma_u^2/\sigma_\sigma^2 \to 0, \bar{\boldsymbol{\beta}}_i^m \to \hat{\boldsymbol{\beta}}_i$, the $i$th country's least squares estimate.

If instead of assuming that $\boldsymbol{\delta}_i$ has a $N[\mathbf{0}, \sigma_\delta^2 (\boldsymbol{X}_i'\boldsymbol{X}_i)^{-1}]$ distribution, we assume that the $\boldsymbol{\delta}_i$'s are independently distributed, with a $N[\mathbf{0}, \sigma_i^2 (\boldsymbol{X}_i'\boldsymbol{X}_i)^{-1}]$ distribution, $i = 1, 2, \ldots, N$, then analysis similar to that presented in connection with (9) yields as the conditional mean of $\boldsymbol{\beta}_i$,

$$\bar{\boldsymbol{\beta}}_i^c = (\hat{\boldsymbol{\beta}}_i + \gamma_i\boldsymbol{\theta})/(1+\gamma_i), \quad i = 1, 2, \ldots, N, \quad (12)$$

where $\gamma_i = \sigma_u^2/\sigma_i^2$. If we further condition on $\boldsymbol{\theta} = \tilde{\boldsymbol{\beta}}$, with $\tilde{\boldsymbol{\beta}}$ given in (8), then

$$\bar{\boldsymbol{\beta}}_i^c = (\hat{\boldsymbol{\beta}}_i + \gamma_i\tilde{\boldsymbol{\beta}})/(1+\gamma_i), \quad i = 1, 2, \ldots, N, \quad (13)$$

which is similar to (11) except that the $\gamma_i$'s are not all equal as is the case in (11).

Upon introducing prior pdfs for $\sigma_u$ and $\sigma_\delta$, or $\sigma_u$ and the $\sigma_i$'s, it is possible to compute the marginal distributons of the $\boldsymbol{\beta}_i$'s, a possibility to be explored in future work - see Miller and Fortney (1984) for interesting computations on a closely related problem. At present, we shall evaluate (13) for various values of $\gamma_i$ and determine the quality of resulting forecasts. That is, the $\gamma_i$-forecast for country $i$ is

$$\tilde{y}_{if} = \boldsymbol{x}_{if}^{'}\bar{\boldsymbol{\beta}}_i^c, \tag{14}$$

with $\bar{\boldsymbol{\beta}}_i^c$ given in (13) and $\boldsymbol{x}_{if}^{'}$ a given input vector.

In summary, we shall use the $\eta$-forecast in (2), the $\gamma$-forecast based on (11) and the $\gamma_i$-forecast in (14) in our forecasting experiments. Also, note that the $\eta$-forecasting approach can be applied to the $\gamma$-forecasts.

### 2.3 Elaboration of the AR(3)LI Model

As mentioned previously, we think that it is advisable to add a variable reflecting world real income growth, denoted by $w_t$, to our AR(3)LI model in (1). Then our equation becomes

$$y_{it} = w_t\alpha_i + \boldsymbol{x}_{it}^{'}\boldsymbol{\beta}_i + u_{it}, \quad i = 1, 2, \ldots, N, \quad t = 1, 2, \ldots, T, \tag{15}$$

where $\boldsymbol{x}_{it}^{'}\boldsymbol{\beta}_i$ represents the constant and other lagged variables in (1) and $\alpha_i$ is the $i$th country's coefficient of the world income growth rate variable, $w_t$. To forecast one period ahead using (15), it is clear that $w_t$ must be forecasted. To do this we introduce the following equation for $w_t$ which will be estimated and used to forecast $w_t$ one period in the future:

$$\begin{aligned} w_t &= \pi_0 + \pi_1 w_{t-1} + \pi_2 w_{t-2} + \pi_3 w_{t-3} + \pi_4 MSR_{t-1} \\ &\quad + \pi_5 MGM_{t-1} + v_t, \quad t = 1, 2, \ldots, T, (16) \end{aligned}$$

where $MSR_t$ is the median of all countries' real stock returns, $MGM_t$, the median of all countries' real money growth rates, $v_t$ a disturbance term, and the $\pi_i$'s are parameters. Thus (16) indicates that we are employing an AR(3)LI model for $w_t$, the rate of growth of world real income. As a proxy for $w_t$, we employ the median of all countries' real output growth rates.

Viewing (15)-(16), it is seen that we have a 'triangular' system. For the future period $f = T + 1$, we have

$$\begin{aligned} Ey_{if} &= Ew_f E\alpha_i + \boldsymbol{x}_{if}^{'} E\boldsymbol{\beta}_i, \quad i = 1, 2, \ldots, N, && (15a) \\ Ew_f &= \boldsymbol{z}_f^{'} E\boldsymbol{\pi}, && (15b) \end{aligned}$$

where $Ey_{if}$ and $Ew_f$ are means of the predictive pdfs for $y_{if}$ and $w_f$, respectively, and $E\alpha_i$, $E\boldsymbol{\beta}_i$ and $E\boldsymbol{\pi}$ are posterior means of the parameters $\alpha_i, \boldsymbol{\beta}_i$ and $\boldsymbol{\pi}^{'} = (\pi_0, \pi_1, \ldots, \pi_5)$, respectively. If the system in (15)-(16)

is *fully recursive* and diffuse prior pdfs for all parameters are employed, $E\boldsymbol{\pi} = \hat{\boldsymbol{\pi}}, E\alpha_i = \hat{\alpha}_i$ and $E\boldsymbol{\beta}_i = \hat{\boldsymbol{\beta}}_i$, where $\hat{\boldsymbol{\pi}}, \hat{\alpha}_i$ and $\hat{\boldsymbol{\beta}}_i$ are least squares estimates - see Zellner (1971, Ch. 8) and Bowman and Laporte (1975). Also, $Ew_f = \boldsymbol{z}_f'\hat{\boldsymbol{\pi}}$ and $Ey_{if} = \boldsymbol{z}_f'\hat{\boldsymbol{\pi}}\hat{\alpha}_i + \boldsymbol{x}_{if}'\hat{\boldsymbol{\beta}}_i, i = 1, 2, \ldots, N$. Thus forecasts under these assumptions are easily computed. If the system in (15)-(16) is not fully recursive, that is, the $u_{it}$'s and $v_t$ are correlated, then the expectations in (15) have to reflect the non-recursive nature of the system. In the present work, we shall use a 'conditional' forecasting procedure which is equivalent to a 2SLS point forecast. That is, the parameters $\alpha_i$ and $\boldsymbol{\beta}_i$ in (15) are estimated by 2SLS, a conditional Bayesian estimate [see Zellner (1971, p. 266)], and these estimates along with a forecast of $w_t$ from (16) are employed to obtain a forecast of $y_{it}$ from (15). Such forecasts will be compared with those that assume that $w_t$'s value in a forecast period is perfectly known, a 'perfect foresight' assumption. In current work, an unconditional Bayesian approach for analyzing (15)-(16) when the $u_{it}$'s and $v_t$ are correlated is being developed.

## 3. DATA

Annual data for eighteen countries employed in our work have been assembled in the main from the International Monetary Fund's International Financial Statistics data base and are available on a diskette from the authors for a nominal fee to cover costs. The output data include annual rates of growth of real output (GNP or GDP), of real stock prices and of real money for each country. In computing rates of growth of real stock prices, an index of nominal stock prices was deflated by an index of the price level for each country. Nominal money, $M_1$, was deflated by a general price index for each country to obtain a measure of real money.

Boxplots of output growth rates, real stock price growth rates and real money growth rates are shown in Fig. 1. It is seen that the median growth rates exhibit a cyclical pattern with that for real stock prices having a considerably greater amplitude than those for output and real money growth rates. Also, as might be expected, the interquartile ranges of the real stock price growth rates are much larger than those of output and real money growth rates. Further, the interquartile ranges for growth rates of real stock prices appear to be slightly smaller in the vicinity of troughs than of peaks for the first half of the sample and the ranges for all three variables tend to be slightly larger in the vicinity of peaks than of troughs in many cases. Last, the plots of the real money and real stock price growth rates give some evidence of a slight lead relative to those of the output growth rates.

It will be noted from the plots in Fig. 1 that there are some apparently outlying points in the data with a number of them present in the data from the nine additional countries' data. These outlying data points are being subjected to close scrutiny and procedures for accommodating outlying data points are being considered for use in future work. In the present

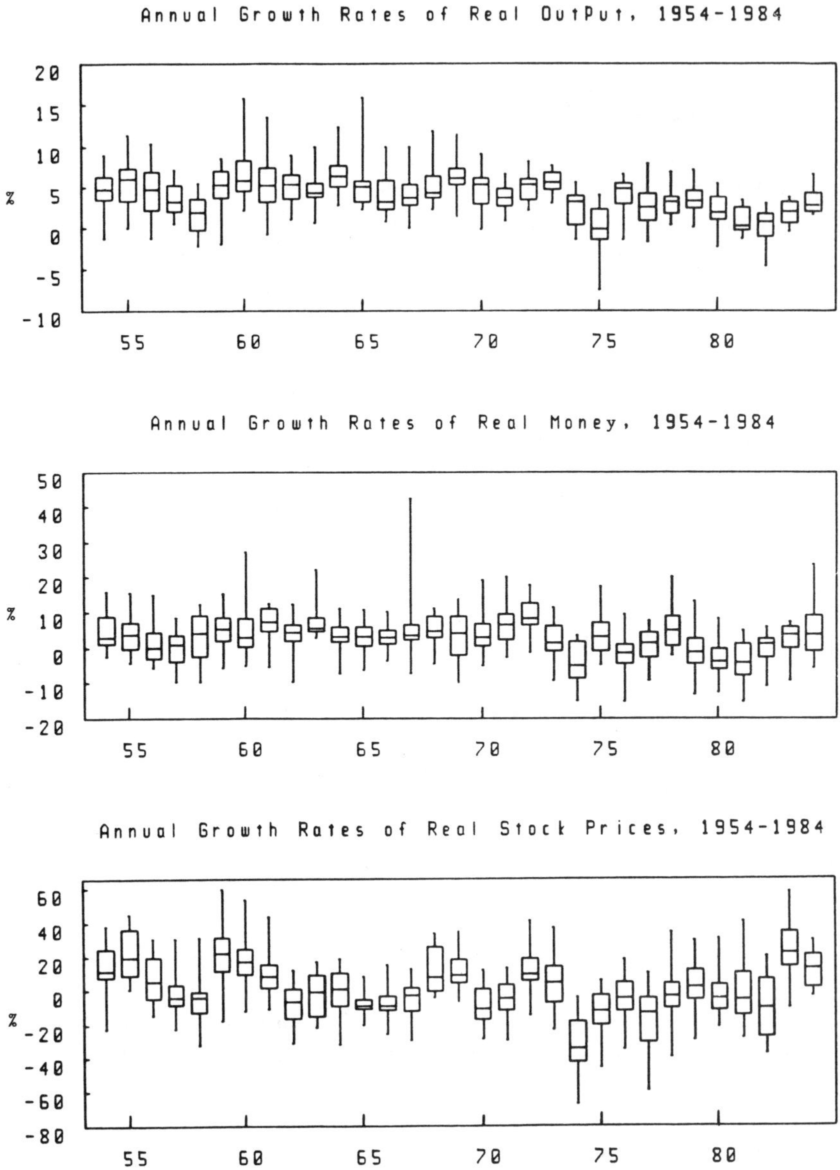


Figure 1. Boxplots of data for eighteen countries, 1954-84.

study, *all data were employed*, including outlying data without any special treatment given to them.

## 4. FORECASTING RESULTS

In this section, we first compare RMSEs of one-year-ahead forecasts, 1974-81 (eight years), with those for the period 1974-84 for our original nine countries using various models and methods. Then for nine additional countries, forecasting results for the period 1974-84 are presented and compared with earlier results. Finally, the effects of introducing a world growth rate variable in our AR(3)LI model in (1a) on forecasting performance will be described.

### 4.1 Forecasting Results for an Expanded Data Set

Shown in Table 1 are the RMSEs of forecast of nine countries for the periods 1974-81 and for 1974-84. Here and elsewhere, all models were re-estimated using data up to the forecast year. In the top panel of Table 1, results for the eight one-year-ahead forecasts, 1974-81, are shown. It is seen that the median RMSE for the AR(3)LI model, 2.23 percentage points, is quite a bit lower than the median RMSEs for the naive models and for the AR(3) model. In addition, shrinkage or pooling techniques applied to the AR(3)LI model led to median RMSEs of 2.22 and 1.78, a very slight reduction in the former case and a somewhat larger reduction in the latter case, from 2.23 to 1.78 percentage points. Use of the $\eta$-shrinkage procedure, with $\eta = 0.5$, led to reduction of RMSEs for seven of the nine countries.

As regards the results for 1974-84, in the lower panel of Table 1, the AR(3)LI model's median RMSE is 2.41 percentage points, a good deal lower than those associated with the naive models and the AR(3) model. Also, the two shrinkage or pooling procedures produced modest decreases in median RMSEs, from 2.41 to 2.26 and 2.31 and in six of nine cases for the $\eta$-shrinkage procedure. In these respects, the forecasting results parallel those obtained for the shorter period, 1974-81. However, note that there is an increase in the median RMSEs for the AR(3)LI model in going from the period 1974-81 to the longer period 1974-84 which also shows up in seven of the nine countries' RMSEs.

In Table 2 RMSEs of forecast are shown for nine additional countries. Here the median RMSE for the AR(3)LI model, 3.33, is somewhat larger than that for the naive models and for the AR(3) model. However, the shrinkage or pooled forecasts based on the AR(3)LI model show substantial reductions in median RMSE's, from 3.33 to 2.01 and 2.32, which are similar in magnitude to those reported for the original nine countries for periods 1974-81 and 1974-84 in Table 1. In Table 2, on comparing the AR(3)LI

Table 1.
Nine Countries' RMSEs of one-year-ahead forecasts, 1974-81 and 1974-84.

| Model ($\hat{y}_t$) | Belgium | Denmark | France | Germany | Ireland | Italy | Netherlands | U.K. | U.S. | Median |
|---|---|---|---|---|---|---|---|---|---|---|
| | | | | (Percentage points) | | | | | | |
| | | | | | (A)1974-81 | | | | | |
| A. NMI [a] | 3.09 | 2.83 | 2.96 | 2.95 | 4.38 | 3.72 | 3.77 | 2.21 | 3.48 | 3.09 |
| B. NMII[a] | 4.25 | 3.73 | 2.43 | 3.26 | 2.06 | 4.88 | 4.04 | 3.91 | 3.60 | 3.73 |
| C. NMIII[a] | 3.23 | 3.48 | 3.05 | 3.87 | 1.88 | 3.90 | 3.74 | 2.95 | 2.81 | 3.23 |
| D. AR(3)[b] | 3.66 | 3.46 | 2.89 | 3.39 | 1.69 | 4.75 | 3.52 | 3.50 | 2.48 | 3.46 |
| E. AR(3)LI[c] | 1.56 | 2.92 | 2.43 | 1.47 | 1.83 | 2.57 | 2.63 | 2.23 | 1.82 | 2.23 |
| 1. Shrinkage(1)[d] | 1.69 | 2.37 | 1.35 | 2.03 | 1.77 | 2.22 | 2.87 | 2.26 | 2.75 | 2.22 |
| 2. Shrinkage(2)[e] | 1.68 | 2.21 | 1.61 | 1.25 | 1.52 | 2.01 | 2.52 | 2.46 | 1.78 | 1.78 |
| | | | | | (B)1974-84 | | | | | |
| A. NMI[a] | 2.59 | 2.78 | 2.65 | 2.69 | 4.02 | 3.27 | 3.32 | 2.30 | 3.79 | 2.78 |
| B. NMII[a] | 3.53 | 3.56 | 2.20 | 2.91 | 2.85 | 4.26 | 3.57 | 3.69 | 3.89 | 3.56 |
| C. NMIII[a] | 3.01 | 3.05 | 3.08 | 3.85 | 2.35 | 3.98 | 3.87 | 2.62 | 3.09 | 3.08 |
| D. AR(3)[b] | 2.98 | 2.96 | 2.47 | 3.10 | 2.29 | 4.34 | 3.35 | 3.21 | 3.01 | 3.01 |
| E. AR(3)LI[c] | 1.73 | 2.73 | 2.52 | 2.28 | 2.80 | 3.40 | 2.41 | 2.32 | 2.14 | 2.41 |
| 1. Shrinkage(1)[d] | 1.96 | 2.26 | 1.66 | 2.00 | 2.14 | 2.45 | 2.53 | 2.39 | 2.79 | 2.26 |
| 2. Shrinkage(2)[e] | 1.81 | 2.37 | 2.07 | 1.94 | 2.31 | 2.73 | 2.50 | 2.63 | 2.03 | 2.31 |

[a] NMI: $\hat{y}_t = 0$; NMII: $\hat{y}_t = y_{t-1}$; NMIII: $\hat{y}_t$ = Past average.
[b] Least squares forecasts from an AR(3) model for each country.
[c] Least squares forecasts from an AR(3) model with leading indicator variables shown in eq. (1a).
[d] Least squares forecasts with use of coefficient estimate in (8).
[e] Use of shrinkage equation in (2) with $\eta = 0.5$.

Table 2.
Nine additional countries' RMSEs of one-year-ahead forecasts, 1974-84.

| Model $\hat{y}_t$ | Australia | Austria | Canada | Finland | Japan | Norway | Spain | Sweden[f] | Switzerland | Med.[f] |
|---|---|---|---|---|---|---|---|---|---|---|
| | (Percentage points) | | | | | | | | | |
| A. NMI [a] | 3.10 | 2.91 | 3.51 | 3.39 | 4.10 | 4.15 | 2.51 | 2.29 | 3.03 | 3.10 |
| B. NMII[a] | 2.90 | 2.82 | 4.13 | 2.34 | 3.12 | 1.98 | 1.86 | 1.87 | 3.91 | 2.82 |
| C. NMIII[a] | 2.75 | 3.11 | 3.31 | 2.88 | 4.98 | 1.76 | 3.56 | 2.43 | 4.48 | 3.11 |
| D. AR(3)[b] | 2.87 | 3.16 | 3.55 | 2.58 | 3.26 | 1.75 | 2.50 | 2.22 | 4.24 | 2.87 |
| E. AR(3)LI[c] | 3.34 | 2.71 | 3.68 | 3.37 | 3.33 | 1.62 | 2.06 | 2.32 | 3.45 | 3.33 |
| 1. Shrink(1)[d] | 2.01 | 1.77 | 2.39 | 2.01 | 2.40 | 1.68 | 1.65 | 2.01 | 2.71 | 2.01 |
| 2. Shrink(2)[e] | 2.32 | 2.12 | 2.92 | 2.42 | 2.51 | 1.52 | 1.81 | 2.29 | 3.25 | 2.32 |

[a,b,c,d,e] See Table 1 for explanations.
[f] Based on ten forecasts, 1974-83.

country RMSEs with the corresponding shrinkage RMSEs, it is seen that there is a reduction of RMSEs in all but one case. Thus for the nine additional countries, shrinkage results in a somewhat greater improvement in forecasting results than in the case of the original nine countries.

In Table 3 summary forecasting results for different models, applied to all eighteen countries' data to forecast year by year for the period 1974-84, are reported. It is seen that the AR(3)LI model's median RMSE is about 13 percent or more below those of the AR(3) and naive model's median RMSEs 2.62 versus 3.00 or greater. Also, the shrinkage or pooled forecasts median RMSEs, 2.14 and 2.32, are 18 and 11 percent lower, respectively, than the AR(3) model's median RMSE. Further, the ranges for the AR(3)LI and shrinkage forecast RMSEs for the eighteen countries are much smaller than those for the AR(3) and naive models.

Table 3.
Summary statistics on forecasting results for eighteen countries, 1974-84.

| Model $\hat{y}_t$ | Median RMSE | Smallest RMSE (% points) | Largest RMSE |
|---|---|---|---|
| A. NMI[a] | 3.07 | 2.29 | 4.15 |
| B. NMII[a] | 3.02 | 1.87 | 4.26 |
| C. NMIII[a] | 3.09 | 1.76 | 4.98 |
| D. AR(3)[b] | 3.00 | 1.75 | 4.34 |
| E. AR(3)LI[c] | 2.62 | 1.62 | 3.68 |
| 1. Shrinkage(1)[d] | 2.14 | 1.65 | 2.79 |
| 2. Shrinkage(2)[e] | 2.32 | 1.52 | 3.25 |

[a,b,c,d,e] See Table 1 for explanations.

In summary, the shrinkage or pooling techniques used earlier in Garcia-Ferrer et al. (1987) for nine countries for the period 1974-81 are effective in reducing overall median RMSEs for an extended forecast period, 1974-84, and for nine additional countries.

The $\eta$-shrinkage forecast RMSEs, reported above, are based on the same value of $\eta = 0.5$ for all countries. It is of interest to see how sensitive forecasting results for individual countries are to variation in the value of $\eta$. In Table 4 one-year-ahead forecast RMSEs are reported for selected values of $\eta$ in (2) for each of the eighteen countries in our sample.

When $\eta = 0$, the forecasts are least squares forecasts for each country based on our AR(3)LI model in (1a). When $\eta = 1.0$, the forecast for each country is the mean of the eighteen countries' least squares forecasts from the AR(3)LI model in (1a). On viewing the median RMSEs at the bottom of Table 4, it is seen that the median RMSEs are 2.43 for $\eta = 0.25$, 2.32

Table 4.
Sensitivity of $\eta$-forecast RMSEs to values of $\eta$ for AR(3)LI model in (1a), 1974-84.[a]

| | Values of $\eta$ | | | | |
|---|---|---|---|---|---|
| | 0 | 0.25 | 0.50 | 0.75 | 1.0 |
| Country | (Percentage points) | | | | |
| Belgium | 1.73 | 1.70[b] | 1.81 | 2.04 | 2.35 |
| Denmark | 2.73 | 2.49 | 2.37[b] | 2.41 | 2.59 |
| France | 2.52 | 2.29 | 2.07 | 1.90 | 1.77[b] |
| Germany | 2.28 | 2.04 | 1.94[b] | 2.01 | 2.24 |
| Ireland | 2.80 | 2.51 | 2.31 | 2.24[b] | 2.29 |
| Italy | 3.40 | 3.04 | 2.73 | 2.46 | 2.27[b] |
| Netherlands | 2.41[b] | 2.42 | 2.50 | 2.65 | 2.87 |
| U.K. | 2.32[b] | 2.44 | 2.63 | 2.89 | 3.20 |
| U.S. | 2.14 | 1.97[b] | 2.03 | 2.31 | 2.75 |
| Australia | 3.34 | 2.80 | 2.32 | 1.92 | 1.67[b] |
| Austria | 2.71 | 2.40 | 2.12 | 1.87 | 1.67[b] |
| Canada | 3.68 | 3.27 | 2.92 | 2.64 | 2.46[b] |
| Finland | 3.37 | 2.86 | 2.42 | 2.10 | 1.95[b] |
| Japan | 3.33 | 2.90 | 2.51 | 2.18 | 1.96[b] |
| Norway | 1.62 | 1.53 | 1.52[b] | 1.57 | 1.70 |
| Spain | 2.06 | 1.86 | 1.81[b] | 1.91 | 2.14 |
| Sweden[c] | 2.32 | 2.28 | 2.29[b] | 2.34 | 2.44 |
| Switzerland | 3.45 | 3.26 | 3.25[b] | 3.42 | 3.75 |
| Median RMSE | 2.62 | 2.43 | 2.32 | 2.21 | 2.28 |
| Min. RMSE | 1.62 | 1.53 | 1.52 | 1.57 | 1.67 |
| Max. RMSE | 3.68 | 3.27 | 3.25 | 3.42 | 3.75 |

[a] The $\eta$-forecast is $\tilde{y}_{if} = \eta\hat{y}_f + (1-\eta)\hat{y}_{if}$, where $\hat{y}_f$ is the mean of the eighteen countries' least squares forecasts and $\hat{y}_{if}$ is the $i$th country's least squares forecast.
[b] Minimum of entries in each row. The median of these RMSEs is 1.96.
[c] Based on ten forecasts, 1974-83.

for $\eta = 0.5$, and 2.21 for $\eta = 0.75$, all below 2.62, the median RMSE for $\eta = 0$. For individual countries, the RMSEs change as the value of $\eta$ changes. For example in the case of Belgium, the RMSEs vary from 1.73 for $\eta = 0$ to 2.35 for $\eta = 1.0$, with a minimum of 1.70 for $\eta = 0.25$. Two countries' minimal RMSEs occur for $\eta = 0$ and $\eta = 0.25$, six for $\eta = 0.50$, one for $\eta = 0.75$, and seven for $\eta = 1.0$. The median RMSE of these minimal values is 1.96, somewhat below that associated with the use of $\eta = 0.5$ for all countries, namely 2.32, or of $\eta = 0.75$, 2.21.

Just as it is of interest to determine the sensitivity of countries' $\eta$-forecasts to the values of $\eta$ employed, it is of interest to determine how sensitive $\gamma$-forecasts for countries are to the value of $\gamma$ employed. The coefficient estimate in (13) permits $\gamma$ to be different for different countries. In Table 5, RMSEs of forecast are reported for each country for selected values of $\gamma$.

Table 5.
Sensitivity of $\gamma$-forecast RMSEs to values of $\gamma$ for AR(3)LI model in (1a), 1974-84.[a]

| | Values of $\gamma$ | | | | | | |
|---|---|---|---|---|---|---|---|
| | 0 | 0.50 | 1.0 | 2.0 | 3.0 | 5.0 | $10^6$ |
| Country | (Percentage points) | | | | | | |
| Belgium | 1.73 | 1.60[b] | 1.62 | 1.69 | 1.74 | 1.81 | 1.96 |
| Denmark | 2.73 | 2.34 | 2.22 | 2.16[b] | 2.16[b] | 2.18 | 2.26 |
| France | 2.52 | 2.17 | 2.01 | 1.87 | 1.81 | 1.75 | 1.66[b] |
| Germany | 2.28 | 1.91 | 1.81 | 1.80[b] | 1.82 | 1.87 | 2.00 |
| Ireland | 2.80 | 2.47 | 2.34 | 2.24 | 2.20 | 2.17 | 2.14[b] |
| Italy | 3.40 | 3.00 | 2.82 | 2.67 | 2.61 | 2.55 | 2.45[b] |
| Netherlands | 2.41 | 2.34[b] | 2.35 | 2.38 | 2.41 | 2.44 | 2.53 |
| U.K. | 2.32 | 2.19 | 2.18[b] | 2.21 | 2.24 | 2.28 | 2.39 |
| U.S. | 2.14 | 1.98[b] | 2.06 | 2.24 | 2.35 | 2.48 | 2.79 |
| Australia | 3.34 | 2.72 | 2.46 | 2.25 | 2.16 | 2.09 | 2.01[b] |
| Austria | 2.71 | 2.35 | 2.19 | 2.03 | 1.96 | 1.89 | 1.77[b] |
| Canada | 3.68 | 3.08 | 2.83 | 2.63 | 2.54 | 2.47 | 2.39[b] |
| Finland | 3.37 | 2.70 | 2.42 | 2.20 | 2.12 | 2.06 | 2.01[b] |
| Japan | 3.33 | 2.89 | 2.71 | 2.57 | 2.51 | 2.46 | 2.40[b] |
| Norway | 1.62 | 1.54[b] | 1.54[b] | 1.56 | 1.58 | 1.61 | 1.68 |
| Spain | 2.06 | 1.85 | 1.76 | 1.70 | 1.68 | 1.66 | 1.65[b] |
| Sweden[c] | 2.32 | 2.18 | 2.12 | 2.07 | 2.05 | 2.03 | 2.01[b] |
| Switzerland | 3.45 | 2.98 | 2.81 | 2.71 | 2.68[b] | 2.68[b] | 2.71 |
| | | | | | | | |
| Median RMSE | 2.62 | 2.34 | 2.22 | 2.21 | 2.16 | 2.13[b] | 2.14 |
| Min. RMSE | 1.62 | 1.54 | 1.54 | 1.56 | 1.58 | 1.61 | 1.65 |
| Max. RMSE | 3.68 | 3.08 | 2.83 | 2.71 | 2.68 | 2.68 | 2.79 |

[a] The coefficient estimate in (13) was employed to compute forecasts. When $\gamma = 0$, the forecasts are least squares forecasts, and when $\gamma = 10^6$, they are produced using the coefficent estimate in (8).
[b] Minimum of entries in each row.
[c] Based on ten forecasts, 1974-83.

A zero value for $\gamma$ yields AR(3)LI least squares forecasts while a very large value for $\gamma$ results in an AR(3)LI forecast based on the pooled coefficient estimate in (8). From the median RMSEs reported at the bottom of Table 5, it is seen that a common value of $\gamma = 5.0$ yields a median RMSE equal to 2.13, not far different from that associated with $\gamma = 10^6$, namely 2.14. However for individual countries, RMSEs show more substantial variation as $\gamma$ assumes different values. For example in the case of Germany, the RMSE is 1.80 when $\gamma = 2.0$, quite a bit lower than the RMSE of 2.00 associated with $\gamma = 10^6$. Ten of the eighteen countries show minimal RMSEs for $\gamma = 10^6$, while the remaining eight countries have minima in the vicinity of $\gamma = 0.5$ to $\gamma = 1.0$ in five cases and of $\gamma = 2$ to $\gamma = 5$ for the remaining three. The median RMSE for these minimal values is 2.01, slightly lower than the median RMSE of 2.14 when $\gamma = 10^6$ is used for all countries. In future work it may be worthwhile to estimate $\gamma$ for each country which will probably produce lower RMSEs for the selected countries, e.g., Belgium, the Netherlands, the U.S., and Norway. Also combinations of $\gamma$-forecast and $\eta$-forecast procedures are under consideration as well as other shrinkage techniques.

### 4.2 Forecasting Using a World Output Growth Rate Variable

As mentioned previously, countries' exports are influenced by world income or output. Thus a world output growth rate variable, denoted by $w_t$, was added to the AR(3)LI model as shown in eq. (15). The variable

Table 6.
RMSEs and MAEs of one-year-ahead forecasts of the medians of eighteen countries' output growth rates, 1974-84.[a]

| Model | RMSE | MAE |
|---|---|---|
| | (Percentage points) | |
| AR(3) | 2.74 | 2.24 |
| AR(3)LI[b] | 1.48 | 1.24 |

[a] The initial estimation period is 1954-73 (20 years). Estimates are updated year by year in the forecast period.
[b] With $w_t$ being median output growth rate in year $t$, the AR(3)LI model is $w_t = \pi_0 + \pi_1 w_{t-1} + \pi_2 w_{t-2} + \pi_3 w_{t-3} + \pi_4 x_{t-1} + \pi_5 z_{t-1} + \epsilon_t$, where, for year $t$, $x_t$ is the median of countries' growth rates of real stock prices and $z_t$ the median of countries' growth rates of real money. This equation was employed to generate one-year-ahead least squares forecasts for each year, 1974-84.

$w_t$ is taken to be the median of the eighteen countries' output growth rates for the year $t$ - see Fig. 1 for a plot of $w_t$ for the years in our sample.

To use (15) in forecasting, it is necessary to forecast $w_t$. The model for $w_t$ in (16) was fitted by least squares, using data from 1954-73, and used to forecast the 1974 value and subsequent values with coefficient estimates updated year by year for the years 1974-84. The RMSEs of these one-year-ahead forecasts and of those yielded by an AR(3) model of $w_t$ are reported, along with MAEs, in Table 6. It is seen that the AR(3)LI model for $w_t$ produced a RMSE of 1.48 and MAE of 1.24, values much smaller than those associated with forecasts from an AR(3) model, 2.74 and 2.24, respectively. The forecasts from these two models for $w_t$, denoted by $\hat{w}_t(1)$

Table 7.
RMSEs of one-year-ahead forecasts of annual real output growth rates employing an AR(3)LI model including the world growth rate, 1974-84.[a]

| Country | AR(3)LI with $w_t^b$ | AR(3)LI with $\hat{w}_t(1)^c$ | AR(3)LI with $\hat{w}_t(2)^d$ | AR(3)LI without $w_t^e$ |
|---|---|---|---|---|
| | (Percentage points) | | | |
| Belgium | 1.54 | 2.80 | 1.79 | 1.96 |
| Denmark | 1.77 | 2.74 | 2.04 | 2.26 |
| France | 1.03 | 1.90 | 1.36 | 1.66 |
| Germany | 0.90 | 2.39 | 1.35 | 2.00 |
| Ireland | 2.82 | 1.57 | 2.54 | 2.14 |
| Italy | 1.57 | 3.08 | 1.82 | 2.45 |
| Netherlands | 1.86 | 2.52 | 2.27 | 2.53 |
| U.K. | 1.87 | 2.67 | 2.21 | 2.39 |
| U.S. | 2.56 | 2.95 | 2.36 | 2.79 |
| Australia | 2.24 | 2.39 | 1.96 | 2.01 |
| Austria | 1.23 | 2.06 | 1.57 | 1.77 |
| Canada | 2.27 | 2.81 | 2.15 | 2.39 |
| Finland | 2.14 | 2.13 | 1.83 | 2.01 |
| Japan | 2.85 | 2.30 | 2.58 | 2.40 |
| Norway | 1.66 | 1.79 | 1.45 | 1.68 |
| Spain | 1.12 | 2.03 | 1.22 | 1.65 |
| Sweden[f] | 1.62 | 2.58 | 1.70 | 2.01 |
| Switzerland | 2.20 | 3.17 | 2.42 | 2.71 |
| Median RMSE | 1.82 | 2.46 | 1.90 | 2.14 |
| Min. RMSE | 0.90 | 1.57 | 1.22 | 1.65 |
| Max. RMSE | 2.85 | 3.17 | 2.58 | 2.71 |

[a] The model employed is shown in (15) and estimated under the assumption that coefficients are the same for all countries.
[b] The value of $w_t$, the median output growth rate or world growth rate, is assumed known in the forecast period, a 'perfect foresight' assumption.
[c] $\hat{w}_t(1)$ is a forecast of $w_t$ from an AR(3) model for $w_t$ - see Table 6.
[d] $\hat{w}_t(2)$ is a forecast of $w_t$ from the AR(3)LI model described in footnote $b$ of Table 6.
[e] Least squares forecasts using AR(3)LI model in (1), with the use of coefficient estimate in (8).
[f] Based on ten forecasts, 1974-83.

for the AR(3) forecasts and $\hat{w}_t(2)$ for the AR(3)LI model forecasts, were used to generate one-year ahead forecasts from (15) for individual countries' output growth rates for 1974-84, using the $\gamma$-forecast with a very large value of $\gamma$. Also, for comparative purposes, a 'perfect foresight' model, one in which it is assumed that $w_t$ is known exactly, was employed to generate forecasts with results given in the first column of Table 7. With the $w_t$ value assumed known in each forecast period, the median RMSE of the annual forecasts for the eighteen countries, 1974-84, is 1.82, a value much lower than those reported in Table 3, and for the AR(3)LI model without the $w_t$ variable it is 2.14, shown in the last column of Table 7. In column 2 of Table 7 are shown forecast RMSEs when $w_t$ was forecasted using an AR(3) model. Since the AR(3) forecasts of $w_t$ are not very good - see Table 6 - the forecasts of country output growth rates based on them are in general not as good as those based on known values of the $w_t$ variable. When the AR(3)LI model in (16) was used to produce forecasts of $w_t$, denoted by $\hat{w}_t(2)$, and these were used to forecast individual countries' output growth rates, the results, as shown in column 3 of Table 7, were much better. The median RMSE associated with these forecasts is 1.90, not far different from the 'perfect foresight' median RMSE of 1.82. Also when $\hat{w}_t(2)$ was employed, sixteen of eighteen countries' RMSEs were reduced relative to the RMSEs for the AR(3)LI model without the world growth variables, shown in column 4 of Table 7. Further, the median RMSE of 1.90, associated with the AR(3)LI world growth rate model, is smaller than all of those shown in Table 3.

### 4.3 Comparisons with OECD Forecast RMSEs

Smyth (1983) has presented a description of the forecasting procedures employed by the Organization for Economic Cooperation and Development (OECD) to produce annual forecasts of seven countries' annual rates of growth of output, 1968-79. The OECD forecasts are derived from elaborate country econometric models and are subjected to judgemental adjustments by individuals not associated with the modeling process, that is by individuals from the OECDs various country desks - see Smyth (1983, p. 37). In Garcia-Ferrer et al. (1987, p. 61ff) comparisons of RMSEs of OECD forecasts for five countries with those provided by AR(3)LI models were presented and discussed. In Table 8, OECD RMSEs of forecast for seven countries, 1968-79, are presented along with forecast RMSEs for the same countries, 1974-84, computed in the present study. While the forecast periods 1968-79 and 1974-84 are somewhat different and different methodologies were employed, it is still of interest to consider the relative forecasting performance of OECD and our forecasts.

From Table 8, it is seen that the OECD forecast RMSEs have a median of 2.12 perentage points with a range of 1.45-4.40. The OECD's RMSE for Japan, 4.40 is quite large. The median RMSEs, in our study range from 2.52 for least squares forecasts from our AR(3)LI model in (1) to 2.15 for the forecasts obtained from our 'AR(3)LI world income' model in (15) using a forecasted value of $w_t$ from (16) and $\gamma$-forecast with $\gamma = 10^6$. The range of the forecast RMSEs in this latter case is 1.35-2.58. The RMSEs in line 3 of Table 8 are smaller than the corresponding OECD RMSEs in five of seven cases. Large reductions were encountered for Germany, Italy and Japan, while smaller reductions appeared for France and the U.K. In the cases of

Table 8.
Comparison with OECD forecast RMSEs.

| Forecasts | Canada | France | Germany | Italy | Japan | U.K. | U.S. | Median | Range |
|---|---|---|---|---|---|---|---|---|---|
| | | | (Percentage points) | | | | | | |
| 1. OECD, 1968-79 | 1.71 | 1.45 | 2.12 | 2.86 | 4.40 | 2.26 | 1.38 | 2.12 | 1.38-4.40 |
| 2. AR(3)LI, 1974-84 | | | | | | | | | |
| a. Least Squares | 3.68 | 2.52 | 2.28 | 3.40 | 3.33 | 2.32 | 2.14 | 2.52 | 2.14-3.68 |
| b. $\gamma$-forecasts[a] | 2.39 | 1.66 | 2.00 | 2.45 | 2.40 | 2.39 | 2.79 | 2.39 | 1.66-2.79 |
| c. $\eta$-forecasts[b] | 2.92 | 2.07 | 1.94 | 2.73 | 2.51 | 2.63 | 2.03 | 2.51 | 1.94-2.92 |
| 3. AR(3)LI with forecasted $w_t$, 1974-84[c] | 2.15 | 1.36 | 1.35 | 1.82 | 2.58 | 2.21 | 2.36 | 2.15 | 11.35-2.58 |

[a] $\gamma = 10^6$ in model for eighteen countries.
[b] $\eta = 0.5$ in model for eighteen countries.
[c] $w_t$, the world output growth rate was forecasted from (16) and used in (15) to produce $\gamma$-forecasts for eighteen countries with $\gamma = 10^6$.

the U.S. and Canada, the OECD RMSEs were smaller, much smaller for the U.S. These comparisons, however, must be qualified since the forecast period and data used by OECD are different from those employed in this study.

## 5. SUMMARY AND CONCLUDING REMARKS

We have presented the results of forecasting experiments for nine countries' annual output growth rates for the periods 1974-81 and 1974-84 and similar results for an additional sample of nine countries for the period 1974-84. In general, the forecasting experiments revealed that methods and models employed earlier in Garcia-Ferrer et al. (1987) worked reasonably well when applied to an extended sample of data and countries. The shrinkage forecasting procedures produced larger reductions in RMSEs of forecast for the nine additional countries than for the original sample of nine countries. The AR(3)LI model, incorporating a world income growth rate variable, forecasted the best and provided forecast RMSEs that compared favorably with the RMSEs of OECD forecasts for a subsample of seven countries that were produced using complex country models and judgmental adjustment.

In future work, we shall extend our forecasting experiment to include forecasts derived from time-varying parameter Bayesian state-space models - see Garcia-Ferrer et al. (1987) and Highfield (1986) for macroeconomic forecasting results obtained with such models. In addition, as stated in Section 1, macroeconomic structural equation systems are under consideration that have reduced form equations for the rate of growth of output that are similar in form to the forecasting equations used in our present and past work. These structural macroeconomic equation systems also yield reduced form equations for several additional variables that will be appraised in future forecasting experiments. With several forecasting equations per country, there will be an opportunity to experiment with a broader range of shrinkage forecasting techniques. Hopefully, this work will yield a set of forecasting equations for each country that yield reasonably good forecasts for a broad sample of countries and for a temporally expanded data set. Further, disaggregation will be studied by modeling components of GNP as fractions or log-fractions of relevant aggregrates. In this way, we hope to 'iterate in' to satisfactory structural models for countries in the structural econometric modeling, time series analysis (SEMTSA) approach put forward in our past work - see Zellner and Palm (1974, 1975), Palm (1983) and Zellner (1979, 1984, 1987).

## ACKNOWLEDGMENT

This research was supported by the National Science Foundation and by income from the H.G.B. Alexander Endowment Fund, Graduate School of Business, University of Chicago. Luis Mañas-Antón, Carla Inclan and Michael Zellner provided valuable research assistance.

## REFERENCES

Bowman, H.W. and Laporte, A.M. (1975) Stochastic optimization in recursive systems with random parameters with an application to control of the money supply, in S. E. Fienberg and A. Zellener, eds., *Studies in Bayesian Econometrics and Statistics in Honor of Leonard J. Savage* (North Holland, Amsterdam) 441-462.

Garcia-Ferrer, A., R.A. Highfield, F. Palm, and A. Zellner (1987) Macroeconomic Forecasting Using Pooled International Data, *Journal of Business and Economic Statistics,* **5**, 53-67.

Highfield, R.A. (1986) Forecasting with Bayesian State Space Models, unpublished Doctoral Dissertation, Graduate School of Business, University of Chicago.

Lindley, D.V. and A.F.M. Smith (1972) Bayes' Estimates for the Linear Model, *Journal of the Royal Statistical Society, Series B*, **34**, 1-41.

Miller, R.B., and Fortney, W.G. (1984) Industry-wide expense standards using random coefficient regression, *Insurance: Mathematics and Economics* **3**, 19-33.

Palm, F.C. (1983) Structural economic modelling and time series analysis: An integrated approach, in A. Zellener, ed., *Applied Time Series Analysis of Economic Data, Proceedings of the Conference on Applied Time Series Analysis of Economic Data*, Oct. 13-15, 1981 (U.S. Government Printing office, Washington, D.C.) 199-233.

Smyth, D.J. (1983) Short-run macroeconomic forecasting: The OECD performance, *Journal of Forecasting,* **2**, 37-49.

Zellner, A. (1971) *An Introduction to Bayesian Inference in Econometrics* (Wiley, New York); reprinted in 1987 (Krieger, Melbourne, Florida)

Zellner, A. (1979) Statistical analysis of econometric models, *Journal of the American Statistical Association,* **74**, 628-651.

Zellner, A. (1983) Applications of Bayesian analysis in econometrics, *The Statistician,* **32**, 23-34.

Zellner, A. (1984) *Basic Issues in Econometrics,* (University of Chicago Press, Chicago, IL)

Zellner, A. (1986) On assessing prior distributions and Bayesian regression analysis with g-prior distributions, in: P. K. Goel and A. Zellner, eds., *Bayesian Inference and Decision Techniques: Essays in Honor of Bruno de Finetti* (North Holland, Amsterdam) 233-243

Zellner, A. (1987) Macroeconomics, econometrics and time series analysis, *Revista Espanola do Economica*, **4**, 3-9.

Zellner, A. and Palm, F.C. (1974) Time series analysis and simultaneous equation econometric models, *Journal of Econometrics*, **2**, 17-54.

Zellner, A. and Palm, F.C. (1975) Time series and structural analysis of monetary models of the U.S. economy, *Sankhyā*, **37**, 12-56.

## COMMENTS

N. Sreenivas Iyengar
Indian Statistical Institute, Bangalore, India

This paper is an extension of an earlier work by Zellner with Garcia-Ferrer, Highfield and Palm (1987), who had established that a third-order auto-regressive model with leading indicators, AR(3)LI, using diffuse prior or least squares estimates, provides reasonable forecasts with small root mean squared errors (RMSE). They had also found that certain shrinkage procedures could give even better results while using the Bayesian approach for forecasting the annual real growth rates in the gross national or domestic products of certain countries in the industrial world.

The GHPZ model consists of one reduced-form equation in which the real GNP growth rate in the year $t, y_t$, is explained in terms of its past values, $y_{t-1}, y_{t-2}, y_{t-3}$, and three leading indicators suitably lagged. The latter include two market variables, (SR) and (GM), denoting respectively, the annual growth rates in real stock returns and annual growth rates in real money supply. The third leading indicator is the world variable (WR), which is defined as 'world return' represented by the median of the countries' real stock returns. Lagged variables $(SR)_{t-1}, (SR)_{t-2}, (GM)_{t-1}$, and $(WR)_{t-1}$ are used as regressors to forecast $y$, making use of the annual time series data (1951-73) for nine countries, taken mainly from the IMF's *International Financial Statistics*. The fitted model is used to forecast growth rates for 1974 and subsequent years with the model re-estimated year-by-year. For the forecasting period (1974-81) the least squares forecasting yields reasonably good forecasts with low RMSEs ranging for 1.47 to 2.92, with a median of 2.23 percentage points, for eight EEC countries and the United States.

The Zellner-Hong (ZH) study further extends the GHPZ model to nine more countries and cover a longer period (1974-84). In their modified model, an additional world variable ($w_t$), which is designed to reflect the world's real income growth, is introduced. As a proxy for $w$ the median of the selected countries' real output growth is used. The current level of $w$, which is endogenous, has to be forecasted for future periods using a suitable reduced-form equation, as in the two-stage least squares approach. This reduced-form equation is assumed to be an AR(3)LI in which the leading indicators are (MSR) and (MGM) lagged by one year. The latter variables represent the medians of the countries' SRs and GMs.

The authors provide valuable analytical insights into the derivation of the shrinkage Bayesian procedures. Their extensive numerical exercises show that the $\eta$- and $\gamma$- shrinkage forecasts using the ZH specification vary between 1.25 and 2.52 with the median of 1.78 percentage points. It may be noted that the corresponding figures in the GHPZ model are slightly higher. In simplicity and performance, the modified ZH model appears to give far better results than the elaborate methods and models employed in the OECD study for the same group of countries.

However, the main focus of the paper is on the techniques and not so much on economic interpretation of their empirical results. In fact, the paper does not address itself to the determination of the factors of growth rate changes and the economic analysis of the individual coefficients, with discussion of the standard errors, the $t$-values, the $\bar{R}^2$ and so on, as in the traditional econometric approach. This is easily understandable, as the purpose of the paper is mainly to demonstrate what Bayesian methodology could do in an important area of economic research, namely making dependable economic forecasts and determining future targets for planning and policy purposes.

In the Bayesian approach, questions usually center round the choice of the loss function and the prior beliefs about the probability density functions (pdf) for the hyperparameters to be estimated. The choice of an appropriate loss function and prior pdf is, however, dictated by necessity and access to the right type of data. Then there is the specification problem, and matching the data with prior specifications could introduce some subjectivity. The use of diffuse prior may be justified in some cases, but this requires stronger assumptions and reliance on judgments based on economic theory. In spite of these well-known limitations, Zellner (1988) has elsewhere argued for the Bayesian approach where the orthodox econometrics cannot provide any solutions. Another common objection to the paper is that it relies solely on one criterion, viz., the RMSE of the forecast. The sampling properties - in large and small samples - are not discussed.

A small sample of countries which are industrially advanced does not make the world, so that the use of terms like 'world growth rate' could create problems of interpretation, particularly in the underdeveloped world. I certainly see a possibility of applying the ZH model to some of the developing countries like India, endowed with good data base, though with certain modification to suit local conditions. I am glad to be told that the authors are themselves planning to undertake such useful extensions to the less developed world.

## REJOINDER

We thank Professor Iyengar for his thoughtful and perceptive comments on our paper. He appreciates very well our objective to show that sophisticatedly simple models and methods can produce more satisfactory results, here forecasting results, than those of more complicated models and methods–see also Zellner, Hong and Gulati (1988) and Zellner, Hong and Min (1989) for evidence on the ability of our models and methods to forecast turning points in rates of growth of 18 countries, 1974-86–approximately 70 percent of 158 turning points forecasted correctly.

On usual coefficient estimates, $t$-values for individual coefficients and goodness of fit measures, which Professor Iyengar mentions, Bayesian posterior means and standard deviations for coefficients and $R^2$'s, approximate posterior means of population $R^2$'s, are presented in Zellner, Hong and Gulati (1988) country by country. For the period 1954-85, the $R^2$'s range from .26 for Norway to .76 for the Netherlands with a median of .56. Further, while the posterior standard deviations for some coefficients were large relative to their corresponding posterior means, the posterior means of coefficients of particular variables tended to have the same algebraic signs country by country. In a pooled relation for the 18 countries, coefficients' posterior means were much larger than their respective standard deviations. It is important to note that usual 5% sampling theory, approximate, asymptotic $t$-tests do not have much power in samples of sizes 20 to 32 observations and thus should not be used mechanically. The sample first order serial correlation coefficients for residuals were found to be small. Also, in Hong (1989), the roots of the AR(3) part of our model were analyzed with the finding that with high posterior probability for each country, there are two complex conjugate roots and one real root. The amplitudes and periods associated with the oscillatory components were found to be reasonable. Thus we have studied many statistical properties of our AR(3)LI model.

On the economic interpretation of our short-run forecasting model, mentioned by Professor Iyengar, we have employed heuristic economic knowledge in formulating our models, an AR(3) component to allow for the possibility of an oscillatory component and a local trend component, leading indicator variables suggested by past studies and monetary theory, etc. In Zellner and Anton (1986), a simple, structural, short-run aggregate supply and demand model is formulated which has a reduced form equation very similar to our AR(3)LI-world income model. In Hong (1989), it is shown that an "IS-LM" macroeconomic model has a reduced form equation in the form of our forecasting model. Thus previous analysis has indicated compatibility of our forecasting model with certain, very simple, structural macroeconomic models. However, much more research and testing is needed to discriminate among alternative structural models.

We have used diffuse prior assumptions in some of our analyses in an effort to let the data "speak for themselves" and also to provide easy comparisons with non-Bayesian results. However, in the $\gamma$- shrinkage approach, more prior information in the form of a hyper-distribution for the parameters is employed and shown to produce better forecasting results for many countries. In Garcia-Ferrer et al. (1987) and Zellner, Hong and Min (1989), more prior information was introduced by assuming that coefficient vectors in our models are time-varying and follow a random walk process which yielded improved forecasting results. By starting simply with fixed parameter models and diffuse prior assumptions, we obtained benchmark forecasting results to compare to those yielded by models incorporating more prior information.

As regards applying our methods and models to data of developing countries, e.g. India, we have forecasting results for Venezuela and India. For forecasts of the Venezuelan annual growth rate of output, 1974-86, the RMSEs of forecast for our AR(3)LI model ranged from 3.3 to 3.7, with the former figure relating to our deletion of 1957-58 data, the time of a revolution and use of a 0-1 dummy variable for the years 1954-64 and 1965-73 and the latter without these modifications to the data for the initial estimation period. For India, with an initial estimation period, 1954-1977, application of the $\gamma$-shrinkage approach produced the results shown in Table R1. With $\gamma = 0$, the case of diffuse prior, least squares forecasts, the RMSE is equal to 4.3 percentage points. With $\gamma = 3.0$ or $\gamma = 5.0$, the RMSE is equal to 3.0. As can be seen in the Table R1, a large forecast error was made in 1979, a year in which the actual Indian growth rate was -4.9 and our forecasts ranged from 2.0 to 2.4. Professor N.S.S. Narayana pointed out in private conversation at our Bangalore meeting that 1979 was a drought year and also that the authorities instituted a tight control of the money supply to counter effects of increased oil prices in 1979. Perhaps we have to include a weather variable in our model for India.

Table R1.
One Year Ahead Forecasts of Indian Output Growth Rates (% points) Using $\gamma$-Shrinkage and Data for 18 Other Countries, 1978-86[a]

| Year | Actual Growth Rate | One Year Ahead Forecasts Using Various Values of $\gamma$ | | | | |
|---|---|---|---|---|---|---|
| | | 0.0 | 1.0 | 3.0 | 5.0 | $10^6$ |
| 1978 | 6.4 | 2.5 | 3.7 | 4.4 | 4.6 | 5.0 |
| 1979 | -4.9 | 2.0 | 2.2 | 2.3 | 2.4 | 2.4 |
| 1980 | 6.3 | 5.2 | 4.3 | 3.8 | 3.6 | 3.3 |
| 1981 | 5.8 | 8.3 | 6.2 | 5.2 | 4.8 | 4.2 |
| 1982 | 3.6 | 4.1 | 3.3 | 3.0 | 2.9 | 2.6 |
| 1983 | 7.5 | 3.6 | 4.0 | 4.2 | 4.3 | 4.4 |
| 1984 | 3.8 | -4.1 | 0.7 | 3.1 | 3.9 | 5.5 |
| 1985 | 6.1 | 1.8 | 3.6 | 4.5 | 4.8 | 5.4 |
| 1986 | 4.3 | 3.1 | 4.9 | 5.8 | 6.1 | 6.7 |
| RMSE | - - | 4.3 | 3.2 | 3.0 | 3.0 | 3.1 |

[a] Forecasts derived from AR(3)LI models for individual countries. The initial estimation period is 1954-1977.

## ADDITIONAL REFERENCES

Hong, C. (1989), Forecasting Real Output Growth Rates and Cyclical Properties of Models, *Core of PhD Thesis*, Department of Economics, U. of Chicago.

Zellner, A., C. Hong and G.M. Gulati (1990), Turning Points in Economic Time Series, Loss Structures and Bayesian Forecasting, in S. Geisser, J. Hodges, S.J. Press, and A. Zellner (eds.), *Bayesian and Likelihood Inference Methods in Statistics and Econometrics: Essays in Honor of George A. Barnard*, Amsterdam: North-Holland 371-393.

Zellner, A., C. Hong and C. Min (1989), Forecasting Turning Points in International Output Growth Rates Using Bayesian Exponentially Weighted Autoregression, Time-Varying Parameter and Pooling Techniques, Technical Report, H.G.B. Alexander Research Foundation, Graduate School of Business, U. of Chicago.

Zellner, A. and L.M. Antón (1986), Macroeconomic Theory and International Macroeconomic Forecasting, Technical Report, H.G.B. Alexander Research Foundation, Graduate School of Business, U. of Chicago.

# 23

# Estimation in the Linear Regression Model with Errors in Variables

## Avanindra N. Bhat [1]

**Abstract**

Bayesian estimation of the unknown parameter vector in the linear regression model, when the independent variables are measured with errors, is considered. Our approach takes into account the indeterminacy of the MLE approach and generalizes the Bayesian approach of Zellner(1970). Conditional on the variance of the error term and on the variance of the prior for the parameter vector, the Bayes estimator depends on these variances through a shrinkage factor. Assuming a proper prior on these parameters, the Bayes estimator is evaluated with respect to the posterior mean of the shrinkage factor.

*Key words and Phrases:* Errors in Variables; Shrinkage factor; Estimation; MLE; Variance components.

## 1. INTRODUCTION

Economists often work with datasets that consist of variables that are not accurately measured. This paper focuses on the problem of estimating the unknown parameter vector in the linear regression context when there is error in measuring the independent variables. Some important examples with this data structure are given in Feldstein (1974), Griliches (1974), Levi (1973, 1977), Nerlove (1971) and Zellner (1970).

A Bayesian estimation procedure, which takes into account the indeterminacy of the MLE approach (Judge et.al.,(1980), p. 514) and generalizes the Bayesian approach of Zellner (1970), is proposed in the paper. A class of informative, proper priors is assumed on the variance of the error term and on the variance of the prior for the parameter vector. The Bayes estimator is known to depend on these unknown variances through a shrinkage factor. Typically, these variances being unknown, the Bayes estimator is not operational and cannot be used to report estimates.

The paper proceeds to obtain the posterior density of the unknown variance components in the context of the model. (This in turn enables us to

---

[1] Jawaharlal Nehru University, New Delhi, India

compute the posterior mean of the shrinkage factor). Such a posterior density has been obtained in the context of the one way analysis of variance models by Hill (1965, 1967), but is new in this context. The Bayes estimator is then evaluated with respect to the posterior mean of the shrinkage factor. This leads to a new estimator of the parameter vector which is known to be admissible.

Section 2 presents the regression model with explicit assumptions that incorporate the measurement errors. The prior distribution assumed on $\beta$ is discussed. The induced prior on $\alpha$ (the reparameterized vector) is derived. In Section 3, alternative estimators of $\beta$ in the context of this model are discussed together with our Bayes estimator of $\beta$. The Bayes estimator of $\alpha$ is explicitly presented. Section 4 derives the posterior distribution of the unknown variance components. This in turn enables us to obtain the posterior mean of the 'shrinkage' factor. (The Bayes estimator depends on this 'shrinkage' factor). We can now evaluate the Bayes estimator with respect to this posterior mean. This leads to a new estimator of $\alpha$ and hence, $\beta$. Conclusions are drawn in Section 5.

## 2. THE LINEAR REGRESSION MODEL

It is only in rare cases that the measurements of the independent variables are absolutely error-free. Generally, economic data derived from government or private publications report only significant digits. These rounding off errors are a likely cause of error in the measurement of variables. In addition, some variables may have inherent difficulties in their measurement.

We will present and consider the problem of errors in variables using the following model as an example:

$$\begin{aligned} \mathbf{Y} &= \tilde{\mathbf{X}}\beta + \epsilon \\ \tilde{\mathbf{X}} &= \mathbf{X} + \mathbf{V}, \end{aligned} \tag{2.1}$$

where,

**Y:** $T \times 1$ vector of observations on the dependent variable.

$\tilde{\mathbf{X}}$: $T \times K$ matrix of the true non-stochastic, independent variables that are not accurately observed.

**X:** $T \times K$ matrix of observed non-stochastic explanatory variables assumed to be distributed independently of $(\beta, \sigma^2)$; $\mathbf{X}'\mathbf{X}$ is generally assumed to be nonsingular.

$\beta$: $K \times 1$ vector of unknown parameters of the model, assumed to have a normal prior, $N[\mu, \sigma_\beta^2 \mathbf{I}]$.

$\epsilon$: $T \times 1$ vector of spherical disturbances assumed to have $N[0, \sigma^2 \mathbf{I}]$ distribution.

**V**: $T \times K$ matrix of measurement errors assumed to have a $N[0, \sigma_v^2 \mathbf{I}]$ distribution.

A multivariate normal prior on $\beta$ is assumed in this analysis. This prior represents the investigator's subjective beliefs about $\beta$, essentially capturing his uncertainty about $\beta$. In the classical framework, this would be a random effects model and the analysis here is in the tradition of random coefficient regression models. The Bayesian analysis proposed here is a multivariate generalization of work by Hill (1967, 1969) in the context of the one way analysis of variance models.

A multivariate transformation that reparameterizes the parameter vector $\beta$ is considered. This transformation is a convenient intermediate step in the estimation of $\beta$.

Description of the transformation:

Consider the model in (1) and assume that $\mathbf{X}'\mathbf{X}$ and $\mathbf{V}'\mathbf{V}$ are positive definite. There exist orthogonal $(K \times K)$ matrices $\mathbf{A}, \mathbf{B}$ such that

$$\mathbf{A}'(\mathbf{X}'\mathbf{X})\mathbf{A} = \mathbf{\Lambda} \text{ and } \mathbf{B}'(\mathbf{V}'\mathbf{V})\mathbf{B} = \mathbf{\Gamma},$$

where $\mathbf{\Lambda} =$ Diag $(\lambda_1, \lambda_2, \ldots \lambda_K), \lambda_i$ are the eigenvalues of $\mathbf{X}'\mathbf{X}$ and $\mathbf{\Gamma}$ is the matrix of eigenvalues of $\mathbf{V}'\mathbf{V}$. The transformation matrix $\mathbf{A}$ depends on $\mathbf{X}$, and $\mathbf{B}$ depends on $\mathbf{V}$ (Both matrices $\mathbf{A}$ and $\mathbf{B}$ are stochastic; $\mathbf{V}$ being unobserved, and so $\mathbf{B}$ is unknown). Using this transformation, the model in (1) is reparametrized as shown below:

The model in (1) is now written as

$$\mathbf{Y} = \mathbf{Z}\alpha + \mathbf{U}\gamma + \epsilon, \tag{2.2}$$

where

$$\mathbf{Z} = \mathbf{XA} \text{ and } \mathbf{Z}'\mathbf{Z} = \mathbf{\Lambda}, \alpha = \mathbf{A}'\beta,$$
$$\mathbf{U} = \mathbf{VB} \text{ and } \mathbf{U}'\mathbf{U} = \mathbf{\Gamma}, \gamma = \mathbf{B}'\beta.$$

The prior assumed on $\beta$ induces a prior on $\alpha$ given $\mathbf{X}$; namely, $\alpha \sim N[\mathbf{A}'\mu, \sigma_\alpha^2 \mathbf{I}]$ conditional on $\mathbf{X}$.

## 3. THE BAYES ESTIMATOR OF THE PARAMETERS

In this section, the focus is on the estimation of $\alpha$ in the reparametrized model (2), which is a convenient intermediate step in the estimation of $\beta$. Similar econometric models have been analyzed in the literature both in a classical as well as a Bayesian framework. It is well known that the OLS estimator in such a context is not only *biased* but *inconsistent*. [See Judge et al. (1980), p. 514]. A maximum likelihood procedure is proposed (Judge et al. (1980), pp. 518-531) which does not yield an explicit solution to the estimation problem due to insufficient information. (The MLE procedure leads to a system of equations which can be solved for the estimators of the different parameters. The problem, however, is that there are more unknowns than the number of equations and hence, the MLE cannot be obtained). For a complete solution to the MLE approach, it is suggested that further information is needed on the variance component $\sigma^2$ and/or $\sigma_v^2$. Thus, a Bayesian approach is natural and justified in this case.

The Bayes estimator of $\alpha$ in (2) is written as

$$\begin{aligned}\hat{\alpha}(\mathbf{B}) &= E[\alpha|\mathbf{y},\mathbf{X},\sigma^2,\sigma_\alpha^2,\delta] \\ &= \mathbf{W}^{-1}(\delta^{-1}\boldsymbol{\Lambda})\hat{\alpha}(o)+[I-\mathbf{W}^{-1}(\delta^{-1}\boldsymbol{\Lambda})]\mathbf{A}'\mu,\end{aligned}$$

where

$$\begin{aligned}\mathbf{W} &= (\delta^{-1}\boldsymbol{\Lambda}+\mathbf{K}), \mathbf{K}=t^{-1}\mathbf{I}, t=\sigma_\alpha^2/\sigma^2, \\ \hat{\alpha}(\mathbf{B}) &= \text{the Bayes estimator of } \beta, \\ \hat{\alpha}(o) &= \text{the OLS estimator of } \beta, \text{ and} \\ \delta &= (1+t\sigma_v^2).\end{aligned}$$

The $i^{th}$ component of this estimator is $\hat{\alpha}_i(\mathbf{B})=\eta_i\hat{\alpha}_i(o)+(1-\eta_i)\mathbf{a}_\mathbf{i}'\mu, i = 1,2,\ldots K$ where

$$\eta_i=\frac{\lambda_i\sigma_\alpha^2}{(\lambda_i\sigma_\alpha^2+\sigma^2)};$$

$\mathbf{a_i}$ is the $i^{th}$ column of $\mathbf{A}$ and $\sigma^2,\sigma_\alpha^2,\delta$ are given. The analysis proceeds in Section 4 to assume prior distributions on $(\sigma^2,\sigma_\alpha^2,\sigma_v^2)$ and obtain the posterior density of $(\sigma^2,\sigma_\alpha^2,\sigma_v^2)$. Further, the posterior mean of $\eta_i$ will be obtained from the posterior density mentioned above. We evaluate $\hat{\alpha}_i(\mathbf{B})$ by taking the posterior expectation with respect to $\eta_i$:

$$\hat{\alpha}_i(\mathbf{B})=E[\eta_i|\text{ data }]\hat{\alpha}_i(o)+(1-E[\eta_i|\text{ data }])\mathbf{a}_\mathbf{i}'\mu.$$

## 4. THE POSTERIOR DISTRIBUTION OF THE VARIANCE COMPONENTS $(\sigma^2,\sigma_\alpha^2,\sigma_v^2)$

In this section, we derive the posterior density of the unknown variance components $(\sigma^2,\sigma_\alpha^2,\sigma_v^2)$. In turn, the posterior density of suitably transformed variance components can be obtained. $E[\eta_i|$ data $]$ is then obtained which enables us to make operational the estimator $\hat{\alpha}_i(\mathbf{B})$ presented in Section 3.

For model (2), the likelihood function is given by

$$L(\alpha, \sigma^2, \delta) \propto (\sigma^2\delta)^{-T/2} \exp[-(\mathbf{y} - \mathbf{Z}\alpha)'(\mathbf{y} - \mathbf{z}\alpha)/2\sigma_\delta^2],$$

where $\alpha$ depends on $\mathbf{X}$ and $\beta$.

Incorporating the prior on $\alpha$, it is easily seen that $\mathbf{y} = \mathbf{Z}\alpha + \mathbf{U}\gamma + \epsilon \sim N[0, \mathbf{\Sigma}]$, where $\mathbf{\Sigma} =$ Cov $(\mathbf{y}|\mathbf{X}, \mathbf{V}) = \sigma^2[\mathbf{I} + t\mathbf{Z}\mathbf{Z}' + t\sigma_v^2\,\mathbf{I}]$. The likelihood function, as a function of $(\sigma^2, \sigma_\alpha^2, \sigma_v^2)$ can be written as the equation (A.1) in the appendix.

A class of informative priors is assumed on the unknown variance components $\sigma^2, \sigma_\alpha^2$ and $\sigma_v^2$ from the inverted gamma family of densities. Assuming that the priors on $(\sigma^2, \sigma_\alpha^2, \sigma_v^2)$ are independent, the joint prior on $(\sigma^2, \sigma_\alpha^2, \sigma_v^2)$ is

$$\rho'(\sigma^2, \sigma_\alpha^2, \sigma_v^2) \propto (\sigma^2)^{-(\lambda+1)}(\sigma_\alpha^2)^{-(\lambda_\alpha+1)}(\sigma_v^2)^{(\lambda_v+1)} \exp[-H_2/2],$$

where

$$H_2 = \left(\frac{c_o}{\sigma^2} + \frac{c_\alpha}{\sigma_\alpha^2} + \frac{c_v}{\sigma_v^2}\right).$$

By Bayes theorem, the posterior density of $(\sigma^2, \sigma_\alpha^2, \sigma_v^2)$ can be written as

$$\rho''(\sigma^2, \sigma_\alpha^2, \sigma_v^2|\text{ Data }) \propto (\sigma^2)^{-(\frac{T}{2}+\lambda+1)}(\sigma_\alpha^2)^{-(\lambda_\alpha+1)}(\sigma_v^2)^{-(\lambda_v+1)}(\delta)^{-T/2}$$
$$\prod_{i=1}^K \left(1 + \tfrac{\lambda_i t}{\delta}\right)^{-1/2} \exp\left(-\tfrac{1}{2}[H_2 + \tfrac{H_1}{\sigma^2}]\right)$$

where $H_1$ is given by equation (A.2) in the appendix.

This posterior density is a generalization of the posterior density derived by Hill (1965, 1967) (of the unknown variance components in the context of the one way analysis of variance models) to the case of the linear regression model with errors in variables.

On further simplification and transforming the $(\sigma^2, \sigma_\alpha^2, \sigma_v^2)$ space to the $(\sigma^2, t, \delta)$ space, the posterior density of $(\sigma^2, t, \delta)$ is obtained. On integrating with respect to $\sigma^2$, the posterior density of $(t, \delta)$ is given by

$$\rho''(t, \delta|\text{ Data }) \propto t^{\lambda_v-\lambda_\alpha-1}(\delta - 1)^{-(\lambda_v+1)} \prod_{i=1}^K (1 + \tfrac{\lambda_i t}{\delta})^{-1/2}$$
$$\exp[-\tfrac{1}{2(\delta-1)}c_v t\; [c_o + \tfrac{c_\alpha}{t} + H_1]^{-(T/2+\lambda+\lambda_\alpha)} \quad (4.1)$$

This posterior density enables us to compute the posterior mean of $\eta_i$ (which depends on $t$ and $\delta$); and hence $\hat{\alpha}(\mathbf{B})$ and $\hat{\beta}(\mathbf{B})$. The following theorem stated (without proof) summarizes the main results of this section.

**Theorem 1**

(a) Given the linear regression model (2.2) together with all its assumptions, the posterior density of $(t, \delta)$ is given in equation (3.1).

(b) $E[\eta_i|\text{ data }] = \int_0^\infty \int_1^\infty \eta_i p''(t, \delta|\text{ Data })dt\, d\delta$ and can be computed by numerical integration methods.

(c) The Bayes estimator of $\beta$ is $\hat{\beta}(\mathbf{B}) = \mathbf{A}'\hat{\alpha}(\mathbf{B})$ where the matrix $\mathbf{A}$ is computed from the matrix $\mathbf{X}$.

## 5. CONCLUSION

The paper considers the linear regression model with measurement errors in the independent variables. A new, operational version of the admissible Bayes estimator is derived. This estimator corresponds to a class of informative priors both on the unknown parameter vector as well as on the unknown variances. The posterior density of the unknown variance components is derived. This in turn enables us to compute the posterior mean of $\eta_i$ by numerical integration methods as a closed form solution is not obtained. The Bayes estimator is evaluated with respect to this posterior mean, which leads to a new estimator of $\beta$ in the context of linear regression models with errors in variables. The sensitivity of this estimator to the choice of the hyper-parameters needs to be explored as that would give us some guidance in choosing the hyper-parameters. Further work needs to be done for writing programs to obtain $E[\eta_i|\text{ data }]$, given the data and the prior parameters.

## APPENDIX

The likelihood function being considered is

$$L(\sigma^2, \sigma_\alpha^2, \sigma_v^2) \propto |\mathbf{\Sigma}|^{-1/2} \exp(-1/2\, \mathbf{y}'\mathbf{\Sigma}^{-1}\mathbf{y})$$

where

$$\mathbf{\Sigma} = \text{ Cov } (\mathbf{y}|\mathbf{X}) = \sigma^2\mathbf{I} + \sigma_\alpha^2\, \mathbf{ZZ}' + \sigma_\alpha^2\, \sigma_v^2\mathbf{I},$$

and

$$|\mathbf{\Sigma}| = (\sigma^2\delta)^T |\mathbf{I} + t\delta^{-1}\mathbf{ZZ}'| = (\sigma^2\delta)^T \prod_{i=1}^{K}(1 + t\delta^{-1}\lambda_i),$$

where $t = \sigma_\alpha^2/\sigma^2$.

Using these results, the likelihood function can be written as

$$L(\sigma^2, \sigma_\alpha^2, \sigma_v^2) \propto (\sigma^2)^{-T/2} \prod_{i=1}^{K} (1 + t\delta^{-1}\lambda_i)^{-1/2} \exp[-\frac{\mathbf{y}'(\mathbf{I} + t\delta^{-1}\mathbf{Z}\mathbf{Z}')^{-1}\mathbf{y}}{2\sigma^2\delta}].$$

This can be further expressed as

$$L(\sigma^2, \sigma_\alpha^2, \sigma_v^2) \propto (\sigma^2\delta)^{-T/2} \prod_{i=1}^{K} (1 + t\delta^{-1}\lambda_i)^{-1/2} \exp[-H_1/2\sigma^2], \qquad (A.1)$$

where

$$H_1 = \frac{1}{\delta}[\mathbf{y}'\mathbf{y} - (t\delta^{-1})\mathbf{y}'\mathbf{Z}(\mathbf{I} + t\delta^{-1}\Lambda)^{-1}\mathbf{Z}'\mathbf{y})]. \qquad (A.2)$$

## ACKNOWLEDGMENTS

The author wishes to thank Prof. Bruce Hill and Prof. E.P. Howrey for their detailed discussions and suggestions on an earlier draft of this paper. I am indebted to Prof. Prem Goel for his valuable and insightful comments on a previous draft of this paper.

## REFERENCES

Feldstein, M.S. (1974), Errors in Variables: A consistent estimator with smaller MSE in finite samples, *Jour. Amer. Stat. Assoc.*, **69**, 990-996.

Fomby, T.B., R. Carter Hill and S.R. Johnson (1984), *Advanced Econometric Methods,* New York: Springer-Verlag.

Griliches, Z. (1974), Errors in variables and other unobservables, *Econometrica,* **42**, 971-978.

Hill, B.M. (1965), Inference about Variance Components in the One Way Model, *Jour. Amer. Stat. Assoc.,* **60**, 806-825.

Hill, B.M. (1967), Correlated Errors in the Random Model, *Jour. Amer. Stat. Assoc.,* **62**, 1387-1400.

Judge, G., W.E. Griffiths, R. Carter Hill and T. Lee (1980), *The Theory and Practice of Econometrics,* New York: John Wiley and Sons.

Levi, M.D. (1973), Errors in the variables bias in the presence of correctly measured variables, *Econometrica,* **41**, 985-986.

Levi, M.D. (1977), Measurement Errors and bounded OLS estimates, *Journal of Econometrics,* **6**, 165-171.

Nerlove, M. (1971), A note on error component models, *Econometrica,* **39**, 383-396.

Zellner, A. (1970), Estimation of regression relationships containing unobservable independent variables, *International Economic Review,* **11**, 441-454.

# 24

# Population Size Estimation with Truncated Data: A Survey


## Saul Blumenthal [1]



**Abstract**

We review the literature for the problem of estimating the size, $N$, of a finite population. Each member of the population has associated with it a value of a random variable $X$, and $X_1$, ..., $X_N$ are independent and identically distributed with density $f(x;\theta)$, $\theta$ being unknown. Only those members of the population with $X$-values in some restricted subset $\mathbf{R}$ of the sample space can be observed. The problems discussed here are more closely related to truncated sampling than to traditional population size estimation.



*Key Words and Phrases:.* Finite population; Size estimation; Truncated data; MLE; Bayes estimates.


## 1. INTRODUCTION

Estimating the size, $N$, of a finite population is a problem which appears in the statistical literature in many variations. The version which I shall be considering is that in which each member of the population has associated with it a value of a random variable $X$, where $X_1$, ..., $X_N$ are independent and identically distributed with density $f(x;\theta)$, $\theta$ is unknown, and only those members of the population with $X$-values in some restricted subset $\mathbf{R}$ of the sample space can be observed, i.e. the data are truncated. With some exceptions, one of which will be noted below, the most widely studied variations of the population size estimation problem cannot be formulated in the truncated data format. Two traditional versions of the population size estimation problem are, estimating the size of an animal population using capture-recapture methods, and estimating the number of categories where each observation must come from one of an unknown number of categories which may have different probabilities of occurence, for instance, estimating the number of words in Shakespeare's vocabulary. One problem which, as formulated by Brookmeyer and Gail (1988), has a structure


[1] The Ohio State University

which is similar to the truncated data problem, but somewhat more complex, is that of estimating the size of the AIDS epidemic. The material to be considered here is more closely related to truncated sampling than to traditional population size estimation. Vcry oftcn, $X$ is non-negative, $\mathbf{R}$ is the interval $(0, \infty)$, and one observes the first $n$ order statistics where $n$ is a binomial random variable with parameters $N$ and $Q(\theta) = P_\theta(X \in \mathbf{R})$. The truncated sampling literature traditionally concentrates on estimating $\theta$, and conditioning on the value of $n$, thus removing the parameter $N$ from consideration. For references to some of the literature on estimating $N$ in this model, see Dahiya and Blumenthal (1986). Two examples illustrate the model.

First, if a population of transistors (say) consists of two homogeneous subpopulations, one being the 'defective' items which fail according to the density $f(x; \theta)$ and the other being the 'good' items which have a failure distribution with all of its weight so far out that the lifetimes could be regarded as infinite, then $N$ represents the number of defectives in a batch of these items. This is sometimes referred to as the "limited failure population" model, and may be regarded as an approximation to the situation where both good and defective items can fail. If this batch is put on life test for a period of time $T$, then the defectives with lives exceeding $T$ cannot be distinguished from the good items. One might wish to estimate how many unfailed defectives are in the batch after the lifetest, or one might want a sequential stopping rule which will assure with a high probability that fewer than k defectives remain at stopping. The sequential estimation procedure is close in its objectives to what is referred to in the reliability literature as burn-in (see Marcus and Blumenthal (1974), and Pan and Blumenthal (1987)). A recent paper in which the limited failure population model is exploited is Meeker (1987). In that paper, the items are assumed to come from a production line and estimation of the fraction of defectives produced is the focus, rather than estimation of $N$ for a single batch. The analytic difficulties are closeley related.

The second example comes from the computer software reliability area. A program is assumed to have $N$ errors. A fairly popular model for time until detection and correction of an error is equivalent to associating with each error an independent $X_i$ with density $f(x; \theta)$. Unlike the transistor problem, this model when appropriate is not an approximation since correct instructions never cause computational errors. The special case where $f(x; \theta)$ is an exponential density with unknown scale parameter has drawn particular attention, and is known as the Jelinski-Moranda (J-M) model (for a discussion of the relevance of this model, see Trachtenberg (1985)). Some approaches to generalizing the J-M model lead to our model with different f(x;q), see Miller (1986) for an extensive discussion of such models, and Raftery (1987) for a Bayesian approach to comparing the fit of such models to the data. For a discussion of generalizations of the J-M model which lead in completely different directions, see Langberg and Singpurwalla (1985).

The program is tested for a period of time $T$ and it is desired to estimate the number of undetected errors which remain. Corresponding to the burn-in procedures mentioned above, there are sequential debugging procedures for errors. These have goals which may be stated a little differently than those natural for burn- in, but share a great deal in common. A recent reference on that topic is Dallal and Mallows (1988).

## 2. METHODOLOGY REVIEW

Regardless of the motivating application, estimation of $N$ and $\theta$ has commanded a lot of attention because the maximum likelihood estimator (MLE) of $N$ is very unstable, i.e. is infinite or changes greatly for small changes in the data. There are two forms of the MLE. The unconditional MLE (UMLE) is obtained by maximizing the likelihood of the observations jointly w.r.t. $N$ and $\theta$, while the conditional MLE (CMLE) comes from treating the n observed values as a fixed size sample from a truncated distribution for the purpose of estimating $\theta$ and then using $n/Q(\hat{\theta})$ to estimate $N$. The strong connection with the truncated sampling literature comes through the CMLE. Note that estimates of $\theta$ which make $Q(\theta)$ close to zero lead to unacceptably large estimates of $N$. For a discussion of the difficulties of estimating $\theta$ for the truncated exponential, see Suich and Rutemiller (1982). Stability of the UMLE is only slightly better than that of the CMLE.

The case of the MLE for exponential $f(x;\theta)$ has been studied in great detail, see for instance Marcus and Blumenthal (1975), Littlewood and Verrall (1981), Joe and Reid (1985b),and Spreij (1985). It is known that infinite estimates of $N$ occur with fairly high probability when both $N$ and Q are small. Yet, the MLE of $N$ is still median biased downward, tending very often to estimate $N$ by the observed value of $n$.

A variety of attempts have been made to come up with more stable estimators of $N$. One approach is to introduce a conjugate prior for $\theta$, and a uniform prior for $N$ and then to find the posterior mode. This gives back the MLE when the prior for q approaches uniformity. See Marcus and Blumenthal (1975), for instance. When a properly chosen prior is used, the resulting estimator is more stable than the MLE but tends to underestimate $N$ rather severely. Other attempts at developing point estimators of $N$ have been by Joe and Reid (1985a,b) who suggested the harmonic mean of the endpoints of a 50% highest relative likelihood interval, and by Raftery (1988a) who computes Bayes posterior mode and median estimates with respect to a Poisson prior for $N$, and a 'natural' conjugate distribution for $\theta$ and the Poisson parameter. A simulation study leads him to conclude that none of the four estimators mentioned heretofore are very good, and that HPD intervals might be a better choice in general. Spreij (1985) briefly discusses using likelihood ratio tests to develop confidence intervals for $N$. Jewell (1986a) develops a predictive distribution for the undetected errors using somewhat different priors from Raftery, and suggests using its mode as a point estimator but does not compare this with any other estimators. The poor performance of all the estimors, even the stabilized ones, is largely due to having small values of $Q$, i.e. of not testing for a sufficient period of time. One attempt at remedying this problem was given by Watson and Blumenthal (1980a), who developed a two stage procedure which estimates $\theta$ at the first stage and chooses the length of the second stage to give $Q$ large enough to achieve a reasonable bound on the asymptotic variance of the $N$ estimator. This sampling plan can be used to improve the asymptotic performance of any estimator and, in fact, will not eliminate the instability of the MLE. Another solution was offered by Nyack (1988) who suggests a "recapture" scheme whereby multiple observations can be collected on the $X$ associated with any computer program error by use of markers in the program. Whether this scheme could be applied in other contexts is not clear, but it does improve the performance of the MLE, where applicable.

The small sample (i.e. small $N$) properties of $N$-estimators tend to be examined by simulation studies. For the various Bayes estimators, large sample properties to the first order, i.e., asymptotic normality, tend to be indistinguishable. To find differences among the estimators asymptotically, it is necessary to look at second and higher order terms in their stochastic expansions, or in expansions for their moments. Using stochastic expansions, see Blumenthal (1977,1982), one of the conjugate distributions for $\theta$ can be singled out as giving a posterior mode with least bias asymptotically. Similar calculations were used by Watson and Blumenthal (1980b) to find an estimator of Bayes modal type which has minimum mean squared error. For large sample comparisons with the other estimators which have been advanced, stochastic expansions should be developed for these estimators.

The properties, either large or small sample, of Bayes posterior means remain to be studied. It could be expected that with a reasonable choice of prior, these would be finite, but would reflect better than the mode or median the long tails of the posterior distribution and avoid the negative bias of these other estimators. Computationally, the posterior mean might be difficult to handle, but not impossible, and why no simulations have appeared comparing small sample properties with its competitors is something of a mystery. Getting stochastic expansions for the posterior mean is more difficult, but should be made feasible by the recent work of Kass, Tierney, and Kadane (1988).

What prior distributions to use for $N$ is also worth looking at. For defective items from a production line, a binomial prior is natural, but the value of p is not likely to be known, creating a heirarchal Bayes problem. In the software problem, there has been a tendency to use a Poisson distribution for $N$. This is computationally tractable, but whether it is justified is not clear. There ought to exist a class of conjugate distributions for $N$, but these have not been explored.

One notable area of overlap between capture-recapture estimation of population size and the model discussed above is the following. In trying to determine the number of particles on a film, for instance, several inspectors may examine the film independently, and particles can be classified according to how many times they were detected. This results in a multinomial distribution with an unobservable category, namely the one consisting of particles which no inspector saw. Estimating the number of particles can then be recast as estimating the number of multinomial trials when **R** consists of one category. Certain models for probability of particle detection lead to the same likelihood and the same estimators as in the capture-recapture problem. Sanathanan (1972a,b,1973) and Jewell (1985b) discuss this relation, and analyze the $N$ estimation problem for various assumptions about detection probabilities. Jewell takes a Bayesian approach to estimation for a single model, while Sanathanan looks at maximum likelihood estimators for several models and procedures for comparing the fit of these models.

It might be noted that if the multinomial distribution has only two categories, i.e., the trials are Bernoulli, and the probability $Q$ of the observable category is not known, then estimation of $N$ is impossible on the basis of one experiment. In this case, a series of Binomial experiments, each consisting of $N$ trials must be conducted in order to estimate $N$. In this sense, the binomial $N$ problem differs from the problems discussed above. Difficulties in getting good estimators are encountered very much like those discussed before, and a considerable literature has developed in recent years. Two recent papers are Hamedani (1988) and Raftery (1988b).

## ACKNOWLEDGEMENTS

Research supported in part by Air Force Office of Scientific Reseearch under Contract AFOSR 89-0357 at The Ohio State University. The U.S. Government is authorized to reproduce and distribute reprints for governmental purposes notwithstanding any copyright notation thereon.

## REFERENCES

Blumenthal, S. (1977), Estimating Population Size With Truncated Sampling, *Commun. in Statist.*, **A6**, 297-308.

Blumenthal, S. (1982), Stochastic Expansions for Point Estimation from Truncated Samples, *Sankhya Indian J. of Statist., Ser. A*, **44**, 436-451.

Brookmeyer, R., and Gail, M.H.(1988), A Method for Obtaining Short-Term Projections and Lower Bounds on the Size of the AIDS Epidemic, *J. Amer. Statist. Assoc.*,**83**, 301-308.

Dahiya and Blumenthal (1986), Population or Sample Size Estimation." In *Encyclopedia of Statistical Sciences Vol. VII, S. Kotz and N. L. Johnson (editors)*,100-110, J. Wiley, New York

Dalal, S.R., and Mallows,C.L. (1988), When Should One Stop Testing Software, *J. Amer. Statist. Assoc.*,**83**, 872-879.

Hamedani, G.G. (1988), Bayes Estimation of the Binomial Parameter n, *Commun. in Statist. Theor. Meth.*, **17**, 1829-1843.

Jewell, W.S.(1986a), Bayesian Extensions to a Basic Model of Software Reliability, *Proc. Internat. School of Physics "Enrico Fermi" : Theory of Reliability*, **93**, 394-404.

Jewell, W.S.(1986b), Bayesian Estimation of Undetected Errors,*Proc. Internat. School of Physics "Enrico Fermi" : Theory of Reliability*, **93**, 405-425.

Joe, H. and Reid, N. (1985a), Estimating the Number of Faults in a System, *J. Amer. Statist. Assoc.*, **80**, 222-226.

Joe, H. and Reid, N. (1985b), On the Software Reliability Models of Jelinski-Moranda and Littlewood, *IEEE Trans. on Reliab.*, **R-34**, 216-218.

Kass,R.E., Tierney,L., and Kadane,J.B. (1988), The Validity of Posterior Expansions Based on Laplace's Method, *Technical Report* Dept. of Statistics , Carnegie-Mellon U., Pittsburgh, PA.

Langberg, N., and Singpurwalla, N.D. (1985), A Unification of Some Software Reliability Models, *SIAM J. Sci. Stat. Comput.*, **6**,781-790.

Littlewood, B. and Verrall, J.L. (1981), Likelihood Function of a Debugging Model for Computer software Reliability, *IEEE Trans. on Reliab.,* **R-30**, 145-148.

Marcus,R., and Blumenthal, S. (1974), A Sequential Screening Procedure, *Technometrics*, **16** ,229-234.

Marcus,R., and Blumenthal, S. (1975), Estimating Population Size with Exponential Failure," *J. Amer. Statist. Assoc.*, **70**, 913-922.

Meeker, W.Q. Jr. (1987), Limited Failure Population Life Tests: Application to Integrated Circuit Reliability, *Technometrics*, **29**, 51-65.

Miller, D.R. (1986), Exponential Order Statistic Models of Software Reliability Growth, *IEEE Trans. on Software Eng.*, **SE-12**, 12-24.

Nyack, T. K. (1988), Estimating Population Size by Recapture Sampling, *Biometrika*, **75** 113-120.

Pan, U.Q., and Blumenthal,S. (1987), Burn-in with Mixed Populations". *Technical Report 372*, Statist. Dept. Ohio State U.,Columbus,OH.

Raftery, A.E. (1987), Inference and Prediction for a General Order Statistic Model with Unknown Population Size, *J. Amer. Statist. Assoc.*, **82**, 1163-1168.

Raftery, A.E. (1988a), Analysis of a Simple Debugging Model, *Applied Statistics*, **37**, 12-22.

Raftery, A.E. (1988b), Inference for the Binomial N Parameter: A Hierarchical Bayes Approach, *Biometrika*, **75**, 223-228.

Sanathanan, L.P.(1972a), Estimating the Size of a Multinomial Population, *Ann. Statist.* ,**43**, 142-152.

Sanathanan, L.P.(1972b), Models and Estimation Methods in Visual Scanning Experiments, *Technometrics*, **14**, 813-830.

L.P.(1973), A Comparison of Some Models in Visual Scanning Experiments, *Technometrics*, **15**, 67-78.

Spreij, P. ( 1985), Parameter Estimation for a Specific Software Reliability Model, *IEEE Trans. on Reliab.* **R-34**, 323-327.

Suich, R. and Rutemiller, H.C. (1982), Point Estimation of the Parameter of a Truncated Exponential Distribution, *IEEE Trans. on Reliab.*, **R-31**, 393-397.

Trachtenberg, M. (1985), The Linear Software Reliability Model and Uniform Testing *IEEE Trans. on Reliab.*, **R-34**, 8-16.

Watson, D.A., and Blumenthal, S. (1980a), A Two Stage Procedure for Estimating the Size of a Truncated Exponential Sample, *Australian J. Statist.* **22**, . 317-327

Watson, D.A., and Blumenthal, S. (1980b) Estimating the Size of a Truncated Sample, *Commun. in Statist. Theor. Meth.*, **A9**, 1535-1550

# 25

# Bayesian Analysis of Co-Integrated Time Series


## Klaus Felsenstein [1]
## Klaus Pötzelberger & Wolfgang Polasek [2]



**Abstract**

Co-integration models have attained considerable interest over the past few years for modeling multivariate economic time series. One of the basic ideas behind these models is the reduction of non-stationary $I(1)$ components of a vector AR process to a linear combination which is a stationary AR process. All approaches which have been proposed so far are classical methods relying heavily on asymptotic theory. This paper proposes a simple Bayesian treatment of a co-integrating model, focussing on bivariate time series. The posterior distribution of the co-integrating vector is derived and the Bayesian analysis, which is exploratory in nature, is demonstrated by a macroeconomic example.



*Key Words and Phrases:.* Cointegration models; Bivariate time series; Vector $AR$ process; Macronomic Data.


## 1. INTRODUCTION

In econometrics, a non-stationary economic time series is denoted by $x_t \sim I(d)$ if the $d$ times differenced time series is stationary. A non-stationary vector time series $x_t$ might have a long-run linear relationship between the components $x_{t,i}, i = 1, \cdots, k$ which can be viewed as a non-stationary trend. We assume that a lower dimensional joint relationship may be reconstructed as a linear combination of the components.

Thus we may hope that a suitable linear combination generates a time series with two components: the non-stationary trend, integrated of order $d$, and a time series integrated of order $d - b$, with $b > 0$. In this paper we assume that the second component may be represented as an autoregressive process (of order $p$, AR($p$)), so that $d = b$. Up to the joint non-stationary trend, the components of the time series are then stationary. Granger (1981) and Granger and Weiss (1983) introduced this concept of non-stationary


[1]Technical University of Vienna

[2]University of Basel

time series representation called co-integration. In order to formalize this approach we give the following

**Definition 1.1**. The vector $\mathbf{x}_t$ is co-integrated with co-integrating vector $\boldsymbol{\alpha} \in R^k$ if the univariate variable

$$z_t = \boldsymbol{\alpha}'\mathbf{x}_t \qquad t = 1, \cdots, n, \tag{1.1}$$

represents an autoregressive process.

Using conjugate priors, the parameter space includes non-stationary AR processes. We focus on the AR structure of the process as generating mechanism of the data, therefore co-integration implies a common generating mechanism of the "stationary" part of the components.

We assume that the order $p$ of the AR($p$) process is known and the choice of $p$ has less impact on the validity of the model than the existence of a co-integrating vector has. Of course, the co-integrating vector $\alpha$ is not unique. In general there exist $r(r < k)$ linear independent vectors, $r$ is called the co-integrating rank of the time series $\mathbf{x}_t$. Especially any multiple of $\alpha$ gives a time series $\{z_t\}$ fulfilling (1.1).

In this paper we restrict the analysis to $k = 2$. Taking the first component of a constant turns out to be a tractable choice of normalisation:

$$\boldsymbol{\alpha} = (1, \tilde{\boldsymbol{\alpha}}). \tag{1.2}$$

The general case, including simulation results, is reported in Felsenstein et al. (1989). Note that for higher dimensional processes our choice of normalisation excludes any co-integrating relationship that does not include the first component. An alternative normalizing condition would be $\alpha\,'\alpha = 1$. This restriction leads to technical difficulties when formulating the co-integrated time series as a submodel of a linear model and leads to rather intractable posterior distributions. Essential properties and structures of co-integrated systems can be found in Granger (1981), Granger and Weiss (1983), Engle and Granger (1987), Johansen (1987) among others. The relationship to error correction models as stated in Granger (1981) increases the importance of models of co-integrated time series for econometric modeling.

## 2. BAYESIAN INFERENCE FOR THE PARAMETER

The underlying autoregressive model can be formulated as

$$\epsilon_t = \theta(B) z_t \qquad \text{with} \quad \epsilon_t \sim N(0, \sigma^2) \tag{2.1}$$

where $B$ is the backshift operator defined as $Bz_t = z_{t-1}$, $\theta(B) = 1 + \theta_1 B + \cdots + \theta_p B^p$ is a (univariate) lag-polynomial of degree $p$, and $\epsilon_t$ are independently distributed Gaussian errors with variance $\sigma^2$. With $\mathbf{x}_t = (x_{t,1}, x_{t,2})$ equation (2.1) reads

$$x_{t,1} + \tilde{\alpha} x_{t,2} + \theta_1 x_{t-1,1} + \theta_1 \tilde{\alpha} x_{t-1,2} + \cdots + \theta_p \tilde{\alpha} x_{t-p,2} = \epsilon_t. \tag{2.2}$$

The aim of the statistical analysis is the estimation of a co-integrating vector $\alpha$ leading to a favourable small model variance $\sigma^2$ and a decision whether a time series is co-integrated or not. $\tilde{\alpha}$, the unknown second component of the co-integrating vector $\alpha$ is treated as a distributed parameter. The maximum likelihood estimate and tests of linear hypothesis are derived in Johansen (1987), and Engle and Granger (1987). In the Bayesian framework the parametric form of the model expressed in the likelihood function and prior information lead to a distribution of $\alpha$ which is the basis of further statistical inference like decisions and parameter estimates.

Bayesian analysis is more or less straightforward if we reparametrize the model (2.2) by the following co-integration condition "Co":

$$\{\beta_0 = \alpha, \quad \beta_i = \theta_i \alpha, \quad \beta_i \in R^2, i = 1, \cdots, p\}. \tag{2.3}$$

Now (2.2) can be written as

$$\epsilon_t = \sum_{i=0}^{p} \beta_i' \mathbf{x}_{t-1} \tag{2.4}$$

We combine all parameters $\beta_i$ into one vector $\boldsymbol{\beta}$ of dimension $2(p+1)$:

$$\boldsymbol{\beta}' = (\beta_0', \beta_1', \cdots, \beta_{p-1}') = (1, \tilde{\boldsymbol{\beta}}') \tag{2.5}$$

which is also normalized because of (1.2). Furthermore, we introduce the $2(p+1) \times 2(p+1)$ moment matrix $\mathbf{M}_t$ for $p$ observations $\mathbf{x}_t$ at time $t$:

$$\mathbf{M}_t = \begin{pmatrix} \mathbf{x}_t\mathbf{x}_t' & \mathbf{x}_t\mathbf{x}_{t-1}' \cdots & \mathbf{x}_t\mathbf{x}_{t-p}' \\ \cdots & \cdots & \cdots \\ \mathbf{x}_t\mathbf{x}_{t-p}' & \cdots & \mathbf{x}_{t-p}\mathbf{x}_{t-p}' \end{pmatrix}. \tag{2.6}$$

Now the likelihood function for $\boldsymbol{\beta}$ has the form

$$\ell(\boldsymbol{\beta}) \propto \exp\left(-\frac{1}{2\sigma^2}\sum_{t=1}^{n} \boldsymbol{\beta}'\mathbf{M}_t\boldsymbol{\beta}\right) = \exp\left(\frac{1}{2\sigma^2}\boldsymbol{\beta}'\mathbf{M}\boldsymbol{\beta}\right) \tag{2.7}$$

with $\mathbf{M} = \mathbf{M}_1 + \cdots + \mathbf{M}_n$. The symbol $\propto$ indicates equality up to a multiple constant. The form of the likelihood function suggests the use of a normal prior for $\tilde{\boldsymbol{\beta}}$. Since normal distributions constitute a conjugate family, the posterior distribution is also a multivariate normal distribution.

Let $\tilde{\boldsymbol{\beta}}$ be distributed according to $\tilde{\boldsymbol{\beta}} \sim N(\boldsymbol{\mu}, \mathbf{P}^{-1})$ with prior mean $\boldsymbol{\mu} \in R^{2p-1}$ and prior precision matrix $\mathbf{P}$. Let $\mathbf{X}$ be the matrix of all observations, then the posterior distribution reads as $\tilde{\boldsymbol{\beta}}|\mathbf{X} \sim N(\boldsymbol{\mu}^\star, \mathbf{P}^{\star-1})$ with posterior moments

$$\boldsymbol{\mu}^\star = \mathbf{P}^{\star-1}(-\sigma^{-2}\mathbf{m} + \mathbf{P}\boldsymbol{\mu}) \quad \text{and} \quad \mathbf{P}^\star = \sigma^{-2}\mathbf{M}_1 + \mathbf{P}. \tag{2.8}$$

The vector $\mathbf{m} \in R^{2p-1}$ and the submatrix $\mathbf{M}_1$ are elements of a partition of $\mathbf{M}$ ($m_{11}$ is a scalar):

$$\mathbf{M} = \begin{pmatrix} m_{11} & \mathbf{m}' \\ \mathbf{m} & \mathbf{M}_1 \end{pmatrix}. \tag{2.9}$$

In many cases the specification of a complete prior will be difficult and hardly reasonable. In this case robust Bayesian methods or even a non-informative prior are recommended. Any (improper) prior distribution with constant density on $R^{2p-1}$ serves as non-informative prior for the given parametrisation. Due to the normal kernel of the likelihood function (2.7), the posterior distribution also remains normal in the non-informative case:

$$\tilde{\boldsymbol{\beta}}|\mathbf{X} \sim N(-\mathbf{M}_1^{-1}\mathbf{m}, \sigma^2\mathbf{M}_1^{-1}). \tag{2.10}$$

The parameters of the normal distribution in (2.8) correspond to the usual linear model given by regressing the first component $\mathbf{x}_{t,1}$ on the remaining components and the lagged variables: $\mathbf{X}_{(-1)} : \mathbf{x}_1 = \mathbf{X}_{(-1)}\tilde{\boldsymbol{\beta}} + \boldsymbol{\mu}$. Given the posterior distribution, we are able to apply the common techniques of Bayesian inference. For example, the posterior mean of (2.10) serves as a point estimate.

We assume fixed starting values and a known variance $\sigma^2$. For unknown $\sigma^2$, a non-informative prior for $\sigma^2$ might be used. These priors have densities proportional to some power of $\sigma^2$, leading to $t$-distributed marginals of $\tilde{\boldsymbol{\beta}}$. Since our method depends on the use of the normal distribution, we replace the unknown $\sigma^2$ by an estimate.

Deriving the posterior distribution for $\alpha$ is straightforward since the first component of $\tilde{\boldsymbol{\beta}}$ coincides with $\tilde{\alpha}$. (The first component of $\alpha$ is one and can be neglected.) Because the posterior distribution of $\tilde{\boldsymbol{\beta}}$ is normal, $\tilde{\alpha}$ is normal after observing the data $\mathbf{X}$. If the assumption of a co-integrated system is correct then the $2p-1$ dimensional distribution of $\tilde{\boldsymbol{\beta}}$ degenerates to a distribution of $\theta$ and $\tilde{\boldsymbol{\alpha}}$ only. Let $\beta_{ij}$ denote the $j$-th component of $\beta_i$ then the condition of the existence of a co-integrating vector (2.1) is

$$\beta_{i2} = \beta_{i1}\ \beta_{o2} = \theta_i\tilde{\alpha}, \quad i = 1, \cdots, p. \tag{2.11}$$

In contrast to classical hypothesis testing we operate with probabilities concerning the hypothesis of co-integration. Under "Co", the posterior distribution should be concentrated on or near points obeying (2.11). A relatively high posterior probability

$$\Pr\left[\ \max_{i=1,\cdots,p} |\beta_{i2} - \theta_i\tilde{\alpha}| < \epsilon |\ \mathbf{X}\ \right]. \tag{2.12}$$

for suitable small $\epsilon > 0$ confirms the validity of the co-integration assumption. The probability in (2.12) is computed numerically, using Gaussian quadrature.

## 3. DISTRIBUTION OF THE PARAMETER UNDER CO-INTEGRATION

We assume that a non-informative prior has been specified. Taken co-integration for granted, further inference should be based on the conditional posterior of $\boldsymbol{\beta}$ given (2.11), denoted by $\boldsymbol{\beta}|\mathbf{X}$,Co . The conditional density (with the same parameters as 2.10) is

$$f(\beta_0, \theta|\mathbf{X}, \mathrm{Co}) \propto 1(\alpha, \theta_1\alpha, \cdots, \theta_p\alpha) \tag{3.1}$$

where $\theta = (\theta_1, \cdots, \theta_p)' = (\beta_{11}, \cdots, \beta_{p1})'$ denotes the AR parameter. To derive the marginal density of $\beta_0|\mathbf{X}$,Co we decompose the moment matrix $\mathbf{M}$ in (2.9) into $(p+1)^2$ square submatrices $\mathbf{M}_{ij} \in R^{2\times 2}$

$$\mathbf{M} = \begin{pmatrix} \mathbf{M}_{00} & \mathbf{M}_{01} & \cdots & \mathbf{M}_{0p} \\ \cdot & \cdot & \cdots & \cdot \\ \mathbf{M}_{p0} & \cdot & \cdots & \mathbf{M}_{pp} \end{pmatrix}. \tag{3.2}$$

The exponent in the likelihood function (2.7) under the constraint $\beta_i = \theta_i \alpha$ becomes (with $\theta_0 = 1$)

$$-\frac{1}{2\sigma^2}\beta'\mathbf{M}\beta = -\frac{1}{2\sigma^2}\sum_{i,j=0}^{p} \theta_i \boldsymbol{\alpha}'\mathbf{M_{ij}}\boldsymbol{\alpha}\ \theta_j. \tag{3.3}$$

Defining the quadratic forms $w_{ij} = \boldsymbol{\alpha}'\mathbf{M}_{ij}\boldsymbol{\alpha}$ to be entries of a matrix $\mathbf{W}$, (3.3) reduces to

$$\frac{1}{2\sigma^2}\beta'\mathbf{M}\beta = -\frac{1}{2\sigma^2}(1, \boldsymbol{\theta})'\mathbf{W}\begin{pmatrix} 1 \\ \boldsymbol{\theta} \end{pmatrix}, \tag{3.4}$$

$\mathbf{W}$ depends upon $\tilde{\alpha}$ and can be compactly expressed by switching from the quadratic $(p+1)\times(p+1)$ partition to the regression-partition as in (2.9):

$$\mathbf{W}\begin{pmatrix} w_{00} & \mathbf{w}' \\ \mathbf{w} & \mathbf{W}_1 \end{pmatrix} = (\boldsymbol{\alpha}'\mathbf{M}_{ij}\boldsymbol{\alpha}), \quad i,j = 0, \cdots, p; \tag{3.5}$$

$\mathbf{W}_1 \in R^{p\times p}$, and $\mathbf{w} \in R^{p\times 1}$. Substituting (3.5) into the quadratic form (3.4) and integrating out $\theta$, we obtain the marginal density of $\tilde{\alpha}$ up to a normalizing constant:

$$f(\tilde{\alpha}|\mathbf{X}, \mathrm{Co}) \propto (\det \mathbf{W}_1)^{-1/2} \exp\left(\frac{1}{2\sigma^2}\left(w_{00} - \mathbf{w}'\mathbf{W}_1^{-1}\mathbf{W}\right)\right). \tag{3.6}$$

Note that $\mathbf{w}'\mathbf{W}_1^{-1}\mathbf{w}$ and $w_{00}$ are functions of $\boldsymbol{\alpha}$. The normalizing constant as well as the expectation of this distribution have to be calculated by numerical integration. In low dimensional problems Gaussian quadrature is appropriate, in higher dimensions Monte-Carlo methods. Because of the symmetry of the likelihood concerning $\boldsymbol{\alpha}$ and $\theta$ we derive the conditional distribution of the AR parameter $\theta$ by similar methods. Let $\mathbf{V}$ be the matrix

$$\mathbf{V} = \sum_{i,j=0}^{p} \theta_i \theta_j M_{ij} = \begin{pmatrix} v_{11} & \mathbf{v}' \\ \mathbf{v} & \mathbf{V}_1 \end{pmatrix}. \tag{3.7}$$

where we use the same decomposition for $\mathbf{V}$ as for $\mathbf{W}$ in (3.5). Integration with respect to the parameter vector $\boldsymbol{\alpha}$ leads to the conditional posterior density of $\theta$ , given the data $\mathbf{X}$ , and the co-integration condition "Co"

$$f(\theta|\mathbf{X}, \mathrm{Co}) \propto (\det \mathbf{V}_1)^{-\frac{1}{2}} \exp(-\frac{1}{2}\left[v_{11} - \mathbf{v}'\mathbf{V}_1^{-1}\mathbf{v}\right]) \tag{3.8}$$

Again this distribution serves as basis for estimation and testing procedures involving the AR parameter which have to be executed numerically. In cases of low dimensions of $\alpha$ and $\theta$ respectively, the densities in (3.8) and (3.6) become relatively easy to handle with.

## 4. A MACROECONOMIC EXAMPLE

A simple macroeconomic model with possible co-integration is the the relationship between real income ($xs$) and real consumption ($cs$) in Switzerland. The joint density of $(\alpha, \theta)$ under co-integration is shown in figure 4.1. The relationship seems to indicate no dependence. The mode of $\alpha$ is near -.20 and the mode of $\theta$ centers around -.89. The marginal densities of $\alpha$ and $\theta$, under co-integration and unrestricted, are shown in figures 4.2 and 4.3, respectively. The density of $\alpha$ for the unrestricted case (2.10) and under co-integration (3.6) is very similar. The density of $\theta$ is quite different for the unrestricted case and under co-integration (3.8): Unrestricted, the mode is around -.9, but under co-integration the mode is much closer to -1 with only a small probablity of undershooting the value -1 (hatched area). The values of $\theta$ which are less than 1 in absolute value correspond to stationary time series. In the Bayesian analysis of AR processes there is always a small probability of non-stationarity, when conjugate priors are used.

To decide whether the two time series are in fact co-integrated, we computed for $0 < \epsilon < 0.15$ the posterior probability of the event co-integration, as given in (2.12). These posterior probabilities are displayed as a function of $\epsilon$ in figure 4.4. The hypothesis of co-integration is supported very strongly, because for $\epsilon = 0.08$ the probability is almost 1. The joint distribution of $(\alpha, \theta)$ under co-integration can be summarized by the following moments:

$$\text{mean}\begin{pmatrix} \alpha \\ \theta \end{pmatrix} = \begin{pmatrix} -.21 \\ -.97 \end{pmatrix} \quad \text{Var}\begin{pmatrix} \alpha \\ \theta \end{pmatrix} = \begin{pmatrix} .05^2 & 0 \\ 0 & .034^2 \end{pmatrix}$$

Skewness $(\alpha, \theta) = (.036, 1.23)$, Kurtosis $(\alpha, \theta)$= (-.1, -1.33).

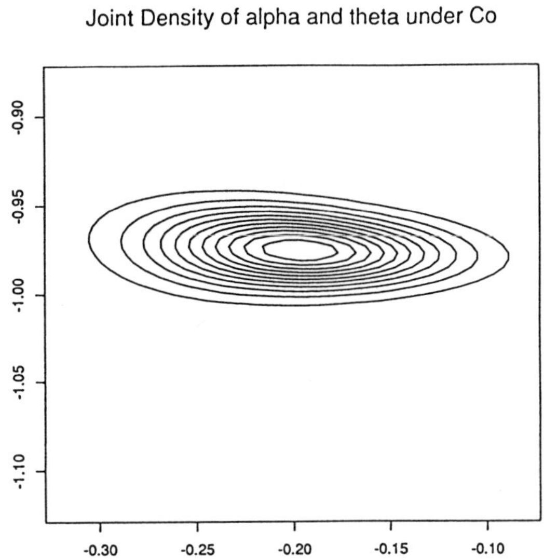


Fig.4.1. CoAR(2,1) model:
Swiss consumption and income

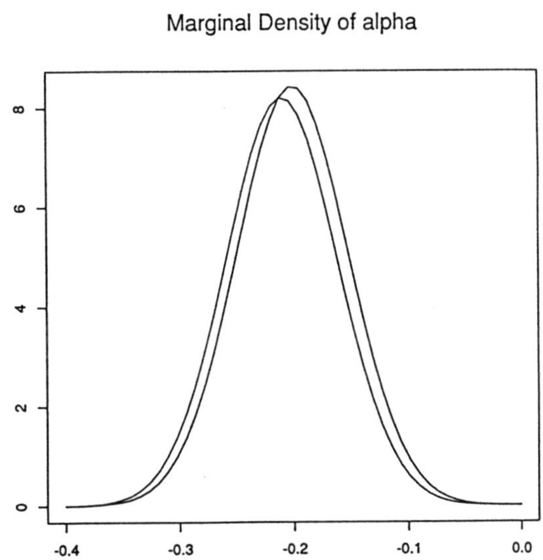


Fig.4.2. Under cointegration
and unrestricted

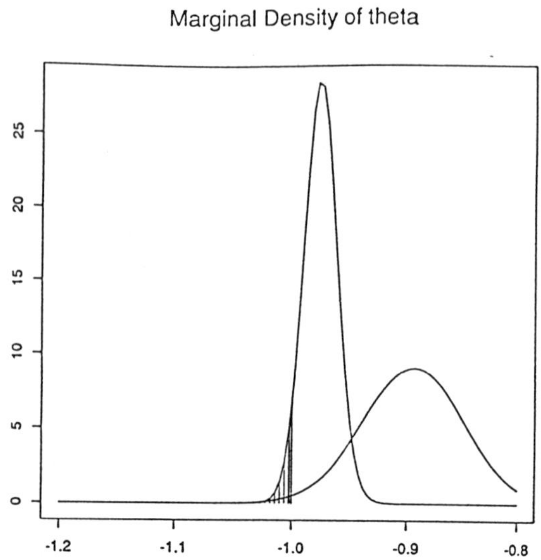


Fig.4.3. Under cointegration and unrestricted

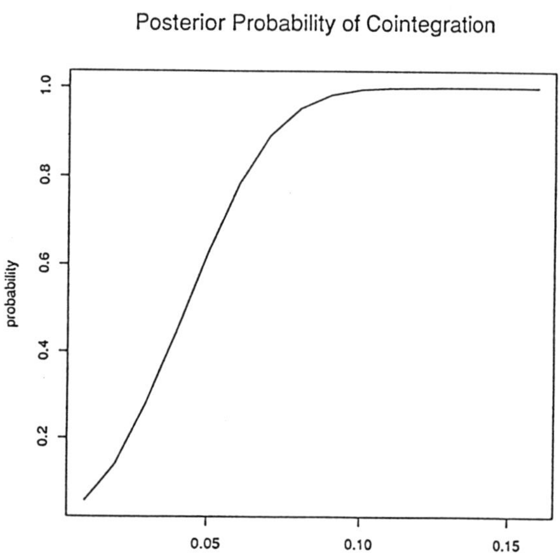


Fig.4.4. Probability (2.12) as a function of ε

## 5. SUMMARY

In this paper we have shown how a Bayesian approach to co-integration can be put into practice. As for many complex linear models, analytically Bayesian solution can be carried out only partly and numerical integration is necessary. The Bayesian approach is conceptually straightforward but numerically demanding. The availability of posterior distributions has the advantage that probability statements about co-integration can be made as in (2.12). This is certainly an advantage compared to the classical approach, as well as the availability of the whole posterior distribution. Since this paper explores the possibilities in two dimensions, further research is needed to explore the higher dimensional cases.

## REFERENCES

Engle, R.F. and Granger, C.W.(1987), Co-Integration and Error Correction: Representation, Estimation and Testing, *Econometrica* **55**, 251-276.

Felsenstein, K., Pótzelberger, K., and Polasek, W. (1989), Bayesian Analysis of Co-Integrated Time Series, Univ. Basel, WWZ-Discussion Papers Nr. 8910.

Granger, C.W.(1981), Some Properties of Time Series Data and Their Use in Econometric Model Specification, *Journal of Econometrics*, 121-130.

Granger, C.W. and Weiss, A. A.(1983), Time Series Analysis of Error-Correcting Models, in *Studies in Econometrics, Time Series and Multivariate Statistics*, New York, Acadamic Press, 255-278.

Johansen, S.(1987), Statistical Analysis of Co-integration Vectors, paper presented at the Econometric Society European Meeting, Copenhagen.

Johansen, S.(1988), The Mathematical Structure of Error Correction Models, Contemporary Mathematics, American Mathematical Society.

Johansen, S. and Juselius, K.(1988), Hypothesis Testing for Co-integration Vectors with an Application to the Demand for Money in Denmark and Finland, preprint No. 2, Inst. of Math. Statistics, Univ. of Copenhagen.

# 26

# A Bayesian Approach to the Measurement of Poverty in India

## N. Sreenivas Iyengar and S.N. Joshi [1] Mallika Gopalakrishna[2]

**Abstract**

For efficient implementation of various anti-poverty measures devised by Indian planners, a suitable definition of poverty is essential. One of the many definitions of poverty is in terms of a cut-off point for the minimum calorie intake level, *MCIL*, fixed exogenously by the nutritionists. A basic problem of economic interest is, however, to determine a threshold for the consumption in monetary terms, which corresponds to MCIL.

In this paper, we depart from the traditional approach which uses conditional measures to estimate the consumption threshold, such as conditional means, median etc., and formulate poverty line as follows: If y is the per capita consumption of a household and x is the corresponding per capita calorie intake, the poverty line is defined as the minimal ($y_*$), which guarantees the availability of the fixed MCIL of ($x_*$) cals with probability arbitrarily close to one. We assume a simple linear relationship between suitably transformed x and y and use the Bayesian approach in estimating the poverty line so defined. Extensive computations using the official National Sample Survey data for the Indian State of Karnataka reveal that poverty is under-estimated by the orthodox methods.

*Key Words and Phrases:*. Poverty line; Consumption; Nutrirional levels; Minimum calorie intake; Calibration; National Sample Survey.

## 1. INTRODUCTION

Poverty is a multivariate concept involving several dimensions. In a purely economic sense it may be regarded as the state where some goods and services essential to maintain an acceptable minimum standard of living are lacking. However a precise definition of poverty will have to differ from

[1]Indian Statistical Institute, Bangalore
[2]Indian Market Research Bureau, Bangalore

society to society, as the concepts of subsistence, deprivation and income adequacy vary.

In India, for instance, poverty is understood and defined as the lack of the very basic essentials required for subsistence. Indian planners have related poverty to nutritional inadequacy. The underlying behavioral assumption here is that the consumption pattern of households will ensure that when the food intake reaches the desired level, other essentials like clothing, housing and health would also be adequately available. Initially, protein was considered to be the crucial nutrient for ensuring nutritional adequacy, but later on it was established that calorie intake was the most crucial criterion to be considered for purposes of measuring poverty. See, Gopalan, et al (1971), Gopalan (1983), and Sukhatme (1977, 1982).

For Indian conditions an ideal requirement of 2400 K cals for the average sedentary male adult has been prescribed. See, Gopalan and Rao (1968) and Indian Council of Medical Research (1980). Calorie norms for other categories were linked to this average.

Calorie intake is not directly and easily available, and further, there may be inter- and intra-individual variations in calorie consumption, quite unrelated to the poverty level. Hence, it becomes important to translate the calorie norm into a more readily observable indicator like the household per capita total consumption in monetary terms. The Planning Commission in India first defines a minimum per capita calorie norm and then takes as the poverty line a central value like the mean, mode or median, of the per capita total expenditure class for which the mean calorie intake exactly matches the fixed norm. Using this procedure, the Planning Commission (1979) estimated the poverty lines for 1973-74 as Rs. 49.09 per month for the rural calorie norm of 2400 K cals daily and Rs. 56.60 for the urban calorie norm of 2100 K cals per day. More recently, a working Group of the Planning Commission has estimated similar poverty lines for later years for different states of India.

There are obvious limitations to this procedure. Firstly, the area covered is too large for either the calorie norm or the poverty line to have much practical meaning. Secondly, for this large area, a general occupational profile was considered, which is necessarily inaccurate. Consequently, applying these poverty lines in a specific region for identification of the poor becomes an exercise fraught with danger. This has been clearly shown for the State of Karnataka in Iyengar and Gopalakrishna (1985).

In Section 2, a novel concept of poverty line is introduced. Section 3 outlines a Bayesian approach to the measurement of poverty so defined, (see Hoadley,1970 for discussion on the need and usefulness of Bayesian approach in a related problem). Section 4 describes the data used and Section 5 gives the main results. A few concluding remarks are offered towards the end.

## 2. POVERTY LINE DEFINED

The two main household variables considered in this analysis are (i) the level of calorie intake ($Y$) and (ii) the level of consumption expenditure ($X$). Instead of per capita calorie intake as used by the Planning Commission, we prefer the concept of *per unit* calorie intake for $Y$, where the unit numeraire is the average sedentary male adult. Different activity levels and the differential requirements thereof are thus taken care of in the derivation of $Y$ itself. Monthly per capita consumption expenditure in current prices is denoted by $X$. The exact nature of relationship between $Y$ and $X$ has to be determined empirically. We may assume a simple proportionate relationship to begin with. But when the relationship is non-linear, it may be possible in some cases to linearize it through suitable transformation of the variables, as is currently done in the analyses of Engel curves from family budget data. For our present purposes a linear double-logarithmic relationship is considered, after a routine examination of the scatter diagram.

Let us write

$$y = \alpha + \beta x + U, \tag{1}$$

where $y = lnY$, $x = lnX$, and $U$ is normally distributed with mean zero and constant variance $\sigma^2$. The problem, then, is to define a level of $x$ related to the minimal calorie norm ($y_*$), such that

$$Pr(y \geq y_*) \geq 1 - \eta, \tag{2}$$

where $\eta$ is an arbitrarily small positive number. The minimal level of $X = e^x$ satisfying (2) may be defined as the $\eta$-poverty line in monetary terms. Naturally, its value depends upon the level of probability ($\eta$) chosen.

In field surveys it is much easier to obtain information on the variable $X$ than on $Y$ from individual households. A household is to be regarded as poor if its $X$-value falls short of the minimum needed to buy $y_*$, the absolutely minimum calorie requirement. Let the poverty line be denoted by $X_*$, i.e., $exp(x_*)$. Obviously, this depends upon the unknown parameters, $\alpha, \beta$ and $\sigma$. Under normality and iid assumptions on $u$'s, it is easy to show that

$$x_* = (y_* - \alpha + \sigma Z_\eta)/\beta = x(\alpha, \beta, \sigma), \text{say}, \tag{3}$$

where $Z_\eta$ is the standard normal deviate corresponding to $\eta$. That is,

$$\eta = \Phi(-Z_\eta) = \frac{1}{\sqrt{2\pi}} \int_{-\infty}^{-Z_\eta} e^{-u^2/2} du. \tag{4}$$

The proof follows from the assumption that $Z = u/\sigma$ is normally distributed with mean zero and standard deviation one and

$$Pr(Z \geq (y_* - \alpha - \beta x)/\sigma) = 1 - \Phi[(y_* - \alpha - \beta x)/\sigma]. \tag{5}$$

For this to be greater than or equal to $1-\eta$, $(y_* - \alpha - \beta x)$ must be greater than $-\sigma Z_\eta$. Hence, the optimal level for $x$ is $(y_* - \alpha + \sigma Z_\eta)/\beta$.

If one ignores the possibility of variation in calorie intake for the same level of per capita expenditure, there is a simpler rule, (see, Kreme and Thorbecke, 1987), for converting $y$ into $x$:

$$x_* = (y_* - \alpha)/\beta. \tag{6}$$

## 3. THE BAYESIAN APPROACH

Consider the statistical problem of determining the unknowns, $\alpha, \beta$ and $\sigma$ in equation (3). To estimate these unknowns, a most appropriate procedure would be to use the standard loss function,

$$L(d, x) = (d - x)^2. \tag{7}$$

The problem is to get a Bayes estimate, $d(\mathbf{y})$, where $\mathbf{y} = (y_1, ..., y_n)$ is the observed vector of the $y$-values in this sample, such that the posterior loss is minimized, i.e., to find $d$ for which $E[(d-x)^2|\mathbf{y}]$ is a minimum. The optimal value of $d$ is given by

$$d_B = y_* E(\beta^{-1}|\mathbf{y}) - E(\alpha\beta^{-1}|\mathbf{y}) + Z_\eta E(\sigma\beta^{-1}|\mathbf{y}), \tag{8}$$

which provides the Bayes estimate for $x_*$. (This is a slightly different application of the "sophisticatedly simple" approach to the control problem as outlined in Zellner(1988). Let us write $\tau = \sigma^{-2}$. To compute the conditional expectations in (8), additional specifications of the priors for $\alpha, \beta$ and $\tau$ are needed. It seems reasonable a priori to assume that $\alpha, \beta$ and $\tau$ are independently distributed as Gamma random variables, with parameters $(\theta_\alpha, \kappa_\alpha)$, $(\theta_\beta, \kappa_\beta)$ and $(\theta_\tau, \kappa_\tau)$ respectively. Further, for the existence of the conditional expectations in (8), we need to assume that $\kappa_\alpha > 0$, $\kappa_\beta > 1$ and $\kappa_\tau > 1/2$. These assumptions may be justified because nutrition is inelastic with respect to total consumption, i. e., $0 < \beta < 1$ and $\alpha > 0$, since nutrition is measured in terms of calorie intake, the units of which are much smaller than units of consumption in terms of value.

The posterior density of $\alpha, \beta$ and $\tau$ given the sample, $\mathbf{y}$, can be written as:

$$\Pi(\alpha, \beta, \tau) \propto L(\mathbf{y}|\alpha, \beta, \tau)\Pi_1(\alpha)\Pi_2(\beta)\Pi_3(\tau), \tag{9}$$

where,

$$L(\mathbf{y}|\alpha, \beta, \tau) = (2\pi)^{-n/2}\tau^{n/2} exp[-(\tau/2)\sum(y_i - \alpha - \beta x_i)^2], \tag{10}$$

and $\Pi_1(\alpha) \sim G(\theta_\alpha, \kappa_\alpha), \Pi_2(\beta) \sim G(\theta_\beta, \kappa_\beta)$, and $\Pi_3(\tau) \sim G(\theta_\tau, \kappa_\tau)$.

Using (9) one can compute the posterior conditional expectations of $\beta^{-1}, \alpha\beta^{-1}$, and $\tau^{-1/2}\beta^{-1}$. Of course, one does not get a simple closed expression for any expectation and, therefore, one has to rely upon numerical integration. The numerical results for some real data are reported in the following section.

## 4. NATIONAL SAMPLE SURVEY DATA

Our analysis is based on ungrouped National Sample Survey (NSS) data relating to households' consumption expenditures in current prices and restricted to the South Indian state of Karnataka. In this survey, which covered the period from October 1973 to June 1974, a stratified two- stage PPS design was employed. The first-stage unit in the urban sector was the household block, while in the rural sector the village formed the first-stage unit. In either case the ultimate unit of observation was the household. The expenditures on various household items pertain to actual and imputed expenditures incurred by the household over a period of 30 days preceding the date of interview (For details, see *NSSO Report No. 240, Government of India*). 620 rural and 369 urban households were included in the self- weighted sample split into two independent and interpenetrating sub-samples.

Information was collected from each sampled household on the following aspects:

(i) quantity and expenditure on all items of consumption during the reference period of thirty days;

(ii) number of adults (male and female) and children in the household; and

(iii) occupation, religion, land owned and social status of the household.

Using this information, the variables, $X_h$ and $Y_h$, were first constructed for each sampled household, $h$. The calculation of $Y_h$ involved certain steps and assumptions. For example, for some households, calorie intake could not be correctly computed and hence the effective sample sizes for this study were only 620 and 367 for the pooled rural and urban samples, respectively. These samples were further stratified into district samples. For a detailed discussion, see Iyengar and Gopalakrishna (1985).

Three simple forms of regression were tried, viz.. linear, semi- logarithmic, and double-logarithmic models, commonly used in traditional family budget analyses. In these specifications it is implied that there is no limit to calorie intake. However, it may be more realistic to assume a Tornqvist form of relationship that incorporates a saturation point for food consumption, no matter how large the per capita total expenditure may be. The Bayesian analysis in Section III would have to be slightly modified in that case. Empirical considerations strongly favored the adoption of the double-logarithmic form, in both rural and urban sectors. In this case, the Bayesian approach yields an estimate of the poverty line in terms of $x_*$, from which $X_*$ $(= e^{x_*})$ can be easily derived.

As noted earlier, our first step would be to make appropriate specifications of the values of the hyperparameters, $(\theta_\alpha, \kappa_\alpha)$ , $(\theta_\beta, \kappa_\beta)$ and $(\theta_\tau, \kappa_\tau)$ , in the Gamma priors. There may be several ways of specifying these quantities, including informed judgments and guesses. In this study, however, we have adopted a crude method using the least squares approximations. For example, consider the hyperparameters $(\theta_\alpha, \kappa_\alpha)$ appearing in the prior distribution of $\alpha$. It is easy to verify that $E(\alpha) = \kappa_\alpha/\theta_\alpha^2$, and that the coefficient of variation $CV(\alpha) = \kappa_\alpha^{-1/2}$. If one assumes that $E(\alpha)$ can be approximated by the pooled least squares estimate of $\alpha$ and use certain arbitrary values for the coefficient of variation, both the parameters $\theta_\alpha$, and $\kappa_\alpha$ get automatically determined. The same argument applies in the case of $(\theta_\beta, \kappa_\beta)$ and $(\theta_\tau, \kappa_\tau)$. In our computations we have assumed different values for the CV's of $\alpha, \beta$ and $\tau$. Three sets of values were used for the values of $\theta$ and $\kappa$ in the assumed prior distributions of $\alpha, \beta$ and $\tau$ to examine the sensitivity of the Bayes estimates to the prior assumptions. However, no separate values were assumed for the rural and urban distributions. These assumptions are set out in Table 1 below.

Table 1: Hyperparameters for Prior Gamma $(\theta, \kappa)$

| Regression | Set I | | Set II | | Set III | |
|---|---|---|---|---|---|---|
| Parameter | $\theta$ | $\kappa$ | $\theta$ | $\kappa$ | $\theta$ | $\kappa$ |
| $\alpha$ | 1 | 2 | 2 | 11 | 5 | 12 |
| $\beta$ | 1 | 3 | 20 | 10 | 12 | 6 |
| $\tau$ | 1 | 2 | 15 | 5 | 6 | 2 |

## 5. MAIN RESULTS

The ordinary least squares estimates of $\alpha, \beta$ and $\sigma$ in rural and urban Karnataka are given in Table 2. Least squares estimates of poverty lines are also given sector-wise (Using (3) with $Y_* = 2400$ and $\eta = 0.05$).

Table 2: Least squares estimates of regression coefficients and poverty lines, Rural and Urban Karnataka, 1973-74

| Parameter | Rural | Urban |
|---|---|---|
| $\alpha$ | 5.2804 | 5.5550 |
| $\beta$ | 0.6078 | 0.4992 |
| $\sigma$ | 0.3089 | 0.2958 |
| Poverty line | 141.66 | 229.93 |
| Sample size | 620 | 367 |

Using the nutrition norm of $Y_* = 2400$ K cals and the hyperparameter-values given in Table 1, the Bayes estimates were obtained after calculating the conditional posterior expectations of $\alpha\beta^{-1}, \beta^{-1}$ and $\sigma\beta^{-1}$, using standard methods of numerical integration, as discussed in Section III. Poverty lines were worked out using expression (8) for three levels of probability, viz. $\eta = 0.05, 0.1$ and $0.2$. The reliability level can be measured by $1 - \eta$. The major results of our computations are summarized in Table 3.

It may be noted that the Bayesian estimates are generally higher than the least squares poverty estimates. However, a strict comparison between the official and other estimates including our own, is perhaps not possible, since their underlying concepts and methods are different. But the widespread impression that poverty in India is grossly underestimated is strengthened by our analysis. The official statistics on poverty put out by the Planning Commission point to a much smaller degree of poverty at national and state levels.

Table 3: Posterior means of $\alpha\beta^{-1}, \beta^{-1}$ and $\sigma\beta^{-1}$
and the Bayesian poverty lines.
Rural and Urban Karnataka, 1973-74

| Parameter | Set I | | Set II | | Set III | |
|---|---|---|---|---|---|---|
| | Rural | Urban | Rural | Urban | Rural | Urban |
| $E(\alpha\beta^{-1}\|\mathbf{y})$ | 8.638 | 11.013 | 8.792 | 11.241 | 8.601 | 10.855 |
| $E(\beta^{-1}\|\mathbf{y})$ | 1.639 | 1.988 | 1.659 | 2.018 | 1.634 | 1.968 |
| $E(\sigma\beta^{-1}\|\mathbf{y})$ | 0.514 | 0.604 | 0.625 | 0.822 | 0.553 | 0.681 |
| Bayesian poverty line $X_*$ (in Rs.) for $1 - \eta =$ | | | | | | |
| 0.95 | 142.85 | 233.95 | 172.05 | 335.99 | 152.30 | 265.11 |
| 0.90 | 118.43 | 187.65 | 136.95 | 248.93 | 124.46 | 206.80 |
| 0.80 | 94.55 | 133.98 | 104.12 | 173.62 | 97.68 | 153.45 |

## 6. CONCLUDING REMARKS

In this paper we have shown that Bayesian methods can be used profitably for estimating the poverty line in monetary terms, which can be readily applied for identifying the poor on the basis of a more acceptable definition of poverty. In this implementation, certain simplifying assumptions about the priors, involving the assumption of independent Gamma distributions with specified hyperparameter values, had to be made. The three sets of specifications used for the hyperparameters are only illustrative but not exhaustive. To get more reasonable values for these parameters requires extensive prior data or expert knowledge. Also, for the application of the Bayesian method, we have assumed the standard squared error loss function. On grounds of simplicity and conventions this could probably be justified. In our Bayesian analysis, we have assumed that the regression parameters $(\alpha, \beta)$ and the error variance $\sigma^2 (= 1/\tau)$ are apriori independently distributed. This assumption can easily be justified using the arguments of Box and Tiao (1973), cf. p. 52.

The official estimates of Rs. 49.09 and Rs. 56.60 as poverty lines for rural and urban areas can be shown to correspond, under specifications in Set I, Table 1, to the the probability level $1-\eta$ equal to 0.67 and 0.76 respectively. In this respect, our present estimates, ensuring greater reliability in terms of the probability levels, appear much better.

The Karnataka data in ungrouped form $(X_h, Y_h)$ are available for 18 rural and 9 urban districts, so that further modelling may be possible, using alternative and perhaps more realistic assumptions. For example, we may do away with the assumption of a common calorie norm and Gamma priors for rural and urban sectors and derive seperate Bayesian estimates for different norms. Also, as stated earlier, we may explore other non- linear forms of relationship such as hyperbolic between nutrition and consumption expenditure, incorporating a saturation parameter for nutrition. Finally, since we believe that $0 < \beta < 1$, it appears more appropriate to use a beta distribution as a prior for $\beta$. We hope to examine the various possibilities and issues in a future study.

## ACKNOWLEDGMENT

The authors wish to thank Prof. Jay Kadane and another referee for their kind comments and suggestions on an earlier version of this paper.

## REFERENCES

Box, G.E.P. and G.C. Tiao (1973). *Bayesian Inference in Statistical Analysis.* Addison-Wesley, Reading, MA.

Gopalan, C. (1983). Measurement of undernutrition - Biological considerations, *Economic and Political Weekly,* **XVIII - 15.** 591.

Gopalan, C., B.V. Rama Sastry and S.C. Balasubramaniam (1971). *Nutritive Value of Indian Foods.* (National Institute of Nutrition, Hyderabad, India).

Gopalan, C. and B.S. Narasinga Rao (1968). *Dietary Allowance for Indians* (Special Report Ser. **#60,** Indian Council of Medical Research, New Delhi, India).

Hoadley, B. (1970) A Bayesian look at Inverse Linear Regression, *Jour. American Statistical Assoc.,* **65,** 356 - 69.

Iyengar, N.S. and Mallika Gopalakrishna (1985). Appropriate criteria for the measurement of levels of living, *Indian Economic Review,* **XX, 2.** 191-229.

Kyereme, Stephen S. and Erik Thorbecke (1987). Food poverty profile and decomposition applied to Ghana, In *World Development,* **15, 9.** 1189-99.

Planning Commission (1979). *Report of the Task Force on Projections of Minimum Needs and Effective Consumption Demand,* Government of India, New Delhi, India.

Sukhatme, P.V. (1977). *Poverty and Malnutrition,* Lal Bahadur Sastry Memorial Lecture, New Delhi, India.

Sukhatme, P.V. (Ed.) (1982). *Newer Concepts in Nutrition and Their Implications for Policy,* Maharashtra Association for the Cultivation of Science Research Institute, Pune, India.

Zellner, A. (1988). Bayesian Inference in Econometrics, *Jour. of Econometrics,* **37.** 27-50.

# 27

# Quantifying Prior Opinion in Length-Biased Linear Mean Natural Exponential Family

**Ramalingam Shanmugam** [1]

**Abstract**

Recently, Shanmugam [3] identified that there are only four members (namely, Binomial, Negative Binomial, Poisson, and Gamma) in a class called Linear Mean Natural Exponential Family (LMNEF). For a length-biased random sample from a LMNEF, we derive (1) Jeffreys' noninformational, (2) Hartigan's asymptotically locally invariant, and (3) conjugate priors, and employ them to drive a Bayesian inference-making engine.

*Key Words and Phrases:*. Length-biased sampling; Linear mean exponential family; Quantifying priors; Noninformative priors.

## 1. INTRODUCTION

Consider a chance mechanism whose real-valued natural observation $X$ is governed by a likelihood function

$$f(x|\theta) = s(x,a)\exp[x\theta - a\psi_\theta] \tag{1}$$

with respect to some counting or Lebesque measure $m(x)$, where $\theta \in \Theta$ is the real-valued unknown natural parameter and the reproductive parameter '$a$' is assumed to be known. From the cumulant generating function $\psi_\theta$, both the population mean

$$E(X|\theta) = \mu_\theta = a\partial_\theta\psi_\theta \tag{2}$$

and variance

$$V(X|\theta) = \partial_\theta\mu_\theta = a\partial_\theta^2\psi_\theta, \tag{3}$$

are obtained, where the notation $\partial_\theta^i$ stands for differentiating $i$ number of items with respect to $\theta$. Also it is easy to show that $V(X|\theta)$ is also a

[1] University of Colorado at Denver

quadratic function of $\mu_\theta$. Furthermore, if the mean $\mu_\theta$ could be expressed in an exponential format as

$$\mu_\theta = a\delta \exp[\alpha\theta - \beta\psi_\theta] \tag{4}$$

with some constants $\delta$, $\alpha$, and $\beta$ not depending on $\theta$, then the class of distributions with the likelihood function $f(X|\theta)$ as in (1) is called Linear Mean Natural Exponential Family (LMNEF) (see Shanmugam [3]). The LMNEF is a subclass of a family called the exponential quadratic variance family. Recently, Shanmugam identified that there are only four members in LMNEF. They are: Binomial (B), Negative Binomial (NB), Poisson (P), and Gamma (G). Their individual values are summarized in Table 1 below.

**Table 1**

Linear Mean Natural Exponential Family

| Distr. | $a$ | $S(X,\alpha)$ | $\psi_\theta$ | $\delta$ | $\alpha$ | $\beta$ | $\Theta$ |
|---|---|---|---|---|---|---|---|
| $B$ | $r>0$ | $\binom{r}{x}$ | $ln(1+e^\theta)$ | 1 | 1 | 1 | $(-\infty,\infty)$ |
| $NB$ | $-r<0$ | $\binom{r+x-1}{x}$ | $ln(1-e^\theta)$ | $-1$ | 1 | 1 | $(-\infty,0)$ |
| $P$ | 1 | $\frac{1}{x!}$ | $e^\theta$ | 1 | 1 | 0 | $(-\infty,\infty)$ |
| $G$ | $r>0$ | $\frac{x^{r-1}}{\Gamma(r)}$ | $-ln(-\theta)$ | 1 | 0 | $-1$ | $(-\infty,0)$ |

The variance stated in (3) could be explicitly written in terms of $\mu_\theta$:

$$V(X|\theta) = V(\mu_\theta) = \frac{\mu}{a}(\alpha - \frac{\beta\mu_\theta}{a}). \tag{5}$$

Also, the population third central moment is

$$M_3 = E[(X-\mu_\theta)^3|\theta] = a\partial_\theta^3\mu_\theta = (\alpha - \frac{2\beta\mu_\theta}{a})\frac{V(\mu_\theta)}{a}. \tag{6}$$

Furthermore, both $\theta$ and $\psi_\theta$ are on a one-to-one relationship with $\mu_\theta$. To see this, we extract from (3) and (5) that

$$\frac{d\mu}{\frac{\mu_\theta}{a}(\alpha - \frac{\beta\mu_\theta}{a})} = d\theta, \tag{7}$$

and use it on the right side integral of

$$\int a\psi'_\theta \, d\theta$$

to get

$$\psi_\theta = \begin{cases} -\frac{1}{\beta}\, ln(\alpha - \frac{\beta\mu_\theta}{a}) & \text{if } \beta \neq 0 \\ \alpha\mu_\theta & \text{if } \beta = 0. \end{cases} \tag{8}$$

Substituting (8) into (4), we recover

$$\theta = \begin{cases} \frac{1}{\alpha}\left[ln\left\{\frac{\mu_\theta}{\alpha\delta(\alpha - \frac{\beta\mu_\theta}{a})}\right\}\right] & \text{if } \alpha \neq 0 \\ -\frac{a}{\mu} & \text{if } \alpha = 0. \end{cases} \tag{9}$$

We now turn to some basics of length-biased sampling. Many times in real life, although a random sample is intended to come from a stated population, we end up having the sample represent a different population due to what is called a length bias in the sampling process. For example, when a random sample on the number of days patients tend to stay in a hospital is intended, the observed data represent not the intended population with the likelihood function $f(X|\theta)$ but instead the one with likelihood function

$$f^*(x|\theta) = \frac{xf(x|\theta)}{\mu_\theta}. \tag{10}$$

For patients with a larger value of $X$, the lengths of stay are more likely to be included in the sample. This phenomenon is called length-biased sampling and is found to be useful in the analysis of skull data and in illustrating the visibility bias in wild life studies, among others. For details, see Rao [2].

For our discussion in this article, we consider a length-biased random sample $X_1, X_2, \ldots, X_n$ which was intended to come from a member of the LMNEF displayed in Table 1, but represents the population with a (weighted) likelihood function

$$f^*(x|\theta) = S^*(x,a)\exp[T(x,\alpha)\theta - (a-\beta)\psi_\theta], \tag{11}$$

where $S^*(x,a) = xS(x,a)/a\delta$ and $T(x,\alpha) = x - \alpha$. Throughout in this writing, a superscript $*$ personifies the length biasedness. Here, a statement calling for caution is needed: the likelihood function $f^*(x|\theta)$ in (11) is no longer a member of LMNEF, nor $(a-\beta)\psi_\theta$ a cumulant generating function. In fact, the mean of the length-biased likelihood function in (11) is

$$E^*(X|\theta) = \mu_\theta^* = \mu_\theta + \frac{V(\mu_\theta)}{\mu_\theta} = \frac{\alpha}{a} + (1 - \frac{\beta}{a^2})\mu_\theta \tag{12}$$

while its variance

$$V^*(X|\theta) = V(\mu_\theta) + \frac{\partial_\theta V(\mu_\theta)}{\mu_\theta} - \left[\frac{V(\mu_\theta)}{\mu_\theta}\right]^2 = (1 - \frac{\beta}{a^2})V(\mu_\theta), \tag{13}$$

where $\mu_\theta$ and $V(\mu_\theta)$ are as given in (4) and (5), respectively. From (12), it is interesting to note that $\partial_\theta \mu_\theta^* = (1 - \frac{\beta}{a^2})\partial_\theta \mu_\theta$ which due to (3) becomes $\partial_\theta \mu_\theta^* = (1 - \frac{\beta}{a^2})V(\mu_\theta)$ and then according to (13)

$$\partial_\theta \mu_\theta^* = V^*(X|\theta). \tag{14}$$

This means the variance is simply the rate of change in the mean function whether or not the length bias occurs in the sampling process.

We are now tempted to pose a question: Is it a simple anomaly or only the tip of the iceberg for many quality preservations under length-biased sampling? In this article, we explore this question from Bayesian viewpoint.

## 2. QUANTIFYING PRIOR OPINION

It is needless to say that the prior distribution plays a crucial role in the Bayesian approach. There are different ways to quantify the prior opinion. In this section, we shall follow (1) Jeffreys' noninformational method, (2) Hartigan's asymptotically locally invariant (ALI) method, and (3) the conjugate method to quantify the prior opinion for a length-biased sample.

### 2.1. Jeffreys' Noninformational Prior

Jeffreys' [1] noninformational prior is locally uniform (a synonimity for complete ignorance) and also data dominated. Jeffreys' prescription for obtaining it is to have it proportional to the square root of Fisher's information $I(\theta) = (E\{[\partial_\theta \ln f(x|\theta)]^2|\theta\})^{-1}$. For a regular sample from a member of LMNEF listed in Table 1, Fisher's information is $I(\theta) = [E\{(x-\mu_\theta)^2|\theta\}]^{-1} = [V(\mu_\theta)]^{-1}$ and hence Jeffreys' noninformational prior $J(\theta)$ due to (5) is

$$J(\theta) \propto \mu_\theta^{-1/2}\left(\alpha - \frac{\beta\mu_\theta}{a}\right)^{-1/2}. \tag{15}$$

Note that $\alpha - \frac{\beta\mu_\theta}{a} > 0$ (see in Table 1) for each member of LMNEF, and hence, $J(\theta)$ is not complex. $J(\theta)$ is sometimes called the reference prior. $J(\theta)$ is a beta distribution of the first kind for binomial, second kind for the negative binomial, and power distributions for the Poisson and gamma, but all in terms of $\mu_\theta$ not $\theta$. Pertaining to a length-biased sample representing a population whose likelihood function is $f^*(x|\theta)$ as in (10), Jeffreys' noninformational prior $J^*(\theta)$ is

$$J^*(\theta) \propto \left(1 - \frac{\beta}{a^2}\right)^{-1/2} J(\theta) \tag{16}$$

The proportionality between $J(\theta)$ and $J^*(\theta)$ is only a constant, free of $\theta$, and it makes sense after all because the parameter $\theta$ is $\theta$ regardless of the sampling scheme (length biased or not). However, the posterior is affected by the length bias.

Looking at the result in (16), we realize that the length-biased statistical experiment $E_X^* = \{X, f^*(x|\theta), \theta \in \Theta\}$ is not always information sufficient in the Fisherian sense for the regular statistical experiment $E_X = \{X, f(X|\theta), \theta \in \Theta\}$; only for gamma population $I^*(\theta) > I(\theta)$ as $\beta < 0$, and hence the information in the length-biased sample is sufficient to make inference on any parametric function of the experiment $E_X$. Of course, the larger the value of "$a$" is, the smaller the difference between the informations $I(\theta)$ and $I^*(\theta)$ where "$a$" is the convolution (or reproductive, as it is sometimes called) parameter. As the ramifications of this inequity in informations are not in tune with the main theme of this article, we shall discuss it elsewhere.

We now turn to Hartigan's method of quantifying prior opinion.

## 2.2 Hartigan's ALI Prior

Hartigan's [4] ALI prior $H(\theta)$ possesses many desirable properties. His prescription to find it is

$$H(\theta) \propto \exp\left[-\int \left\{\frac{E[\partial_\theta \ln f(X|\theta)\, \partial_\theta^2 \ln f(X|\theta)]}{E[\partial_\theta^2 \ln f(X|\theta)]}\right\} d\theta\right] \tag{17}$$

provided the expected value of the influence function is zero. For the LMNEF, the influence function $\partial_\theta \ln f(X|\theta) = X - \mu_\theta$ has zero expected value. Because $\partial_\theta^2 \ln f(X|\theta) = -V(\mu_\theta)$, independent of the observation $X$, expression (17) yields $H(\theta) \propto 1$ which is a flat and improper prior.

In the length-biased case also, the influence function $\partial_\theta \ln f^*(X|\theta) = X - \mu_\theta - \frac{V(\mu_\theta)}{\mu_\theta}$ has zero expected value. Also, $\partial_\theta^2 \ln f^*(X|\theta) = -\left(1 - \frac{\beta}{a^2}\right) V(\mu_\theta)$ is independent of $X$, expression (17) again yields a flat and improper prior $H^*(\theta) \propto 1$.

We now discuss conjugate priors which are considered to be convenient building blocks in Bayesian inference making.

## 2.3 Conjugate Priors

Conjugate prior is one whose posterior mean is linear in the natural observation. The prescription to find it is focused on the sample size $n$ and the sufficient statistic $T = T(x_1, \ldots, x_n)$ for the parameter $\theta$ in the likelihood function $\prod_{i=1}^n f(x_i|\theta)$. By mimicking the likelihood function with $n$ and $t$ replaced by prior parameters $n_0$ and $x_0$, the conjugate prior is obtained. For LMNEF listed in Table 1, the sufficient statistic for the parameter $\theta$ is $T = n\overline{x}$ where $\overline{x}$ is the sample mean. Hence the conjugate prior has a proper probability density

$$C(\theta) = K(n_0, x_0, a) \exp[n_0 x_0 \theta - n_0 a \psi_\theta] \tag{18}$$

with respect to $d\theta$ where $K(n_0, x_0, a)$ is a normalizing constant. This conjugate prior is known to possess posterior linearity

$$E[\mu_\theta | X = x] = bx + d \tag{19}$$

for some constants "$b$" and "$d$". In fact, Diaconsis and Ylvisaker [5] used (19) as a definition to characterize the conjugate prior in (18).

In the case of the length-biased density $f^*(x|\theta)$ in (11), the prescription for finding the conjugate prior is not quite applicable as $n\overline{x}$ is not a sufficient statistic. The density $f^*(x|\theta)$ does not have the reproductive property. By simply mimicking the format in (11), we suggest using

$$C^*(\theta) = K^*(x_0, n_0, a) \left(\mu_\theta e^{a\psi_\theta}\right)^{-n_0} e^{n_0 x_0 \theta} \tag{20}$$

as a pseudo-conjugate prior density where $n_0$ and $x_0$ are its parameters and $K^*(x_0, n_0, a)$ as the normalizing constant to make $C^*(\theta)$ proper. In Section 3, we shall explore whether a posterior linearity like (19) exists.

## 3. POSTERIORITY

In practice, researchers are often interested in the mean $\mu_\theta$ of the intended population rather than the parameter $\theta$ or the mean $\mu_\theta^*$ of the actually studied population. Hence, we focus our discussion on the Bayes estimator $\tilde{\mu}_\theta$ with respect to a minimum squared loss: posterior mean of $\mu_\theta$.

When $J(\theta)$ in (15) is used, the posterior density of $\theta$ given the complete sufficient statistic $\overline{x} = \sum n_i/n$ is beta distribution of first kind for binomial, second kind for negative binomial, gamma for Poisson, and inverted gamma for gamma, but all in terms of $\mu_\theta$ not $\theta$. They all yield Bayes estimator $\tilde{\mu}_\theta = E(\mu_\theta|\overline{x})$ which converges to the sample mean $\overline{x}$ as the sample size increases, with their posterior risk $V(\mu_\theta|\overline{x})$ approaching zero. Likewise, using Jeffreys' noninformational prior, $J^*(\theta)$ in (16) under a length-biased sample, we realize that $E^*(\mu_\theta^*|\overline{x}) = \int \left[\mu_\theta + \frac{V(\mu_\theta)}{\mu_\theta}\right] P^*(\theta|\overline{x})\, d\theta = \overline{x} + \frac{1}{2n}$ which is the Bayes estimator of only $\mu_\theta + \frac{V(\mu_\theta)}{\mu_\theta}$ instead of $\mu_\theta$. It suggests then under a length biased sampling, one should rather use $(\overline{x} + \frac{1}{2n} - 1)$ for Poisson, $\left[\frac{na}{n(a+1)-1}\right] \overline{x}$ for gamma, $\left[\frac{a}{n(a-1)+1}\right] [n(\overline{x} - 1) + \frac{1}{2}]$ for binomial or negative binomial distributions as Bayes estimator for $\mu_\theta$.

We then assess the Euclidean distance

$$d = [\{E^*(\mu_\theta|\overline{x}) - E(\mu_\theta|\overline{x})\}^2 + \{V^*(\mu_\theta|\overline{x}) - V(\mu_\theta|\overline{x})\}^2]^{1/2} \qquad (21)$$

in its estimation and posterior risk spaces. The gap in the risk space shrinks as the sample size $n$ increases and eventually is zero. But, in the estimation (mean) space, it is not so, as $E^*(\mu_\theta|\overline{x})$ and $E(\mu_\theta|\overline{x})$ never become equal, no matter how large the sample size $n$ might be. Even for a large sample, there is a finite amount of shift, and it is

$$d \simeq \begin{cases} 1 & \text{if } \beta = 0, \alpha = 1 \\ \overline{x}/(a+1) & \text{if } \beta = -1, \alpha = 0 \\ \sqrt{[(a-\overline{x})/(a-1)]^2} & \text{if } \beta = 1, \alpha = 1, a > 0, \text{ or } a < 0. \end{cases} \qquad (22)$$

Expression (22) indicates that for a fixed convolution parameter, $a$, in the case of negative binomial, and gamma populations, the asymptotic distance $d$ is smaller whenever $\overline{x}$ is smaller but in the case of binomial, $d$ increases as $\overline{x}$ decreases. In the case of Poisson, it is a constant shift.

When Hartigan's ALI prior is used, the shift $d$ in both estimation and posterior-risk spaces for a large sample size turns out to be the same as expression (22). This then implies that it makes no difference only in the asymptotic sense whether Jeffereys' noninformational or Hartigan's locally invariant prior is used as far as the shift '$d$' is concerned.

With conjugate prior $c(\theta)$ in (18), and the pseudo-conjugate prior $C^*(\theta)$ in (20), in all cases, it is seen that the shift $d$ is again the same. Hence, we summarize our results in a theorem.

**Theorem 1** The asymptotic Bayes estimator $E^*(\mu_\theta|\overline{x})$ of the mean $\mu_\theta$ based on a length biased sample from the linear mean exponential family stated in (1) is $d$ units less than the asymptotic Bayes estimator $E(\mu_\theta|\overline{x})$ of the mean $\mu_\theta$ based on the non-length biased sample from the same family. The value of $d$ is unity in Poisson, $\overline{x}/(a+1)$ units in gamma and $\sqrt{[(a-\overline{x})/(a-1)]^2}$ in binomial or negative binomial cases, where $\overline{x}$ is the sample mean.

## ACKNOWLEDGMENT

The author thanks Department of Mathematics and College of Liberal Arts and Sciences at the University of Colorado for travel support to present this paper at the Indo-U.S. Bayesian Workshop in Bangalore, India, on December 23, 1988. The author dedicates this work to his beloved wife Malarvizhi for her understanding, affection, and a great companionship.

## REFERENCES

[1] H. Jeffreys (1961). *Theory of Probability*, London: Oxford University Press.

[2] C. R. Rao (1987). Weighted distributions arising out of methods of ascertainment: What population does a sample represent? Chapter 24 in *A Celebration in Statistics: The ISI Centenary Volume* edited by A. C. Atkinson and S. E. Fienberg, New York: Springer Verlag Press.

[3] R. Shanmugam (1987). Asymptotic homogeneity tests for mean exponential family distributions, *J. StatPlanning and Inference* 23, 227@–241.

[4] J. A. Hartigan (1964). Invariant prior distributions, *AnnMathStat* 35, 836@–845.

[5] P. Diaconis and D. Ylvisaker (1979). Conjugate priors for exponential families, *Annals of Statistics* 7, 269@–281.

# Subject Index

# Author Index

**Indo-U. S. Workshop, Bangalore, India**
**December 19 - 23, 1988.**

**Bayesian Analysis in Statistics and Econometrics**

## PARTICIPANTS FROM OUTSIDE INDIA


Basu, Asit P.
Department of Statistics
University of Missouri,
Columbia, MO 65211

Berger, James O.
Department of Statistics
Purdue University,
W. Lafayette, IN 47907

Bhat, A.N.
Department of Statistics
University of Georgia,
Athens, GA 30602

Blattenberger, Gail
Department of Economics
University of Utah,
Salt Lake City, UT 84112

Blumenthal, Saul
Department of Statistics
The Ohio State University,
Columbus, OH 43210

Dalal, Siddhartha R.
Bell Communications Research,
435 South Street
Morristown, NJ 07960

Delampadi, Mohan
Department of Statistics
Univ. of British Columbia,
Vancouer, B.C., Canada V6T 1W5

DuMouchel, William
BBN Software INC.,
10 Fawcett Street
Boston, MA

Epstein, Leonardo D.
Department of Statistics
Carnegie Mellon University,
Pittsburgh, PA 15213

Fienberg, Stephen E.
Department of Statistics
Carnegie Mellon University,
Pittsburgh, PA 15213

Geisser, Seymour
School of Statistics
270 Vincent Hall
University of Minnesota,
Minneapolis, MN 55455

Ghosh, Malay
Department of Statistics
University of Florida,
Gainsville, FL 32611

Goel, Prem K.
Department of Statistics
The Ohio State University,
Columbus, OH 43210

Gupta, Shanti S.
Department of Statistics
Purdue University,
W. Lafayette, IN 47907

Highfield, Richard
Graduate School of Management
Cornell University,
559 Malott Hall
Ithaca, NY 14853

Hill, Bruce
Department of Statistics
University of Michigan,
Ann Arbor, MI 48109

Kadane, Joseph B.
Department of Statistics
Carnegie Mellon University,
Pittsburgh, PA 15213

Meeden, Glen
Department of Statistics
Iowa State University,
Ames, IA 50011

Mittal, Yash
Department of Statistics and O.R.
VPI & State University,
Blacksburg, VA 24061

Polasek, Wolfgang
Institute of Stattistics &
Econometrics
University of Basel,
Petersgraben 51,
CH 4051 Basel,
Switzerland

Press, S. James
Department of Statistics
University of California,
Riverside, CA 92521

Raj, Baldev
School of Business and Economics
Wilfrid Laurier University,
Waterloo, ONT, Canada N2L 3C5

Rao, C. R.
Department of Statistics
Penn State University,
Pond Lab
University Park, PA 16802

Schervish, Mark
Department of Statistics
Carnegie Mellon University,
Pittsburgh, PA 15213

Sethuraman, Jayaram
Department of Statistics
Florida State University,
Tallahassee, FL 32303

Shanmugam, Ram
Department of Mathematics
University of Colorado,
Denver, CO 80204

Singpurwalla, Nozer
Department of Operations Research
George Washington University,
Washington, DC 20052

Waiker, Vasant B.
Department of Math. & Stat.
Miami University,
Oxford, OH 45056

Waternaux, Christine
Department of Biostatistics
Harvard School of Public Health,
677 Huntington Ave.,
Boston, MA 02115

Zellner, Arnold
Graduate School of Business
University of Chicago,
1101 E. 58th Street
Chicago, IL 60637

**Indo-U. S. Workshop, Bangalore, India**
**December 19 - 23, 1988.**

**Bayesian Analysis in Statistics and Econometrics**

## PARTICIPANTS FROM INDIA


K. G. Aneja
Senior Scientist
Indian Agricultural Statistics
Research Institute (IASRI)
Library Avenue, Pusa Campus
New Delhi 110 012

P. Balasiddamuni
Lecturer in Statistics
Sri Venkateswara University
Tirupati 517 502

Ashok K. Bansal
Department of Statistics
University of Delhi
Delhi 110 007

D. Basu
Indian Statistical Institute
203 Barrackpore Trunk Road
Calcutta 700 035

Rajinder Singh Bawa
Professor and Head
Punjab School of Economics
Guru Nanak Dev University
Amritsar 143 005

P. R. Brahmananda
ICSSR National Fellow and
Honorary Visiting Professor
Economic Analysis Unit
Indian Statistical Institute
8th Mile, Mysore Road
Bangalore 560 059

Chandrahas
Indian Agricultural Statistics
Research Institute (IASRI)
Library Avenue, Pusa Campus
New Delhi 110 012

Anoop Chaturvedi
Department of Mathematics
and Statistics
University of Allahabad
Allahabad 211 002

S. Ganga
Scientific Officer
Post Box No: 1242
Central Power Research Institute
Bangalore 560 012

H. N. Geetha
Research Student
High Voltage Engineering Dept.
Indian Institute of Science
Bangalore 560 012

J. K. Ghosh
Director
Indian Statistical Institute
203 Barrackpore Trunk Road
Calcutta 700 035

Ramesh Chand Goel
Medical Education Cell
Dean's Office
Postgraduate Institute
Chandigarh 160 012

Mallika Gopalakrishna
19 Andree Road
Bangalore 560 027

M. A. Gopinath
Indian Statistical Institute
8th Mile, Mysore Road
Bangalore 560 059

Anwar Hassan
Postgraduate Department of
Mathematics and Statistics
Kashmir University
Hazaratbal
Sprinagar 190 006

N. S. Iyengar
Professor and Head
Economic Analysis Unit
Indian Statistical Institute
8th Mile, Mysore Road
Bangalore 560 059

S. N. Joshi
Stat-Math Unit
Indian Statistical Institute
8th Mile, Mysore Road
Bangalore 560 059

Vaman Balbhim Kale
Department of Statistics
University of Poona
Ganeshkhind
Pune 411 007

Gopal K. Kadekodi
Secretary
Indian Econometric Society
Institute of Economic Growth
University Enclave
Delhi 110 007

Balwant R. Kandoi
Department of Economics
Panjab University
Chandigarh 160 014

Attar Ali Khan
Faculty of Veterinary Sciences
and Animal Husbandry
Sher-e-Kashmir University of
Agricultural Sciences
Nowshera. Srinagar 190 011

Sudhakar Kunte
Department of Statistics
University of Poona
Ganeshkhind
Pune 411 007

Ram Lal
27-B Allahabad Agricultural
Institute
Allahabad 211 007

A. Loganathan
Research Student
Department of Statistics
Annamalai University
Annamalainagar 608 002

S. M. Manjunath
Department of Statistics
Bangalore University
Bangalore 560 040

R. N. Mishra
Department of Statistics
Patna University
Patna 800 005

T. S. Krishnan Moothathu
Reader in Statistics
University of Kerala
Kariavattom
Trivandrum 695 581

N. S. S. Narayana
Indian Statistical Institute
8th Mile, Mysore Road
Bangalore 560 059

V. R. Padmawar
Stat-Math Unit
Indian Statistical Institute
8th Mile, Mysore Road
Bangalore 560 059

B. K. Pal
Statistical Quality Control,
Indian Statistical Institute
8th Mile, Mysore Road
Bangalore 560 059

B. N. Pandey
Reader in Statistics
Banaras Hindu University
Varanasi 221 005

Sudha Ganesh Purohit
1943 Madiwale Colony
Pune 411 030

C. Raghavendra Rao
School of Mathematics & CIS
University of Hyderagad
Hyderabad 500 001

I. K. Ravichandra Rao
Documentation Research and
Training Centre (DRTC)
Indian Statistical Institute
8th Mile, Mysore Road
Bangalore 560 059

R. N. Rattihalli
Department of Statistics
Shivaji University 416 004

Somnath Ray
Statistical Quality Control,
Indian Statistical Institute
8th Mile, Mysore Road
Bangalore 560 059

S. G. A. Rizvi
Department of Statistics
Aligarh Muslim University
Aligarh 202 001

Divakar Sharma
Department of Mathematics
Indian Institute of Technology
Kanpur 208 016

Randhir Singh
Senior Professor
Indian Agricultural Statistics
Research Institute (IASRI)
Library Avenue, Pusa Campus
New Delhi 110 012

Umesh Singh
Reader in Statistics
Banaras Hindu University
Varanasi 221 005

Koduri Srinivas
School of Mathematics & CIS
University of Hyderabad
Hyderabad 500 001

M. B. Srinivas
Research Student
High Voltage Engineering Dept.
Indian Institute of Science
Bangalore 560 012

T. Srivenkataramana
Department of Statistics
Bangalore University
Bangalore 560 056

Alladi Subramanyam
Department of Mathematics
Indian Institute of Technology
Powai
Bombay 400 076

K. Surekha Rao
Economic Analysis Unit
Indian Statistical Institute
8th Mile, Mysore Road
Bangalore 560 059

L. N. Upadhyaya
Indian School of Mines
Dhanbad 826 004

H. J. Vaman
Department of Statistics
Bangalore University
Bangalore 560 056

P. Vellaiswamy
Department of Mathematics
Indian Institute of Technology
Kanpur 208 016

Vol. 1: R.A. Fisher: An Appreciation. Edited by S.E. Fienberg and D.V. Hinkley. XI, 208 pages, 1980.

Vol. 2: Mathematical Statistics and Probability Theory. Proceedings 1978. Edited by W. Klonecki, A. Kozek, and J. Rosinski. XXIV, 373 pages, 1980.

Vol. 3: B.D. Spencer, Benefit-Cost Analysis of Data Used to Allocate Funds. VIII, 296 pages, 1980.

Vol. 4: E.A. van Doorn, Stochastic Monotonicity and Queueing Applications of Birth-Death Processes. VI, 118 pages, 1981.

Vol. 5: T. Rolski, Stationary Random Processes Associated with Point Processes. VI, 139 pages, 1981.

Vol. 6: S.S. Gupta and D.-Y. Huang, Multiple Statistical Decision Theory: Recent Developments. VIII, 104 pages, 1981.

Vol. 7: M. Akahira and K. Takeuchi, Asymptotic Efficiency of Statistical Estimators. VIII, 242 pages, 1981.

Vol. 8: The First Pannonian Symposium on Mathematical Statistics. Edited by P. Révész, L. Schmetterer, and V.M. Zolotarev. VI, 308 pages, 1981.

Vol. 9: B. Jørgensen, Statistical Properties of the Generalized Inverse Gaussian Distribution. VI, 188 pages, 1981.

Vol. 10: A.A. McIntosh, Fitting Linear Models: An Application of Conjugate Gradient Algorithms. VI, 200 pages, 1982.

Vol. 11: D.F. Nicholls and B.G. Quinn, Random Coefficient Autoregressive Models: An Introduction. V, 154 pages, 1982.

Vol. 12: M. Jacobsen, Statistical Analysis of Counting Processes. VII, 226 pages, 1982.

Vol. 13: J. Pfanzagl (with the assistance of W. Wefelmeyer), Contributions to a General Asymptotic Statistical Theory. VII, 315 pages, 1982.

Vol. 14: GLIM 82: Proceedings of the International Conference on Generalised Linear Models. Edited by R. Gilchrist. V, 188 pages, 1982.

Vol. 15: K.R.W. Brewer and M. Hanif, Sampling with Unequal Probabilities. IX, 164 pages, 1983.

Vol. 16: Specifying Statistical Models: From Parametric to Non-Parametric, Using Bayesian or Non-Bayesian Approaches. Edited by J.P. Florens, M. Mouchart, J.P. Raoult, L. Simar, and A.F.M. Smith, XI, 204 pages, 1983.

Vol. 17: I.V. Basawa and D.J. Scott, Asymptotic Optimal Inference for Non-Ergodic Models. IX, 170 pages, 1983.

Vol. 18: W. Britton, Conjugate Duality and the Exponential Fourier Spectrum. V, 226 pages, 1983.

Vol. 19: L. Fernholz, von Mises Calculus For Statistical Functionals. VIII, 124 pages, 1983.

Vol. 20: Mathematical Learning Models — Theory and Algorithms: Proceedings of a Conference. Edited by U. Herkenrath, D. Kalin, W. Vogel. XIV, 226 pages, 1983.

Vol. 21: H. Tong, Threshold Models in Non-linear Time Series Analysis. X, 323 pages, 1983.

Vol. 22: S. Johansen, Functional Relations, Random Coefficients and Nonlinear Regression with Application to Kinetic Data, VIII, 126 pages, 1984.

Vol. 23: D.G. Saphire, Estimation of Victimization Prevalence Using Data from the National Crime Survey. V, 165 pages, 1984.

Vol. 24: T.S. Rao, M.M. Gabr, An Introduction to Bispectral Analysis and Bilinear Time Series Models. VIII, 280 pages, 1984.

Vol. 25: Time Series Analysis of Irregularly Observed Data. Proceedings, 1983. Edited by E. Parzen. VII, 363 pages, 1984.

Vol. 26: Robust and Nonlinear Time Series Analysis. Proceedings, 1983. Edited by J. Franke, W. Härdle and D. Martin. IX, 286 pages, 1984.

Vol. 27: A. Janssen, H. Milbrodt, H. Strasser, Infinitely Divisible Statistical Experiments. VI, 163 pages, 1985.

Vol. 28: S. Amari, Differential-Geometrical Methods in Statistics. V, 290 pages, 1985.

Vol. 29: Statistics in Ornithology. Edited by B.J.T. Morgan and P.M. North. XXV, 418 pages, 1985.

Vol 30: J. Grandell, Stochastic Models of Air Pollutant Concentration. V, 110 pages, 1985.

Vol. 31: J. Pfanzagl, Asymptotic Expansions for General Statistical Models. VII, 505 pages, 1985.

Vol. 32: Generalized Linear Models. Proceedings, 1985. Edited by R. Gilchrist, B. Francis and J. Whittaker. VI, 178 pages, 1985.

Vol. 33: M. Csörgo, S. Csörgo, L. Horváth, An Asymptotic Theory for Empirical Reliability and Concentration Processes. V, 171 pages, 1986.

Vol. 34: D.E. Critchlow, Metric Methods for Analyzing Partially Ranked Data. X, 216 pages, 1985.

Vol. 35: Linear Statistical Inference. Proceedings, 1984. Edited by T. Calinski and W. Klonecki. VI, 318 pages, 1985.

Vol. 36: B. Matérn, Spatial Variation. Second Edition. 151 pages, 1986.

Vol. 37: Advances in Order Restricted Statistical Inference. Proceedings, 1985. Edited by R. Dykstra, T. Robertson and F.T. Wright. VIII, 295 pages, 1986.

Vol. 38: Survey Research Designs: Towards a Better Understanding of Their Costs and Benefits. Edited by R.W. Pearson and R.F. Boruch. V, 129 pages, 1986.

Vol. 39: J.D. Malley, Optimal Unbiased Estimation of Variance Components. IX, 146 pages, 1986.

Vol. 40: H.R. Lerche, Boundary Crossing of Brownian Motion. V, 142 pages, 1986.

Vol. 41: F. Baccelli, P. Brémaud, Palm Probabilities and Stationary Queues. VII, 106 pages, 1987.

Vol. 42: S. Kullback, J.C. Keegel, J.H. Kullback, Topics in Statistical Information Theory. IX, 158 pages, 1987.

Vol. 43: B.C. Arnold, Majorization and the Lorenz Order: A Brief Introduction. VI, 122 pages, 1987.

Vol. 44: D.L. McLeish, Christopher G. Small, The Theory and Applications of Statistical Inference Functions. VI, 124 pages, 1987.

Vol. 45: J.K. Ghosh (Ed.), Statistical Information and Likelihood. 384 pages, 1988.

Vol. 46: H.-G. Müller, Nonparametric Regression Analysis of Longitudinal Data. VI, 199 pages, 1988.

Vol. 47: A.J. Getson, F.C. Hsuan, {2}-Inverses and Their Statistical Application. VIII, 110 pages, 1988.

Vol. 48: G.L. Bretthorst, Bayesian Spectrum Analysis and Parameter Estimation. XII, 209 pages, 1988.

Vol. 49: S.L. Lauritzen, Extremal Families and Systems of Sufficient Statistics. XV, 268 pages, 1988.

Vol. 50: O.E. Barndorff-Nielsen, Parametric Statistical Models and Likelihood. VII, 276 pages, 1988.

Vol. 51: J. Hüsler, R.-D. Reiss (Eds.), Extreme Value Theory. Proceedings, 1987. X, 279 pages, 1989.

Vol. 52: P.K. Goel, T. Ramalingam, The Matching Methodology: Some Statistical Properties. VIII, 152 pages, 1989.

Vol. 53: B.C. Arnold, N. Balakrishnan, Relations, Bounds and Approximations for Order Statistics. IX, 173 pages, 1989.

Vol. 54: K.R. Shah, B.K. Sinha, Theory of Optimal Designs. VIII, 171 pages, 1989.

Vol. 55: L. McDonald, B. Manly, J. Lockwood, J. Logan (Eds.), Estimation and Analysis of Insect Populations. Proceedings, 1988. XIV, 492 pages, 1989.

Vol. 56: J.K. Lindsey, The Analysis of Categorical Data Using GLIM. V, 168 pages, 1989.

Vol. 57: A. Decarli, B.J. Francis, R. Gilchrist, G.U.H. Seeber (Eds.), Statistical Modelling. Proceedings, 1989. IX, 343 pages, 1989.

Vol. 58: O.E. Barndorff-Nielsen, P. Blæsild, P.S. Eriksen, Decomposition and Invariance of Measures, and Statistical Transformation Models. V, 147 pages, 1989.

Vol. 59: S. Gupta, R. Mukerjee, A Calculus for Factorial Arrangements. VI, 126 pages, 1989.

Vol. 60: L. Györfi, W. Härdle, P. Sarda, Ph. Vieu, Nonparametric Curve Estimation from Time Series. VIII, 153 pages, 1989.

Vol. 61: J. Breckling, The Analysis of Directional Time Series: Applications to Wind Speed and Direction. VIII, 238 pages, 1989.

Vol. 62: J.C. Akkerboom, Testing Problems with Linear or Angular Inequality Constraints. XII, 291 pages, 1990.

Vol. 63: J. Pfanzagl, Estimation in Semiparametric Models: Some Recent Developments. III, 112 pages, 1990.

Vol. 64: S. Gabler, Minimax Solutions in Sampling from Finite Populations. V, 132 pages, 1990.

Vol. 65: A. Janssen, D.M. Mason, Non-Standard Rank Tests. VI, 252 pages, 1990.

Vol. 66: T. Wright, Exact Confidence Bounds when Sampling from Small Finite Universes. XVI, 431 pages, 1991.

Vol. 67: M.A. Tanner, Tools for Statistical Inference: Observed Data and Data Augmentation Methods. VI, 110 pages, 1991.

Vol. 68: M. Taniguchi, Higher Order Asymptotic Theory for Time Series Analysis. VIII, 160 pages, 1991.

Vol. 69: N.J.D. Nagelkerke, Maximum Likelihood Estimation of Functional Relationships. V, 110 pages, 1992.

Vol. 70: K. Iida, Studies on the Optimal Search Plan. VIII, 130 pages, 1992.

Vol. 71: E.M.R.A. Engel, A Road to Randomness in Physical Systems. IX, 155 pages, 1992.

Vol. 72: J.K. Lindsey, The Analysis of Stochastic Processes using GLIM. VI, 294 pages, 1992.

Vol. 73: B.C. Arnold, E. Castillo, J.-M. Sarabia, Conditionally Specified Distributions. XIII, 151 pages, 1992.

Vol. 74: P. Barone, A. Frigessi, M. Piccioni, Stochastic Models, Statistical Methods, and Algorithms in Image Analysis. VI, 258 pages, 1992.

Vol. 75: P.K. Goel, N.S. Iyengar (Eds.), Bayesian Analysis in Statistics and Econometrics. XI, 410 pages, 1992.